P9-BTN-823

PEOPLE and POLITICS in URBAN SOCIETY

Volume 6, URBAN AFFAIRS ANNUAL REVIEWS

PEOPLE and POLITICS in URBAN SOCIETY

Edited by
HARLAN HAHN

Volume 6 URBAN AFFAIRS ANNUAL REVIEWS

SAGE PUBLICATIONS / BEVERLY HILLS-LONDON

Printed in the United States of America

For information address:

SAGE PUBLICATIONS, INC.
275 South Beverly Drive
Beverly Hills, California 90212

SAGE PUBLICATIONS LTD
St George's House / 44 Hatton Garden
London E C 1

International Standard Book Number 0-8039-0168-2

Library of Congress Catalog Card No. 72-77744

FIRST PRINTING

CONTENTS

PEOPLE
and POLITICS in
URBAN SOCIETY

Volume 6, URBAN AFFAIRS ANNUAL REVIEWS

1

Reassessing and Revitalizing Urban Politics: Some Goals and Proposals

HARLAN HAHN

□ ONE OF THE MOST FUNDAMENTAL PREREQUISITES for the investigation of urban society is a thorough understanding of the politics that affects communities. Although major interest in the study of urban affairs usually has been focused on the numerous difficulties that confront cities, perhaps equal or greater attention might be devoted to the political processes through which those problems may be solved. Efforts to achieve important objectives, such as the eradication of poverty, inadequate financial resources, and racial or ethnic conflict, as well as attempts to improve the aesthetic quality of urban life, ultimately could depend upon the capacity of governmental bodies to cope with the responsibilities that are imposed upon them. Unless political agencies are equipped to handle such momentous challenges, the possibility of attaining those goals may be negligible. As a result, politics can be found at the core of the many social, economic, and even psychological problems that afflict cities.

This focus on urban politics seems to be especially appropriate in the year of a critical national election. As many voters visit the polls to cast their ballots in a presidential campaign, as well as other state and congressional races, their attention may be focused not only on the broad societal issues that dominate those contests, but also on the urgent requirements of local areas. Although relatively few mayors, city councilors, and other local officials normally are chosen in a presidential election, widespread public interest in urban issues appears to bear implicit testimony both to the importance of metropolitan areas in national politics and to the significance of government efforts in solving urban problems. By whatever definition one may choose to employ, the bulk of the American population now lives in urban areas. Hence, popular concern about issues that have their primary impact upon localities might appear to be natural and unexceptional. Perhaps more significantly, however, the urgent needs of the cities can no longer be neglected or ignored. Many of the controversies that have arisen in municipalities are so inescapable and so crucial that they outrival other questions as topics of national debate. Furthermore, most of the problems that emerge in cities are essentially *political* problems, and they require political action for their amelioration or solution. The basic importance of urban issues in national politics and of politics in the solution of urban problems, therefore, seems to form an essential foundation for the examination of cities.

POPULAR VIEWS OF URBAN POLITICS

Despite the importance of this subject, however, conventional accounts of urban issues often seem to treat local politics as an exercise in civic activity rather than as an essential process for the resolution of conflicting demands and for the alleviation of municipal needs. In large measure, this orientation may be founded upon the same motivations that arouse voter interest on election day; namely, the intense and overriding desire to secure a final solution to urban problems. Many specialists in urban affairs and other commentators consider politics merely as another obstacle to be averted in their efforts to achieve this ultimate objective. As a result, governments are perceived primarily as a means of implementing—or of obstructing—policies and programs that have been fully perfected before they are submitted to the political process. Members of the public and political leaders seldom are viewed as performing a constructive task in the formulation of urban policies; their role is usually regarded as inhibiting or destructive.

Although many discussions seem to reflect this essentially apolitical—or even anti-political—approach to the solution of urban problems, perhaps few would deny the need at least for some method of transmitting public concerns to responsible government officials. As a result, politics frequently is characterized as creating or establishing a basic linkage between the residents of a community and their elected representatives. The importance of this perspective must not be obscured. Political institutions do provide a vital means of relating popular grievances and aspirations to governmental authorities. In fact, a principal objective of this volume is to examine and to explore the association between the public and political decision makers.

Yet, this approach to urban politics also appears to be incomplete. Many urban specialists and supposedly disinterested or enlightened citizens appear to regard politics as a ritualistic obligation. They may submit their proposals to the public primarily to fulfill the requirements of consultation. The basic function of community residents, therefore, is treated like that of any other consultant. Advice may be freely rejected or accepted, without any firm obligation to modify prepared plans according to the wishes of the people. Similarly, the role of public officeholders frequently is described as providing a necessary source of legitimation. Public policies cannot be accepted or implemented until they have been sanctioned by local politicians.

The basic difficulty with this orientation is that it does not allow for significant inputs either by the public or by government leaders. According to this view, if a controversy arises, the major duty of urban experts is to stand aside from the political process and to allow the principal combatants in the battle to resolve the issue by themselves. In many circumstances, of course, the proponents of urban policies may find it necessary to temper or to amend their plans according to the requirements of political expediency. This intervention usually is regarded as an action that may detract from—rather than contribute to—the substantive merits of the original proposal. Many students of urban affairs seem to neglect the possibility that participants in the political struggle might be able to offer constructive improvements and suggestions that would not have been discovered without a full review by the decision-making process.

Perhaps more important, this perspective seems to treat politics as a somewhat sterile and impersonal mechanism for communicating public demands to elected representatives and for holding those representatives accountable to the people. Hence, the basis of government is contained in the machinery providing for political representation and elections. Although this view represents a highly simplistic and mechanistic description of government, perhaps its basic flaw is reflected in a failure to comprehend the broad impact of political behavior upon public policy. The submission of plans to the political process is not an exercise undertaken solely because of the constraints of the law; it also may contribute substantially to the enrichment of urban programs. Solutions to metropolitan problems often may be found *because* of rather than *in spite* of politics. Moreover, the nature of local governance seems to raise some fundamental issues concerning the welfare of society generally.

THE SIGNIFICANCE OF LOCAL POLITICS

Perhaps one of the major reasons for the efforts by social scientists to stimulate increased interest in the study of communities rests upon the importance of urban politics in the evaluation of democratic theory. Local governments at least provide an opportunity for more direct contact between the public and political leaders than their state or national counterparts. Presumably if governments are truly responsive to the wishes of the people, they are most likely to be found at the grass-roots level. Consequently, a great deal of attention has been focused on attempts to compare and contrast the operation of local politics with the classic tenets of democracy.

In large measure, this interest may explain the long-standing preoccupation with the study of community decision-making as well as the fervent debate that developed between so-called "elitist" and "pluralist" schools of thought. Early studies, which indicated the existence of a "community power structure" that dominated the city, seem to contain some ominous implications for democratic society. Clearly, if local decision-making was controlled by a small group of elite influentials, the prospects of finding genuine democracy in national politics did not appear to be encouraging. As a result, those investigations inspired many other researchers to employ similar as well as different methods of examining community power and influence.

Perhaps the principal challenge to the "elitist" position was presented by the "pluralists," who discovered that decisions on separate local issues frequently were determined by different individuals. While the latter studies did not identify active popular involvement by all segments of the community in each local controversy, they offered a sharp attack on the notion of a monolithic power structure and they suggested that diverse persons might be granted an opportunity to exercise decisive influence, depending upon the nature of the issue involved. Subsequently, much of the initial conflict between the two opposing positions developed into a methodological debate between the relative merits of so-called "reputational" and "decisional" techniques of studying community decision-making. Actually, appropriate research strategies for identifying the role of elites either at the national or the local level have not yet been developed or refined.

The methodological difficulties that plague those studies, however, probably should not be allowed to overshadow their theoretical significance. The influence in local decision-making exerted either by self-appointed elites or by residents broadly representative of the general population might be an appropriate concern of all persons interested in the preservation of a democratic society. Cities are defined by their boundaries, and they are confined to a relatively limited geographic area. If for no other reason, therefore, one might expect the principle of proximity to encourage a closer collaboration between the average citizen and public officials in urban areas than in more distant or remote governmental jurisdictions. Although the quagmire encountered in research on community power and influence has stimulated many investigators to explore alternative approaches to the subject, this emphasis illustrates the importance of studying local politics as a means of probing broader social issues.

Two other distinguishing characteristics of municipalities also seem to demonstrate the central role of research on local decision-making in the development of social or political theory. Initially, cities are classified by their size. Although the definition of urban areas may range from 2,500 to 250,000 or more, most population levels employed to demarcate cities are relatively high. Since a large aggregation of people usually implies increasing public demands for governmental services, one might expect to find more active competition for political resources in cities than in rural or suburban areas. In addition, urban areas are characterized by population density, which is commonly associated with a heterogeneous rather than a homogeneous population. The diversity of political interests in cities also might tend to promote intense struggle for political rewards as well as a dispersal of influence and authority.

The game of urban politics may be played by a disparate set of contestants within a sharply confined arena. As a result, it ought to be a lively sport, engaging the time and energy of a large number of participants and perhaps even outrivaling other political contests as a source of interest and attention. Any evidence that local politics is not a thriving activity, therefore, may contain some serious connotations for political competition at higher levels of government.

Yet, this description of urban politics seldom is realized within communities. Local issues frequently fail to arouse active concern and involvement, either among large segments of the population or among responsible decision makers. In some respects, therefore, the disinterest expressed by many observers in local decision-making might seem to be justified. City politics may not attract a wide following simply because it does not appear to be a particularly exciting or stimulating sport.

PERSONALIZING POLITICAL FAILURES

Perhaps more important, the explanation for the relatively emaciated state of political life at the local level usually is attributed to the principal actors in the governmental process. Members of the public frequently are described as ignorant or apathetic concerning local issues, and city politicians often are accused of failing to provide effective or inspiring leadership in the performance of their duties. As a result, the defects of urban policies commonly are ascribed to the characteristics and capabilities of participants in local government rather than to the nature of the political process itself.

Indeed, some support for this attitude might be interpreted as emerging from a series of studies of state and local politics which found that policy outcomes were more closely related to demographic or social and economic attributes than to political variables such as party competition and electoral behavior. Although this research has not yet completely unraveled the complex set of forces that mediate the development of policy, it appears to suggest that the composition of the population–and the talent or sophistication located in different geographic areas–as well as the economic resources available, may have a determinative effect upon the capacity of local governments to contend with the social problems that beset them. Since political bodies seldom have been totally successful in meeting those challenges, the personnel encompassed by local politics–as well as the economic support that they muster to finance government programs–usually are blamed for such failures.

This emphasis on the participants in local politics, including those who might potentially become involved, also has major implications for other aspects of the study of urban decision-making. From this perspective, the presence of a small ruling clique in a city, for example, may be ascribed either to the self-serving motives of the elite or to the failure of other members of the community to become actively engaged in efforts to strip this clique of its power and to broaden the basis of influence. Elitism in cities is depicted as arising both from the desire of some individuals to control the community and from the reluctance of the citizenry to alter the prevailing arrangement of power. On the other hand, the discovery of an essentially pluralistic decision-making process may be ascribed to the interest of both public officials and private citizens in different local issues and to their willingness to become politically active concerning those matters. Similarly, the failure of even larger numbers of people to become politically involved in communities bearing the characteristics of pluralism might be explained by their indifference or inertia. In many respects, therefore, the attributes and abilities of social or economic influentials, government officials, and members of the public may be regarded as accounting for the existence of elitism or pluralism at the local level. Moreover, such factors frequently receive more attention than the general configurations of the structure of influence in cities which may either facilitate or inhibit the participation of each of those groups in local politics.

This preoccupation with personal characteristics also may be related to the widespread neglect of community politics by those who are principally concerned with the solution of urban problems. In many accounts, the failure to devise appropriate remedies for the needs of the cities is blamed upon the incompetence or the ineptitude of the public and political leaders. Although this orientation seems to impose the primary obligation for the absence of effective solutions to local problems upon the major actors in urban politics, the same emphasis also is employed to demonstrate the irrevelance of the political process. Many writers appear to assume that the lack of ability of participants in community politics may prevent them from making a significant contribution to the formulation of urban policy. As a result, the principal responsibility for the development of policy is shifted from political activists to the urban expert and the detached observer of metropolitical affairs. Both interpretations, however, seem to yield essentially similar results. In either event, the failures of urban policy are attributed to personalized factors rather than to the institutions of community politics.

STRUCTURAL INFLUENCES ON COMMUNITY POLITICS

This tendency to neglect the institutional influences that shape local politics may have critical implications for the investigation of urban problems. Political actions, like all other forms of human behavior, are not performed in a vacuum; they are profoundly affected by the organizations or the settings in which they occur. In fact, the close interdependence of environmental or structural characteristics and public activity that has assumed a prominent position in many theoretical concepts, especially in the field of urban studies, is a striking finding verified by extensive behavioral research. This relationship may exert a significant impact upon policy-making as well as other aspects of urban life. Nonetheless, relatively little attention is devoted even to such basic issues as the political effects either of elitist and pluralist structures of influence or of the organizational features of city governments. By stressing an essentially anthropomorphic view of urban decision-making and by neglecting to emphasize the institutional contraints and opportunities that mold political behavior, some observers seem to overlook one of the most salient and significant elements in city politics.

Perhaps more important, the failure to recognize or to observe the importance of institutional influences may create serious problems in the study of politics at the local level. Although the basic framework of local governments may be among the most prominent factors that must be noted by researchers, other features of the community, such as the prevailing structure of influence, the distribution or stratification of social status within the population, and similar characteristics also might account for a sizable proportion of the variance in political activity. In fact, the terms institutions, organizations, and structures are employed in this essay to include those attributes as well as other regularized patterns of personal interaction in communities rather than merely the formal machinery of government outlined in legal codes and city charters.

Since it has been demonstrated that environmental and instructional factors exert a major impact upon personal activities, the researcher who does not display a sensitivity to this issue may be left in an awkward position concerning the interpretation of his findings. Presumably, results that emerge from political research could be attributed either to the variables under investigation or to the institutional milieu in which they were examined. While it may be difficult–if not impossible–to delineate all of the means by which institutions may affect urban politics, research on this subject at least might be expected to include an explicit recognition of their impact. The investigation of social and political phenomena may be incomplete without a clear understanding of the context in which they appear. Contextual factors can exert a strong effect upon political behavior, and they must be acknowledged by social scientists. By failing to appreciate the significance of institutional or structural factors in city politics, some investigators seem to have contributed to the confusion rather than the clarification of research on this subject.

In many studies, the institutional attributes of the political process at the local level are treated as "givens," which seem to lie beyond the legitimate purview of the researcher. As a result, attention is narrowly confined to the particular form of behavior under study, and few social scientists attempt to examine the structural conditions that may augment or diminish that behavior. To some extent, of course, this focus might appear justifiable. The institutional arrangements that mold political life in the community are determined by legal dictates and by complex social forces which the researcher frequently cannot control. Social scientists hardly could ask cities to alter their social and political structures to suit the requirements of a research design.

But even this excuse may fail to comprehend the vast amount of change that occurs in the organizational features of city politics.

Although perhaps the most resilient political institutions are those established by a written framework such as the United States Constitution, even this document is constantly subject to subtle changes that may affect the foundations of the governmental apparatus. In addition, state and local laws that outline the organization of municipal governments are even more susceptible to changes that may affect the ability of persons to participate in community politics and to develop appropriate solutions for urban problems. Other community organizations and institutions also display highly variable traits. Political structures are not fixed and immutable; they are dynamic and variable. Thus, the tendency to emphasize the permanence of the institutional components of local politics does not appear to provide a sufficient rationale for their neglect in research on urban decision-making.

Although the importance of institutional or structural influences apparently has not received adequate notice in much contemporary research, this fact was clearly understood by the municipal reformers who altered the course of urban history in America at the turn of the century. By advocating the adoption of major changes in the form of municipal government such as nonpartisanship, at-large elections, the city manager plan, and similar proposals, the reformers attempted to alter the nature of urban politics for many years to come. Fundamentally, they sought to institutionalize their own concept of the "public interest," or "the welfare of the community," defined in a manner that would be compatible with their personal values and styles of life. Consequently, in those communities in which their efforts were successful, public policies that have filtered through the political process probably also have borne the imprint of their initial goals. In seeking to attain a convergence between the formal mechanisms of urban government and their own policy preferences, the reformers may have been able to exert an impact upon city policies that outlived the period of their activity.

Perhaps one of the most intriguing aspects of community politics is that there have been few subsequent efforts to gain specific policy objectives by altering the formal process of decision-making. Although many researchers and other observers have lamented the plight of urban society, their diagnoses seldom have resulted in the advocacy of concrete institutional or structural changes. In part, perhaps this tendency has resulted from the desire of social scientists to avoid the prescription of normative cures. Although many political scientists, sociologists, and other members of the academic community were involved in the reform movement that swept American cities in the early twentieth century, subsequent generations of scholars have attempted to maintain a more detached view of the problems of cities. One source of this reticence may have been related to the outcome of the earlier reform movement. Since the reformers obviously were not successful in correcting all of the ills that plague urban areas, many persons were understandably hesitant about attempting this approach again.

THE STUDY OF INSTITUTIONAL CHARACTERISTICS

However, there may be another, and perhaps more compelling, reason for this position. At the present time, social and political researchers frequently fail to recognize—let alone to understand—the precise impact of institutional and structural arrangements upon political activity and policy outcomes within communities. In addition, cities have not yet begun to explore, or to exhaust, the variety of ways in which the governmental process might be organized to produce different political results. Any attempt to prescribe institutional solutions for urban problems, therefore, might be premature and potentially dangerous.

Growing pressures on social scientists to make value-laden judgments about the institutions that affect political life in the community obviously entail serious risks. By yielding to the temptation of offering recommendations for the reform or the reorganization of political structures, the researcher is exposed to the danger not only of losing his own objectivity but also of providing advice that may prove to be inaccurate because it is not based upon sufficient evidence. Furthermore, the injection of personal values into social analysis could deprive subsequent investigators of the opportunity to replicate or to verify the results of prior studies.

Yet it is frequently difficult to escape any hint of normative concerns in research on social or political behavior. Both in the selection of this topic for study and in the interpretation of the implications of his findings, the researcher is frequently compelled to unmask his own values; he must state why he considers his problem an important subject for investigation and what he feels his conclusions mean for progress either in the discipline or in society as a whole. Behavioral research and a normative concern about the social and political effects of different institutional arrangements, therefore, may not necessarily be incompatible. Without a clear understanding of the ways in which people actually behave, it might be impossible to devise institutions that would maximize the opportunities for people to participate in the formulation of policy or to achieve other political objectives. Similarly, without an acute appreciation of the impact of structural conditions upon personal conduct, it may be equally difficult to compile a comprehensive inventory of the total range of social or political behavior.

Perhaps more important, the failure to take cognizance of structural influences upon political conduct could result in a serious distortion of research findings. Any investigation that neglects to consider those factors may reflect a relatively constricted perspective. As abundant evidence demonstrates, the organizations and institutions that provide a vehicle for political activity often may have a powerful–even a determinative–impact upon both the nature and the form of that activity. Unless structural characteristics are included in the appraisal of social or political data, that research may simply be incomplete.

One of the most fundamental reasons for the consideration of institutional factors in social science research, however, is related to a basic problem in behavioral investigations. Although quantitative studies have contributed greatly to an expanded knowledge of human conduct, they do not appear to have made an equivalent contribution to an understanding of the possibilities for social change. By focusing exclusively upon an examination of the way things are, social scientists are deterred from considering the way things ought to be. Hence, researchers are vulnerable to the charge that the primary impact of their studies has been to strengthen the influence of the status quo. This criticism may be especially pertinent because of the earlier failure of many social scientists to examine the behavioral effects of social or political structures. Studying behavior only within the context of existing institutional arrangements, those investigators may project an image of conservatism. In addition, the effects of their findings frequently appear to retard rather than promote the process of institutional change.

One method of avoiding this bias might be provided by placing an increased emphasis on the institutional determinants of behavior. The examination of human conduct in a variety of institutional contexts, for example, could provide the researcher with an opportunity to explore the extent to which his findings are affected by structural characteristics. Enhanced attention also might be appropriately devoted to the study of social change. Some investigators might profitably concentrate on the analysis of local areas in which patterns of behavior are changing to determine if there are comparable changes in the social or political structure of those communities, while other researchers may examine urban areas that are undergoing structural transitions to ascertain the extent to which such alterations produce major shifts in the behavior of local residents. Perhaps most important, however, a simple recognition of the fact that institutions frequently shape personal conduct may offer an important means of correcting—or at least of reducing—the tendency of behavioral research to be interpreted as inhibiting social change or progress.

The neglect of institutional influences on social and political behavior also has resulted in another, and perhaps even more critical, bias in quantitative research. Since behavioral scientists usually focus on the examination of people rather than institutions, they often are inclined to adopt the same perspective in evaluating their research findings. Consequently, any evidence of defects or faults in the political system also are ascribed to personal rather than to institutional attributes. This orientation is especially evident in the study of local politics. The results of empirical investigations frequently are cited in support of the general tendency to personalize the blame for political failures and to denigrate the role of major participants in the governmental process within cities. On the basis of evidence gathered by social scientists, citizens are described as apathetic and ill-informed, and political leaders are characterized as parochial and unimaginative. Hence, such statements often are employed to deprecate the function of both the public and political decision makers. Little is expected from either group in the formulation of public policy, because it is presumed that they have little to contribute.

INSTITUTIONAL CONSTRAINTS ON POLITICAL BEHAVIOR

What those conclusions generally overlook, however, is that the options available both to the public and to government officials often are limited by institutional factors. Members of the public, for example, may be prevented from playing a more constructive role in the policy-making process by a series of structural barriers. Initially, they seem to be confronted by some serious deficiencies in the system by which information is transmitted to the community. Important facts may not be disseminated for a variety of reasons. Moreover, even if such material is made available to the interested citizen, it may escape his notice amid the host of stimuli that compete for his time and attention. Those considerations, of course, tend to place members of the public at a severe disadvantage. Unless they are equipped with adequate information, citizens can hardly be expected to make an optimal contribution to the decision-making process.

Furthermore, persons who have succeeded in gathering the necessary arsenal of facts usually encounter another series of obstacles that may impede the effective expression of their sentiments. Opportunities for the public to exercise political influence typically are confined to actions such as communicating with legislative representatives, voting in elections, participating in interest group or political party activities, and engaging in public demonstrations or protests. Although those vehicles for political participation might appear to be impressive, all of them entail substantial costs for the average citizen and none of them can guarantee success. Conceivably, alternative structures might be created to provide more appropriate mechanisms for political expression than those offered by existing institutions.

Similarly, the effectiveness of public officeholders also may be curtailed by the organizational contexts in which they must operate. In fact, each of the restrictions on the transmission of communications to the people also may hamper the ability of decision-makers to acquire needed information both from the public and from other sources. Limitations that affect the quality of the information available to government officials may constitute one of the most basic restrictions on their capacity to achieve major improvements in public policy. In addition, political leaders often may be hindered in their ability either to make correct decisions or to enact innovative proposals by other restraints such as pressures from party organizations, the necessity of cooperating with other public officials, and restrictions on the resources at their disposal. While the existing machinery of government has not yet become totally inoperative, it is not necessarily the most effective apparatus that could be constructed either to facilitate the expression of public sentiments or to achieve permanent solutions to major social problems. In failing to account for either the structural restraints that inhibit political behavior or for the potential effects of alternative institutional arrangements, social scientists may have committed a serious error by underestimating the political capabilities of participants in the governmental process.

Although both members of the public and political decision makers have been criticized for their apparent inability to fulfill standards of performance established for them, perhaps the most vehement criticism has been directed at the public. The mythical average citizen commonly is condemned for his indifference to public issues, his lack of information about them, and his unwillingness to become involved in political controversies.

In part, this characterization may be symptomatic of an implicit bias that is prevalent among social scientists. Research on three topics, which will be discussed in greater detail below, might be cited to illustrate this orientation. Studies of alienation, for example, usually focus on the estrangement of the general population; few of them seek to examine the extent to which political leaders are alienated from the public. Similarly, acts of public authorities seldom are investigated by researchers attempting to probe the causes that may trigger protest and violence. Even in the study of public policy, the failure of legislative representatives to represent the wishes of their constituents usually is not considered as one of the factors that may contribute to their inability to solve pressing problems. (Although such speculation is fraught with peril, it is interesting to consider the possibility that the strongly anti-populist orientation of many researchers may reflect the natural interest of a highly educated elite, which is clinging tenaciously to declining social status, in maintaining a wide distance from those whom it regards as socially and intellectually inferior.)

Perhaps an even more basic explanation of the political failures of the public, however, might be found in the lack of institutions designed to maximize the effects of public pressure. Many established channels of political expression such as voting, party activity, lobbying, and protests may be regarded as meaningless or ineffective by the average citizen. The prospects of achieving either significant political benefits or intense personal satisfaction from participation in those activities may be regarded as so low that they do not justify the expenditure of time and energy. This, of course, does not necessarily imply that the individual also would refrain from participating in the decision-making process if more convenient, meaningful, or effective channels of political expression were created. On the contrary, he might be willing—and even eager—to become involved in political structures that are designed specifically to suit his interests and needs.

As another example, consider the case of the person who never avails himself of the opportunity to vote in an election. His refusal to exercise the franchise might be based upon a number of reasons rather than apathy or a lack of information concerning political issues. He may feel that the choices presented to him in an election do not provide a sufficient opportunity to express his political preferences, or he may believe that elections are simply an inadequate means of registering public sentiments. In any event, his opinions seldom are heard or studied by government decision makers. (Perhaps only if he should happen to be included in a random sample drawn for a major opinion survey would his attitudes be recorded, and even then they may be ignored because of his failure to participate in elections.) In fact, some social scientists might argue that nonvoting because of a belief that elections are meaningless does not constitute an act of political behavior worthy of careful study.

Yet, the prevalence of this conviction among sizable proportions—or even a majority—of the population could have serious political repercussions. As citizens become increasingly disenchanted with electoral institutions, they may seek more direct—and potentially more disruptive—means of communicating their grievances. Moreover, a reluctance to vote in elections might not denote a desire to abstain from all forms of political participation. The nonvoter may not refrain from communicating his attitudes to his friends, to public opinion pollsters, or to candidates who visit him during campaigns. Efforts to determine the will of the people by conducting personal interviews with community residents, for example, could tend to stimulate increased public participation in politics. Although this method might appear to be less practical than elections, it is conceivable that citizens could be furnished with other appropriate mechanisms for communicating their views on political issues.

Existing institutions of government provide those who wish to become involved in political decision-making with numerous constraints and opportunities. In general, however, the constraints imposed by political structures may appear to overshadow the opportunities for effective political participation. Elections, for example, permit voters to record their preferences at the ballot box, but they are perceived by large segments of the population as an inadequate or inappropriate form of political expression. If the goal of encouraging increased public involvement in politics is considered one of the basic objectives of democratic governments, attention might be appropriately devoted to the invention of other institutions that would supplement elections as a means of reflecting popular sentiments. Similarly, government officials frequently find that the authority bestowed upon them is not sufficient to permit the discharge of their responsibilities. Efforts to redesign or reorganize political institutions, therefore, might promote more active and more effective participation in all stages of the decision-making process.

INVESTIGATING STRUCTURAL CHANGE

Social scientists might perform two critical functions in this reexamination of political institutions. Initially, they could devote expanded attention to the study of common forms of political activity that have evolved without specific support or recognition from legal and governmental structures. In the course of their everyday lives, many people engage in activities that may have little direct impact on government but that have important political implications and consequences. A complete understanding of all those actions might tend to stimulate major improvements in the design of governmental structures. Conceivably, for example, this information could be utilized in the development of plans for restructuring government institutions to promote increased public participation in the policy-making process or to achieve other political objectives. In this way, it may also be possible to achieve a closer correspondence between informal patterns of social or political conduct in everyday life and the formal institutions of government.

In addition, increased research might be focused on social or political behavior within different institutional contexts. This emphasis seems to be especially appropriate in the study of local politics. Communities exhibit a wide range of variations, and they possess disparate social and political structures. Thus, efforts to identify the forms of political behavior that emerge in cities with diverse structural characteristics might become a major goal in the study of urban affairs. In this endeavor, of course, the researcher may be greatly aided by the multiplicity of institutional variables at his disposal and by the opportunities they afford for comparative analysis.

In part, this need is being recognized by a growing number of social scientists. As many of the chapters in this volume indicate, there is an increasing tendency among urban researchers to focus on the comparative study of city politics, both within a single country and between different nations. In contrast to the case studies that dominated the field for many years, this trend promises to extend the generality of the findings uncovered by social and political research. Yet, there also may be major value in studying communities that have unusual or divergent institutional arrangements. By comparing the nature of political activity in cities with unique structural characteristics and in municipalities that maintain the predominant institutional patterns, social investigators may gain some important insights concerning the institutional determinants of political behavior.

In addition, such evidence may be of great value in any attempts to redesign political institutions in accordance with the requirements of modern society. Obviously, it would be difficult—if not impossible—to prescribe appropriate social or political structures for cities without a clear understanding of the effects that institutions exert upon personal conduct. Guided by this information, however, the social analyst may be in an improved position to offer recommendations concerning institutional changes or reorganization.

While this responsibility undoubtedly places urban investigators in a controversial and demanding role, the interpretation of research findings is an important duty that has broad theoretical as well as social implications. Just as the results of earlier studies of political behavior often were seen as reflecting the inadequacies and the limitations of major participants in local politics, the conclusions of research concerning the behavioral impact of structural characteristics might be employed to locate or to identify critical faults in existing political institutions. Moreover, the needs and demands that are emerging in cities appear to be so urgent and so compelling that researchers may be remiss in their obligations if they fail to allow pertinent information to be used in efforts to solve those problems.

Perhaps one of the most critical challenges in the field of urban affairs is the need for a bold attitude toward experimentation both in the design of political institutions and in the study of city politics. Communities might be encouraged to experiment with different decision-making structures, and researchers may be induced to locate quasi-experimental settings in which they could examine the impact of institutional influences upon political behavior. Undoubtedly, such attempts would be marked by many failures as well as an occasional success. But there seems to be an increasing interest in organizational forms and innovations among municipal officials as well as social scientists. While the tasks of either group will not be easy, the mounting problems of urban areas seem to suggest the necessity of new and radical approaches to the solution of those problems.

SOCIAL RESEARCH AND INSTITUTIONAL CHANGE

Despite those difficulties, the need for fundamental changes in city politics seems to be increasingly imperative and inescapable. Many commentators are raising basic questions not only about the relative merits of political bodies at the local level but also about the adequacy of governments to cope with mounting demands. In fact, each of the contributors to this volume was presented with a series of questions reflecting those concerns: How are political interests and needs brought to the attention of government officials? Are the existing channels of political activity such as elections, interest groups, and other forms of expression adequate to handle public demands, especially by powerless and disadvantaged segments of the community? In a period of mounting polarization, how are opposing and conflicting political desires reconciled or resolved equitably? Can political institutions acquire the resources and the will necessary to satisfy the demands that confront them? Although the studies reported in this volume do not necessarily represent an effort to deal with—or to answer—*all* of those questions, each of the following chapters contains some important implications concerning them. The task of devising political institutions that are appropriate to the needs and requirements of contemporary society has not yet been accomplished, but it is a problem to which a large number of social scientists and other observers are devoting increased attention.

ALIENATION

In large measure, the impetus for this concern probably can be traced to a number of other problems that have emerged in cities. For many years, as an example, sociologists, psychologists, and political scientists have been preoccupied with the problem of alienation in modern life. According to a classic view, the tendency of man to feel estranged from his environment is intensified by an urban milieu. In the alien and impersonal world of the city, people are supposedly more apt to experience a sense of disaffection than in the comfortable and supportive environs of the small town or village. Although this perspective is undergoing extensive revision as new information becomes available, the continuing presence of severe symptoms of alienation cannot be denied. Its manifestations are evident in phenomena that range from the commission of violent and destructive crimes to the defeat of issues endorsed by civic activists in municipal referenda. Alienation not only constitutes a serious problem with potentially devastating effects upon society, but it also reflects an issue of fundamental concern to social scientists. In fact, attempts to cope with the concept of alienation could be described as one of the most basic theoretical goals of sociology as well as other academic disciplines. As a result, an extensive amount of time and effort is being focused on the task of grappling with the social, economic, and political consequences of alienation.

In seeking to understand and to solve the problems that arise from a widespread sense of alienation, many social scientists eventually were induced to expand their range of interest. Although alienation initially was characterized by some researchers as an essentially psychological trait that emanated from the individual and that could be alleviated most effectively by dealing with persons on an individual basis, others began to realize that alienation also might have external or environmental sources. Instead of regarding alienation as a deviant or abnormal state of mind, an increasing number of scholars gradually were persuaded to recognize the possibility that people who experienced a sense of alienation might have a plausible basis for their feelings. Actually, the organizations and institutions provided by an urbanized society may be more conducive to the growth of estrangement and disaffection than to the development of rapport and security. Thus, attention naturally was shifted from individual cures to the goal of restructuring institutions so as to minimize alienation. While this trend has led to an emphasis in sociology on the creation of social structures that might diminish the gap among all members of society, it has also resulted in a concentration in the political sphere on efforts to reduce the distance between the public and government decision-makers. Ultimately, therefore, the traditional focus on the behavioral effects of alienation has prompted many social scientists to acquire a normative concern about the institutional context in which alienation has been studied.

VIOLENCE

The same tendency also has become evident in research on violence. As many citizens became alarmed at the riots and disorders that erupted in cities during the decade of the sixties, a number of behavioral scientists began to devote their attention to research on those events. Subsequent developments seem to prove an important–and a painful–lesson for social investigators. While the predominant reaction of the public to urban violence was shock and disbelief, many social scientists experienced an acute sense of embarrassment about their inability either to offer a comprehensive explanation for the riots or to suggest remedies that might prevent their recurrence. As a result, their role was preempted by some public leaders who ascribed the violence to individualistic sources and to deviant or atypical members of society. Slowly and painstakingly, researchers amassed enough evidence to refute those interpretations. The riots did not appear to be the product of outside agitators, organized conspiracies, known criminals, or other deviants in society. Instead, they seemed to reflect the basic needs and aspirations of broad segments of the population.

Although divergent theories of urban violence have remained the subject of sharp contention, at least some researchers not only have rejected the individualistic explanations of rioting, but they also have placed increasing emphasis on social and political factors that may have caused the disorders. In a basic sense, the riots seem to denote a massive expression of public protest. In part, the violence in American ghettos, for example, represented an appeal to the conscience of the white public and political leaders. In addition, however, rioting may have been sparked by the failure of deprived minorities to achieve their objectives through existing political structures such as elections, interest group activity, the legal system, peaceful demonstrations, and verbal persuasion. Fundamentally, therefore, the outbreak of violence appeared to indicate the collapse of traditional mechanisms for resolving conflict which compelled political authorities to mobilize physical force to suppress the disorders. Moreover, during the course of the rioting, the participants in the violence seem to establish their own rules of conduct on the streets in defiance of the laws that normally govern society.

Consequently, the analysis of urban riots has prompted some investigators to consider the possibility of reshaping social and political institutions to grant minority groups an increased opportunity to participate in the decision-making process without resorting to the ultimate threat of violence. Research on urban rioting, therefore, has tended to follow a path that is comparable to trends in other areas of investigation. With the change from a focus on the personal characteristics of rioters to an emphasis on the social and political structures that had failed to deter violent protest and that comprised the major targets of attack during the rioting, social scientists were forced to consider questions about how those institutions might be redesigned to satisfy the legitimate needs of minorities and to avoid future violence. At the same time, they were compelled to examine not only the behavior of rioters but also the values implicit both in existing and alternative institutional arrangements of power.

PUBLIC POLICY

An additional source of this emphasis on the examination of political structures and on the exploration of possibilities of institutional change has been reflected in the growing interest in the study of public policy. Despite the adoption of expansive government programs designed to ameliorate persistent urban crises such as poverty, welfare, housing, crime, education, unemployment, and transportation, those problems continue to plague the cities. Hence, the failure of major political efforts to eradicate many of the most basic ills of urban areas has inspired some social scientists to devote increasing attention to the analysis of both the content of public policies and the process by which they are formulated. Although this field of investigation also was influenced for many years by a tendency to personalize the faults of policies by assigning the blame to inexperienced administrators or to ungrateful recipients of government aid, renewed interest in this subject generally has been characterized by an increased willingness to assess structural characteristics not only of the agencies that are responsible for the enactment and administration of policy but also of the populations that they are mandated to serve.

Emerging research on this subject also has not been restricted to a consideration of the behavior of policy makers in reaching decisions and to the factors that may influence their choice of policy preferences. Many investigators have been prepared to make personal judgments that involve fundamental social or political values and to evaluate the merits or flaws of policy proposals. While those studies have not yet produced concrete proposals to reorganize the policy-making process, they have contributed significantly to an enhanced awareness of the normative implications of political action.

DECENTRALIZATION

Another important example—both of the growing interest in restructuring political institutions and of changes that might be implemented at the local level—is reflected in the expanding movement for decentralization or "community control" in urban neighborhoods. While this plan may contain some of the traditional problems associated with decentralization such as the dispersal of authority and the difficulty of mobilizing for prompt action, it at least represents a relatively new and innovative effort to solve some of the most persistent problems in city politics. By placing political authority at the neighborhood rather than the municipal or the metropolitan level, the advocates of "community control" hope to promote increased public participation in urban politics. Implicit in the proposal, therefore, is a realization that the means by which political institutions are organized in cities may affect both the nature and the outcome of the decision-making process.

The drive toward decentralization in urban areas not only reflects an emphasis on the institutional components of political behavior, but it also suggests the value of structural innovations in the study of urban politics. Despite the theoretical importance of this issue to democratic governments, for example, there is little empirical evidence to indicate that people are more likely to participate in relatively small political units than in higher levels of government. By examining political involvement in neighborhoods that have implemented decentralization and by contrasting these results with comparable studies of activity in city-wide organizations, social scientists might begin to develop some inferences concerning the effect of structural characteristics upon political behavior. Perhaps the results of such studies–by indicating whether people are more apt to participate in decentralized or centralized institutions, for example–may have important normative implications, depending upon the value that one attaches to public participation. Still, the investigation of political structures or institutions does not require the researcher to sacrifice his scientific objectivity. Nor does it compel him to abandon his theoretical interests and to respond only to immediate social problems. As this example reveals, the effects of structural characteristics upon political behavior is a critical question that has important theoretical as well as practical implications.

RACE RELATIONS

The mounting interest in attempts to restructure political institutions and the escalating problems of the cities also have been reflected in research on race relations. As many of the studies in this volume reveal, race has become a central fact of urban life; and it may be impossible to examine urban politics without a careful analysis of the conflicting aspirations of black and white citizens. Yet, as several of the chapters also indicate, issues concerning minority groups in urban areas seem to be undergoing some marked changes. For many years, the principal focus in the investigation of racial or ethnic relations was devoted to the attempts of deprived minorities to achieve integration into a predominantly white society and to their efforts to secure improvements in housing, education, employment, and other areas. Increasingly, however, the attention of researchers is being shifted to topics such as the establishment of predominantly black communities and the emergence of black candidates for prominent positions of local leadership. In part, this change may have been occasioned by major social and demographic trends. As burgeoning numbers of black and other minority groups migrated to the cities, this movement created the possibility either of the development of all-black communities or of the election of black officials.

In addition, however, this focus also may reflect an increasing concern about institutional or structural influences that affect racial and ethnic relations. Since the pervasive effects of racial segregation and discrimination traditionally have prevented minority groups from gaining equitable representation in decision-making bodies, many minority leaders have sought to attain political influence either by winning control of major elective offices in cities where they comprise a sizable proportion of the population or by developing independent sources of political power within separate municipalities. Although it is difficult to predict which of those strategies eventually will be more successful, both approaches seem to represent a relatively determined effort to achieve increased control of governmental institutions.

FEDERALISM

The efforts by ethnic or racial minorities as well as other segments of the population to acquire an expanded role in the decision-making process, however, also may be deterred by still another feature of governmental institutions. At least in the United States, and in several other countries, community politics has been strongly affected by the principles of federalism and by the imposition of several other layers of government upon the decision-making process at the local level. In fact, state and national governments commonly have played a more determinative role in shaping the nature of urban areas than local institutions. Since political agencies at the state and national level have been able to amass greater financial resources and power, they frequently have seemed to force local governments into a subordinate position. This trend seems to contain several critical implications, including one consequence of special importance to minority populations. As a result of inadequate local resources, black and other ethnic groups who achieve power either in central cities or in separate, incorporated enclaves may suffer enhanced frustration about their inability to effect major change in the community. Without extensive support, especially from the federal government, cities—under black as well as white leadership—may be incapable of solving the problems that confront them.

Although the intervention of state and national governments in urban affairs may tend to reduce the ability of local residents to influence city policies, it is undoubtedly one of the most salient and significant facts affecting community politics in America. As a result, most of the sections of this volume conclude with a study examining the implications of various forms of political behavior at the national level. The expanding federal role in metropolitan areas not only seems to underscore the importance of governmental institutions in the solution of urban problems, but it also seems to raise some interesting implications for the study of urban politics. By assuming that the solution to urban problems cannot be found locally, this approach seems to reinforce the tendency to diminish the importance of community residents and city officials in the formulation of public policy.

SUMMARY

The trend toward increasing federal involvement in the solution of urban problems, therefore, appears to be consistent with many traditional perspectives on city politics. Although the study of community decision-making long has occupied a crucial position in the evaluation of democratic theory, many commentators have expressed a critical view of the major participants in the political process at the local level. Members of the public and local decision makers seldom have been credited with contributing significant inputs to the development of urban policies and programs. As a result, this orientation has tended to elevate the status of the so-called "urban expert," or the detached observer of urban affairs, in the policy-making process.

Furthermore, there has been a common tendency to personalize the blame for political failures by emphasizing the ignorance or the ineptitude of both citizens and urban politicians. Although some studies of social or political behavior have been cited to support this interpretation, relatively little attention has been devoted by researchers or other commentators to the structural influences that may shape political activity at the local level. A careful review of institutional characteristics, however, seemed to indicate that those attributes frequently may exert a constricting effect upon the choices available to both community residents and public officials.

Increasingly, therefore, as a result of research on such diverse topics as alienation, urban violence, public policy, decentralization, and race relations, a growing interest has emerged in the possibility of restructuring political institutions to encourage the enhanced participation of all segments of the population in the policy-making process. Consequently, a need also has seemed to emerge for increased experimentation both in the design of political institutions and in the study of urban politics. In particular, efforts might be devoted to the creation of political structures that would enhance the opportunities for local citizens to become actively involved in community politics. Although the people frequently have been the brunt of sharp criticism, their collective wisdom may be one of the most important resources available in the effort to alleviate the crisis of the cities.

Part I

THE EXPRESSION
OF PUBLIC SENTIMENTS

THE EXPRESSION
OF PUBLIC SENTIMENTS

One of the most critical features of urban politics is reflected in efforts of the public to communicate their wishes to government officials. Before political leaders can respond to the needs and demands of the people, they must gain an accurate understanding of popular grievances and aspirations. As a result, each of the chapters in this section examines the means by which people, acting as individuals, seek to express their sentiments to public officeholders.

Initially, individuals may contact public officials directly. In a survey of white and black residents of Milwaukee, Eisinger examines the characteristics of persons who make such contacts, the types of contacts that are made, and the political leaders most likely to be contacted. His conclusions indicate that the communications received by government officials from direct contacts usually emanate from white citizens of relatively high social status who also participate in other political activities.

Communications between the public and political decision makers, however, not only may occur through direct contacts, but they also may be transmitted by intermediaries. In a study of interpersonal linkages, Bockman and Hahn employ standard reputational analysis as well as snowball and random sample surveys to trace the relationship between local elites and members of the community.

The results of this investigation—and especially the discovery of a communication network which encompassed minorities as well as white segments of the community—seemed to indicate that an important method of transmitting information has developed at the local level without specific institutional support.

Perhaps the most common vehicle for the expression of public views is provided by elections. In the investigation of elections involving candidates for public office, Pettigrew explores an important emerging development in urban politics, namely, the entrance of black candidates into campaigns for high elective offices. His study of voting in Gary, Cleveland, Los Angeles, and Newark suggests that resistance to black mayoral candidates might be attributed to a social psychological sense of relative deprivation rather than to a simple white "backlash."

An even more direct opportunity for voters to record their preferences on policy issues, however, is offered by public referenda. Hence, two chapters in this section assess voting behavior in local referenda. In the study of a referendum on a mass transportation system in the Atlanta area, Ippolito, Levin, and Bowman employ concepts such as self-interest, public-regardingness, and confidence in local government to analyze both white and black attitudes concerning that issue. In addition, the broad implications of referendum votes are explored in a study by Clubb and Traugott, based upon a national survey, which indicates that participants in referenda seem to reflect the characteristics of politically active segments of the population rather than the attributes of alienation.

2

The Pattern of Citizen Contacts with Urban Officials

PETER K. EISINGER

□ ONE MEANS BY WHICH INDIVIDUAL citizens express policy preferences and communicate grievances directly to public officials, namely personally initiated contacts with government personnel, is a subject which has commanded little attention among social scientists. Citizen contact with people in government is an important dimension of the representation relationship: contact is a demand for representation in that the contactor asks in effect that an official act on behalf of his concerns.

The patterns of such personal contact are manifold. Some citizens make no contacts whatsoever, but among the substantial number who do, the targets and concerns vary widely. Given the variety of contact patterns, it follows that some citizens are represented through the device of contact more than others or differently than others. That some citizens seize this opportunity for representation

AUTHOR'S NOTE: *This research was made possible by support from the Governmental and Legal Processes Committee of the Social Science Research Council and by funds granted to the Institute for Research on Poverty at the University of Wisconsin by the Office of Economic Opportunity pursuant to the Economic Opportunity Act of 1964. The conclusions are the sole responsibility of the author. I wish to thank Ira Sharkansky for his comments and Robert Neis for his assistance in processing the data.*

while others do not and that citizens pursue contacts with different officials at different levels of government make the study of the patterns of contact an enterprise of considerable importance in understanding how and to whom representative political systems respond.

Contacting government officials is not always an easy or a rewarding business. Contemplating the task evokes visions of Kafkaesque corridors and frosted glass doors in anonymous buildings. The problems of making contact with bureaucrats often approach the absurd. For example, to complain in New York City about the sale of unsanitary food one must telephone the Department of Health. But to notify the proper official that one has become ill from the unsanitary food he bought, the citizen must call a different number at the same department. And if the food is unsanitary because it has remained on the shelf too long, the hapless shopper must call still a third number, this one at the Department of Consumer Affairs, Consumer Complaints Division. Such chaotic arrangements have been common in New York City: when John Lindsay became mayor, he discovered that three different bureaus and agencies were responsible for dealing with problems of residential water supply, depending upon whether one had insufficient water, no hot water, or no water at all (Lindsay, 1969: 79).

Not only must the citizen confront a baroque structure in his search for the proper place to lodge a complaint, but once he finds the responsible authority, his chances of getting redress are not always good. To cite one extreme but not uncommon example, in the first 59 years of existence of the Milwaukee Fire and Police Commission, a board empowered to hear citizen complaints about police conduct, no individual patrolman has ever been tried as a result of a citizen accusation. The laws of probability make it unlikely that not one complaint in this century of difficult police-citizen relationships was sufficiently justified to warrant a trial.

Complaining to public agencies about unjust or inadequate public or private services is only one form of contact the individual citizen might initiate with officials in government. A more common type of contact involves sending an expression of personal opinion regarding public policy or a plea for action to one's elected representative or executive. Here the target for contact is easy to identify. Yet even in this case the probability of satisfaction may be low, depending on the official contacted and the nature of the message sent, a fact to which those who have sent messages protesting the conduct of the Asian war can readily attest.

The examples cited above warrant the tentative conclusion that for citizens to attempt to contact their government often means to embark on a difficult and frustrating course. To do so requires a substantial effort of will, great persistence, and limited expectations of gratification. Yet it is striking—in spite of all this—that great numbers of citizens do manage to contact their government about one thing or another, and even greater numbers are able to contemplate doing so in the event they have something to complain about or communicate. Nevertheless, there is a self-selection process at work in the individual contact process, and, as a result, certain types of people are more visible to public officials than others. That contact involves difficulty—and thus prevents many from attempting contact—and that it serves as one type of feedback which public officials receive from their constituents make the patterns of contact inherently interesting for the political scientist as well as for the policy maker. It is the broad purpose of this article to explore these patterns in some detail as they occur among the inhabitants of one American city.

To begin with we should be clear about the meaning and dimensions of the notion of citizen contact. In this paper we are concerned with contact with government officials initiated by individuals acting in their capacity as private citizens and not as group spokesmen. The extent to which individuals contact officials in their various governments was determined in a survey conducted in Milwaukee during the summer of 1970. The respondents, residents of the city age 18 and over, were selected by block cluster techniques. The sample was stratified by race such that black respondents comprised a much larger proportion of the total than they did of the population. In all, 313 white interviews and 241 black interviews were completed, constituting two separate samples of each racial group. In each case the race of the interviewer and that of the respondent were matched.

Respondents were asked if, "within the last couple of years," they had personally written a letter, sent a telegram, or spoken to any of a variety of different types of public officials on a list which was handed to them. These officials ranged from the President of the United States to the mayor of Milwaukee to "any person at all who works for the City of Milwaukee." If any contact was made, the interviewers determined the identity of the target and the nature of the contact.

Our first task will be to discover who the contactors are and how they differ from the noncontactors. Then we shall explore the various dimensions of contact—the nature, the content, and so on, of each instance of contact. Finally, we shall discuss the implications of the patterns of contact for the idea of political representation.

THE BASIC PATTERNS OF CONTACT

Americans are a self-confident people when it comes to assessing their individual capacity to influence political affairs—or at least they were ten years ago. Almond and Verba (1965: 188) reported at that time that a substantial majority of citizens in this country believed not only that they would bring influence to bear on political officials under certain circumstances, but that they stood a high chance of success in doing so.

Yet if the Milwaukee respondents possess this same high sense of "subjective competence," their behavior does not reflect it in full measure. The number of people who have actually initiated any sort of contact with government officials at any level in this sample is relatively low. Not only are those who make contact a distinct minority of the city's residents, but they are also overwhelmingly white. In certain crucial ways, the contactors are unrepresentative of the city's population.

Of the 313 whites in the sample, 103 (32.7%) had made a total of 254 contacts with all levels of government, or nearly 2.5 contacts per contactor. Among the 241 blacks, however, only 27 had initiated contact with public officials (11.1%), and 23 of those had made only one contact. The total number of black contacts amounted to 32, a figure low enough to inject considerable caution into the succeeding discussion.

Milwaukeeans, especially those who are black, are not confident that they could gain the ear of the local government even if they tried. While Almond and Verba (1965: 188) found in their nationwide sample that approximately 65% thought they could successfully influence *local* government, only 20% of the Milwaukee blacks and 36% of the whites agreed with the following statement:

"If a group of people have problems here in Milwaukee, it's pretty easy to get somebody in the city government to listen to them." Fewer than half the blacks (44.8%) but nearly three-quarters of the whites (72.8%) said that they would feel free to talk to someone in the government about getting something done for their neighborhood.

As with any form of political self-assertion, contacting public officials is associated with social class indicators (Lane, 1959: 67). The strength of the relationships of selected socioeconomic and demographic variables and contact among the Milwaukee respondents is generally not remarkable except for education. The gamma correlation between education and contact for whites is .46, while for blacks it rises to .70. Income and contact are related for blacks (.47) but not for whites. Relationships between contact and age, occupation, and length of residence in Milwaukee are virtually absent. Such figures indicate that the educated population in the city is overrepresented among contactors of both races.

If a control for education is introduced, the data indicate that education is more important as a support for blacks in making contacts than it is for whites. Controlling for education requires combining the two racial samples, a procedure which does not allow us to generalize to the larger population, since blacks are now overrepresented. Nevertheless, the findings are suggestive.

The relationship between race and contact among those with less than a high school education is –.70 (gamma), while for those with a high school diploma it is –.57. For respondents with some college experience, the gamma score is –.27. The strongest relationship occurs among those of limited education.

The negative value of the coefficient suggests that uneducated whites are more likely to contact public officials than are uneducated blacks. The strength of the relationship diminishes as we go up the educational scale, but at each level blacks appear to labor under a disability that cannot be attributed to educational achievement. We may draw two conclusions from these figures. First, it would seem that education is more important in predicting black contact than it is for predicting white contact. To put it another way, education provides a more important resource for blacks: it serves as a means for overcoming the disability of race. Education is less important for whites. Being white confers an advantage in attempting contact. We may guess that these racial differences are related to the degree of confidence an individual brings with him in his dealings with society.

Thus, uneducated whites find it easier to contact government than uneducated blacks, making the latter group badly underrepresented among the contactors. Second, race appears to have an independent effect, apart from education, in predicting contact. Those blacks who have achieved high educational status are still less likely to have contacted government than are whites at the same level.

To summarize, the voices of the relatively well-off in terms of socioeconomic status are communicated disproportionately to public officials through the process of individually initiated contact. And within the population, the better-off whites make contact in greater proportion than the better-off blacks.

THE DIMENSIONS OF CONTACT

For our purposes we may identify at least five important dimensions of any instance of contact:

(1) the nature of the contact;

(2) the content of the contact;

(3) the referent for whom the contact is made;

(4) the level of government at which contact is made; and

(5) the target at whom contact is directed.

(1) *The nature of the contact.* Empirically, there are several different types of contact with public officials. We have already had occasion to note that one type is the grievance or complaint. Such contacts may be included in a broader category which we may call *request contacts.* Another broad category is *opinion contacts.*

The first category covers all those cases in which the object of the contact is to provoke action on the part of the target *by providing him an opportunity to act.* That is to say, in such instances the contact provides the occasion for action, if the target so desires. There are three subcategories of request contacts. In many cases this form of contact involves a *complaint* about some injustice or difficulty the contactor has suffered or anticipates suffering. He seeks to have the injustice rectified by public authorities or he calls upon them to obviate the threat. Examples would be the complaint about a landlord's refusal to provide customary services or the grievance lodged when one believes he has been mistreated by the police.

Another form of request contact is the communication seeking *help* or a favor or a service from a public official or agency to which the contactor believes he or some referent is rightfully entitled. Seeking information about zoning laws or making known the need for a traffic light on a busy corner are examples of this type of request contact. This communication is not occasioned by the sense that one has suffered injustice.

Finally, we may classify as a form of request contact those communications directed to a public official asking him to "do something" about a problem that is viewed by the contactor in generalized terms. Such a contact is not designed as a comment on what the contactor believes is the preferable option among an array of options already being considered by an official. Nor is it an expression of opinion regarding some explicit aspect of existing policy. Thus, a request contact of this type would be the call to do something about job discrimination or ghetto problems or pollution. This form of contact is void of references to particular options open to the official; rather its purpose is to call attention to an area of concern in the hopes of initiating action. Unlike the other two types of request contacts, an appeal to do something offers only the vaguest sort of mandate or opportunity for the public official to act. The opportunities supplied by a complaint or a request are much more explicit.

The second broad category of contacts includes communications expressing an opinion. In one category of *opinion contacts* the individual is engaged in the act of throwing his weight on the side of one of several options for action already explicitly open to the public official. A decision is pending or action is being considered. To ask one's congressman to vote against a new weapons system is an apt example.

A second type of opinion contact is the *comment* on an existing state of affairs or the communication of support or opposition for actions already taken. Congratulatory messages and pledges of support for a position taken fall in this group.

Opinion contacts are generally *reactions* to policy decisions or possibilities, while request contacts are designed to *initiate* consideration of policy or problems. To make an opinion contact is not to present the target of the contact with an opportunity to act. Rather

TABLE 1

TYPES OF CONTACT BY RACE AND LEVEL OF GOVERNMENT

	Whites						Blacks					
	All Levels of Government		City Government Targets		County, State and Federal Targets		All Levels of Government		City Government Targets		County, State and Federal Targets	
	%	n	%	n	%	n	%	n	%	n	%	n
Request contacts	43	(108)	67	(61)	29	(47)	84	(27)	93	(13)	78	(14)
Opinion contacts	52	(146)	27	(25)	66	(108)	16	(5)	7	(1)	22	(4)
Uncodable	5	(13)	5	(5)	5	(8)	–	–	–	–	–	–
Totals[a]	100	(254)	99	(91)	100	(163)	100	(32)	100	(14)	100	(18)
Complaints	15	(38)	36	(33)	3	(5)	22	(7)	21	(3)	22	(4)
Ask for help	15	(39)	22	(20)	12	(19)	25	(8)	29	(4)	22	(4)
Do something	12	(31)	9	(8)	14	(23)	38	(12)	43	(6)	33	(6)
Urge action on pending decision	14	(37)	5	(5)	20	(32)	9	(3)	7	(1)	11	(2)
Comment	38	(96)	22	(20)	46	(76)	6	(2)	–	–	11	(2)
Uncodable	5	(13)	5	(5)	5	(8)	–	–	–	–	–	–
Totals	99	(254)	99	(91)	99	(163)	100	(32)	100	(14)	99	(18)

a. Percentages do not always total 100 due to rounding errors.

it is to express an opinion about actions taken or pending or considered when such action is not dependent upon the stimulus of the citizen's contact. Hence, we may classify the nature of citizen communication to public officials as request contacts or opinion contacts, depending upon whether they provide an opportunity for the target or whether they offer support or opposition for an opportunity already before or taken by a target.

In Milwaukee the two races exhibit very different patterns in the nature of their contacts. Looking at instances of contact made at all levels of government, we find that whites made opinion contacts more frequently than request contacts. Of the contacts that were codable (only 5% were not), 52% communicated opinions while 43% were requests. Blacks, however, divided their contacts overwhelmingly in favor of request contacts (84%).

Table 1 provides a breakdown of the different types of contact by race at various levels of government.

Several figures are of special interest in this table. The number of white contacts which fall in the "comment" category under the heading "all levels of government" would seem to indicate both a degree of political awareness and self-confidence not manifest in the black sample. To send a public official one's comment on some aspect of existing public policy or to communicate one's judgment about a position taken by that official presupposes that the contactor keeps abreast of public affairs to some extent. In addition, to send such a communication requires a certain degree of self-confidence, a sense that one's evaluation is worth a public official's time. That blacks are poorly represented in this category hints at the relative absence of these qualities in the Milwaukee sample.

A second figure of note is the concentration of blacks in the "do something" category. The plea to do something about a problem understood in general terms is the most ambiguous of requests. The target of such a contact is hard-pressed to know exactly what the contactor wants, and this makes a satisfying response difficult. While the number of black contacts is extremely small, necessitating a certain caution in interpreting the patterns we find, it is nevertheless interesting to speculate that if the same pattern prevails in the larger black population, it may do much to explain why public officials often appear unresponsive to black demands. Some well-intentioned officials simply may not know what is being asked of them. On the other hand the ambiguity of many individually initiated black contacts may also provide an excuse for officials not to respond. For some officials it may be convenient not to understand exactly what is being asked of them.

The table also breaks down the type of contact according to whether the target was a city official or an official at another level of government. It is apparent that on this dimension blacks relate similarly to local government and the other levels. That is, their contacts with all officials and agencies are prompted by needs to which they expect a public response. Contact of this sort plays neither a supportive nor nonsupportive role for public policy makers seeking citizen opinions.

The data also show that individuals of neither race generally stir themselves to take publicly articulated opinion positions in city conflicts over public policy. When factions and interests coalesce on the various sides of an issue, as they inevitably must, there appears to be little spontaneous expression of support or opposition by members of the mass public. When conflicts occur in the various arenas of city government, individual citizens appear largely in the role of spectators rather than participants in the pageant.

In conflicts at other levels of government, whites actively take sides through their expression of support, opposition, or comment. As a summary judgment about this table, however, it can be said that blacks in no way appear to make a significant contribution to the body of public opinion at any level of government, at least insofar as that opinion is generated by individuals acting in their private capacity rather than as group spokesmen.

(2) *The content of the contact.* Another dimension of citizen contacts is their content. What are people moved to communicate to public officials? What sorts of problems provoke spontaneous expressions of opinion?

To make contact is to provide an important kind of feedback for public officials. We know, for example, that congressmen rely heavily on contact initiated by constituents, despite its admittedly unrepresentative quality. Bauer et al. (1963: 434, 436) write:

> To our surprise, we found many congressmen looking to mail and personal contacts as sources of information on vital issues. . . . Visitors and telephone callers . . . are listened to as indicators of feeling back home.

Other scholars found that a high proportion of a sample of upper-level administrators in the federal government depended on newspaper comment, letters of complaint, and clients' grievances for their information about constituency attitudes (Friedman et al., 1966: 196). And Olson (1969) discovered in his study of citizen grievance letters received by the governor of Wisconsin that such communications are perceived by the state chief executive as some indication of agency performance and provide him with a control device to wield against agencies accordingly.

In short, the content of citizen contacts presumably serves in large measure to shape the pictures of constituency opinion and needs that public officials form, and these contacts help to set the agenda of priorities by informing officials of areas of concern and sensitivity.

An analysis of the content of citizen request contacts in Milwaukee, however, throws into some doubt the utility of using individually generated contacts as indicators of special problems, except perhaps in extraordinary cases. At the most general level one may say that what normally concerns contactors in Milwaukee, regardless of race, is the quality and scope of public services. Poor sanitation pickup, inadequate traffic control, the absence of decent housing, and the failure of the city to regulate citizen behavior (e.g., landlords who violate housing codes or juveniles who vandalize neighborhood shops) are the types of concerns which comprise the major portion of citizen contacts at this level. While Gellhorn (1966: 164) found that complaints about housing dominated the content of citizen letters to a New York City borough president and Campbell and Schuman (1968: 40) found that dissatisfaction over park and recreation facilities outranked negative feelings about other services in fifteen large cities, the Milwaukee whites complained most frequently about faulty garbage collection and snow removal. Data for black service-related contacts are too scanty to discern patterns.

Yet in general for both races the request contacts range fairly evenly across the spectrum of city services, providing little sense of areas of special concern. As a source of information, a tabulation of these complaints and pleas for help or service would provide little aid for officials responsible for assessing the performance of particular municipal services or agencies. What does emerge simply is that a moderate level of dissatisfaction exists, voiced by a small minority of the city's residents. One suspects that if large numbers of people are ever moved to contact a city agency about its service, it is likely that such contact must be prompted by a crisis. (The case comes to mind of the failure of New York City to plow the streets of Queens after a snowstorm in 1968, stranding commuters in their homes, an oversight which stimulated an outpouring of angry mail.)

As noted in Table 1, blacks almost exclusively made request contacts, regardless of the level of government, while whites tended to send expressions of opinion or comment to officials outside local government. The content of the white opinion contacts primarily concerned the war in Asia and the Middle East crisis. Other controversial issues–gun control, birth control, sex education, the ABM system, and state aid to parochial schools–elicited messages from the white sample in relatively even measure. The distribution of the white opinion contacts seems a better gauge than that of the request contacts for measuring the degree of public concern and for spotting areas of sensitivity.

(3) *The referent for whom the contact is made.* When people contact public officials, they do so on behalf of some referent. That is, they hope by their contact to elicit some decision or action by a target in relation to someone or in order that someone will benefit or pay.

In coding the instances of contact we may identify three principal referents: the individual or his primary groups, the secondary or social group, and the community at large. For example, a contactor may complain to government about something done to him or that he wishes done for him or his family for which some public agency or official has responsibility. Many of the Milwaukee contacts of this type concerned requests to legislators for military deferments, pleas for help in obtaining Social Security and veterans' benefits, and requests for information on licensing and zoning laws. In these cases the citizen pursued the contact on behalf of his own interests or those of this family.

A citizen may also initiate contact, still in his capacity as an individual, to request that government do something to or for a particular social group (e.g., black people, the poor, young people), a client group (e.g., for welfare mothers, veterans), or a neighborhood.

Finally, he may make contact to elicit action for the community at large, defined in this case as the entire body of the undifferentiated citizenry of the jurisdiction in which the public official serves. To call for lower state sales taxes, to support the ABM system, to decry the state of community relations, or to seek action on pollution are all examples of contacts in which the community–city, state, or nation–is the referent. That is, whatever response is elicited, it will affect everyone.

The idea of contactors' referents is important theoretically because contact is a demand for attention. By making contact the citizen directs attention to his referent. If demands are never made on behalf of some possible referent, then that group or individual is unlikely to occupy a very prominent place in a public official's mind. For example, it is reasonable to speculate that some neighborhoods house vociferous contactors whose referent is often the neighborhood. These are the good citizens who notice dangerous street intersections, monitor street-cleaning operations, and complain to the police about the lax handling of juvenile loiterers. It is often necessary to call such problems to the city's attention in order to have them treated. As a result of frequent contact, officials will develop images of vocal neighborhoods, as Lindsay has done with regard to the areas of Queens whose streets were not plowed. These vocal neighborhoods can attract attention in the press and perhaps cause political trouble for elected officials. In other neighborhoods the population is silent and passive, and thus invisible. Those neighborhoods are never referents of individual citizen contacts.

The notion of contact referents is also important in that it invites a very limited test of the Banfield and Wilson (1963) public-regarding versus private-regarding ethos theory. Briefly, the theory posits two competing views as to whose interests the city government is designed to serve, those of the community or public or those of particular subcommunities or groups within the city. The public-regarding view is the province of the white middle class, while the private-regarding view is generally held by people of lower-class immigrant stock. While numerous problems of measurement and conceptualization attend this theory (Hennesy, 1970; Wolfinger and Field, 1966), it suggests in its largest outlines that different groups have very different notions as to the purposes to which government might appropriately be put. What the Milwaukee data provide is an opportunity to compare whites and blacks in this regard according to the referents for whom the two racial groups seek benefits from government.

Table 2 shows that both races almost invariably contact city government on behalf of some group or individual referent. The bulk of contacts with local government reflect little concern for broader, communitywide interests at the city level. Except in the few instances in which contactors called for city officials to do something

TABLE 2
CONTACT REFERENTS

	White				Black			
	City Government Only		All Levels of Government		City Government Only		All Levels of Government	
Referents	%	n	%	n	%	n	%	n
Individual, primary group	38	(35)	26	(65)	21	(3)	22	(7)
Secondary group, neighborhood	42	(38)	28	(71)	79	(11)	66	(21)
City, nation, community at large	11	(10)	28	(72)	–	–	6	(2)
Uncodable	9	(8)	18	(46)	–	–	6	(2)
Totals	100	(91)	100	(254)	100	(14)	100	(32)

about community relations or about pollution, contacts were made primarily on behalf of neighborhood or family. If we collapse group and individual concerns into the private-regarding category, as have Banfield and Wilson (1963: 46), we find that there is little difference between the races. Insofar as the referents of contacts measure people's notions of which interests government ought to serve, the Milwaukee sample is mostly private-regarding at the city level.

The pattern changes when the referents of contacts made at all three levels of government are taken into account. Whites achieve more balance among the three types of referents, but still remain weighted on the private-regarding side. Blacks almost exclusively remain private-regarding: government at all levels is perceived as an instrument to serve group and individual interests rather than some notion of the general or community interest.

Such a finding is not surprising: blacks comprise a somewhat more cohesive social category than do whites. While black communities show cleavages similar to those found in white communities, blacks are still an easily identifiable group. Blacks also inhabit segregated neighborhoods, tend to cluster at the low end of the socioeconomic scale, and comprise large portions of the client groups dependent upon supportive services administered by the various levels of government (e.g., welfare, antipoverty, public housing, manpower programs). Blacks also believe that their neighborhoods receive poorer treatment from the city in comparison with white neighborhoods. Poverty, discrimination, and heavy dependence on public services may do much to explain black tendencies to contact government on behalf of group and individual rather than community referents.

(4) *The level of government at which contact is made.* A fourth dimension of citizen contact of some concern involves the level of government to which contact is directed. Americans express greater interest in national and local politics than in state politics and tend to follow more faithfully the activities at the former two levels (Jennings and Zeigler, 1970: 525). Indeed, the Milwaukee data show that the high salience of national and local politics is associated with a much greater propensity for both races to make contact here rather than at the state level. State government is especially free of the sort of scrutiny represented by individually initiated citizen contacts brought to bear by blacks. Nearly half of both the white (44%) and the black (47%) contacts were made at the national level, while 36% of the white contacts and 44% of the black were made at the city level. State targets account for only 18% of white contacts and 3% of those made by blacks.

(5) *The target at whom contact is directed.* A final dimension of citizen contacts concerns the specific targets of the contactors. Hypothetically, a citizen making contact pursues his objective with a minimal expenditure of personal resources. A calculation to this effect, which includes in the equation an attempt to ensure the highest probability of success with a minimal expenditure of time and energy, leads the citizen in the vast majority of cases to contact elected rather than bureaucratic officials. Well over 80% of the contacts made by both blacks and whites were directed at elected figures.

Such a strategy is a rational one. To contact an elected official is by far the easier course. The elected official is normally more visible than the civil servant or the appointed administrator. Much less information must be gathered as to where to direct a contact. In addition, if a citizen request requires action which only a bureaucrat may handle, then by contacting one's elected representative first, the citizen may hope to expedite such action by having this official intercede on his behalf. Indeed, to contact an elected official in the hopes that he will intercede with the bureaucracy is often to enlist a powerful ally in one's bout with the government. To call upon the bureaucracy alone may be to go naked and unarmed into the lion's den.

On occasion the contactor may call upon an elected official simply to enlist his aid as a pathfinder by relegating to him the difficult task of discovering which bureaucratic agency has the authority to satisfy the contactor.

There is some evidence that contacting elected officials rather than administrators increases the chances of success and speed of response in having a request contact satisfied. Olson found that the governor was much more likely to respond favorably to citizen requests (and more promptly) than were agency administrators, even, apparently, if the requests were similar in content (Olson, 1969: 746). In a related vein, Gellhorn (1966: 136-137) reports that complaints which pass through governors' offices usually receive prompter attention by higher-level officials than if they had been sent to administrators.

If elected officials are more responsive than bureaucrats, there is a reasonable explanation. The former depend for their jobs on a satisfied constituency, and in satisfying those who make contact, the elected official is less bound than the bureaucrat by standards of professional public administration. The congressman who makes a special plea for a constituent about to be inducted into the military has greater freedom to devise a rationale for making an exception than the bureaucrat in the draft board has, bound as he is by the necessity for impersonal, uniform administration. If this is the case, then elected officials more than bureaucrats lend the political system flexibility, the obvious possibilities for administrative discretion notwithstanding.

In Table 3 the tendency to contact elected officials is graphically illustrated.

When citizens contact elected officials, it is clear from this table that they tend to contact their representatives to the various legislative bodies rather than the more visible chief executives. This pattern holds true at each level of government. While people are normally more aware of the identity and behavior of their chief executives than that of their representatives, the data here indicate a greater willingness to rely upon the latter.

At all levels citizens virtually ignore the bureaucracies. In view of the arguments made in favor of contacting elected officials, this is rational behavior. There is some evidence, however, to indicate that the likelihood of contacting a bureaucracy depends upon the nature

TABLE 3
TARGETS OF CONTACT

	White		Black	
	%	n	%	n
All targets				
Elected official	84	(213)	88	(28)
Bureaucratic official	14	(36)	6	(2)
Other, unspecified	2	(5)	6	(2)
Total	100	(254)	100	(32)
Specific targets				
President	6	(15)	–	–
U.S. representative	18	(45)	28	(9)
U.S. senator	20	(50)	19	(6)
U.S. bureaucrat	1	(2)	–	–
Governor	4	(10)	3	(1)
State legislator	13	(34)	–	–
State bureaucrat	1	(2)	–	–
Mayor	3	(7)	6	(2)
Alderman	20	(52)	31	(10)
City bureaucrat	12	(32)	6	(2)
Other	2	(5)	6	(2)
Total	100	(254)	99	(32)

of the issue. When all the respondents, both contactors and noncontactors, were asked first whom they would contact about getting a traffic light in their neighborhood, a majority of both races named elected officials. But when they were later asked whom they would contact about getting better police protection, a majority named various officials in the police bureaucracy. The willingness or ability to contact bureaucratic officials is probably a function of the salience of the particular bureaucracy and the ease with which people can identify the locus of bureaucratic responsibility. The source of responsibility for traffic lights and stop signs is relatively obscure, while that for the police is not.

A curious variation on this finding appears when contactors and noncontactors are compared. Contactors of both races—but especially those who are black—reveal a greater tendency to predict that they would contact elected officials than do the noncontactors. That is, in both the case of the traffic light and that of better police protection, noncontactors are more likely than the contactors to predict that they would call upon bureaucratic officials. Such a pattern seems to reflect a kind of innocence on the part of the noncontactors: their assumption is that to get action, one goes to the source of

TABLE 4

REFERENTS OF CONTACTORS BY SPECIFIC TARGETS

	White				Black			
	Individual and Group Interests		Public or Community Interests		Individual and Group Interests		Public or Community Interests	
	%	n	%	n	%	n	%	n
President	1	(1)	14	(10)	–	–	–	–
U.S. representative	12	(16)	25	(18)	29	(8)	50	(1)
U.S. senator	15	(20)	29	(21)	18	(5)	50	(1)
U.S. bureaucrat	1	(1)	1	(1)	–	–	–	–
Governor	4	(5)	4	(3)	3	(1)	–	–
State legislator	12	(17)	11	(8)	–	–	–	–
State bureaucrat	1	(2)	–	–	–	–	–	–
Mayor	2	(3)	6	(4)	7	(2)	–	–
Alderman	31	(42)	3	(2)	36	(10)	–	–
City bureaucrat	21	(29)	7	(5)	7	(2)	–	–
Totals	100	(136)	100	(72)	100	(28)	100	(2)

Uncodable: 46 Uncodable: 2

administration. Contactors, however, apparently acquainted with the difficulties of finding the locus of bureaucratic responsibility, prefer to go to elected officials.

Table 3 shows that the degree of contact with individual citizens varies for different types of public officials. In Table 4 data are presented which suggest at least for whites that the referent of the contactor determines to some extent the target of his contact. City aldermen and bureaucrats are more likely than any other officials to be the recipients of contacts made on behalf of individual or group interests. National political figures–especially senators and representatives–are most likely to be the targets of contact made on behalf of public or community interests.

In summarizing the basic patterns of contact, the following points stand out:

(1) Contactors are better off than noncontactors in terms of social well being. In addition they are most likely to be white.

(2) Citizens tend to make request contacts in the city rather than opinion contacts.

(3) Complaints in the city range across the spectrum of public services, offering little aid in spotting areas of special concern.

(4) In the city, contactors of both races—but especially blacks—are private-regarding, at least insofar as we have measured this orientation by the use of the contactors' referents.

(5) The races differ little in their tendency to contact local and national government targets rather than state officials.

(6) Citizens tend to contact elected officials rather than bureaucratic officials at all levels of government, although *predictions* of contact seem to vary with the problem.

POLITICAL PARTICIPATION AND CONTACT

Contacting public officials is a form of political participation (Milbrath, 1965: 18). All the forms of contact are ways of making demands for a particular allocation of rewards and resources in the society. Taken as an isolated act, an instance of contact poses little threat to a public official or to the society. Failure to meet the demands posed by any particular contactor seldom results in the loss of the official's job, nor does failure to satisfy a single contactor generally have grave consequences for the social order.

Yet to understand the act of contact as an isolated form of individual political participation would be a mistake. Indeed, contact reflects expectations which can be backed up by the use of political sanctions inherent in the act of contact itself. Contact, as we have conceived it, is a solitary act. Rather, contactors possess political resources as any other citizen which may be used as sanctions against unresponsive public officials. These include the vote, party and campaign efforts, protest tactics, appeals to higher authority, and in some cases the ability to create unfavorable publicity. If contactors never used their political resources, public officials would be under little constraint to satisfy their individual demands. What will be made clear in this section is that the contactors are members of the politically active stratum in the society and that contactors are more active in politics genenerally than noncontactors.

To say that contactors participate in other forms of political activity is not to argue that they explicitly seek to back up the demands they make as individual contactors through political action. Rather it is to argue that contact is simply one aspect in a syndrome of behavior. Experiences in one form of participation may, of course, determine action taken in another form: the man who is rebuffed by the alderman to whom he complains may not vote for him for that reason in the next election. But the relationship between contact and behavior is probably not always so direct; contact does not always precede participation or cause participation in other forms of activity.

What is clear, however, is that contactors provide the public official with one indication of what some activists are thinking and working for. For one thing, the reason for the contact is a potential rallying point. Less specifically, the contactor may be a bellwether, a representative of the politically attentive and active citizenry. Insofar as elected officials are dependent on a satisfied and supportive constituency and insofar as bureaucrats attempt to serve a clientele satisfactorily, the voice of the contactor is important *because* he is a political activist.

In every case in Table 5, there is a substantial positive relationship between contacting public officials and participation in conventional electoral and party politics. What the simple coefficients of association obscure, however, is the extent to which contactors are active in each of the forms of participation in comparison with the noncontactors. While the exigencies of space preclude showing each table, it is sufficient to note the average percentage difference between contactors and noncontactors in each instance of political activity is 19.3 for whites and 31.7 for blacks. That is, on the average, one-fifth fewer noncontactor whites and one-third fewer noncontactor blacks participate in any given act of political participation. For both races contactors are substantially overrepresented in the politically active stratum.

TABLE 5
RELATIONSHIP BETWEEN CONTACTING PUBLIC OFFICIALS AND POLITICAL PARTICIPATION (gamma)

	White	Black
Vote for President 1968	.57	.85
Vote for mayor 1968	.21	.69
Talk to anyone about presidential election	.60	.63
Talk to anyone about mayoral election	.57	.52
Persuade someone to vote for any candidate	.47	.64
Help in an election campaign	.74	.87
Give money to party or candidate	.46	.61

Voting and donating money may be done anonymously, but talking to people about an election, persuading people to vote in a particular way, and helping out in a campaign are necessarily group or social activities. The latter forms of participation offer opportunities for interpersonal influence on a face-to-face basis. Although we have no way of knowing how much conventional political activity is a function of experiences in contacting government, it is still probable that impressions gained in contact carry over into other areas of political behavior. (The reverse is also true: political activity provides incentives and opportunities to initiate contact.) The point to make is that contactors are active in these social forms as well as in the more private ones. That this is the case indicates the potential importance to an elected public official of the way he handles contacts over time. If he is unresponsive in general, he not only alienates people who are likely to vote, but also people who are likely to be in a position to influence others. In regard to the white population this seems especially important. Among the whites the average percentage difference between contactors and noncontactors is greater for the group context political activities (22.8) than for the private ones (14.3). White contactors differ from white noncontactors *especially* in their tendency to participate in politics in social settings. However, for blacks there are no such differences between contactors and noncontactors when we control for social and private forms of political participation.

Contactors also demonstrate a greater familiarity than the noncontactors with the personnel of Milwaukee city government, thereby indicating not only a higher awareness of the details of local government but also the ability to identify potential targets of contact. If the energy involved in gathering information as to where to make a contact represents one of the costs of this form of activity, then contactors appear to possess a distinct advantage over noncontactors. Respondents were shown a list with the names of the mayor, the president of the common council, the school superintendent, the head of the city welfare department, the chief administrator of the city housing authority, and the head of the Model Cities agency. They were then asked what job each person held. Except for the case of the Model Cities director, where only one person in the entire sample was able to identify him correctly, contactors of both races invariably identified the names with greater accuracy than the noncontactors. Black contactors were more aware than were the white contactors of the identity of the heads of welfare and housing, while whites were more accurate in placing the president of the common council. Black competence here may probably be explained by the racial makeup of the clientele of the two bureaucratic agencies. The obscurity of the president of the common council for black citizens (over half the total white sample placed him correctly while only 6% of the black sample could do so) is less easily explained. The fact that the man represents a white ethnic constituency and is not outspoken on racial matters undoubtedly are contributing factors.

Contactors, then, comprise a significant portion of the city's politically active stratum. Contact is a part of the entire syndrome of political activity. As such it is possible to argue that the contact experience and participation in other forms of activity interact to influence, lead into, or reinforce one another. The public official is not dealing in most cases with an isolated individual when he responds to an individually generated contact. Rather he has come into contact with a citizen who demonstrates a willingness to participate in group forms of political activity and who exhibits an awareness of government that signifies the potential to make more contacts.

CONCLUSIONS

Individually initiated citizen contact with public officials may best be understood as an aspect of the relationship inherent in the structure of political representation. In her book, Hanna Pitkin (1967: 209) writes that representing "means acting in the interest of the represented, in a manner responsive to them." A citizen contact, understood in the context of this definition, is a demand on the part of the represented for consideration of some interest in which he has a perceived stake. Contact is made necessary in cases in which political society has delegated to government the authority to regulate, protect, or advance certain interests.

When the individual citizen is unable to further or protect his interests by his solitary private initiatives, he calls upon those who have the publicly accorded power and resources to do so. Some interests concern the welfare of the individual citizen which he cannot promote or protect by his own efforts. When a landlord refuses to make repairs required by building safety codes or when a draft board is unsympathetic about a special need to avoid military service, the only means a citizen has for redress is to appeal to those who represent his interests. Other interests are of a collective nature. Individuals identify in varying degrees with a variety of social collectivities, ranging on occasion from a racial group to a neighborhood to the nation-state. For each of these collectivities the individual identifies what he conceives to be certain interests and preferred states of being. In many cases the disposition of the interests of these various groups are dependent upon government behavior. Frequently the citizen delegates the task of protecting these interests before the councils of government implicitly or explicitly to group spokesmen, but at other times he seeks to foster what he conceives his group interests to be by his own initiative. Such action may occur, of course, in conjunction with attempts by his group spokesman to pursue the same ends. Thus, the citizen may petition the President for an end to the war at the same time that spokesmen for an organization to which he belongs are doing so. In short, individually initiated citizen contact is one means of activating the representation relationship between the individual and those to whom he, as a member of political society, has granted authority to act in regard to interests which the citizen cannot foster by his own devices.

This study of citizen contacts provides some data on who takes advantage of this opportunity for representation and suggests at the same time some elaboration of a theory of representation. On the first point the most outstanding pattern revealed by the data is the differential rate at which blacks and whites initiate individual contacts with public officials.

Not only are whites much more likely to make contact, but they are also more likely to do so in a sustained fashion. Many among the one-third of the white sample who had made contact did so several times, while blacks were largely one-time contactors. To the extent that visibility in the public official's eye is a function of frequent contact, whites, in their capacity as individual private citizens, are more visible than blacks. Other factors, to be sure, contribute to visibility: the severity of the problems people face and the extent to which these impose costs on the society are two related factors. Blacks in the city suffer more severely than do whites in matters affecting life style and opportunities, and this deprivation does much to gain public attention. In addition, there exist a number of organizations to protect racially defined interests, and these have few acceptable or strong counterparts in the white community. *Blacks as a group,* then, are not without visibility. Yet black private citizens are comparatively silent about their individual or collective condition. What direct communication is done with public officials is apparently left to organization spokesmen. In contrast, white individuals are vocal: in sheer volume white demands for attention through the device of citizen contact literally drown out those of individual blacks.

For both races there are moderate positive relationships between the tendency to initiate contact on the one hand and education and conventional political activism on the other. Politically, contactors constitute a segment of the population with considerable resources, experience, and willingness to participate actively.

The data on education suggested (but did not necessarily demonstrate) some degree of independent effect of race on propensity to initiate contact. To the extent that blacks do not contact public officials *because* they are black, the explanation lies in a host of factors. These would certainly include the fear of being rebuffed or refused a hearing because of racial prejudice, the lack of black public officials to whom one might appeal, and the sense that appeals are futile.

To summarize, those who use individual contacts as a means of activating the representational relationship are likely to be middle-class whites who are active in politics. Contactors represent a favored segment of society, both in socioeconomic and racial terms, and it is their concerns and their versions of problems to which public officials are exposed through this means of communication.

One conclusion that may be drawn relative to a theory of representation is that only an unrepresentative few take advantage of the opportunities for representation through the device of citizen contact. Yet the data provided here take us beyond this observation to more fertile areas. In a recent article on the political representation function of city councils, Kenneth Prewitt and Heinz Eulau (1969: 427) speak of "the unresolved tension between the two main currents of contemporary thinking about representational relationships."

> On the one hand, representation is treated as a relationship between any one individual, the represented, and another individual, the representative—an *inter-individual* relationship. On the other hand,

> representatives are treated as a group, brought together in the assembly, to represent the interest of the community as a whole—an *inter-group* relationship.

While Prewitt and Eulau do not entirely reject the interindividual formulation, they embrace the notion of representation as an intergroup phenomenon as the more crucial to an empirical theory of representation. Representation in this sense is understood as a system property: thus, they write (1969: 428), "representation as well as other variables we consider are group rather than individual properties; thus we make statements about governing bodies and not individual public officials." The intergroup relationship is one in which the governing group responds to or represents politically organized viewpoints among the citizenry (Prewitt and Eulau, 1969: 430). As we have seen, however, individual contacts are also demands for representation and the contactors themselves are not necessarily organized.

What an analysis of individual citizen contacts, understood as demands for representation, suggests is that the tension between these views of the representational relationship is resolvable and that both views are important to an understanding of representation. The pattern of contacts studied here indicates that citizens call upon different public figures at different levels of government to perform different sorts of representational tasks. There is, in other words, an implicit *division of the labor of representation,* and this division may be understood in part in terms of the individual-collective formulations.

To understand how this is the case, it must be made clear first that all of the public officials dealt with in this study as targets of contact have a representative function. The Prewitt-Eulau conception of representation focuses narrowly on city councilmen. As Edward Muller (1970: 1150) points out, "legislative bodies are not the only loci of representational linkages in political systems." The presidency—and by extension, other chief executives—and the bureaucracies at all levels also perform representative functions in the sense that Pitkin defines representation.

That there exists a division of labor simply in terms of whom private individuals contact among representatives is clear from a review of the data. People tend to contact elected officials rather than bureaucrats. Among elected officials, they generally call upon legislators rather than executives at each level of government. Insofar as *individual* citizens demand representation, they do so mainly through legislative representatives. Executives and bureaucrats perform representative functions, but their representational relationship, we may infer, is largely with group spokesmen. Bureaucracies must respond to organized clientele groups, and chief executives must deal with a variety of interest collectivities as well as attempt to represent some conception of the public interest.

Not only do individuals tend to contact certain types of public officials more frequently than others, but they also make different kinds of contacts according to their target. Opinion contacts are directed largely at national and state government officials, while request contacts dominate the communications to city officials. The role of private citizens is to contribute some measure of support or opposition for representatives at the two superior levels of government, while at the local level it is more to initiate opportunities for representation.

Finally the data showed that people making contact on behalf of individual or group referents communicate primarily with local officials, while those concerned with community referents call upon national officials.

Individual contacts with public officials are inter-individual demands for representation of certain interests as defined or conceived by private citizens. Some of the interests on whose behalf citizens make contact are group or public interests, but the essence of the representational relationship is still the link between the individual and his target, not between the collective represented and the representative body. The notion of individual contact makes clear that the interindividual component of the representational relationship as well as that component in which representative bodies respond to organized demands are both operative in any structure providing for political representation. Some officials at certain levels are called upon more frequently than others to perform the task of interindividual representation, while others represent primarily public or organized group interests. By understanding this division of labor of the representation task, the tension between the two views of representation is largely resolved. Representation is a complex job, requiring responsiveness to both group and individually generated conceptions of critical interests. Different officials are asked to pay heed to interests defined by these different sources within the public.

The notion of a division of labor appears more important for whites than for blacks. To the extent the interests of the latter are represented at all, it is mainly through contacts made by organized group spokesmen. The existence of individually forged avenues for representation points to a means of gaining attention which blacks appear to have left largely unexploited.

REFERENCES

ALMOND, G. and S. VERBA (1965) The Civic Culture. Boston: Little, Brown.

BANFIELD, E. and J. Q. WILSON (1963) City Politics. Cambridge, Mass.: Harvard Univ. Press.

BAUER, R., I. POOL, and L. A. DEXTER (1963) American Business and Public Policy. New York: Atherton.

CAMPBELL, A. and H. SCHUMAN (1968) Racial Attitudes in Fifteen American Cities. Ann Arbor, Mich.: Survey Research Center, Institute for Social Research.

FRIEDMAN, R. S., B. KLEIN, and J. H. ROMANI (1966) "Administrative agencies and the publics they serve." Public Administration Rev. 26 (September): 192-204.

GELLHORN, W. (1966) When Americans Complain: Governmental Grievance Procedures. Cambridge, Mass.: Harvard Univ. Press.

HENNESY, T. (1970) "Problems in concept formation: the ethos 'theory' and the comparative study of urban politics." Midwest J. of Pol. Sci. 14 (November): 537-564.

JENNINGS, M. K., and D. ZEIGLER (1970) "The salience of American state politics." Amer. Pol. Sci. Rev. 64 (June): 523-535.

LANE, R. (1959) Political Life. New York: Free Press.

LINDSAY, J. (1969) The City. New York: New American Library.

MILBRATH, L. (1965) Political Participation. Chicago: Rand McNally.

MULLER, E. (1970) "The representation of citizens by political authorities: consequences for regime support." Amer. Pol. Sci. Rev. 64 (December): 1149-1166.

OLSON, D. J. (1969) "Citizen grievance letters as a gubernatorial control device in Wisconsin." J. of Politics 31 (August): 741-755.

PITKIN, H. (1967) The Concept of Representation. Berkeley: Univ. of California Press.

PREWITT, K. and H. EULAU (1969) "Political matrix and political representation: prolegomenon to a new departure from an old problem." Amer. Pol. Sci. Rev. 63 (June): 427-441.

WOLFINGER, R. and J. O. FIELD (1966) "Political ethos and the structure of urban government." Amer. Pol. Sci. Rev. 60 (June): 306-326.

3

Networks of Information and Influence in the Community

SHELDON BOCKMAN
HARLAN HAHN

□ A PRIMARY METHOD OF COMMUNICATION between political decision makers and other members of the community is the acquisition of information from interpersonal sources. Through daily conversation with family, trustworthy friends, and knowledgeable leaders, individuals fashion their attitudes and values about local events. Although this information often is transmitted by the mass media, personal communications still play a crucial role in this process (Katz and Lazarsfeld, 1955).

Within the democratic tradition the critical nexus between power, influence, and information is of fundamental concern. Government officials are required to be responsive to the interests of the community. On the other hand, to secure an informed citizenry, a democratic society must assure a constant flow of information, provide for accessible leaders, and guarantee independent sources of political knowledge through a free press as well as other forms of communication. As a result, there must be a constant exchange of information between political elites and other members of the

AUTHORS' NOTE: *The authors wish to acknowledge the support of this project by an Urban Crisis grant from the University of California funded by the Ford Foundation. Appreciation also is expressed to Miss Aylene Waggoner and Mr. George Naso for their assistance in this project.*

community. Decision makers are obliged to justify their decisions and policies to the people, and they are held accountable by direct elections and referenda.

Although there have been numerous studies of the relationship between public officials and other citizens through conventional mechanisms such as voting (Campbell et al., 1960), political parties (Key, 1964), and interest groups (Truman, 1951), relatively little attention has been devoted to the importance of interpersonal forms of communication between political leaders and the general population. On the other hand, recent research on community decision-making has suggested that many critical political communications occur outside the institutional framework of government (Dahl, 1961; Hunter, 1953; Bachrach, 1967; Mills, 1956). Thus a major need has emerged for a synthetic analysis which would integrate the study of community influentials with the analysis of communication patterns. As Clark (1968: 48-49) has noted, "A more rigorous theory of decision-making than exists at the present . . . will have to integrate, on theoretical and empirical levels, both the power and communications approaches to decision-making. Any contribution toward such a synthesis would be strategic for research."

While the importance of interpersonal contacts has been recognized in the dissemination of information from the mass media (Katz and Lazarsfeld, 1955), perhaps interactions between political elites and citizens are of equal or greater significance in the formulation of community policy. The nature of political controversies frequently may be shaped by the information transmitted by influentials to local residents and by communications that flow from the citizenry to community elites. One of the few theoretical perspectives that can be employed to confront this problem is social circle analysis (Kadushin, 1968). A major contribution of this approach is embodied by the concept of linkages that weld together diverse individuals through interpersonal communications.

While the value of social circle analysis has been recognized for the study of various social and political organizations (Kadushin, 1968; 1966), little effort has been made to exploit its potential for explaining the linkages between political influentials and other members of the community. By tracing the patterns of interaction between political elites and others with whom they discuss local issues, it may be possible to integrate research on communication and decision-making and to probe some crucial questions concerning the relationship between political leaders and the populace.

This approach, of course, implies that communications between elites and nonelites may either be transmitted directly or through intermediaries. In either event, the exchange of information between decision makers and the community at large is of critical significance for the operation of a democratic society because it indicates the opportunities for access to both the influential and the average citizen. Moreover, such interpersonal contacts may not be random or haphazard; in fact, there may be an identifiable network that facilitates communications between local residents. As Kadushin (1968: 691) has noted, a critical unresolved question is: "What is the precise nature of links, if any, and what is the total structure of the linkage system." One major objective of this research, therefore, is to examine interpersonal communications reported by decision makers and to relate their sources of information to the persons mentioned by the general population.

Moreover, the relationship between elites and other members of the community is of special concern to minority groups in a democratic society. Frequently deprived of formal representation in government, denied the resources necessary to secure a majority, and often systematically excluded from extensive interaction by virtue of a legacy of discrimination, black and Mexican-American citizens may encounter formidable obstacles in their efforts to affect political decisions. The nature of minority group involvement in the flow of information could be a major prerequisite for understanding the political implications of racial and ethnic relations in a democratic society. As a result, another major purpose of this research is to explore the role of minority groups in the transmission of communications in the community.

RESEARCH STRATEGIES IN THE COMMUNITY

The community selected as the setting for this research, which was designated by the pseudonym Spectra, is a small town located in Southern California. Although the population of this city was slightly more than 4,000 in 1970, it is a commercial center for a valley area with a total population of approximately 23,000. One of the principal reasons for the choice of Spectra as a research site was the presence of a substantial number of black and Mexican-American residents. Three issues were chosen as a focus for this study: a proposed rerouting of a major highway through the community, a city council election involving the unsuccessful challenge of two white incumbents by a black candidate, and a planned unification of the school district. Although each of the issues contained important implications for black and Mexican-American citizens, none of the issues aroused strong feeling either among the minority communities or the white population. As a result, the study centered on relatively routine issues of low saliency which are the most common types of issues emerging in any community.

Spectra also appeared to offer an unusual opportunity to experiment with new methodological techniques for the examination of communications and decision-making. Although limitations on the resources available for this research necessitated the choice of a community of this size as a pilot study, the procedures and technique developed in this project could easily be applied to more numerous and more extensive urban areas.

A modified reputational analysis coupled with an issue-specific approach was used to identify political elites in Spectra.[1] Decision makers were selected on the basis of information supplied by 31 organizational leaders who were interviewed separately after a complete enumeration had been made of all organizations in the community. Not only were white organizational leaders randomly selected to ensure the representation of diverse institutions within the community, but all leaders of black and Mexican-American organizations were interviewed to elicit the selection of major decision makers in the minority populations. Organizational leaders were free to offer as many responses as they wished in selecting the issues and the decision-makers that formed the basis for subsequent procedures.[2] The decision-making structure of the community was defined from the pool of names produced by the questions concerning both general and issue-specific influence.

Initially, a top decision maker was defined as a person who received eight or more nominations from organizational leaders in response to five questions concerning both general influence and influence exerted on specific issues. In subsequent interviews, the top decision makers also were asked: "Which people in Spectra do you regard as generally most influential in the local issues which confront this community?" Key decision makers were defined as persons who received ten or more nominations in response to this question from the original group of top decision makers. An additional group of top decision makers was selected on the basis of receiving three or more nominations from key decision makers. While key decision makers received from 13 to 28 nominations from other top decision makers, no other member of the elite circle obtained more than six selections. Finally, all black and Mexican-American leaders who acquired any nominations from either organizational leaders, top decision makers, or key decision makers were included within the decision-making elite. No black or Mexican-American leaders, however, met the criteria necessary for designation as a key decision maker.

The second phase of this study employed the use of snowball sampling to map communication linkages in Spectra.[3] Although the standard reputational techniques may appear to resemble snowball sampling, in this research an expanded version of the method was used not only to identify major political elites but also to trace patterns of communication between and among elites and other residents of the community.

First, all top decision makers were interviewed and asked to select every other person with whom they discussed the three local issues. The persons identified by the top decision makers, in turn, were interviewed and requested to choose another group of individuals with whom they discussed local issues. This procedure was repeated for an additional two stages of interviewing. At each stage, approximately 25% of the persons mentioned most frequently as sources of information about local issues were interviewed. A total of 112 persons were interviewed in the snowball sample. Despite the difficulties posed by the size of the sample and the community, the procedure apparently was successful in identifying most persons who played a critical role in the transmission of information concerning local issues. In fact, this technique permitted the discovery of a communication network which formed the major conduit for the flow of information in the community.

A basic feature of snowball sampling consisted of a series of questions designed to elicit the sociometric choice of persons with whom the respondent discussed local issues. As Kadushin (1968: 693) has noted, a major problem in the study of power and influence is "to construct an open-ended sociometric, rather than a sociometric of a closed system."[4] To achieve this objective, sociometric items were devised which allowed the respondent unlimited choice in the selection of persons who provided information about local issues. All respondents were asked in a series of items to name individuals whom they would go to for information or who would come to them for information about the three local issues. Furthermore, to secure reports on informal conversations in Spectra or elsewhere, a series of memory probes was designed to provide a basis for the recall of discussions about local issues in a variety of social settings including the family, the job, churches, clubs, social gatherings, stores, restaurants, barber or beauty shops, talks with friends and neighbors, telephone conversations, and similar situations. Included among the questions was the major query employed by Katz and Lazarsfeld (1955: 140) to isolate opinion leaders.[5]

The third major basis of this study consisted of a cross-sectional survey of 240 residents of the community. This represented a random sample of 20% of the dwelling units in Spectra. The questionnaire used in this survey included all of the sociometric items, memory probes, and other identical questions used in the snowball sample. As a result, the procedure permitted an examination of the convergence between the network emanating from the top decision makers and the sources of information mentioned by the general population.

THE COMMUNICATION STRUCTURE

The reputational technique yielded 35 influentials for Spectra of whom ten were recognized by other decision makers as key decision makers for the community. All of the ten key decision makers were white. Six were businessmen; of the remaining four, two were lawyers, one was an editor, and the other person was a city engineer. Both political parties were represented among the key decision makers. Among the 35 top decision makers, there were ten nonwhites—seven blacks and three Mexican-Americans.

The snowball sample yielded four distinguishable groups based upon the number of links separating decision makers and all other members of the community. A linkage was defined as a direct connection between two persons resulting from their discussion of local issues. Hence, persons communicating indirectly through one intermediary were separated by two linkages, and those joined by two intermediaries were three links removed from each other.

Group one was composed of all the top decision makers, including key decision makers. Every decision maker in group one had a direct contact with at least one other top decision maker. Group two comprised 175 persons who were not top decision makers, but they all had direct contact with top decision makers. Group three consisted of 107 people who were two links removed from top decision makers. Thus, a top decision maker would contact one intermediary to communicate with a member of group three. The 45 members of group four were three links removed from the decision-making elite.

Variations existed not only in the size of the groups but also in the rate of interpersonal communications between the groups. The mean number of persons mentioned as sources of information by members of the groups indicated that frequency of interaction was directly related to linkages separating them from the key decision makers. Increasing proximity to the decision-making elite was related to enhanced discussion. Among the key decision makers, the mean number of sociometric choices was 17.5; top decision makers selected an average of 13.9 persons; group two members mentioned on the average 7.0 individuals. The equivalent means for groups three and four were 6.5 and 4.55, respectively. The findings denoted an early indication that influence and communication were enmeshed in a pattern that might have important implication for community decision-making.

Since the population of Spectra was divided among whites, blacks, and Mexican-Americans, one of the major objectives of this research was to ascertain the amount of communication among members of those three ethnic groups. To achieve this goal, the sociometric choices of the top decision makers were examined to determine their ethnic characteristics. The 25 white leaders selected 143 individuals of whom 87% were white, 7% black, and 6% Mexican-American. Strikingly, the choices of the three Mexican-American leaders were similar. Of the 22 nominations made by Mexican-American leaders, only 5% reflected ties with other Mexican-Americans; 13% of the choices indicated links with the black community. An overwhelming 81% of the people selected by Mexican-American leaders were white. The equivalent distribution for the black leaders revealed that 69% of their choices were directed at individuals in the white community, 24% consisted of other blacks, and only 7% of the sociometric choice went to Mexican-Americans. Although black leaders were somewhat more likely than Mexican-American decision makers to select members of their own ethnic group, their nominations still reflected the tendency of minority leaders to rely upon whites primarily as sources of information concerning local issues. Consequently, black and Mexican-American residents of the community were underrepresented among those persons who have a direct access to top decision makers. Apparently, either minority leaders had not developed extensive contacts within their own communities or the exigencies of political reality required minority leaders to devote inordinate attention to the sentiments of the white community. In any event, the net effect of this informational flow has been to exclude minority groups from full participation in a decision-making process which was dominated by the white community.

Furthermore, individuals chosen by persons who had direct access to the top decision makers failed to reflect any significant representation of minority groups. In the entire snowball sample stemming from nominations made initially by white top decision makers, 83% of the respondents mentioned whites; 2% referred to Mexican-Americans, and 15% to blacks. Only 10% of the persons mentioned originally by black leaders cited other blacks, with 88% of their choices going to whites and 2% encompassing Mexican-Americans. Among persons whose nominations were initiated by Mexican-American decision makers, only 3% of the selections were Mexican-Americans, while 88% were directed at whites and 9% mentioned blacks.

Table 1 confirms the tendency of all ethnic communities to nominate whites. More than 90% of whites, blacks and Mexican-Americans selected whites as sources of information concerning the issues. On the other hand, few whites reciprocated by mentioning members of the two minority groups. There was a tendency for blacks to nominate other blacks, but few blacks nominated Mexican-Americans. Moreover, only a small number of Mexican-Americans nominated blacks. Especially striking, however, was the failure of Mexican-Americans to mention other Mexican-Americans as sources of information. While the findings suggested little exchange of information between the two minority communities, they also indicated that the white community may have had a disproportionate impact on the information available to the minority populations. Since attitudes and opinions tend to be shaped by available information, the white community may have exerted a strong influence on the positions taken by black and Mexican-American residents. The ability of either black or Mexican-American citizens to develop a distinctive viewpoint toward local issues may

TABLE 1
PERSONS SELECTED AS SOURCES OF INFORMATION BY ETHNIC GROUPS

	Ethnic Group		
	White	Black	Mexican-American
Whites			
Number of persons selecting whites	174	38	29
Percentage of persons selecting whites	99	93	91
Number of whites selected	1183	132	114
Blacks			
Number of persons selecting blacks	37	34	9
Percentage of persons selecting blacks	21	83	28
Number of blacks selected	65	114	13
Mexican-Americans			
Number of persons selecting Mexican-Americans	19	4	8
Percentage of persons selecting Mexican-Americans	11	10	25
Number of Mexican-Americans selected	25	8	21

have been diluted by the apparent one-way flow of information between the dominant white population and the two minority groups.

Although the random sample of the community identified a large number of people who participated in the discussion of local issues, a substantial proportion of each of the three ethnic communities failed to mention anyone as a source of information. In the white community, approximately 30% were unable to name any person with whom they had discussed the issues. In the black community, a comparable proportion–approximately one-third–also failed to mention any other person. Despite the use of extension probes designed to assist the respondent in recalling conversations about such issues, the data indicated that three out of every ten members of the white and black communities were isolated from any discussion of the issues facing Spectra. On the other hand, 53% of the Mexican-American sample could not identify any other persons with whom they had discussed the three issues. In comparison with the white and black segments of the population, there appeared to be little effort to involve the Mexican-American community in conversations about these issues.

Despite the fact that the public level of controversy concerning the three issues in Spectra was less intense than conflict that has erupted in many other communities (Coleman, 1957), the issues apparently provoked a relatively large amount of interest throughout the community. At least two-thirds of the white and black respondents were related to other persons through direct conversation about the issues. Even in the Mexican-American community, nearly half the residents had discussed the issues with other individuals. Thus the findings did not seem to support the prevalent image of massive apathy concerning community issues. They did, however, raise some interesting questions regarding the extent to which conventional forms of political activity such as voter turnout, attendance at community meetings, and similar forms of organized behavior express the actual level of political interest within the community. Since traditional measures of local political involvement usually have neglected interpersonal communications, they may have omitted a critical dimension of political activity at the local level.

THE COMMUNICATION NETWORK AND THE COMMUNITY

A major purpose of this research was to relate the communication network determined by the snowball sampling procedure to the cross-sectional survey of the community. All of the sociometric choices produced by the snowball sample were defined in this study as the *communication network* emanating from the top decision makers. Persons within this interlocking system of linkages seemed to have an increased opportunity to occupy a critical position in the transmission or relay of information between elites and other members of the community. The concept of a communication network was of theoretical importance because it provided a means of relating the flow of information to decision-making structures and of relating major influentials to other members of the community. An analysis of a local communication network offered an opportunity not only to determine the proportion of the general population linked to the network but also the proportion of the population which relies on independent sources of information. Although there may be a variety of strands or clusters within the communication network, this total web of communications constituted a critical linkage between elites and other members of a community.

When the sociometric choices of the random sample population were examined, the data revealed that approximately nine-tenths of the respondents selected one or more persons within the communication network. Although the snowball sampling did not produce closure by exhausting all possible sociometric choices, only one in ten people in the general community mentioned anyone who was not previously identified in the communication network. The finding, therefore, suggested several critical ramifications concerning the relationship between the public and decision-making processes in the community.

One of the most significant implications of the convergence between the sociometric choices of the community and the communication network is that the flow of information regarding community issues is highly structured. Interaction between members of the community does not appear to be random or haphazard. Given the size of the community and the unbounded opportunities for the respondent to mention anyone he wished, the propensity of the respondents to choose informants from the limited pool encompassed by the communication network must be regarded as extraordinary and significantly greater than the results that might have been obtained by chance. Consequently, there also seemed to be a general recognition of individuals who occupy a critical role in the dissemination of information throughout the community.

In addition, the overlap between the communication network and the informants selected by the general community suggests a high potential for the mobilization of opinion and the monitoring of information. Alternative or independent sources of information are not likely to be available to the community. However, there is an extensive opportunity for the top decision makers to co-opt or manipulate the public by the selective release of information and by maintaining a constant surveillance over the community. On the other hand, this configuration may permit widespread access to the top decision makers, and it may heighten their sensitivity to emerging public sentiments. In any event, the extensive overlap between the sociometric choices of the cross-sectional sample and the communication network suggests the existence of well-defined channels for the transmission of information.

The data revealed that these channels permeated the two minority communities as well as the white population. Ninety percent of the white population selected informants in the network. The corresponding figure for the black and Mexican-American communities were 83% and 90%, respectively. While a somewhat larger percentage of the black respondents selected informants outside the communication network, more than four-fifths of the black citizens who reported discussing the issue were linked to the network. As subsequent evidence will indicate, the sociometric choices of the black population unrelated to the network failed to disclose the existence of an independent informational structure. Although the communication network may operate differentially within each of the ethnic communities, the findings suggested that there was at least a potential vehicle for communications between top decision makers and nearly all segments of the population.

Of particular interest in this study was the nature of the connection between the general population and the communication network. One method of exploring this problem entailed an examination of the "single strongest connection," or the tie that involved the fewest possible linkages between the respondent and the top decision makers. If, for example, a member of the community mentioned three informants in groups two, three, and four, his single strongest connection was considered to be the link with group two which involved only one intermediary between him and the top decision makers. Although the range of nominations varied from one to thirteen, the median number of nominations was one. Thus the single strongest connection and the number of nominations were frequently equivalent.

Although the sociometric choices of the general population encompassed all of the groups in the communication network, three-fourths of the single strongest connections were linked either to group two or to the top decision makers. An examination of the nominations made by each of the three major ethnic communities also revealed some interesting patterns. While 49% of the whites reported a direct linkage to the top decision makers, the corresponding figures for the black and Mexican-American communities were 62% and 66%, respectively. Perhaps the stronger impetus for contact between the two minority groups and the top decision makers arose from a heightened sense of urgency about the solution of local problems. Since black and Mexican-American residents confront especially pressing and imperative needs, they may be inclined to contact the top decision makers directly rather than to rely upon more indirect and problematic means of conveying their demands.

A more intensive examination of the sociometric choices directed at the top decision makers indicated some important differences among the three ethnic populations. While an equivalent percentage—32%—of the choices made by each ethnic group within the decision-making elite were directed at key decision makers, twice as many Mexican-Americans and blacks mentioned other top decision makers. Among those who nominated decision makers, 30 and 34% of the black and Mexican-American citizens, respectively, mentioned other top decision makers rather than key influentials as compared to 17% of the whites. Furthermore, more than half the black citizens selected top decision makers who were members of their own ethnic group. On the other hand, white and Mexican-American selections were directed at top decision makers from the white population. The tendency of blacks to choose black decision makers while Mexican-Americans failed to utilize Mexican-American decision makers may reflect different political strategies for influencing the outcome of community issues. Apparently the black community regarded black members of the top decision-making group as appropriate brokers for their interests. On the other hand, the tendency of Mexican-Americans to approach white decision makers may not indicate a lack of Mexican-American leadership. Instead, it may simply denote a belief that contact with white decision makers is an effective alternative means of influencing a decision-making process dominated by whites. In any event, the two minorities generally rely on top decision makers as spokesmen for their needs and aspirations. The absence of black and Mexican-American representation among key decision makers seems to impose sharp restraints upon the techniques employed by minority groups to affect community decisions. Moreover, the distinct tactics employed by the two minority groups apparently do not include any concerted efforts to cooperate or to form coalitions between the two minorities.

While this examination of contacts with both the communication network and the top decision makers has been primarily concerned with the issue of accessibility, an additional objective of this study was to investigate those communications which might serve as the basis for representation. As a result, the distinction between perceived and actual communication appeared to be of special theoretical importance in the study of contacts between members of the community and top decision makers. For the limited purpose of assessing communications between the general population and the top decision makers exclusively, a potential for representation was operationalized as actual rather than perceived communication. The finding revealed that for all three ethnic communities approximately one-half of all contacts between local residents and top decision makers were actual rather than perceived. Perhaps more important, however, six of the seven interactions between black citizens and black decision makers actually occurred. Black decision makers, therefore, appeared to be in the position to play a more effective role in representing the wishes of the black community than decision makers from the other two ethnic groups.

Although the primary focus of this study was concerned with linkages between members of the community and persons in the communication network, another important goal of this research was to account for the total number of sociometric choices offered by the respondents. Conceivably, individuals may have been related not only to the communications network revolving around top decision makers, but they also may have been linked to independent clusters that could serve as sources of information outside the network. As a result, an additional analysis was made of the total range of people mentioned by the cross-sectional sample. This investigation revealed that 70% of all the sociometric choices in the community were directed at persons in the communication network. Most residents of the community, therefore, were related to the communication network not only by a single linkage but also by multiple connections. Given the thousands of people who could have been selected in Spectra and the surrounding area, the overwhelming tendency to select persons from the limited pool in the communication networks seemed impressive. This evidence appeared to validate not only the communication network, but also its centrality in the flow of information in the community.

Moreover, the high proportion of sociometric choices located within the communication network reflected both an aggregate and an individual relationship. As Table 2 indicates, individuals nominating relatively few persons were as likely to choose someone within the communication network as individuals who mentioned many persons. The strong linkages between members of the community and the communication network, therefore, were not an artifact of the number of sociometric choices made by the respondents. In general, nearly two-thirds or more of the sociometric choices within the three ethnic communities referred to persons within the

TABLE 2
PERCENTAGE OF SELECTIONS IN THE COMMUNICATION NETWORK BY ETHNIC GROUP AND NUMBER OF SELECTIONS OFFERED

	Number of Selections Offered	Total Number of Selections	Percentage of Selections in the Communication Network
Whites	Low (1-4)	153	73 (91)
	High (5-13)	149	73 (109)
Blacks	Low (1-4)	59	75 (44)
	High (5-13)	126	58 (73)
Mexican-Americans	Low (1-4)	28	64 (18)
	High (5-13)	94	70 (58)

communication network. Black respondents who made a large number of selections, however, tended to nominate a somewhat larger proportion outside the network. In part, this may have reflected the tendency for blacks who were highly involved in discussing local issues to interact frequently with other members in their own community.

An examination of persons outside the communication network chosen by blacks as well as other respondents failed to disclose any indication of a cluster of individuals which might comprise an autonomous information structure. Presumably, the existence of independent sources of information would be reflected by individuals outside the communication network who were selected by several other persons in the community. This proposition was tested by employing the concept of *communication leaders.* For the purposes of this study, communication leaders were operationally defined as individuals who received four or more sociometric choices from people either in the snowball or the random sample. As a result, communication leaders were identified, and an effort was made to determine whether or not they were members of the communication network. The results revealed that 94%, or 79 of the 84 communication leaders could be located in the network. Of the five persons outside the communication network who were labeled as communication leaders, four received the minimum number of sociometric selections and one received five selections. On the other hand, communication leaders within the network received as many as 94 sociometric choices, and approximately two-thirds of those persons acquired six or more choices. The absence of major independent sources of discussion, therefore, indicated that the flow of information in the community seemed to occur within the limited confines of the communication network.

OPINION LEADERS AND THE COMMUNICATION NETWORK

Perhaps one of the most significant studies concerning the diffusion of information in the community was the research by Katz and Lazarsfeld (1955). The major findings of this seminal treatise delineated the pivotal role of opinion leaders in transmitting information from the mass media to other persons in what was termed the "two-step flow" of communication. Although this research has been acclaimed by social scientists as a classic statement describing the process by which information is disseminated at the local level, relatively little attention has been devoted to exploring the role of the opinion leader in community decision-making. In particular, few subsequent efforts have been made to examine the function of opinion leaders in a broad social milieu. One explanation for this neglect might be traced to an initial preoccupation with the mass media which is reminiscent of the shortcoming of the early voting studies (Lazarsfeld et al., 1948). Consequently, attention has been diverted from any systematic treatment of opinion leaders in the transmission of information that does not originate from the mass media. The concept of the "two-step flow" of communication, therefore, seldom has been applied to the everyday conversations that form an integral part of local decision-making and of the social fabric of the community.

Opinion leaders do not act in isolation; they are embedded in a social structure. While the original formulation of the two-step flow hypothesis emphasized the importance of primary groups in the dissemination of information, it failed to stress the impact of other factors in the transmission of information such as the nature of community power structures. Not only do opinion leaders facilitate the relay of information from the mass media, but they are enmeshed in a social context which exposes them to numerous sources of information and which shapes their activities in the community. The contacts of opinion leaders are not haphazard; in fact, they may be affected by a broader system of the dissemination of information which has been identified in this paper as a communication network.

In this study a question similar to the one used by Katz and Lazarsfeld (1955: 140) was employed to identify opinion leaders. This question asked, "Do you know anyone in [Spectra] who keeps up with [each specific issue] and whom you can trust to let you know what is really going on?" Although this item was not utilized in the compilation of sociometric choices, it was asked in both the snowball and random samples. In order to ensure the maximum possible coverage, all persons receiving a single selection were considered to be opinion leaders in Spectra. In addition, opinion leaders nominated by the random sample of the community were distinguished from those nominated by the snowball sample. This procedure was adopted to facilitate the examination of the association between opinion leaders and the communication network.

In the survey which encompassed a cross-section of the community, a total of 165 opinion leaders were selected by the respondents. When the convergence between opinion leaders and the communication network was investigated, the data revealed that 61% of the opinion leaders were included in the network. Perhaps of even greater significance was the finding that only two of the opinion leaders located outside the network received more than a single vote. (Both persons who were exceptions to this statement each received only two mentions.) The adoption of a more stringent standard than the one-mention criterion, therefore, would have placed nearly all the opinion leaders in the network. Hence, the results failed to suggest any evidence of a structured system of transmitting information outside the communication network. On the contrary, the findings clearly demonstrated that opinion leaders perform their vital functions almost exclusively within the parameters defined by the communication network.

An examination of the selections made by the ethnic communities also produced some noteworthy patterns. While the white community nominated 82 opinion leaders, black and Mexican-American residents nominated 60 and 23, respectively. Whites, therefore, chose nearly as many opinion leaders as the two minority communities combined. The disproportionate representation of whites among the opinion leaders became even more apparent when the nominations of each of the three ethnic groups were investigated separately.

Of the opinion leaders chosen by the white community, 92% were white. Only 6% were blacks and 2% were Mexican-Americans. The seven black or Mexican-American opinion leaders chosen by whites received a total of merely nine selections. Moreover, while several of the white opinion leaders with a single nomination were not located in the communication network, all seven of the black and Mexican-American opinion leaders mentioned by whites were a part of the network. The selection of black or Mexican-American opinion leaders even by a single member of the white community, therefore, seemed to be based upon their distinctive position in the network rather than upon accidental or personal circumstances.

The white community also supplied a disproportionate number of the opinion leaders mentioned by the two ethnic minority communities. Two-thirds of the opinion leaders chosen by the black residents were white. In addition, whites constituted approximately 70% of the opinion leaders chosen by Mexican-Americans.

The selection of black and Mexican-American opinion leaders by those two populations also suggested some important implications for relations between minority groups. Black citizens comprised one-third of the opinion leaders reported by the black community; not a single Mexican-American was mentioned by a black respondent. On the other hand, Mexican-Americans mentioned twice as many black opinion leaders than they chose Mexican-Americans. Less than 10% of the opinion leaders selected by Mexican-Americans were other Mexican-Americans. But black opinion leaders comprised approximately 21% of the Mexican-American selections. The evidence seems to indicate that Mexican-Americans were apparently more inclined than blacks to rely upon the leadership provided by the other minority community in the dissemination of information.

Blacks were also more likely than Mexican-Americans to choose top decision makers from their own ethnic community as opinion leaders. Mexican-Americans selected seven top decision makers as opinion leaders. None of them was Mexican-American. Moreover, no opinion leader received more than a single nomination from the Mexican-American community. By contrast, when black respondents chose opinion leaders, 21% of their total nominations were directed at black top decision makers. Only 12% of the total selections of opinion leaders by black respondents were white top decision makers. In addition, 40% of the total opinion leader selections by white respondents were white top decision makers. Among the white and black populations, there was a strong tendency to mention top decision makers from their own communities as opinion leaders.

In general, a relatively high percentage of opinion leaders selected by all three ethnic communities were centered around the top decision makers. Thirty-five percent of the 234 total choices of opinion leaders mentioned by the three communities consisted of top decision makers. Top decision makers were more likely to serve as opinion leaders than any other group in the community. Furthermore, as the number of linkages separating each group from the top decision makers increased, the number of opinion leaders in a group decreased. A preponderance of opinion leaders were located at or near the center of community decision-making. Thus a large proportion of the residents of the community apparently acquired information about local issues directly rather than through intermediaries. This process of communication may be identified as a one-step flow rather than the two-step flow emanating from the mass media as postulated by Katz and Lazarsfeld (1955). In addition, many people who selected opinion leaders in groups three and four were part of an informational flow that encompassed three or four steps and several intermediaries rather than only one opinion leader. The evidence therefore suggests that at the local level the original formulation of the two-step flow of information proposed by Katz and Lazarsfeld might be modified or extended to include a variety of means by which information is transmitted throughout the community.

Although most opinion leaders selected by the cross-section of the population were located in the communication network, persons in the latter group may not have chosen other members of the communication network as opinion leaders. Since different methods were employed to locate members of the communication network and to identify opinion leaders, it was possible to conduct a separate examination of the opinion leaders mentioned by persons in the snowball sample. This investigation seemed to have some important theoretical implications. The selection of opinion leaders within the communication network—by members of the network themselves as well as by the general population—would seem to indicate a relatively high degree of closure in the transmission of information throughout the community. On the other hand, if individuals within the network relied upon persons outside the network as opinion leaders, this might indicate either that the concept of opinion leader denoted a separate and distinct dimension or that the communication structure was more open than previous evidence had indicated.

Most of the opinion leaders selected in the snowball sample also were located in that network. Eighty-seven percent, or 154 of the 179 opinion leaders selected in this investigation, consisted of other persons in the communication network. (When the total number of selections was considered, 93% were within the network.) Furthermore, none of the opinion leaders chosen outside of the network received more than a single mention. Opinion leaders seemed to be found almost exclusively within the communication network. The transmission of information in this community, therefore, appears to resemble a closed rather than an open communication structure.

Moreover, persons within the communication network mentioned more opinion leaders than did other members of the community. Whereas the mean number of opinion leaders chosen by the general population was approximately one, individuals within the communication network selected an average of nearly two opinion leaders with whom they discussed local issues. Membership in the communication network appeared to connote increased sources of interpersonal information.

In general, however, members of the two minority groups seldom played a major role as opinion leaders for the communication network. No Mexican-American was selected by a member of the communication network as an opinion leader. Although some black opinion leaders were chosen, they constituted only 6% of all opinion leaders selected by the snowball sample. Of the thirteen black opinion leaders mentioned, only one—a top decision maker—received more than a single choice. Hence, black and Mexican-American citizens were accorded more sociometric choices than nominations as opinion leaders by the snowball sample. Even though members of the two minority communities were included in the dissemination of information throughout the network, they seemingly have not achieved the status of trust necessary to act as opinion leaders.

Individuals who were placed at relatively high positions in the communication network—and who may have been accorded a large amount of status—tended to receive more selections as opinion leaders than did those who were further removed from the top decision makers. Eighty-two percent of the total nominations were aimed at either the top decision makers, or at persons who were directly linked to them (group two); 58% focused explicitly on the top decision makers. All ten of the key decision makers were selected as opinion leaders, receiving an average of ten nominations each. Slightly more than half the remaining decision makers also were chosen as opinion leaders, and each of them also received an average of ten nominations. On the other hand, 11% of nominations were referred to individuals in either groups three or four. Just as the general population often chose top decision makers as opinion leaders, members of the communication network also relied on top decision makers as opinion leaders. The evidence provided by both sample populations, therefore, indicated a need to revise or modify the two-step flow hypothesis to include additional configurations in the transmission of information concerning community issues. While the flow of communications occasionally takes a form that involves one or more intermediaries, the most common pattern seems to be a direct one-step transmission in which top decision makers also serve as opinion leaders for a large segment of the community.

Opinion leaders, however, were not the only persons who may play a pivotal role in the transmission of information. While the concept of the opinion leader may be useful in isolating persons who serve as trusted sources of information, much of the routine discussion that shaped the flow of communications in the community is based simply upon the frequency of interpersonal contact. As previously noted, individuals who engaged in a high degree of interaction were defined in this study as *communication leaders.* Unlike the criterion used to select opinion leaders, which required only a single choice, communication leaders were identified by the amount of contact with other persons. This procedure yielded 84 communication leaders of which 83% were white. In addition, the convergence between communications leaders and those classified as opinion leaders was 70%. Although many communication leaders occupied a position of trust in the discussion of local issues, there was not a perfect correspondence between communication and opinion leaders.

A partial explanation for this apparent discrepancy was related to the distribution of communication leaders among the various groups. While the greatest concentration of opinion leaders was found among top decision makers, the highest incidence of communication leaders occurred in group two. Twenty-two of the communication leaders were top decision makers; forty of them were located in group two. This difference has interesting implications. In fact, the evidence suggested that communication leaders may play a more crucial role in the two-step flow of information between elites and the other local residents than those persons who acquired the trust that qualified them as opinion leaders.

SUMMARY AND DISCUSSION

Although the study of communications often has been viewed simply as an effort to depict the means by which information is transmitted within social groups, it also may have major ramifications for democratic theory. In addition to such traditional concepts as voting, legislative representation, and political accountability, the effective operation of a democratic community depends upon the routine flow of interpersonal information between and among political leaders and other citizens.

In this research, the delineation of sources of information emanating from top decision makers disclosed a highly structured and well-defined communication network. Furthermore, this network encompassed an overwhelming share of the individuals with whom other members of the community discussed local issues. Community influentials not only occupied a position at the apex of the decision-making structure, but they also served as a focal point for the transmission of information about local issues. This convergence seemed to indicate a strong association between influence and information that may either hinder or enhance the expression of popular sentiments.

From one perspective, the involvement and centrality of top decision makers in the web of communications seems to suggest a strong potential for control. Political elites may be in a position both to mobilize community sentiment and to restrict the circulation of opposing viewpoints. An examination of other possible sources of information in the community such as opinion leaders and communication leaders failed to detect any independent or autonomous system for the transmission of information about local affairs.

In addition, the general configuration of the flow of information seemed to denote a marked opportunity for the cooptation of dissatisfied or disaffected residents. In Spectra, the two minority communities apparently were enmeshed in the communication network which was dominated by whites. Although the participation of black and Mexican-American citizens in this network might have enabled them to promote their interests, it also may have imposed constraints on the nature of the objectives they sought.

On the other hand, this research can be seen as suggesting that the faults of democracy may be located in political institutions rather than among the people. Although many social scientists and critics of democracy have emphasized the reluctance of people to become involved in traditional modes of political activity such as elections, voluntary organizations, and campaigns, they have usually neglected or ignored the vast amount of political interest expressed in everyday conversations. Unlike other forms of behavior, no organized structures have been established to facilitate the communication of political ideas through interpersonal sources. Moreover, this study suggests that an informal but extensive structure for conveying political information has evolved without explicit institutional support at the local level. The data uncovered through the use of new methodological techniques in this research, therefore, seem to underscore the importance of interpersonal communications as a crucial linkage between the public and political leaders in a democratic society.

NOTES

1. For an incisive discussion of the advantages and disadvantages of reputational analysis, see Clark (1968: 72-81) and Presthus (1964: 422-432). Also see Clark (1968: 77-78; 471) for a general discussion of the issue-specific reputational method.

2. Many of the items in this schedule were purposely designed to avoid any assumptions about either the existence or the possible size of a decision-making elite. In this manner, an attempt was made to circumvent the major pitfalls that have marred prior research employing the reputational analysis of community power structures.

3. For a discussion of techniques of snowball sampling, see TenHouten et al. (1971).

4. For another approach to this problem, see Mayer (1966).

5. The questionnaire also included a number of sociopsychological scale items, standard SES and demographic questions, a series of items from Almond and Verba (1963), and questions which handled the other sources of information such as the mass media. The findings from this material and their relationship to interpersonal informational networks and the elite structure will be reported in subsequent papers being prepared by the authors.

REFERENCES

ALMOND, G. A. and S. VERBA (1963) The Civic Culture. Princeton, N.J.: Princeton Univ. Press.

BACHRACH, P. (1967) The Theory of Democratic Elitism. Boston: Little, Brown.

CAMPBELL, A., P. E. CONVERSE, W. E. MILLER, and D. E. STOKES (1960) The American Voter. New York: John Wiley.

CLARK, T. N. [ed.] (1968) Community Structure and Decision-Making: Comparative Analysis. San Francisco: Chandler.

COLEMAN, J. S. (1957) Community Conflict. New York: Free Press.

DAHL, R. A. (1961) Who Governs? New Haven: Yale Univ. Press.

HUNTER, F. (1953) Community Power Structure. Chapel Hill: Univ. of North Carolina Press.

KADUSHIN, C. (1968) "Power, influence and social circles: a new methodology for studying opinion makers." Amer. Soc. Rev. 33 (October): 685-699.

--- (1966) "The friends and supporters of psychotherapy: on social circles in urban life." Amer. Soc. Rev. 31 (December): 786-802.

KATZ, E. and P. F. LAZARSFELD (1955) Personal Influence. New York: Free Press.

KEY, V. O., Jr. (1964) Politics, Parties, and Pressure Groups. 5th edition. New York: Thomas Y. Crowell.

LAZARSFELD, P. F., B. BERELSON, and H. GAUDET (1948) The People's Choice. New York: Columbia Univ. Press.

MAYER, A. C. (1966) "The significance of quasi-groups in the study of complex societies," pp. 97-122 in M. Banton (ed.) The Social Anthropology of Complex Societies. New York: Praeger.

MILLS, C. W. (1956) The Power Elite. New York: Oxford Univ. Press.

PRESTHUS, R. (1964) Men at the Top. New York: Oxford Univ. Press.

TENHOUTEN, W. D., J. STERN, and D. TENHOUTEN (1971) "Political leadership in poor communities: applications of two sampling methodologies," pp. 215-254 in P. Orleans and W. R. Ellis, Jr. (eds.) Race, Change, and Urban Society. Urban Affairs Annual Review, vol. 5. Beverly Hills, Calif.: Sage Publications.

TRUMAN, D. B. (1951) The Governmental Process. New York: Alfred A. Knopf.

4

When a Black Candidate Runs for Mayor: Race and Voting Behavior

THOMAS F. PETTIGREW

INTRODUCTION

☐ NOVEMBER 7, 1967 WITNESSED THE BEGINNING of a new era in American politics, the beginning of significant black entry into the political decision-making of urban America. Democrat Carl Stokes, a black state legislator, defeated Republican Seth Taft by less than one percent of the votes to become Mayor of Cleveland. And on the same day, Democrat Richard Hatcher, a black city councilor, defeated Republican Joseph Radigan by an equally narrow margin to become Mayor of Gary, Indiana.[1]

Since that historic election day, Stokes has won reelection in 1969, Kenneth Gibson has become Mayor of Newark, New Jersey, in 1970, and Hatcher has won reelection in Gary in 1971. Moreover, two other black candidates narrowly lost in mayoralty bids in major cities: Thomas Bradley in Los Angeles in 1969 and Richard Austin in

---

AUTHOR'S NOTE: *The research reported here was made possible by a basic grant from the Rockefeller Foundation, with subsidiary grants from the Metropolitan Applied Research Center, the Ottinger Foundation, the Joint Center for Urban Studies of MIT and Harvard, and the Anti-Defamation League of B'nai B'rith. I wish to express my indebtedness for the many contributions of*

Detroit in 1969.[2] And there will be many more competent blacks running for mayor in major cities. Indeed, there are serious black bids being made for the top office in other cities such as Boston, Baltimore, and New Haven together with preparations for such bids in cities from Atlanta (Vice-Mayor Maynard Jackson in 1973) to Los Angeles (Bradley again in 1973). Clearly, the 1970s will be the decade of major racial progress in this area that was opened up in 1967.

These races offer ideal opportunities for research to students both of politics and race relations. There have now been enough of these critical elections, with the promise of many more to come, to warrant careful study. What uniformities emerge across these varied cities when a black candidate runs for mayor? Who are the whites who vote for a competent black for high office? And what are some of the political conditions that make it possible for a black aspirant to succeed?

Another feature of these mayoralty races makes them particularly attractive for research in race relations. The office of the mayor, like those of governor and President, involves being "the captain of the ship." This contrasts with the offices to which a disproportionate share of blacks have been so far elected (city council member, legislator, and lower executive posts) where blacks are either a minority in a larger body or in a lower post below the top executive position. Running for "captain of the ship" is obviously, then, a far more rigorous test of the white voter's racial prejudices and behavior.

Finally, social science must begin to study this phenomenon now if it is to learn about the racial and political dynamics that undergird its evolution. Before too many years have passed, the election of competent black mayors will have become normative; and the chance to gather baseline data and to observe its acceptance over time by initially threatened whites will have been lost. Witness the manner in which the election of President John F. Kennedy in 1960 practically removed the issue of membership in the Roman Catholic church as a dominant national political concern.[3] Interestingly, the considerable discussion of various Democratic Party possibilities to run for President in 1972 has virtually ignored as irrelevant their religious affiliations.

---

my colleagues on the project: Robert T. Riley of Harvard, J. Michael Ross of MIT, and Reeve Vanneman of Harvard. In addition, I want to thank Mrs. Shirley Rosess, Professor David Sears of UCLA, and Mrs. Eve Weinberg and her capable staff at the National Opinion Research Center of the University of Chicago.

Consequently, I initiated a research project in 1968 to study the campaigns of black mayoralty candidates throughout the nation. So far, this project has focused on the 1967 and 1971 races of Mayor Hatcher in Gary, the 1969 race of Mayor Stokes in Cleveland, the 1969 race of Bradley in Los Angeles, and the 1970 race of Mayor Gibson in Newark. Both precinct analyses and surveys of probability samples of registrants have been utilized in each of the four cities. This article presents the first preliminary overview of the project's work. First, brief descriptions and discussions of the elections will be provided. Then we shall explore in more detail a number of the uniformities and dynamics that have been uncovered so far across all of the elections.

FIVE MAYORALTY ELECTIONS

Gary, 1967. The project began its work with extensive aggregate analyses of the vote by precinct for Richard Hatcher for Mayor of Gary in November 1967. Next we conducted during October of 1968 a survey of 257 white adult males who were registered to vote and who were representative of nine varied white precincts of Gary. The auspicious timing of this survey right before the 1968 presidential election allowed us to study simultaneously the support for Governor George Wallace in one of the northern communities where he ran strongest (the Wallace data are reported in Pettigrew, 1971).

A year prior to our survey, Mayor Richard Hatcher had won a narrow victory over Joseph Radigan in a bitter contest that required U.S. Department of Justice intervention to ensure its being held fairly. Both our survey and aggregate data agree on the following election statistics:

(1) in 1967, the black percentage of the total Gary electorate was between approximately 40 and 44%;

(2) turnout for both whites and blacks was unusually large, with well over 80% of those registered of each race voting;

(3) Hatcher received virtually all the black votes but only about 15% of the white votes (including Spanish-speaking voters who generally supported Hatcher);

(4) since mayoralty elections in Gary are partisan, Hatcher was the nominee of the Democratic Party, and about 65% of the city's whites

TABLE 1

ATTITUDES TOWARD DESEGREGATION AND LAW AND ORDER AND THE 1967 WHITE GARY VOTE FOR HATCHER (in percentages)

Anti-Desegregation Attitudes[a]	Low			Medium			High		
Concern Over Law and Order[b]	**Low**	**Medium**	**High**	**Low**	**Medium**	**High**	**Low**	**Medium**	**High**
1967 Gary Mayoralty Vote For:									
Hatcher	77.8	20.0	0.0	18.2	14.3	12.9	0.0	18.2	0.0
Radigan	22.2	80.0	100.0	81.8	81.0	87.1	100.0	72.7	100.0
Don't know, did not vote, other	0.0	0.0	0.0	0.0	4.8	0.0	0.0	9.1	0.0
Total %	100.0	100.0	100.0	100.0	100.1	100.0	100.0	100.0	100.0
(n)	(18)	(20)	(13)	(11)	(21)	(31)	(6)	(11)	(47)

a. "Anti-desegregation attitudes" are measured by responses to seven items such as: "Negroes shouldn't push themselves where they're not wanted" (agree); "White people have a right to keep Negroes out of their neighborhoods if they want to, and Negroes should respect that right" (agree); "Do you think white students and Negro students should go to the same schools or to separate schools" (separate schools).

b. "Concern over law and order" is measured by responses to the following three items: "Buses aren't safe these days without policemen on them" (agree); "When looting occurs during a riot, police should shoot to kill" (agree); and "Safety on the streets is the most important issue facing America today" (agree).

> are registered Democrats, this 15% figure means that over three-fourths of the white Democrats in Gary in November 1967 voted against their party's nominee. In other words, for the vast majority, race overcame party identification. Of course, this phenomenon reflects the fact that the Lake County Democratic Committee—the local political machine which Hatcher opposed—openly supported the Republican candidate.

Who, then, were the rare white voters for Hatcher in 1967? In capsule form, they were disproportionately found among Jewish Americans and upper-status, college-educated Democrats. Not only were they significantly less anti-Negro, but they were far less concerned over "law and order" issues. As Table 1 shows, attitudes toward desegregation and concern over law and order are positively correlated, of course, but, surprisingly, each independently accounts for approximately equal percentages of the variance in voting for Hatcher in 1967. Note, for example, that at the underlined extremes, 78% of the 18 respondents who both favor desegregation and are not concerned over "law and order" report having voted for Hatcher compared to none of the 47 respondents who oppose desegregation and are concerned over law and order.

This brief sketch of the 1968 Gary findings will be modified in detail by the more complex analyses now being prepared. But let one example serve here to illustrate the interesting complexities of the data. Annual family income does not relate to either the Hatcher or Wallace white votes in a linear fashion. Both our precinct and survey data reveal that the poorest whites vote slightly more for Hatcher and less for Wallace than the lower-middle-income white voters. Thus, those between $6,000 and $12,000 annual income form the core of Hatcher's opposition and Wallace's support in the Gary electorate (this is best expressed statistically as a quadratic function centered around $9,000 as follows: $(\$9{,}000 - \text{reported income})^2$, where the larger this squared discrepancy, the larger the Hatcher vote and the smaller the Wallace vote). Other findings from this early work in Gary are described below.

Cleveland. Mayor Carl Stokes, a former state legislator from a black district of Cleveland, first ran for mayor in 1965 as an Independent. He lost, though he received fully 11% of the white vote. In 1967, he ran again; this time, he nosed out the incumbent in the Democratic Party primary and went on to gain a narrow victory over the Republican nominee, Seth Taft, in November 1967. In the latter race, he received, by our calculations, about 19% of the white vote. With only a two-year term, he had to run for reelection in 1969 and again won in both the primary and the partisan election. In the latter race, he received about 22% of the white vote. In outline, then, the Stokes elections for mayor in 1967 and 1969 are characterized by the following statistics:

(1) in 1967 and 1969, the black percentage of the total Cleveland electorate was between approximately 37 and 40%–slightly smaller than that of Gary, thus making Stokes somewhat more dependent upon white support than Hatcher;

(2) turnout for both whites and blacks was unusually large–especially in 1967, with over 70% of those registered of each race voting;

(3) Stokes received all but a scattering of the black votes in each election, while, as mentioned above, he slowly increased his white percentage from 11 to 19 to 22%. It should be noted, however, that Stokes lost a number of his key campaign staff members between 1967 and 1969, and this apparently explains some of the slight reduction in his black support in 1969. Yet his slightly greater white support made up the difference and set up his narrow victory margin in 1969;

(4) just as in Gary, mayoralty elections in Cleveland are partisan; Stokes in 1967 and 1969 was the nominee of the Democratic Party, and a strong majority of the city's whites are registered Democrats. Yet once again the small percentages of 19 and 22 of the white vote indicate that most of the white Democrats in Cleveland in both November of 1967 and 1969 voted against their party's nominee. And just as in Gary, Stokes has been opposed by the local county Democratic committee–though not as openly as in Gary.

The profile of the white voter for Stokes resembles what we found in Gary. Our surveys in Cleveland were conducted with 488 white registrants and 400 black registrants in the late spring of 1969, with the same white respondents being reinterviewed in November and December of 1969 after the election (the retrieval rate was over 80% of the original white sample). Both surveys, together with the aggregate analyses of the voting pattern by precincts, found the upper-status, college-educated Democrats to typify the white support for Stokes. Few Jewish voters still reside in the central city, but those few Jewish respondents in our sample did tend to favor Stokes disproportionately.

Another difference with the Gary data concerns ethnicity. In the Indiana city, our survey revealed that those ethnic whites who did not live in ethnic enclaves were the most likely to reject Hatcher and accept Wallace—precisely counter to our theoretical expectations. But in the Ohio city, our expectations were verified; that is, Polish Americans, Hungarian Americans, Czech Americans, and so forth, were most likely to vote against Stokes and for Wallace in neighborhoods which were largely ethnic enclaves for one group. We are now testing for this effect in our Newark data. But what caused this reversal between Gary and Cleveland? We have one promising lead at present; and it concerns the lower-middle-class white voter again. Put simply, this class and income phenomenon of the greatest resistance to black candidates centering in the lower-middle-status ranks appears to be far more critical than the ethnic factor. Thus, in Gary the lower-middle-class whites center in mixed ethnic areas while in Cleveland they are largely found in the ethnic enclave—and this seems to have caused our reversal of ethnic findings between the two cities.

Los Angeles. In the spring of 1969, Thomas Bradley, a city council member, became the first serious black contender for Mayor of Los Angeles. In April 1969, he ran in the nonpartisan race against incumbent Mayor Samuel Yorty and over a half-dozen other candidates of widely varying positions and characteristics. Bradley led this first race with an impressive 42% of the total vote, and entered the runoff against Yorty, who came in second with 28% of the total vote. In June, however, Bradley could not significantly increase his percentage and lost to Yorty by a decisive 55,000 vote margin (52.1 to 47.9%). We conducted two surveys during this period, the first of 300 white registrants between the two elections and the second following the runoff involving the reinterviewing of the same respondents (again our follow-up response rate was 80%). The vast size of Los Angeles, however, necessitated one change in the sampling design from the studies conducted in Gary and Cleveland: namely, the sampled precincts were drawn exclusively from two areas of the sprawling metropolis—West Los Angeles and the northern half of the San Fernando Valley area, which included heavily Jewish strongholds of Bradley's support as well as areas of largely lower-middle status where Bradley tended to run weakest.

Drawing upon our aggregate and survey data as well as those of colleagues who also studied these Los Angeles elections (Sears et al., forthcoming; Hahn and Almy, forthcoming), we believe the Bradley-Yorty runoff election in 1969 is characterized by the following statistics:

(1) in 1969, the black percentage of the total Los Angeles electorate was only about 15 to 18%–a situation qualitatively different from the three other cities studies;

(2) turnout for both whites and blacks was unusually large in the runoff–as in Gary and Cleveland–with over 70% of those registered of each race voting. The total voting percentage of 76% set a city record for mayoralty elections;

(3) Bradley received all but a scattering of the black votes in each election, while securing in the runoff a white percentage of roughly 38 to 41% (including Spanish speakers, a majority of whom voted for Yorty)–the highest degree of white support yet obtained by a black candidate for mayor of a major city. Our data suggest that this result largely reflects in order of importance: (a) widespread discontent with the Yorty administration of the previous eight years; (b) qualitatively less anti-Negro prejudice and perceived racial threat in Los Angeles; (c) Bradley's success with Jewish-American voters, whose concentration in Los Angeles is second only to that of New York City; and (d) a well administered and financed campaign by Bradley;

(4) mayoralty elections in Los Angeles are technically nonpartisan and both Yorty and Bradley are identified as members of the Democratic Party. Bradley, however, ran to some extent as a Democrat, while Yorty's widely publicized breaks with the Democratic party made him most popular among white Republicans. Our surveys suggest the whites of Los Angeles are split almost evenly between the two major parties. Bradley led slightly among white Democrats but trailed almost one to three among white Republicans: his greatest strength among Democrats were those who labeled themselves "liberals," roughly four-fifths of whom indicated support of Bradley in our survey.

The question thus becomes: what happened to Bradley's white voting base between the two elections? The black vote remained solidly behind Bradley and significantly increased in turnout in June from the earlier April race. But the white vote increased even more between the two elections, and this suggests that part of our answer lies in the "out-from-under-the-rocks" phenomenon. That is, many whites who do not routinely vote in civic elections–and did not in fact vote in the April 1969 election–were attracted to the polls in June 1969 and voted overwhelmingly for Samuel Yorty. Similar instances of this phenomenon have been noted by this project in the Boston races of Mrs. Louise Hicks and the presidential races of Governor George Wallace. And it is consistent with a vast political research literature on nonvoters who are repeatedly found to be more conservative, authoritarian, and anti-Negro as a group than voters.

But the "out-from-under-the-rocks" appearance at the polls of whites who typically do not vote is only part of the explanation of Mayor Yorty's triumph. A detailed analysis of our post-runoff interviews reveals that many of the whites who initially favored Bradley were in fact more anti-Yorty than pro-Bradley. The election placed them in a harsh avoidance-avoidance conflict between a Mayor they did not like and a challenger whose race presented a threat. Moreover, those whites who did shift from Bradley to Yorty were especially upset over campus unrest, a revealing survey fact that coincides with the student violence at UCLA and San Fernando State College on the weekend immediately before the June election on Tuesday. It is tempting to speculate, then, that Bradley's 52 to 48% defeat can also be attributed in part to these campus disturbances which harmonized so completely with the Mayor's campaign charges against Bradley as "one of *them* who cannot control *them*" and as a leader who would unnecessarily restrain the important operations of the Los Angeles Police Department. At any rate, we know that there were many "undecided" white voters late in the campaign, that there was a white voter shift toward Yorty late in the campaign not unlike the Truman-Dewey presidential race of 1948, and that only the last survey made (by the Field Organization) detected this trend in a telephone survey.

Finally, the profile of the white voter for Bradley is one that the Gary and Cleveland results had prepared us to expect. Jewish voters were over twice as likely to favor the black candidate as other white voters; those who identified themselves as either Democrats or Independents were over twice as likely to favor Bradley as Republicans; the young (21-40) were twice as likely to favor Bradley as the old (51+); and those with some college training were twice as likely to favor Bradley as those with a limited education (11 years or less). Unlike Gary and Cleveland interviewees, respondents in Los Angeles of all types did not typically evince the more blatant forms of racism; thus, they overwhelmingly rejected notions of biological inferiority, of sanctioned racial discrimination and segregation, and of the fairness of treatment of Negroes in America today. But the somewhat more subtle and symbolic forms of racism—"most Negroes who receive welfare . . . could get along without it if they tried"—is reflected in our Los Angeles data and does differentiate between Bradley and Yorty supporters, a critical point to which we shall return.

Newark. In many ways, Newark and its black-white mayoralty election resembles Gary and its 1967 election: in the shadow of a powerful metropolis, it is a rundown city with a large black population, profound social problems, and a history of ethnic machine politics. Kenneth Gibson, a former city council member and an unsuccessful candidate for mayor in 1966, wrote civic history by becoming the first black mayor in 1970. After leading comfortably in the initial nonpartisan contest in April, he ran up an impressive 56% to 44% victory over the entrenched incumbent, Hugh Addonizio. In addition to aggregate analyses of precinct votes, we conducted four surveys in Newark: 300 whites and 200 blacks were interviewed prior to the first election, and 240 of the same whites (an 80% retrieval rate again) and 200 different blacks were interviewed in May 1970 between the initial election and the runoff. Our data suggest the following electoral statistics:

(1) in 1970, the black percentage of the total Newark electorate was approximately 45 to 48%;

(2) turnout for both whites and blacks was high for Newark in the June runoff election, though our estimate of 58% of the registered Negroes voting is by far the lowest black turnout of our four cities;

(3) in the final race, Gibson received about 97% of the black vote, though Addonizio had been expected by local observers to attract from 8 to 10% (mostly the families of city workers, and the like). But Gibson received only about 16 to 17% of the white vote. This figure seems to be the crude order of magnitude a black candidate, not running as a "machine man," can expect in white support in a city such as Gary, Cleveland, and Newark where anti-Negro feelings are relatively high and no serious challenge for mayor has previously been made by a black candidate. Moreover, this figure holds up in this instance even when the white incumbent is undergoing trial and conviction for civic corruption;

(4) in broader perspective, the Gibson campaign constitutes the opposite extreme from the highly financed and polished campaign of Bradley in Los Angeles. Gibson did not make the all-out effort for white votes

> made by Bradley and Stokes nor did he mobilize the black community in the fashion of the other three candidates. But he benefited enormously from the federal trial against his opponent held during the campaign as well as from the active campaigning in the white areas of the former fire chief and defeated white candidate for mayor, John Caufield (after election, Gibson reappointed the popular Caufield as fire chief). And, not incidentally, Gibson entered the race with the highest black registration percentage of any of the four black aspirants.

Our preliminary analysis of the two white surveys suggests that their results replicated the previous studies on who are the white voters for a black mayoralty candidate. Virtually all of Newark's previously extensive Jewish community have now departed the city. But Gibson's white support came disproportionately from the ranks of the young, the college-educated, the higher-income respondents, Democrats, and the least racially bigoted.

Gary, 1971. Two years after our initial study of Gary in October of 1968, the project returned to the city to investigate the mayoralty election shaping up for 1971 in which incumbent Mayor Richard Hatcher would try to win reelection. Two surveys were conducted, one of a probability sample of 192 black registrants and the other of a probability sample of 291 white registrants.

Hatcher faced two serious challengers: City Council President John Armenta and Lake County Coroner Alexander Williams. Armenta is a Mexican-American with some following throughout the city. Dr. Williams is a well-known black politician and physician who was generally regarded as the candidate of the local "machine"—the Lake County Democratic Committee. The mass media took considerable interest in Gary's mayoralty primary, held on May 4, 1971, largely because Mayor Hatcher's major rival was another black. Some media observers mistakenly thought that the Mayor was in serious trouble, that Williams would effectively split the black vote and make it possible for either Williams or Armenta to upset the incumbent.

The surveys painted a different picture. And unlike Los Angeles, there were no sharp shifts in sentiment at the close of the primary campaign so that our electoral predictions proved accurate. The surveys indicated that the overwhelming majority of black voters would continue to support Hatcher; and that Hatcher would keep or increase his modest 1967 white support while the entry of white candidates into the race would sharply erode the white support of Williams. The primary returns gave Hatcher a decisive 58% of the vote with Williams a distant second and the other candidates out of the running. The survey data together with the precinct data from the election suggest the following statistics:

(1) by 1971, the black percentage of the total Gary electorate had risen from 1967 several percentage points to approximately 42 to 46%. But in a Democratic Party primary, the registered blacks of Gary may slightly outnumber registered whites;

(2) the black turnout was again large, though it may not have reached the remarkable record of 1967. White turnout, however, was down in many areas, unlike 1967. It appears that for a sizable minority of whites an avoidance-avoidance conflict provided by two black candidates led to withdrawal despite the efforts of the machine to achieve a major white turnout for Williams;

(3) Hatcher probably received more than 90% of the black votes, with precinct totals in the black areas of 518-55 and 471-7 not uncommon. His white support increased from about 15% in the 1967 final election to about 22% in this 1971 primary. These results are close to what the project's surveys indicated five months earlier;

(4) Hatcher's black following is so extensive that few demographic distinctions can be made, though the survey of blacks suggested his greatest black strength is found among those with characteristics similar to Hatcher himself: males, the well educated, home owners, and those born in or near Gary. His limited white following is concentrated among males, the well educated, English-Scots and Germans, and those residing in their neighborhoods for less than three years. Not surprisingly, too, the Mayor's white voters are more liberal than other Gary whites on a wide range of issues.

In choosing three adjectives to describe Mayor Hatcher, his supporters—both black and white—see him as "intelligent, honest, and progressive." Anti-Hatcher whites see him as "intelligent, out-for-himself, and prejudiced." This last description—"prejudiced"—is one that occurs often in the four cities when resistant whites describe their black mayoralty candidate. It refers to the fear that many whites share that powerful blacks in high office will prove antiwhite and openly discriminate against whites in jobs, services, and taxes. Some observers might see in this phenomenon a simple case of projection. Yet it is a genuine and widespread fear that aspiring black candidates throughout the country will have to face. One might expect this fear to recede in time, particularly when the black mayor proves effective and provides little or no validity to the fear. Yet in the two years between the 1968 and 1970 Gary surveys, we detect scant reduction of this particular racial anxiety in spite of Hatcher's able leadership.

Blacks in Gary share similar racial attitudes with other urban blacks in the North. They overwhelmingly reject violence, support integrated schools, and favor liberal governmental policies in general. They are unusual in one interesting respect: presumably as a result of Hatcher's electoral successes and administration, they appear to believe in the political system as a principal means for needed racial change somewhat more than comparable blacks in other cities.

Whites in Gary, however, express an extreme degree of racial animosity, even when compared with such other predominantly lower-middle-class cities as Cleveland and Newark. Furthermore, this racial prejudice is directly related to attitudes toward and voting for Hatcher in both 1967 and 1971 (as shown in Table 1). This severely limits the degree of electoral penetration the Mayor can achieve in the city's white community.

Nevertheless, the black and white voters of Gary have a surprising degree of consensus regarding the major local issues. Leading the list by far is concern over crime followed by attention to sanitation, street repairs, and parks and recreational facilities. High levels of dissatisfaction are recorded among both races over present police protection and recreational facilities, among whites over the public schools, and among blacks over garbage collection. This considerable overlap in interests across the races gives the black candidate the opportunity to seek white and black votes with the same platform. And it provides him the opportunity to emphasize issues which many whites place over racism in importance—a point which we shall emphasize in the next section.

TENTATIVE ANSWERS TO INITIAL QUESTIONS

Data on these mayoralty races shed light on two sets of initial questions; one set focuses specifically upon fundamental issues in American race relations, the other upon fundamental issues in social psychology. The mayoralty elections per se, then, provide a salient, real-life situation in which to pursue these matters.

Issues in American Race Relations

How salient will "race" be in mayoralty elections? Is racial bigotry (or "individual white racism" in today's fashionable parlance) operating in a relative vacuum apart from a person's other values and beliefs? Is there a pattern among white voters across cities of the greatest support of and greatest resistance to competent black mayoralty candidates?

Not surprisingly, our data reveal that "race" has been extremely salient in all of the elections studied. Indeed, the fact that a leading candidate for mayor is Negro becomes a dominant feature for both white and black voters, overwhelming in importance political party identification in Gary and Cleveland and scandals in incumbent administrations in Los Angeles and Newark. Thus, the voting turnouts were unusually high for mayoralty races in all four cities (save for whites in Gary in 1971 when the leading opponent was also black); respondents spoke freely of the influence of race on their voting intentions; and our survey measures of prejudiced attitudes related highly and consistently with voting for the white candidates. We have already seen in Table 1 the predictive value of anti-Negro attitudes in the 1967 mayoralty race in Gary.

Yet the salience of race in these political contests cannot be interpreted as evidence for the widely voiced assumption that "white racism" is almost impossible to combat because it operates in a virtual personality vacuum. Said in this manner, of course, this assumption appears psychologically absurd; accordingly, this project has utilized as a guiding hypothesis that racist attitudes are in constant interaction with other attitudes and values and must take their place in a dynamic hierarchy of what is most important to the individual. We believe our data fully support this hypothesis, though the hypothesis directly challenges much of the loose popular analysis of racial prejudice which followed the Kerner Commission Report on Civil Disorders in 1968.

Consider responses to such an open-ended query as: "Is there anything that Mayor Hatcher could do that would cause you seriously to consider voting for him in the next election?" Among the Gary respondents who would not vote for Hatcher for mayor as of October 1968, only about 30% were unable to suggest reasonable actions. Interestingly, most of these respondents were supporters of George Wallace for President; they were, in short, racists who tended to put their anti-Negro beliefs high up in their value hierarchy. But

70% did supply reasonable actions they cared about, most of them local matters that touched their daily lives directly. "If Hatcher would close that stinking city dump at the end of the street," blurted out one interviewee, "then I'd vote for him even if he is black." Or, in our terms, he is stating: "Though I am a racist, I value the closing of the local dump more than I do my racial beliefs." Similarly, two lower-status white precincts in Cleveland which had given Stokes only minimal support in 1967 became pro-Stokes in 1969 after he had opened a much desired playground in the area. We believe this phenomenon is critical to an understanding of American race relations and its future, and we plan to explore it further in future analyses and studies. One theoretical application of it has already been made to the general problem of racism and mental health (Pettigrew, 1972).

Our brief review of results of the elections in the four cities has already indicated the tentative answer to our third race relations question: consistent patterns emerge across Gary, Cleveland, Los Angeles, and Newark of where the chief support for and resistance to competent black mayoralty candidates lie within the white electorate. Where there are significant concentrations of Jewish voters, as in Gary and Los Angeles, they are especially conspicious among the supporters. And in all the cities, the college-educated, the young, the least prejudiced, and the least concerned about student unrest and crime in the streets tend to be disproportionately found among white supporters. It also appears that Democrats are more likely to favor black candidates than Republicans; but this finding is rendered ambiguous by the fact that all four black candidates are highly identified Democrats.

The source of greatest white opposition to black political aspirations appears even more clearly. It is not centered among the poorest and least educated, but rather among the lower-middle class. This pattern is sharpest for Wallace support, but it is also to be found in the white votes against all four black mayoralty candidates. One of the dynamics we have uncovered which undergirds this special resistance of the urban lower-middle class is a keen sense of deprivation relative to both blacks and white-collar workers. We shall probe this social psychological phenomenon in the next section.

Issues in Social Psychology

No other subject has so dominated the attention of the discipline of social psychology as attitude change. Yet much of the research literature on this subject has at least three glaring weaknesses. First, the attitudes influenced in experiments seldom involve deeply rooted, emotionally charged sentiments that typify racial bigotry among white Americans. Second, these experiments have rarely linked the changed attitudes to nonverbal, concrete acts of behavior. And, third, these experiments have generally treated attitudes as simply responses to a particular object (e.g., "Negroes") rather than toward an object in the context of a particular situation (e.g., a particular Negro as the mayor of my city). Seen from this perspective, then, our studies present a unique opportunity for testing attitude change in a real-life situation that corrects for all three of these weaknesses in the present research literature in social psychology.

The discussion above of racial attitudes not existing in a psychological vacuum is, of course, directly relevant to these concerns. In addition, the success of Mayor Yorty's campaign to communicate a message which effectively aroused white fears of a black mayor is of prime importance to social-psychological theories of attitude change. Since neither the Mayor personally nor his fear-provoking message were "popular" among most white voters in Los Angeles, this example neatly illustrates a widely recognized phenomenon in the laboratory: namely, the effectiveness over time of a salient message from a deprecated source.[4] We plan to explore in more detail our before-and-after survey data relevant to this point both within this theoretical context and that of the avoidance-avoidance paradigm.

Also important to social-psychological theory are the effects upon individual whites and blacks of having personally voted for a black mayor who goes on to win the office. We predict for whites, especially if they publicly discuss their vote with others, that this act will lead to more positive racial views in the classic manner of salient behavioral commitment. This notion requires more detailed analysis of our survey data; but the aggregate data are reassuringly consistent. Thus, Mayor Stokes' white total rose slowly with every race from 11% to 19% and 22%. And Mayor Hatcher's white vote also appears to have risen somewhat between 1967 and 1971.

The project has given most of its attention to date, however, on the general theoretical area of social evaluation theory.[5] This theoretical area is built on the basic notion that human beings learn about themselves largely through comparisons with others. A wide range of molecular theories and concepts can be subsumed under this more general theory, such as Festinger's social comparison theory, Hyman and Merton's reference group theory, Lenski's status inconsistency theory, Thibaut and Kelly's concept of comparison level, Homans' concept of distributive justice, and Stouffer's concept of relative deprivation. It is this last concept, relative deprivation, that has proven unusually useful in our analyses of white voters (Pettigrew, 1971, 1967).[6]

We have developed over the course of our twelve surveys an elaborate battery of measures of relative deprivation, the fullest we believe yet attempted in survey research. Eight basic questions are asked requiring 49 different responses; and though this sounds complex, even poorly educated respondents typically had no trouble supplying meaningful data. In addition to the standard Cantril (1965) self-anchoring ladder items, we ask about the respondent's economic gains over the past five years and his satisfaction with them. More important, we obtain comparative ratings of his economic gains relative to eight critical groups: white-collar workers, blue-collar workers, Negroes, professionals, whites, unskilled laborers, people in this neighborhood, and people in the suburbs.

Two general trends across the two races and all four cities are of interest. First, the average ratings assigned the economic gains of the eight groups by our respondents are quite accurate. This suggests that the social science dogma that Americans are relatively unaware of their social class structure deserves serious questioning. Second, there exists a broad resentment in these cities of the economic gains of white-collar workers in general and professionals in particular. Large numbers of the respondents of varying background from Newark to Los Angeles believe, for instance, that "professionals in America today have gained more economically in the past five years than they are entitled to."

Our interest in the role of relative deprivation in racial voting was initiated by an array of amazingly consistent relationships noted between a single relative deprivation item and support for Governor George Wallace in Gary in 1968. As previously reported (Pettigrew, 1971), Table 2 shows how agreement with the straightforward statement—"In spite of what some say, the lot of the average man is

TABLE 2
RELATIVE DEPRIVATION AND WALLACE SUPPORT, GARY, 1968 (in percentages)

	Supporters			
	Wallace	Nixon	Humphrey	Total
"In spite of what some people say, the lot of the average man is getting worse, not better."[a]				
Agree (118)	41.5	33.1	25.4	100
Disagree (122)	18.9	49.2	32.0	100
Union Members				
Agree (76)	47.3	30.2	22.5	100
Disagree (63)	27.0	33.3	39.7	100
Nonmembers				
Agree (40)	27.5	40.0	32.5	100
Disagree (59)	10.2	66.1	23.7	100
Religion				
Protestants				
Agree (53)	50.9	34.0	15.1	100
Disagree (45)	22.2	53.7	24.1	100
Roman Catholics				
Agree (53)	34.0	34.0	32.1	100
Disagree (51)	17.6	43.1	39.2	100
Social-class Identification				
Close to the working class				
Agree (49)	57.1	18.4	24.5	100
Disagree (36)	25.0	47.2	27.8	100
Not close to the working class				
Agree (25)	36.0	36.0	28.0	100
Disagree (20)	30.0	20.0	50.0	100
Close to the middle class				
Agree (27)	25.9	44.4	29.6	100
Disagree (40)	7.5	60.0	32.5	100
Not close to the middle class				
Agree (15)	26.7	60.0	13.3	100
Disagree (23)	21.7	56.5	21.7	100
Total sample (245)	29.8	42.0	28.2	100

a. This item was originally introduced in Srole (1956).

getting worse, not better"—predicted Wallace voting intentions within a number of relevant social controls. Moreover, as Table 3 indicates, the item's predictive value is independent of anti-Negro prejudice despite their positive relationship.

Once we measured relative deprivation with a battery of items beginning with the Cleveland surveys, we soon learned that the most

TABLE 3
ANTI-NEGRO PREJUDICE, RELATIVE DEPRIVATION, AND WALLACE SUPPORT, GARY, 1968 (in percentages)

	Supporters			
	Wallace	Nixon	Humphrey	Total
"The lot of the average man is getting worse."				
High Anti-Negro Prejudice				
Agree (59)	52.5	20.3	27.1	100
Disagree (34)	23.0	42.3	34.7	100
Moderate Anti-Negro Prejudice				
Agree (38)	36.8	44.7	18.5	100
Disagree (38)	26.3	44.7	28.9	100
Low Anti-Negro Prejudice				
Agree (18)	27.7	44.6	27.7	100
Disagree (58)	12.1	55.2	32.8	100
Total sample (245)	29.8	42.0	28.2	100

effective approach was through use of the scheme shown in Table 4. This scheme builds on the theoretical analysis of relative deprivation advanced by Runciman on the basis of his social class studies in the United Kingdom (Runciman, 1966). Table 4 is formed with two pieces of information: how each respondent views his own economic gains over the past five years in relation (1) to his ingroup (his class or racial category) and (2) to the relevant outgroup (e.g., white-collar workers for the blue-collar respondent, or blacks for the white respondent). Type A respondents are *doubly gratified,* for they feel they have been doing as well or better than both their ingroup and outgroup. Type B are the critical respondents, for they feel

TABLE 4
FOUR TYPES OF RELATIVE DEPRIVATION AND GRATIFICATION

		Personal Economic Gains Compared to Outgroup ("white-collar workers" or "Negroes")	
		Equal or Greater Than	Less Than
Personal economic gains compared to ingroup ("blue-collar workers" or "whites")	Equal or greater than	A. doubly gratified	B. fraternally deprived
	Less than	C. egoistically deprived	D. doubly deprived

fraternally deprived. This is Runciman's term for his key group in British class-deprivation research. They feel they have kept up with or even surpassed the gains of their own group but that they have slipped behind those of their outgroup. Consequently, their deprivation is fraternal in that it is their group as a whole which is seen as losing ground in comparison with the outgroup, and they are likely to perceive this situation as unfair.

By contrast, Type C consists of individuals who sense their gains to have been less than those of their ingroup but at least equal to those of their outgroup; they are therefore termed by Runciman as the *egoistically deprived.* Finally, and least interesting, are the *doubly deprived* respondents of Type D who feel they have lost ground to both their ingroup and outgroup. These individuals are typically older and often retired; their fixed incomes probably have in fact been surpassed by younger groups generally.

Both social class and racial comparisons, using the scheme of Table 4, have been found to be particularly important. The class comparisons contrast blue-collar versus white-collar workers; and since our respondents are largely blue-collar themselves, this means blue-collar workers form the ingroup and white-collar workers the outgroup for our purposes. The racial comparisons, of course, contrast whites versus blacks in economic gains.

Tables 5 and 6 demonstrate how *one* of these two types of comparisons operates differently as a predictor of Wallace support in Cleveland and Gary. Fraternal *class* deprivations relate to Wallace voting or preferences in both cities, especially for such meaningful subsamples as working-class identifiers and non-Democrats (Tables 5

TABLE 5
CLASS DEPRIVATION AND 1968 WALLACE VOTE IN CLEVELAND (in percentages)

	Class-Deprivation Type			
	A. Doubly Gratified	B. Fraternally Deprived	C. Egoistically Deprived	D. Doubly Deprived
Entire Cleveland sample[a] (n=301)	16	31	15	13
Just those who identify themselves with "working class"[a] (n=154)	11	41	23	15

a. Those who did not vote in the 1968 presidential election are omitted.

TABLE 6
CLASS DEPRIVATION AND 1972 WALLACE PREFERENCE IN GARY, 1970 (in percentages)

	Class-Deprivation Type			
	A. Doubly Gratified	B. Fraternally Deprived	C. Egoistically Deprived	D. Doubly Deprived
Entire Gary sample (n=288)	12	24	15	17
Non-Democratic party identifiers subsample only (n=92)	9	47	13	18

and 6). Some, but not all, of these differences, however, are traceable to background differences of the four class-deprivation types. Whites who feel fraternally deprived in class terms are disproportionately concentrated among those of medium income and education who are younger, full-time, working-class members of labor unions. These respondents, of course, are precisely the ones we have isolated in other analyses as especially prone to being pro-Wallace and against black candidates. Yet controls for these factors, as in the earlier Gary analysis of Table 3, reduce but do not remove the predictive value of the relative class-deprivation measures. This fact strongly implies that fraternal class deprivation acts as a mediator of some, though not all, of the special lower-middle-class component of the Wallace phenomenon in the North. In sharp contrast, fraternal race deprivations do not effectively predict Wallace leanings in either city. This suggests, together with other evidence (Pettigrew, 1971), that the Wallace appeal in northern cities had a strong populist as well as racist flavor.

Perceived racial deprivations become important, however, for predicting white support of black mayoralty candidates. Table 7 provides these consistent and dramatic results across the four cities. Note that the fraternally deprived on race report less willingness to vote for, and a more negative image of, the black candidate in every instance. The background differences among the four racial-deprivation types are similar to those among the four class-deprivation types noted above, though they are less extensive. Controls for these background variables do not substantially affect the relationships shown in Table 7.

TABLE 7

RACIAL DEPRIVATION AND THE REACTIONS OF WHITES TO BLACK MAYORALTY CANDIDATES

(in percentages)

	Racial-Deprivation Type			
Reactions to Black Candidates	A. Doubly Gratified	B. Fraternally Deprived	C. Egoistically Deprived	D. Doubly Deprived
Mayoralty Voting				
For Stokes versus Perk, Cleveland, 1969[a]	31	12	49	29
For Bradley versus Yorty,				
LA primary vote, 1969	26	17	34	30
Runoff preference, 1969	51	30	46	46
Runoff vote, 1969	35	21	52	42
For Gibson versus Addonizio, Newark, 1970	19	14	29	20
For Hatcher versus Williams, etc., Gary primary, 1971	17	7	30	15
Candidate Image (% favorable)[b]				
Stokes, 1969	57	33	64	50
Bradley, 1969	65	44	71	49
Gibson, 1970	25	18	27	36
Hatcher, 1970	35	17	36	29

a. For Democrats only, since this was a partisan final election.

b. The respondents were each presented a printed card with 12 adjectives from which three were chosen as the most descriptive of the black candidate. Half the adjectives were favorable in tone (e.g., intelligent, honest) and half were unfavorable (e.g., out-for-himself, prejudiced). The favorable percentages provided here represent those whites who chose 3 favorable adjectives in the cases of Stokes and Gibson, and 2 or 3 favorable adjectives in the cases of Bradley and Hatcher.

CONCLUSION

Despite the wide variations in political systems, political personalities, and urban contexts, a number of important phenomena appear to be replicated across cities when a black candidate runs for mayor. It is not surprising in the United States of the late twentieth century that racial considerations become paramount in these contests. And public opinion survey data over the years prepare us for the findings on which white voters are most likely to support the black candidate: Jewish Americans, the young, the college-educated, and the self-designated "liberals."

The base of major resistance in the white electorate to black aspirants, however, is more subtle. As with the "George Wallace" phenomenon in the North, the lower-middle-class whites are conspicuous in their opposition even in comparison with lower-class whites. This phenomenon has been much discussed and distorted in popular analysis, and variously labeled "the backlash of the ethnics," "the revolt of the silent majority," and other catchy but largely meaningless phrases. Initial analyses of the data described in this chapter point to a specific social-psychological condition that appears to be an important motivational source of this phenomenon. Fraternal relative deprivation as perceived by workers themselves acts in these data as both a cause and mediator of lower-middle-class white anger directed at blacks in northern cities. It is relative deprivation and not absolute deprivation; and it is relative to the perceived gains of black Americans as a group compared to those of white Americans as a group. By no means is this social-psychological factor a complete explanation of the larger lower-middle-class phenomenon of special resistance to racial change. But fraternal relative deprivation does seem to attain an importance that cannot be ignored in any broader explanation.

NOTES

1. Just to top off the day, Mrs. Louise Day Hicks, a leading northern opponent of racial change, lost by the convincing margin of 54% to 46% on her bid to become Mayor of Boston on November 7, 1967. Working from a too-simple "white backlash" theory, the mass media had some difficulty the next day accounting for the results of the three elections.

2. We are emphasizing serious black aspirants in major cities. This omits, for example, recent black mayoralty campaigns in Houston, Texas, and Minneapolis, Minnesota, that failed to come into serious contention with the white victors. It also omits the recent increase in black mayors in such small towns as Chapel Hill, North Carolina, and Fayette, Mississippi.

3. According to Gallup surveys, in 1937 only 64% of American voters would consider voting for a "well-qualified" candidate of their own party who was a Roman Catholic for President. By 1960, on the eve of Kennedy's campaign, this figure had risen only 7% to 71%. But by 1961 it had climbed dramatically to 82%; and by 1969 it had reached 88% (compared to 86% for a Jewish candidate, 65% for a black candidate, and only 53% for a woman candidate; see Gallup Opinion Index, 1969: 2-7). This shift appears to be yet another example of the well-established "fait accompli" effect (Pettigrew, 1971).

4. This phenomenon is generally known as "the sleeper effect."

5. This theoretical area has been sketched out and discussed in Pettigrew (1967).

6. The theory has also been usefully applied to black Americans as well (Pettigrew, 1971, 1967).

REFERENCES

CANTRIL, H. (1965) The Patterns of Human Concerns. New Brunswick, N.J.: Rutgers Univ. Press.

Gallup Opinion Index (1969) 46 (April).

HAHN, H. and T. ALMY (forthcoming) "Ethnic politics and racial issues: voting in Los Angeles." Western Pol. Q.

PETTIGREW, T. F. (1972) "Racism and mental health of white Americans: a social psychological view," in C. Willey, B. Kramer, and Brown (eds.), Racism and Mental Health. Pittsburgh: Univ. of Pittsburgh Press.

--- (1971) Racially Separate or Together? New York: McGraw-Hill.

--- (1967) "Social evaluation theory: convergences and applications," in D. Levine (ed.), Nebraska Symposium on Motivation. Lincoln: Univ. of Nebraska Press.

RUNCIMAN, W. G. (1966) Relative Deprivation and Social Justice. London: Routledge & Kegan Paul.

SEARS, D. O., D. R. KINDER, and R. T. RILEY (forthcoming) "The good life, 'white racism,' and the Los Angeles voter." Amer. Pol. Sci. Rev.

SROLE, L. (1956) "Social interaction and certain corollaries: an exploratory study." Amer. Soc. Rev. 21: 709-716.

5

Self-Interest and Referendum Support: The Case of a Rapid Transit Vote in Atlanta

LEWIS BOWMAN
DENNIS S. IPPOLITO
MARTIN L. LEVIN

REFERENDA, SELF-INTEREST, PUBLIC INTEREST, AND SYSTEM SUPPORT

□ IN THE UNITED STATES, thousands of referenda are presented to the voters annually. Some of these are concerned with important policy questions and are therefore vigorously contested (Hamilton, 1970: 125), especially referenda involving civil rights, metropolitan government, education, and similar issues. As a manifestation of direct democracy, these types of referenda properly raise questions about the bases of electoral behavior, particularly the relative effects of self-interest and public interest considerations.

When faced with a referendum issue involving significant increases in public expenditures, for example, a voter can respond to a variety of factors.[1] Previous studies have indicated the varying effects of factors such as race, social class, and even party in referendum voting (Salisbury and Black, 1963; Jennings and Zeigler, 1966; Hahn, 1970b). It is generally assumed, however, that perceptions of self-interest significantly influence referenda in general, and fiscal referenda in particular (Hamilton, 1970: 130-131; Wilson and Banfield, 1964: 876-877). This assumption is made even though the rationale for the referendum procedure leans heavily on the prior assumption that voters can and will acquire the necessary information and evaluate the appropriate considerations in seeking the "public interest."

This being the case, two seemingly contradictory but nevertheless fundamental assumptions are often made about voting behavior. First, self-interest, which Downs (1957: 27) defines as "rational behavior directed toward primarily selfish ends," is considered a critical factor. Second, electoral behavior apparently admits of some degree of altruism. As Downs (1957: 27) states, "In reality, men are not always selfish, even in politics. They frequently do what appears to be individually irrational because they believe it is socially rational–i.e., it benefits others even though it harms them personally."

Several formulations can be utilized in examining the self-interest component of voting, but one formulation is especially intriguing, since it attempts to "explain" the altruistic component of voting behavior. Sometimes referred to as ethos theory, it suggests that some voters are "public regarding" and others "private regarding" when they weigh proposed public expenditures (Wilson and Banfield, 1964; Banfield and Wilson, 1963: 35-46). Moreover, it hypothesizes that:

> voters in some income and ethnic groups are more likely than voters in others to take a public-regarding rather than a narrowly self-interested view of things–i.e., to take the welfare of others, especially that of "the community" into account as an aspect of their own welfare. We offer the additional hypothesis that both the tendency of a voter to take a public-regarding view and the content of that view . . . are largely functions of his participation in a subculture that is definable in ethnic and income terms [Wilson and Banfield, 1964: 885].

Previous studies have questioned the utility of the public regarding thesis (Hahn, 1970a; Wolfinger and Field, 1966), but there has been little effort to employ survey data in such investigations (Hamilton, 1970: 124). Despite the obvious difficulties of assessing voter "motivations," the implications of the public regarding thesis relating to the efficacy of the referendum procedure and to ethnic and class distinctions are sufficiently important to warrant a much more specific approach to the bases of voter choice. As Hamilton (1970: 131) states:

> In a tax levy "election," does the voter ponder in the booth whether the community benefits are commensurate with the cost, or does he weigh his personal benefit against the increment of his property tax?

The following analysis investigates this question.

Public and Private Regardingness

Drawing upon Hofstadter's description of the contrasting political ethos characterizing "Yankee" and "immigrant" voters (Hahn, 1970a, 1970b), Banfield and Wilson (1963: 41) define public regardingness as an "emphasis upon the obligation of the individual to participate in public affairs and to seek the good of the community 'as a whole'. . . ." While seeking the good of the community as a whole represents a form of self-interest, it is considered to be enlightened self-interest or self-interest broadly conceived (Wilson and Banfield, 1964: 885).

The contrary approach to public affairs is described as "ethnic" or private regarding, with the individual essentially interested in the advantages or benefits which he or members of his family will receive if a particular policy is adopted. Banfield and Wilson (1963: 38-44) classify this approach as self-interest narrowly conceived.

The public and private regarding approaches have been utilized in other studies to examine attitudes toward forms of local government (Wolfinger and Field, 1966) and toward public spending, particularly expenditure referenda (Wilson and Banfield, 1964). With respect to forms of local government, the public and private regarding models conflict over such matters as governmental consolidation, nonpartisan versus partisan elections, city manager versus mayor-council governments, and at-large elections versus ward or district constituencies—in other words, over what is sometimes described in the municipal reform literature as "professionalized" as opposed to "politicized" local government. In examining these and related propositions of ethos theory, Wolfinger and Field (1966: 324-326) found them to have limited applicability.

It can be argued, however, that attitudes toward public spending and expenditure referenda, rather than attitudes toward forms of local government, constitute a more appropriate focus for ethos theory. This is because certain referenda are more readily translatable into both narrow and broad conceptions of self-interest. As Wilson and Banfield (1964: 876) note:

> Local bond and other expenditure referenda present such situations: it is sometimes possible to say that a vote in favor of a particular expenditure proposal is incompatible with a certain voter's self-interest narrowly conceived. If the voter nevertheless casts such a vote and if there is evidence that his vote was not in some sense irrational or accidental, then it must be presumed that his action was based on some conception of "the public interest."

The method and difficulty of operationalizing "self-interest narrowly conceived" will, of course, vary with the issues involved in a given referendum. It is much easier, for example, to draw the relevant distinctions in cases where the benefits and costs are differentially apportioned within the community. Indeed, it is especially important that some measure be available for potential costs. Persons who perceive no immediate potential costs from a proposal are not faced with a perceived potential sacrifice–that is, the incompatibility between self-interest and the public interest with which the public regarding thesis is concerned. An example of this is non-property owners participating in referenda where the proposed financing is through increases in property taxes.

The difficulties of operationalization notwithstanding, political ethos theory provides a suggestive way of approaching political cleavages in metropolitan areas. It states that the most important cleavages in these areas–such as conflicts between the central city and the suburbs, and between racial and class groups–"tend to cut across each other and, in general, to become one fundamental cleavage separating two opposed conceptions of the public interest" (Banfield and Wilson, 1963: 35).

System Support

In addition to the manner in which variables such as race and class affect ethos theory, the concept of system support provides further refinement. The theory of system support focuses upon an individual's attitudes toward government and the implications of these attitudes for community governance. System support can be operationalized in terms of an individual's evaluations of governmental outputs or services and his emotional attachments to government.

David Easton (1965) has developed the concept of support by differentiating between types of support and objects of support. The types of support, specific and diffuse, differ in their reliance upon governmental outputs. Specific support "flows from the favorable attitudes and predisposition stimulated by outputs that are perceived by members to meet their demands as they arise or in anticipation" (Easton, 1965: 273). Diffuse support, on the other hand, "is independent of the effects of daily outputs. It consists of a reserve of support that enables a system to weather the many storms when outputs cannot be balanced off against inputs of demands" (Easton, 1965: 273).

The objects of support include the authorities or incumbents, political institutions and procedures, and the political community. The potential consequences of fluctuations in support become increasingly more serious as the object of support becomes more encompassing. Where support for the authorities declines, for example, the continued processing of citizen demands may become increasingly more difficult. Lack of support for political institutions and procedures or lack of support for the political community, however, may represent a threat to stability and maintenance (Easton, 1965: 157).

The concept of support has been used in analyzing attitudes toward forms of government, particularly proposals for governmental reorganization. In his study of a merger proposal in Nashville, Hawkins (1966: 413) found that voters who were dissatisfied with governmental services were more likely to support reorganization than were voters who were satisfied. In another study, Roth and Boynton (1969) used "communal ideology" as a form of diffuse support and analyzed its impact on attitudes toward local government merger.

Since an expenditure referendum involves either expanding current governmental services or instituting new services, it might be expected that elements of specific or diffuse support would be equally useful in analyzing attitudes toward such a referendum. In particular, an individual's evaluation of current services and his confidence in the adequacy of future governmental performance represent elements of support which are relevant considerations in questions of expansion or innovation. Thus system support might be an additional factor affecting public regarding behavior. For example, an individual who stands to benefit from a proposed program might oppose it because his potential costs outweigh the benefits he expects. He might also oppose it because he has little confidence that the government proposing the program has the ability to carry it out successfully. For an individual who does not perceive any personal benefit from a proposed program, but who would share in the costs if it were adopted, considerations about past or future governmental performance would seem even more important.

THE STUDY

On November 5, 1968, the voters in the city of Atlanta, DeKalb County, and Fulton County, Georgia, rejected in referenda a proposal to finance the construction of a rail rapid transit system for the Atlanta metropolitan area. As generally occurs after a rejection of this sort, a variety of explanations for the voters' action were advanced. These involved such disparate factors as the type of design (exclusive rail rapid transit), the method of funding (dependent largely upon property taxes), the differentials in planned services (particularly between high-income white areas and black areas), and the degree of citizen participation in the planning stage. In any case, seven years of planning, a large investment in planning costs, and rather tortuous battles with state government came to an abrupt and–to many leaders in the white community at least–an unexpected end.

Proponents of issues like rapid transit usually attempt to place them in as nonpolitical a context as possible. If the Atlanta campaign was typical, however, the efficacy of such attempts would appear to be limited. In Atlanta, the traditional urban cleavages–suburb versus central city; white versus black; middle class versus working class–surfaced during the campaign and, according to some observers, were key factors in the final result. Indeed, the racial and class cleavages received particular attention, since previous studies had found the controlling voter coalition in Atlanta to be composed of blacks and "northside" (upper-class and middle-class) whites (Jennings and Zeigler, 1966; Rooks, 1970). This traditional alliance, however, appeared to dissolve on the rapid transit issue.

In this study, the relationships among variables measuring public regardingness, system support, race, class, and community will be investigated in an attempt to explain the structuring of the vote on this referendum, and to provide insight into the relationship of these variables among the electorate generally.

The Data

The data reported here are from a survey conducted in the city of Atlanta, Fulton County, and DeKalb County, Georgia, in the spring of 1969.[2] Personal interviews were conducted with a sample drawn from the official lists of registered voters.[3] This paper reports only on the respondents in the city of Atlanta and DeKalb County for whom the relevant data are complete. Fulton County respondents have been excluded because they receive some services from the city of Atlanta, and some from the county government. Potentially this confounds comparisons of the evaluation of government services and confidence in government variables defined below. The Atlanta and DeKalb data do, however, permit a central city (Atlanta) versus suburban (DeKalb County) comparison. Within the Atlanta sample, a sufficient number of black respondents are available to examine differences in attitudes among blacks and whites. The number of blacks in the DeKalb sample was not sufficiently large to allow analysis.

Operational Definitions

Public regardingness. Since public regardingness involves support "of a particular expenditure proposal [which] is incompatible with a certain voter's self-interest narrowly conceived" (Wilson and Banfield, 1964: 876), the possibility of an individual's being public regarding will vary with specific referenda. This is because categorization depends upon the "benefits" and "costs" perceived by the voter.

On the rapid transit or MARTA (Metropolitan Atlanta Rapid Transit Authority) referendum, the proposal involved financing a rail rapid transit system through increases in property taxes. Therefore, two factors—*potential use by* and *potential cost to* the respondent—appeared particularly appropriate in distinguishing between potentially public regarding individuals and others.[4]

As Figure 1 indicates, there were four possible use-cost categories for each individual. The distinction utilized here focuses on the

		Respondent Anticipates Property Tax Increases to Finance MARTA	
		Yes	No
Respondent anticipates using MARTA for work or other purposes	Yes	Use + Costs (compatible self-interest)	Use + No Costs (compatible self-interest)
	No	No Use + Costs (Incompatible self-interest)	No Use + No Costs (compatible self-interest)

Figure 1: POTENTIAL PUBLIC REGARDING VOTERS

element of self-interest. Those persons who anticipated costs but did not intend to use the system and who still supported the MARTA proposal could be said to be acting against their self-interest narrowly conceived. For these persons, the element of sacrifice was present, and their support for the MARTA proposal could be reasonably interpreted as acting out of some conception of the public interest. This group has therefore been categorized under "incompatible self-interest," and, hence, is potentially public regarding.

Individuals in the remaining cost-benefit categories, on the other hand, did not have a similar incompatibility between self-interest and the public interest. Individuals who did not anticipate tax increases, for example, could support MARTA without sacrifice, whether or not they thought they would use the system. Those who anticipated a tax increase but who also intended using the system presumably could support MARTA and still serve some degree of self-interest. These three groups, then, have been categorized under "compatible self-interest," since support for MARTA can reasonably be interpreted as compatible with their self-interest.

The operational definition used here does, of course, present some problems. Some voters may have little or no information about a proposal. Other voters may take into consideration indirect benefits or costs, such as the effect on property values. Perhaps most important, the definition assumes that support for the MARTA proposal, in this case, is in the public interest. The proposal was indeed presented to the public in these terms, which perhaps mitigates the problem, but this type of arbitrary if necessary dichotomy between self-interest and public interest does not leave much room for principled opposition to a proposal. These problems, of course, will affect most attempts to specify public regardingness, and they indicate some of the limits on empirical testing of this concept.

Specific support: evaluation of services. As noted above, support theory involves a distinction between specific and diffuse support, with the former being directly related to governmental outputs. Therefore, respondent evaluations of local government services were used to operationalize the concept "specific support." Each of the respondents was classified as a "high evaluator" or a "low evaluator" on the basis of his evaluation of twelve services performed by local government.[5] "High evaluators" (the equivalent here of high specific support) included those who said that local government was doing a "very good" or a "good" job in nine or more services. "Low evaluators" (the equivalent here of low specific support) included those who assigned a "poor" or a "very poor" rating to four or more services. This cutting point was chosen on the basis of the mean number of negative evaluations for the Atlanta and DeKalb respondents.

Mixed support: confidence in local government. Diffuse support is independent of the effects of daily outputs, and given the difficulty of operationalizing such a support variable, the "confidence" variable used here will represent "mixed support"–that is, support which is affected by governmental outputs but which also has an emotional content.

Respondents were classified as to their confidence in local government in the following manner. They were first asked to list the major public problems which they thought affected their local area. They were then asked whether they expected their local government to be "very successful," "somewhat successful," "not very successful," or "not successful at all" in solving the problems which they had mentioned. Those who expected local government to be "very successful" or "somewhat successful" were classified as having "high confidence," while those who expected local government to be "not very successful" or "not successful at all" were classified as having "low confidence."

The confidence measure, then, has elements of specific and diffuse support. It is tied to the respondent's assessment of specific policy problems. A recent paper using these data found that confidence in government also was related to evaluations of current services (Levin and Ippolito, 1970). Respondents who evaluated current services positively tended to have higher confidence in future governmental performance than did those who were dissatisfied with current services. The confidence measure incorporates as well, however, factors which are elements of diffuse support, notably judgments about and feelings toward local government. These factors may serve to reinforce or to balance attitudes emerging from evaluations of governmental services (Levin and Ippolito, 1970: 8-10).

FINDINGS

Community, Race, and Education

In Atlanta and DeKalb, support for the MARTA proposal was quite similar, with approximately one-half the respondents in each area reporting that they favored the proposal.[6] Within Atlanta, support was greatly affected by race, with 57% of the whites as opposed to 39% of the blacks having favored MARTA. Fifty-four percent of the DeKalb whites supported MARTA. Among the Atlanta whites and DeKalb whites, it would appear the central city and suburban distinctions had little effect (as indicated earlier, a similar comparison among blacks was not possible).

The effect of education upon support, moreover, was significant among whites but not among blacks.[7] As Table 1 indicates, Atlanta and DeKalb whites with high education tended to be considerably more favorable toward MARTA than did whites with low education. Education explained 29% of the variation in MARTA support among Atlanta whites and 19% of the variation among DeKalb whites. Among Atlanta blacks, however, support was virtually the same in both educational subgroups, approximating in each case the degree of support exhibited by whites with low education.

These findings tentatively support the Wilson-Banfield hypothesis. It is apparent that race and social class (as measured by education) affected MARTA support in the hypothesized direction. A substantial majority of whites with high education, regardless of community, tended to support MARTA. A majority of whites with low education and a majority of blacks, regardless of education, opposed MARTA. If one accepts the MARTA proposal as being in

TABLE 1
SUPPORT FOR MARTA, BY COMMUNITY, RACE, EDUCATION, AND CONFIDENCE IN LOCAL GOVERNMENT (mixed support)

	Favoring MARTA					
	Atlanta Whites		Atlanta Blacks		DeKalb Whites	
	%	n	%	n	%	n
Education[a]						
High education	70	(151)	38	(56)	61	(366)
Low education	41	(119)	40	(144)	42	(211)
a' education=	.29[b]		.02[c]		.19[b]	
Confidence (mixed support)						
High confidence	61	(148)	44	(108)	61	(318)
Low confidence	55	(101)	27	(67)	47	(230)
a' confidence=	.06[c]		.17[b]		.14[b]	
Education and Confidence						
High education						
High confidence	73	(78)	54	(28)	68	(210)
Low confidence	69	(61)	17	(23)	53	(132)
Low education						
High confidence	49	(68)	41	(80)	48	(103)
Low confidence	35	(40)	33	(43)	41	(90)
a' education=	.28[b]		.02[c]		.16[b]	
a' confidence=	.08[c]		.18[b]		.12[b]	

a. Low education is defined here and in following tables as high school or less. High education is formal education beyond high school.

b. Statistically significant from zero beyond the 1% level.

c. Not statistically significant from zero beyond the 5% level.

the public interest, these limited data show clinic and class distinctiveness as far as public regardingness is concerned.

System Support

It is somewhat surprising that specific support (evaluation of local government services) did not appreciably affect support for MARTA among either whites or blacks. As might be expected, specific support was affected by race. Within Atlanta, for example, approximately two-thirds of the whites reported high evaluations of government services, while a majority of blacks reported low evaluations. Specific support among DeKalb whites was somewhat lower than that among Atlanta whites. In each subgroup, the level of support for MARTA was slightly higher among those with high evaluations as compared to those with low evaluations, but the effects of evaluation were not significant.

Mixed support (confidence in local government) did affect attitudes toward MARTA. As Table 1 indicates, support for MARTA in each of the three groups was greater among respondents with high confidence than among those with low confidence. While the direction of the effect was consistent for Atlanta whites, the effect was not significant. The effect of the confidence variable was greatest among Atlanta blacks, explaining 17% of the variation in MARTA support. For DeKalb whites, the confidence variable accounted for 14% of the variation in MARTA support.

The combined effects of education and confidence show some interesting patterns. Among Atlanta whites, education continued to exert an independent effect, accounting for 28% of the variation in MARTA support. The effect of the confidence variable, however, remained not significant, although the differences by confidence among those with low education were greater than the differences observed among those with high education.

For DeKalb whites, the effects were somewhat different. First, education and confidence had independent effects, explaining 16% and 12%, respectively, of the variation in MARTA support. Further, unlike the results for Atlanta whites, confidence had a greater effect among DeKalb whites with high education than it did among those with low education. Among Atlanta blacks, differences in MARTA support occurred by educational level where none had appeared in the zero-order relation, and the direction of these differences was not consistent with that observed for whites. In particular, highly educated blacks who reported high confidence also had the highest support for MARTA (as was the case for whites), but highly educated blacks with low confidence had the lowest support for MARTA of any subgroup. Thus education "intensified" the effect of the confidence variable among blacks, but it did not have an independent effect.[8]

It would appear, therefore, that the emphasis of the public regarding thesis upon ethnic and class differences requires qualification at this point. The expected relationships were found among Atlanta whites. However, among DeKalb whites mixed support did affect MARTA attitudes, and the extent of the effect of the confidence variable was greater among those with high education. Moreover, the confidence variable had a very decided effect among Atlanta blacks, particularly those with high education. In terms of

MARTA support, the distinctiveness of blacks with high education vis-à-vis whites with comparable education is attributable to a large extent to the impact of the confidence variable among highly educated blacks.

Self-Interest and Public Regardingness

A minority of respondents in each of the three groups can be categorized as having "incompatible self-interest." This includes 22% of the Atlanta whites, 28% of the Atlanta blacks, and 31% of the DeKalb whites. These respondents constitute what might be termed the "potential public regarding" portion of the sample–i.e., only the individuals in these groups who actually did support MARTA could be considered public regarding.

As Table 2 indicates, self-interest considerations had a significant effect in both Atlanta and DeKalb. Among both blacks and whites, support for MARTA was much higher for those with "compatible self-interest" than for those with "incompatible self-interest."[9] Indeed, self-interest accounted for 31% of the variation in MARTA support among blacks, 36% among DeKalb whites, and 37% among Atlanta whites, indicating a consistently strong effect.

Moreover, among Atlanta and DeKalb whites, the self-interest distinction explained considerably more of the variation in MARTA attitudes than did education (see Table 3). In DeKalb, for example, 34% of the variation was accounted for by self-interest, as opposed to 16% for education. Education once again had no significant effect among Atlanta blacks, but self-interest had a strong, positive effect on support for MARTA. The importance of the confidence variable

TABLE 2
SUPPORT FOR MARTA, BY COMMUNITY, RACE, AND SELF-INTEREST

	Favoring MARTA					
	Atlanta Whites		Atlanta Blacks		DeKalb Whites	
	%	n	%	n	%	n
Self-Interest						
Compatible self-interest	65	(166)	40	(94)	63	(312)
Incompatible self-interest	28	(47)	9	(35)	27	(138)
a' self-interest=	.37[a]		.31[a]		.36[a]	

a. Statistically significant from zero beyond the 1% level.

TABLE 3
SUPPORT FOR MARTA, BY COMMUNITY, RACE, SELF-INTEREST, EDUCATION, AND CONFIDENCE IN LOCAL GOVERNMENT

	Favoring MARTA					
	Atlanta Whites		Atlanta Blacks		DeKalb Whites	
	%	n	%	n	%	n
Education						
High education						
Compatible self-interest	73	(104)	33	(24)	68	(211)
Incompatible self-interest	41	(17)	14	(14)	36	(73)
Low education						
Compatible self-interest	52	(60)	43	(70)	54	(99)
Incompatible self-interest	20	(30)	5	(20)	17	(59)
a' education=	.21[a]		.002[c]		.16[a]	
a' self-interest=	.32[a]		.33[a]		.34[a]	
Confidence (mixed support)						
High confidence						
Compatible self-interest	65	(99)	43	(46)	66	(177)
Incompatible self-interest	40	(20)	16	(19)	38	(58)
Low confidence						
Compatible self-interest	69	(55)	26	(38)	62	(117)
Incompatible self-interest	18	(22)	0	(11)	21	(71)
a' confidence=	.02[c]		.16[a]		.08[b]	
a' self-interest=	.40[a]		.26[a]		.35[a]	

a. Statistically significant from zero beyond the 1% level.
b. Statistically significant from zero beyond the 5% level.
c. Not statistically significant from zero beyond the 5% level.

(mixed support) among blacks is also emphasized in Table 3. It explained 16% of the variation in MARTA attitudes, as opposed to 26% accounted for by self-interest. For whites, particularly Atlanta whites, confidence had only a minor effect. Thus it appears that the confidence variable has an effect among blacks which is similar (in both direction and magnitude) to that of education among whites. But it is also clear that considerations of self-interest are the most important explanatory variable for each group.

SUMMARY AND CONCLUSIONS

The findings presented here are suggestive in several ways. First, and most important, it would appear that the public regarding thesis is of very limited utility in explaining a vote such as MARTA. If a rigorous operational definition of self-interest is employed, only a small portion of the electorate can be potentially public regarding on this type of issue. And, as our findings indicate, support for MARTA was considerably lower among the potentially public regarding electorate than it was among individuals with a positive (compatible) self-interest involvement. Among whites and blacks, for example, a majority of the potentially public regarding group actually opposed MARTA.

On the other hand, the importance of self-interest in a referendum of this type is apparent. It is true that class (as measured here by education) has an effect among whites, but its explanatory value is considerably less than that of self-interest. Even among whites with high education, the differences in MARTA support attributable to self-interest were considerable.

Second, mixed support (confidence in local government) has a considerable effect among blacks, particularly blacks with high education, but a small effect among whites when education is controlled. Support for MARTA, for example, was virtually the same for whites and blacks with low education, ranging from 40% of the Atlanta blacks to 42% for the DeKalb whites. Moreover, while mixed support has little explanatory value for the whites when self-interest is controlled, among blacks the relationship between mixed support and support for MARTA is maintained even when self-interest is controlled.

It should be noted, however, that while the effect of the confidence variable is virtually nonexistent among whites when positive self-interest is involved, it has some importance for whites in the potentially public regarding group. Thus mixed support may have marginal impact on whites who do not have a positive self-interest at stake.

Given the publicity which transportation problems receive in the Atlanta area, an issue such as the MARTA proposal is inevitably associated (whatever the merits of the proposal) with community progress and the public interest. It would appear, however, that campaigns for issues of this sort cannot wholly rely upon the altruism of the electorate, even among those Wilson and Banfield consider to be public regarding individuals. And among blacks, the difficulty is even greater, for their political behavior is strongly affected by generalized attitudes toward government. Support from whites can apparently be obtained through self-interest considerations, regardless of class. For blacks, self-interest considerations are also important, but the disaffection of a considerable number of blacks must also be considered.

It would appear evident, then, that an issue such as rapid transit is extremely political. It is not susceptible to a "professionalized" or nonpolitical approach. Success with this and similar issues in the future will probably depend as much upon the political astuteness with which they are handled as upon the technical proficiency of their design.

NOTES

1. The analyses of referenda voting can generally be categorized into three groups. The first type of analysis emphasizes the specific issues involved in a particular referendum. Thus, the referendum becomes the focus for analysis. The second approach deemphasizes the specific issues involved, treats the referendum or referenda as simply a useful species of voting, and focuses on the effect of traditional political variables–party, race, social class–on this type of voting (Hahn, 1970a; Jennings and Zeigler, 1966; Salisbury and Black, 1963). The third approach is similar to the second in focus, but the issue involved is of such general interest and importance (e.g., fair housing referenda) that substantial attention is paid to the issue at stake, if not to the very specific features of the referendum (Hahn, 1968, 1970b; Hamilton, 1970; Wolfinger and Greenstein, 1968). Our study is appropriately categorized as an example of the second approach.

2. The authors acknowledge the support of the Institute of Public Administration (New York) and the Urban Mass Transportation Authority, U.S. Department of Transportation, which provided funding for the survey, and the assistance of Professors Jack Hopkins and William Pendleton who participated in the larger study from which these data are drawn.

3. The total sample size of 1,400 was allocated among the three jurisdictions in proportion to the number of registered voters residing in each place. Precincts were then selected in terms of probabilities weighted by the number of registered voters residing therein. A systematic sample of voters in the selected precincts was then drawn. Some systematic substitutions were allowed and 1,321 interviews were completed, with all respondents interviewed by members of their own race.

4. The authors wish to acknowledge the assistance of Professor William Hulbary who suggested the operational definition used here. Professor Hulbary is not, however, responsible for any shortcomings in the explanation or use of the operational definition.

5. The list of services included: (1) parks and recreation; (2) housing; (3) education; (4) law enforcement; (5) roads; (6) fire protection; (7) planning and zoning; (8) race relations; (9) providing jobs; (10) water supply; (11) garbage and trash disposal; and (12) health and hospital services.

6. Recalled attitude toward the MARTA proposal on election day has been used in this study even though recall data on the respondents' votes are also available. Attitude was used rather than vote to increase the number of respondents in each table. In the total sample, 108 persons did not have a definite attitude position relating to MARTA. Of the remainder, 131 did not vote on the proposal but expressed an attitude pro or con. However, of the 1,213 expressing an opinion, only 17 reported voting contrary to their opinion.

7. The "measure of effect" employed here was developed by James S. Coleman (1964: 189-210) and designed to be used with attribute data. Basically, it provides a measure of the percentage of variation in a dichotomous dependent variable which can be attributed to each of the independent variables. To account for differences in the size of the bases in the cross-tabulation, a weighted measure of effect was also developed which weights the contribution of each observed percentage difference in the calculations of the effect measure by the inverse of its variance. This procedure was used here.

8. This particular data configuration has been referred to as the "intensifier phenomenon." According to Coleman (1964: 224), it was first identified by Professor Paul Lazarsfeld in a seminar at Columbia University's Bureau of Applied Social Research (a review of the major findings of this seminar may be found in Suchman and Menzel, 1955: 148-155). In developing his measures of effect, Coleman (1964: 228-229) emphasized the fact that his measures are apt to prove misleading if employed on intensifier attributes. Instead, Coleman suggested as a rough measure of the intensifier effect an index which might be termed the "amplification factor." For the data with education, this value is 8.98; i.e., high education amplified the effect of confidence in government upon MARTA attitudes 8.98 times.

9. Within the "compatible self-interest" category, support was higher in the "use + no costs" group as compared to the "use + costs" group, especially among blacks. The "no use + costs" and the "no use + no costs" groups have quite similar levels of support among whites, but show larger differences among blacks. In the case of the "no costs" groups, however, the number of cases is quite small, particularly among blacks. More than four-fifths of these respondents who could be classified were in the "no use + costs" or "use + costs" categories, regardless of race.

REFERENCES

BANFIELD, E. C. and J. Q. WILSON (1963) City Politics. New York: Vintage Books.

COLEMAN, J. S. (1964) Introduction to Mathematical Sociology. New York: Free Press.

DOWNS, A. (1957) An Economic Theory of Democracy. New York: Harper & Row.

EASTON, D. (1965) A Systems Analysis of Political Life. New York: John Wiley.

HAHN, H. (1970a) "Ethos and social class: referenda in Canadian cities." Polity 2 (Spring): 295-315.

--- (1970b) "Correlates of public sentiments about war: local referenda on the Vietnam issue." Amer. Pol. Sci. Rev. 64 (December): 1186-1198.

--- (1968) "Northern referenda on fair housing: the response of white voters." Western Pol. Q. 21 (September): 483-495.

HAMILTON, H. D. (1970) "Direct legislation: some implications of open housing referenda." Amer. Pol. Sci. Rev. 64 (March): 124-137.

HAWKINS, B. W. (1966) "Public opinion and metropolitan reorganization in Nashville." J. of Politics 28 (May): 408-418.

JENNINGS, M. K. and H. ZEIGLER (1966) "Class, party, and race in four types of elections: the case of Atlanta." J. of Politics 28 (May): 391-407.

LAZARSFELD, P. F. and M. ROSENBERG, [eds.] (1955) The Language of Social Research. New York: Free Press.

LEVIN, M. L. and D. S. IPPOLITO (1970) "Some implications of evaluational orientations for confidence in local government among registered voters." Paper presented at Forty-Second Annual Meeting of the Southern Political Science Association.

ROOKS, C. S. (1970) The Atlanta Elections of 1969. Atlanta: Voter Education Project.

ROTH, M. and G. R. BOYNTON (1969) "Communal ideology and political support." J. of Politics 31 (February): 167-185.
SALISBURY, R. and G. BLACK (1963) "Class and party in partisan and nonpartisan elections: the case of Des Moines." Amer. Pol. Sci. Rev. 57 (September): 584-592.
SUCHMAN, E. A. and H. MENZEL (1955) "The interplay of demographic and psychological variables in the analysis of voting surveys." in P. F. Lazersfeld and M. Rosenberg (eds.) The Language of Social Research. New York: Free Press.
WILSON, J. Q. and E. C. BANFIELD (1964) "Public-regardingness as a value premise in voting behavior." Amer. Pol. Sci. Rev. 58 (December): 876-887.
WOLFINGER, R. E. and J. O. FIELD (1966) "Political ethos and the structure of city government." Amer. Pol. Sci. Rev. 60 (June): 306-326.
WOLFINGER, R. E. and F. I. GREENSTEIN (1968) "The repeal of fair housing in California: an analysis of referendum voting." Amer. Pol. Sci. Rev. 62 (September): 753-769.

6

National Patterns of Referenda Voting: The 1968 Election

JEROME M. CLUBB
MICHAEL W. TRAUGOTT

□ INSTRUMENTALITIES OF DIRECT DEMOCRACY have a well-established but ambiguous place in American political life. At every general election American voters have the opportunity to pass judgment on numerous public issues; and at primary, special, and local elections, a wide variety of propositions are also included on the ballot. In the 1968 general election well over three hundred statewide referenda, initiatives, and amendments to state constitutions were at issue. Furthermore, if estimates presented by one scholar are employed, it appears that as many as fifty local propositions may have been voted upon in 1968 for each statewide issue (Hamilton, 1970: 125). In other words, over 16,000 local issues were probably voted upon in the course of that year.

The initiative and referenda, and other devices of direct democracy are, of course, by no means phenomena only of contemporary years. Tentative and partial exploration of the historical record suggests that 12,000-15,000 propositions have been voted upon in statewide elections during the twentieth century alone. Once again, if this estimate is projected to include propositions voted upon in local elections, the number of policy issues that have been subjected

AUTHORS' NOTE: *We wish to thank Warren Miller for his careful reading of an earlier draft and for his valuable suggestions.*

historically to popular vote would appear substantially larger indeed. Obviously, the voting record on these issues provides a potentially important means to trace shifts in public attitudes toward policy issues across time. Knowledge of contemporary voting behavior on referenda, however, is not yet adequate to satisfactorily support the use of these historical materials.

Indeed, despite their ubiquity, the role and significance of these devices of direct democracy in the shaping of government and public policy has not been fully or systematically assessed. We know, however, that the institution of a variety of public policy innovations at the state and local levels often depends upon constitutional amendments, the passage of which requires popular ratification. In many states, for example, tax revisions, including institution of state and local income taxes, which have vital significance for the solution of pressing social and governmental problems, require popular ratification of amendments to state constitutions. The degree to which public education, to cite but a single further example, is dependent upon procedures of direct democracy has not been fully or systematically assessed. It is probably fair to say, however, that most observers are aware of numerous instances in which institution or expansion of instructional programs, increases in salaries, expansion of facilities, construction of buildings, and consolidation of school districts in the interest of economy and improved instruction have depended upon popular approval. And the list of such policies could probably be almost indefinitely extended.

Although a substantial literature on the subject exists, generalizations bearing upon the nature and mechanics of the popular response to referenda, initiatives, and the like, have not been satisfactorily integrated into more general knowledge of mass electoral behavior. The findings produced by empirical studies are often conflicting, and interpretations range from those which stress alienation and irrational negativism as major elements in referenda voting, to those which emphasize rational self-interest and "other-regardingness" as motivational factors determining the responses of referenda voters (see, e.g., Gamson, 1961; Horton and Thompson, 1962; McDill and Ridley, 1962; Wilson and Banfield, 1964; Hahn, 1970). Investigations of referenda voting, at least in recent years, have tended to focus rather heavily upon occasional and spectacularly controversial issues–the fluoridation and open-housing referenda of the 1950s and 1960s are cases in point–but less spectacular, because more common, issues have received relatively little attention. At least in part as a consequence of this relative emphasis, little is known of the ways in which information bearing upon school and tax referenda, for example, is communicated to the electorate and of the factors which shape and determine the popular response. The electoral response to referenda thus constitutes a lacunnae in theories of popular voting behavior, and, of more practical and immediate importance perhaps, public officials, school administrators, and others responsible for vital local facilities are without guidance in presenting their cases to the electorate or in interpreting and assessing the results of electoral decisions.

Our principal goal in this investigation is descriptive in nature: we are interested primarily in identifying the characteristics of referenda voters as compared with those who do not vote on such propositions. We are also concerned with two conceptual schemes which might be used to explain and interpret referenda voting. The first view emphasizes the alienation of referenda voters and stresses the negativism characteristic of referenda voting. At the risk of over-simplification and possible distortion, this view suggests that some referenda provoke a response on the part of those afflicted with a sense of powerlessness and mistrust of government. Although these individuals may not ordinarily participate in partisan elections, they sometimes express their resentment and alienation by voting negatively on particular referenda issues (see, e.g., Gamson, 1961; Horton and Thompson, 1962; McDill and Ridley, 1962; Aberbach, 1969).

The alternative view, we believe, is implicit in the literature of partisan voting behavior. Much of this literature leads to the expectation that referenda voters are likely to include a disproportionate number of better-educated individuals of higher socio-economic status and income, who are more politically involved, better informed, and marked by a stronger sense of political efficacy than those who do not participate in referenda (the basic work in this connection is Campbell et al., 1960). This view suggests, in other words, that referenda voters are unlikely to include large numbers of individuals who do not participate in other elections, and who are marked by characteristics that are normally associated with political alienation. Obviously, these conceptualizations are not necessarily or completely contradictory. The "alienated voter model" does not assume that all referenda provoke a negative or any other response on the part of alienated voters. Rather, only some, and perhaps primarily local, referenda are seen as provoking such a response.

The data upon which the study is based are the aggregate state-level returns of statewide referenda in the 1968 general election, and information collected in the national survey conducted by the Survey Research Center of the University of Michigan in conjunction with that election.[1] These data allow the alienated voter model, if such a voting pattern exists, to be placed in a more adequate national perspective marked by a common context, that of a national campaign. Furthermore, by combining the two categories of data, we feel that this perspective is still further enhanced. The data employed in this study, however, do not allow investigation of the voting response on specific referenda. We view the study as a preliminary report and expect in the future to extend the investigation to include referenda voting in the 1970 and 1972 general elections as well as on more purely local issues.

THE ELECTORAL RESPONSE

Number and variety were the most outstanding, and the most obvious, characteristics of the issues voted upon in the 1968 election. Referenda were included on the ballot in forty-four of the fifty states, and the issues voted upon ranged from ferry construction in Alaska, through race tracks in North Dakota, and eighteen-year-old voting in several states, to modification of the government of specific counties in Alabama. Modification and adjustment of state and local government, including such matters as tenure, succession, mode of selection, and responsibilities of public officials, were the most common issues voted upon. Money issues in one form or another, such as tax revision, special levies and bond issues, ran a close second. And it is obvious, of course, that the issues included on the ballots varied widely in substantive importance.

Even a cursory examination of the aggregate vote at the state level on these issues suggests several rather broad characteristics of referenda voting. In those states in which several referenda were on the ballot it was usually not the case that all received either a favorable or a negative vote. Rather, in most cases, there was considerable variation in level and direction of the vote, suggesting, of course, that the response of at least some voters varied from issue to issue and was neither consistently negative nor positive. Nevertheless, the overwhelming majority of the referenda on the ballots received a favorable response: 233 out of 320 referenda, or approximately 73%, passed. This figure is somewhat lower than Carter and Savard's (1961) finding with regard to the nationwide response to referenda on school bonds and taxes for the period 1948 to 1959, when 84% of all bonding proposals and 90% of all tax proposals passed. It is quite consonant, however, with Bone's (1968) finding for all referenda in the 1960s. Although some individual states did have much higher success ratios than others, in the most gross aggregate sense, we did not find widespread negativism in the 1968 vote. In fact, just the opposite case appeared to be true.

Voter turnout on referenda was consistently lower than the total vote for candidates to major national and state offices. Here again, however, there was considerable variation both within and between states. An indication of this latter variation is provided by comparison of turnout on those referenda in each state on which turnout was highest. Voter turnout on these propositions, computed as a proportion of estimated eligible population (U.S. Bureau of the Census, 1968), ranged from a high of 74% in Utah, to a low of 9% in South Carolina. The average referendum turnout was 40.1%. In these terms, turnout was on the whole highest in the states of the far West and, as might be expected, lowest in the southern states. On the other hand, when the turnout on these same referenda is computed as a proportion of the total vote for candidates to major offices, a somewhat different pattern emerges. Turnout on these referenda calculated as a proportion of the total vote for President ranged from a low of 30.5% in South Carolina to a high of 93.3% in Utah. The comparable figures based on races for the House was somewhat higher ranging from a low of 32.4%, again in South Carolina, to 113.5% in Mississippi. The latter proportion, of course, reflects uncontested House races in that state. Average turnout on these referenda calculated as a proportion of the total vote for President and House was 69.6% and 76.7% respectively. Moreover, the dropoff between the vote for major offices and referenda turnout tended to be less sharp in the southern states than in other regions even though turnout levels in the latter regions were considerably higher.

These turnout figures present a somewhat confusing picture. At first glance they confirm the expectation of low participation in referenda voting. They do, however, show a somewhat higher rate of referenda voting when the vote on referenda is compared with voting for President and congressmen. The 1968 survey data reveal a similar

TABLE 1
VOTER PARTICIPATION IN STATES IN WHICH STATEWIDE REFERENDA WERE HELD[a] (in percentages)

	Voted for Referenda	Did Not Vote for Referenda	Total
Did not vote	–	24.6	24.6
Voted for referenda only	.2	–	.2
Voted for Congress	.4	.6	1.0
Voted for President	3.1	5.7	8.8
Voted for President and Congress	36.5	28.9	75.4
Total	40.2	59.8	100.0
n=	(468)	(696)	(1,164)

a. Cell entries are based upon responses to questions asked of postelection respondents about their vote for President, U.S. representative, and referenda. Respondents for whom no postelection interview was obtained or who resided in states in which no statewide referenda appeared on the 1968 general election ballot are omitted.

pattern. As Table 1 shows, slightly over 40% of the respondents who had an opportunity to vote on statewide referenda in 1968 reported doing so, a figure which accords closely with average turnout level of 40.1% in all referenda observed in the preceding paragraph. In comparison, two out of three respondents reported that they had voted for a House candidate and almost three out of four respondents reported that they had voted for a presidential candidate. On the other hand, virtually all respondents who indicated that they had voted on referenda also reported that they had voted for candidates for President or Congress. Only three respondents reported voting for referenda only, and nine out of ten referenda voters also cast ballots for both the offices of President and representative.

One of the basic arguments offered by progressive reformers during the early twentieth century in support of the institution of the referenda, initiative, and other mechanisms of direct democracy was, of course, the view that the traditional devices of representative democracy did not give the citizen an adequate voice in public affairs. Numerous citizens, it was argued, had become disaffected with the political system because of the corruption and the frequent failures of elected representatives to adequately express the voter's wishes. Thus, direct democracy was seen as a means to restore confidence and restrain elected representatives, and it is probably fair to say that at least some contemporary intellectuals also hold generally similar views. Indeed, the alienated voter model of referenda voting, in at least some of its formulations, suggests that referenda voters are likely to include significant numbers of individuals who are disenchanted with the mechanics of representative democracy and the party system, and who frequently do not participate in partisan elections. In fact, Table 1 provides no indication that referenda voters were more dissatisfied with the mechanisms of representative government than were non-referenda voters. Neither does it indicate a groundswell of response to referenda beyond the response to partisan races on the ballot.

In 1968 there was significant variation in participation between referenda on the same ballot, and, here again, the aggregate vote does not suggest a consistent "knee-jerk" response on the part of the voters. As might be expected, the length of the ballot apparently affected participation on referenda. Among the forty-four states that included statewide referenda on the ballot, the average number of referenda was seven, with a maximum of fifty and a minimum of one. Voter participation was inversely related to ballot position, with referenda appearing relatively later in the ballot receiving, on the average, a lower total vote than referenda that preceded them. The correlation between ballot position and turnout ($r = -.48$) indicates, however, that this relationship, although strong, was by no means perfect. Moreover, when Louisiana with its fifty referenda was excluded from computation as a deviant case, the correlation between turnout and ballot position dropped to a more modest −.30.

As suggested elsewhere, turnout on referenda varied significantly from state to state and from one issue to the other within the same state. If it were assumed that higher turnout on referenda was the result of increased participation by the disaffected, it might be expected that negative voting would increase as participation levels increased. In fact, we found little or no correlation between turnout and negative voting, measured as percent not corrected for reversals of wording, in the aggregate returns. The correlation between the two variables for all 320 referenda was −.15 and −.07 when the fifty Louisiana referenda were excluded. Similarly, there was little indication that the response to referenda that appeared relatively late in the ballot was more strongly negative than the response to those that appeared toward the beginning of the ballot. The correlation between ballot position and negative voting for all referenda was .22 and −.07 when the Louisiana referenda were excluded. Thus the aggregate vote provides little evidence of association between either higher turnout or longer ballots and negative voting.

Taken in total, examination of these gross indicators of the electoral response to statewide referenda in 1968 provides no immediate or obvious evidence of negativism or alienation on the part of referenda voters. The participating electorate was small, but almost all referenda voters also voted for candidates in the major partisan races. The large majority of the referenda passed, and, although there was some association between turnout and ballot position, little or no association was found between either turnout or ballot position and negative voting.

CHARACTERISTICS OF REFERENDA VOTERS

The 1968 survey data provide an opportunity to compare the characteristics of respondents who reported that they had voted on referenda with those who indicated that they had not done so. In Table 2, respondents are compared in terms of several broad socioeconomic characteristics. In this case, respondents who reported that they did not vote in the 1968 election are grouped in the first category, and those who reported voting for either President or representative, or for both, are grouped in a second. All respondents who reported that they had voted on referenda are included in the third category whether or not they also reported voting for President and representative. As will be recalled from Table 1, however, virtually all referenda voters reported that they had also voted in at least one partisan race, and a slightly smaller proportion reported voting for both President and representative. Thus the classification groups respondents roughly in terms of their levels of political interest and activity as reflected by the extent of their electoral participation. Only respondents who resided in states in which statewide referenda were on the ballot are included in this and the following tables.

When compared in these terms, referenda voters often differed quite visibly from non-referenda voters. Although these differences are sometimes small when considered individually, their consistency when taken in total lends them a significance that might not otherwise be appreciated. As Table 2 indicates, referenda voters were even more overwhelmingly white than were the groups who indicated that they had not voted on referenda. Of the several groups, respondents who reported voting on referenda included a larger number of relatively high education attainment (35% with some college education or more), while the proportion of less well-educated was smallest. Similarly, the income levels of referenda voters tended to be higher than in the case of the other groups. Forty-two percent of the referenda voters had family incomes of $10,000 per year or more as opposed to only 16% of nonvoters and 30% who voted only for partisan races. Conversely, referenda voters included a smaller proportion of members of low-income families; only 22% of the referenda voters had family incomes below $6,000 as compared to 41% of those who voted only for offices. As might be expected in

TABLE 2
SOCIOECONOMIC CHARACTERISTICS OF REFERENDA AND NON-REFERENDA VOTERS (in percentages)

	Voted For:		
	Nonvoters	Offices Only	Referenda
Race			
White	85	87	92
Negro	13	12	7
Other	2	1	1
n=	(286)	(409)	(469)
Education			
Eight grades or less	37	23	13
High school	44	51	52
Some college or more	19	27	35
n=	(285)	(409)	(469)
Income			
Less than $6,000	57	41	22
$6,000-$9,999	26	29	36
$10,000 or more	16	30	42
n=	(278)	(399)	(462)
Subjective Social Class			
Working class	67	55	48
Middle class	28	36	36
Upper-middle class	5	9	16
n=	(278)	(400)	(449)
Place of Residence			
Large cities	23	26	25
Suburbs and small cities	30	29	40
Rural and outlying areas	47	45	35
n=	(286)	(409)	(469)

view of these differences, a smaller proportion of the referenda voters classified themselves as members of the working class than did non-referenda voters in the other two categories, and a larger proportion classified themselves as members of the middle or upper-middle class. The consistency of these differences as one moves across the table from left to right is of interest. Each category of respondents is relatively more highly educated, or higher on subjective social class, and includes a larger proportion of people of high incomes than the category to its left.

Referenda voters also differed in terms of their place of residence. In this case, respondents were classified in three categories: the first category includes residents of the central areas of large urban areas; the second includes residents of smaller cities surrounding larger cities and of more remote smaller cities. The third category includes residents of rural and outlying areas. As Table 2 indicates, the suburbs and smaller cities were better represented among respondents who reported that they had voted on referenda than were rural and outlying areas, and the central cities were least well represented.

Again, these differences are remarkably consistent from one descriptive and voting category to the next. Taken in total they suggest that decisions on public policy made through statewide referenda in 1968 disproportionately reflected the views of residents of the suburbs and small cities, of the highly educated, and of those of relatively high incomes and of higher social class in terms of their own self-classification. To put the matter somewhat differently, in 1968, statewide referenda constituted an even less effective means to communicate with, and receive communications from, the poor and the poorly educated, blacks, and residents of large cities and rural areas than did other electoral mechanisms.

In view of the differences observed above, a variety of other differences between referenda and non-referenda voters are quite predictable. Here again, these differences are impressively consistent. In the first place, the survey data indicate that referenda voters paid more attention to and, it may be inferred, had more information about the political events and the campaign of 1968 than had other respondents. As a group, respondents who indicated that they had voted on referenda read newspapers more regularly, listened to more radio programs, read magazines more extensively, and observed more television bearing upon the campaign than did other groups of respondents (see Table 3). There was, of course, considerable variation in attention from one media type to the other. Although

TABLE 3
MEDIA USAGE FOR POLITICAL INFORMATION BY REFERENDA AND NON-REFERENDA VOTERS (in percentages)

		Voted For:	
	Nonvoters	Offices Only	Referenda
Read Newspaper Articles About the Election			
Regularly	21	35	52
Often	7	16	12
From time to time	14	22	17
Once in a great while	30	7	5
Did not read	48	20	14
n=	(280)	(399)	(463)
Listened to Radio Programs About the Election			
A good many	11	12	15
Several	13	21	13
Just one or two	10	12	13
Did not listen	66	56	58
n=	(281)	(405)	(465)
Read Magazine Articles About the Election			
A good many	3	9	14
Several	7	12	14
Just one or two	10	11	21
Did not read	80	68	51
n=	(282)	(406)	(466)
Watched Television Programs About the Election			
A good many	34	40	51
Several	34	36	33
Just one or two	15	14	9
Did not watch	18	10	7
n=	(283)	(405)	(468)

referenda voters made heavier use of all media than other respondents, their greater attention to what might be termed the "active" media—newspapers and magazines—was particularly pronounced. Differences were less marked between referenda voters and other respondents in attention to radio and television which require only passive listening and watching. The direction and consistency in these differences in attention to the media are, of course, in keeping with and lend additional weight to the variations observed in preceding paragraphs.

Differences in attention to the political offerings of the media of communication, however, may indicate little more than access to political information and a generalized interest in the 1968 campaign. Indeed, these variations could reflect no more than the heavy media coverage given the campaign coupled with the general reading

TABLE 4

INTEREST IN PUBLIC AFFAIRS: REFERENDA AND NON-REFERENDA VOTERS (in percentages)

	Voted For:		
	Nonvoters	Offices Only	Referenda
Follow Public Affairs			
Most of the time	19	29	47
Some of the time	24	34	32
Only now and then	23	21	14
Hardly at all	34	16	7
n=	(285)	(408)	(469)
Attention to International and World Affairs			
Great deal	28	31	37
Some	47	55	53
Not much	25	14	11
n=	(188)	(340)	(435)
Attention to National Affairs			
Greal deal	24	36	46
Some	57	54	48
Not much	19	10	6
n=	(188)	(340)	(435)
Attention to State Affairs			
Greal deal	21	34	47
Some	58	54	45
Not much	21	12	8
n=	(188)	(341)	(436)
Attention to Local Affairs			
Great deal	24	36	47
Some	49	48	42
Not much	27	16	11
n=	(187)	(341)	(436)
Interest in School Board			
Very high	9	12	24
Moderately high	21	38	35
Moderately low	22	26	22
Very low	48	24	19
n=	(285)	(408)	(469)
Interest in Statewide Propositions[a]			
Interested	–	8	100
Not interested	–	64	0
Don't know if interested	–	2	0
No propositions on the ballot	–	26	0
n=		(367)	(469)

a. The following question was asked only of respondents who reported voting in the 1968 general election. "How about propositions on the ballot in November? Were there any state, county, or local propositions on the ballot for people to vote on? (If YES) Was there one proposition you were particularly interested in?"

and television watching habits associated with higher income and education levels. We do find that referenda voters who made extensive use of newspapers and magazines included a disproportionate number of individuals of higher income. The survey data, however, allow a somewhat more precise assessment of the level and nature of political interest on the part of referenda voters as compared with respondents who indicated that they had not voted on referenda. As Table 4 indicates, a significantly larger proportion of referenda voters than of respondents who did not vote on referenda indicated that they followed public affairs most of the time while a much smaller proportion said that they paid little attention to these matters. Similar, although smaller, differences were also characteristic of levels of interest in public issues ranging from world to local affairs and from the national level of government to the local school board.

These characteristics, of course, do not necessarily suggest any particular interest in referenda, and voting on such issues could have been in some measure merely the obligatory act of the good citizen unaccompanied by any specific information, motivation, or interest. Our data on this point are not extensive, but it is worth noting that all referenda voters could recall the presence on the ballot of referenda and all indicated specific interest in particular propositions. A significant proportion of the non-referenda voters in states in which statewide referenda were at issue, on the other hand, expressed no interest in any particular referenda or incorrectly responded that no propositions were included on the ballot (see Table 4).

It appears, then, that referenda voters in 1968 constituted a small but select segment of the active electorate in terms of education, income, self-perceived social status, and level of political information and interest. It is not surprising, therefore, that referenda voters tended to be more independent in voting behavior, at least in some senses of the word independent, and more heavily Republican than were the other groups of respondents. As Table 5 suggests, referenda voters were more likely to split their tickets in voting in state and local races than were respondents in other categories. This difference did not signify, however, that referenda voters were indifferent to the political parties or that their partisan identification was consistently less strong than those of respondents who indicated that they had not voted on referenda. As Table 5 also indicates, most referenda voters identified more or less strongly with one or the

TABLE 5

PARTY IDENTIFICATION ON REFERENDA AND NON-REFERENDA VOTERS (in percentages)

		Voted For:	
	Nonvoters	Offices Only	Referenda
Party Identification			
Strong Democrat	15	23	22
Weak Democrat	34	26	22
Independent Democrat	11	9	9
Independent	14	8	8
Independent Republican	7	8	10
Weak Republican	12	15	18
Strong Republican	7	11	11
n=	(272)	(408)	(468)
Democratic identifiers	60	58	53
Independents	14	8	8
Republican identifiers	26	34	39
Vote for State and Local Offices			
Straight ticket	–	58	48
Split ticket	–	42	52
n=		(383)	(451)

other of the major parties, although this proportion was but little larger than the component of party identifiers in the other groups. Democratic identifiers were apparently less likely to vote on referenda than were those who identified with the Republican Party. These differences are in keeping, of course, with the higher proportion of marginal and poorly socialized voters among Democrats than among Republicans. Even so, and for what it means, these differences suggest that on the whole the views of Democrats in 1968 were less well-represented in referenda voting than were the views of Republicans.

Taken in total, the data presented above do not provide any obvious reason to suspect that referenda voters included a disproportionate number of the politically alienated or of individuals marked by a sense of political powerlessness, or to believe that referenda voting was primarily motivated by a desire to strike a direct blow at an unjust and unresponsive political Establishment. To the contrary, the overwhelming majority of referenda voters indicated that they had voted for candidates for President and for Congress, and most referenda voters identified with one or the other of the political parties. As we have also seen, referenda voters included a larger proportion of individuals of higher socioeconomic status and tended to be better informed about and more interested in politics and public affairs than were respondents who reported that they had not voted on referenda. In other words, as a group, referenda voters included a disproportionate number of individuals marked by characteristics that are usually associated with a sense of political power.

To this point, however, the data presented have been at best indirectly and inferentially related to these latter issues. Here again, the available data allow a somewhat more direct approach to questions related to alienation and powerlessness. The 1968 survey included questions relevant to the sense of political efficacy of respondents, to their trust in government and governmental officials, and to their assessment of the responsiveness of government. On the basis of responses to these questions, three indices–political efficacy, government responsiveness, and trust in government–were constructed to compare respondents who reported voting on referenda with those who indicated that they had not done so. The indices have been employed in various past studies and are described elsewhere (see Robinson et al., 1968: 187, 632, 633).

As originally formulated, the sense of political efficacy was defined as "the feeling that individual political action does have, or can have, an impact upon the political process, i.e., that it is worthwhile to perform one's civic duties. It is the feeling that political and social change is possible, and that the individual citizen can play a part in bringing about this change" (Campbell et al., 1954: 187). Various studies have found that political participation in partisan elections tends to be associated with a high sense of political efficacy. If it is assumed that referenda sometimes provoke a response on the part of alienated voters characterized by a sense of political powerlessness, it might be expected that a larger proportion of referenda voters than of other voters would score low in political efficacy. In fact quite the reverse appears to have been the case in 1968 (see Table 6). As one moves from nonvoters, to voters in partisan races only, to referenda voters, increasingly, larger proportions of each group have high scores on the political efficacy index and smaller proportions have low scores. Almost one-half the referenda voters were marked, in terms of the index, by a high sense of political efficacy as compared to one-fourth of the nonvoters.

The second index is intended to measure attitudes about the responsiveness of government "to the people and to the presumed instruments of their will—political parties and the electoral process" (Robinson et al., 1968: 632). Here again, alienated voters would not be expected to find the government highly responsive and could be expected to have low scores on the index. Differences between referenda voters and the other categories of respondents as measured by this index were not as sharp as in the case of the political efficacy measure. Even so, consistent differences are observable. The group of respondents who reported that they had voted on referenda included a larger proportion of individuals who expressed confidence in the responsiveness of government and a smaller proportion with low scores. Respondents who voted only for partisan offices were less confident in the responsiveness of government, and nonvoters were the least confident of the three groups.

The final index employed is concerned with respondents' trust in government. The questions upon which the index is based were relevant to respondents' overall trust in government, their view as to whether government is run for the benefit of all or a few interests, and their assessment of governmental employees in terms of

TABLE 6
ATTITUDES TOWARD POLITICS AND THE POLITICAL SYSTEM OF REFERENDA AND NON-REFERENDA VOTERS
(in percentages)

	Voted For:		
	Nonvoters	Offices Only	Referenda
Political Efficacy			
Low	39	21	12
Medium	37	42	40
High	24	37	48
n=	(266)	(393)	(455)
Government Responsiveness			
Low	36	28	23
Medium	40	42	46
High	24	29	31
n=	(220)	(364)	(438)
Trust in Government			
Low	26	21	18
Medium	57	58	66
High	17	20	16
n=	(246)	(362)	(432)

intelligence, capacity to handle money, and honesty (see Robinson et al., 1968: 633). Alienated voters, again, would be expected to have low scores on this index. Differences between referenda voters and other respondents are even less clear than those measured by the two indices discussed above. However, the proportion of referenda voters who expressed low trust in government was smaller than the corresponding proportion of nonvoters and voters for partisan offices. On the other hand, a somewhat smaller number of the referenda voters than of respondents in the other categories expressed a high level of trust in government. These differences were faint, however, and most referenda voters scored in the middle category on this index. Taken in total, the index does not suggest that referenda voters were any more or less trustful of government than respondents who did not vote on referenda.

In terms of these indices, then, referenda voters appear as more confident of their own capacity to affect the political process and, to a somewhat larger degree, more inclined to see government as responsive than were nonvoters or those who voted only for partisan offices. On the other hand, they also tended to be as mistrustful of government and government employees as the other categories of respondents. In view of these characteristics, and those observed in preceding pages, referenda voters appear as a well-informed and well-politicized segment of the electorate that was attuned to perceived dishonesty and waste in government and able to respond to these perceptions at the ballot box.

ISSUE SALIENCE

As suggested elsewhere, the data employed in this study are by no means completely adequate to effectively explore a number of important aspects of referenda voting. The propositions voted upon in 1968 differed from state to state and varied significantly in terms of their importance and controversiality. Thus it is certainly possible to assume that particular issues attracted the interests of more voters and of voters marked by different characteristics than did other issues. The sample, however, does not allow us to explore the impact of a particular referendum within a given state. It is, of course, possible to group referenda into more general issue categories, but local contexts and orientations and the nature of specific propositions within these categories would vary widely from state to state. Thus this procedure would be of limited utility as a means to investigate the effects of particular issues upon referenda voting, or to assess the characteristics of voters attracted to the polls by highly salient or controversial propositions.

To gain additional information bearing upon the latter question, we have employed the aggregate vote to classify referenda in terms of their apparent interest and controversiality. A variety of factors in addition to the intrinsic interest of the issues at stake can influence turnout on referenda, including traditional or historical participation patterns, the presence on the ballot of a highly popular candidate, and hotly contested candidate races. To minimize the effects of such nonissue-related factors, we have classified referenda on the basis of relative turnout. Referenda characterized by high relative turnout as compared with turnout on other referenda on the same ballot or turnout in congressional or presidential races were classified in one category; those on which relative turnout was low were classified in a second category. These categories were again subdivided in terms of the closeness of the electoral contest, producing four categories of referenda. Referenda marked by high relative turnout and close contests have been termed salient; those marked by high relative turnout but on which the contest was not close have been termed consensual; and those characterized by low relative turnout and close contests have been termed special issues. Finally, referenda on which relative turnout was low and the margin of decision was wide have been classified as routine issues.[2] On the basis of this procedure, we have compared respondents from states that included on the ballot referenda in the several categories. This procedure does not, of course, eliminate the effects of the non-issue-related factors mentioned above, and it has a number of other imperfections. Even so, it provides a suggestive means to explore the characteristics of the electorate that voted on issues that were probably of higher salience as compared with the electorate that decided the outcome of referenda of apparent lower salience.

We have not attempted to report the results of these comparisons in detail. In general, however, our comparisons indicate that referenda voters in the four salience categories were similar in their collective characteristics. That is, referenda voters in all four categories included a larger proportion of individuals of higher income and education, of higher subjective social class, and who indicated higher levels of political interest and information than did non-referenda voters. Most referenda voters in all four categories identified more or less strongly with one or the other of the political parties. The distribution of party identifiers was approximately the same from one category to the other as was the proportion of independents. Similarly, referenda voters in the four categories were on the whole as confident in government and their own political capacities as those who did not vote on referenda. In short, our effort to control on issue salience does not suggest that the differences between referenda voters and other respondents observed in the preceding pages were associated with the salience or controversiality of the issues at stake.

As we have noted elsewhere, some of the literature on referenda voting suggests that some referenda provide an occasion for the negatively inclined and the politically disaffected to express their views. From this perspective it might be expected that participation votes among various social and economic groups would differ from one category of referenda to the other. More specifically, it might be expected that referenda in our salient category, those with high relative turnout and contested outcomes, would be marked by higher participation levels on the part of groups with characteristics that have sometimes been associated with political alienation and disaffection than would referenda in the other categories. In fact we do find (Table 7) that participation rates among low-income, poorly educated, working-class, black and urban respondents were higher on referenda in the salient category than on other referenda. On the other hand, we also find that participation rates of all groups tended to be higher on these referenda than on referenda in the other categories, although the differences were small in some cases. Participation rates on salient referenda, moreover, were highest among whites and suburban respondents and among those of higher subjective social class, income, and educational attainment.

The data also suggest considerable discrimination on the part of these latter groups. As Table 7 indicates, suburban respondents, whites, and those of higher social and economic status, participated more consistently on referenda in all four categories than did other respondents. Their participation votes were highest, however, on referenda in the salient and special categories. The latter referenda, it will be recalled, were marked by low relative turnout and by closely contested outcomes. Taken in total, then, the data suggest that the more affluent, the better-educated, and those of higher social status

TABLE 7

PARTICIPATION RATES OF SELECTED SOCIAL AND ECONOMIC GROUPS ON REFERENDA IN SALIENCE CATEGORIES

	Salient		Popular		Special		Routine	
	%	n	%	n	%	n	%	n
Place of Residence								
Urban	54	(72)	32	(63)	38	(37)	37	(117)
Smaller cities and suburbs	65	(89)	47	(45)	53	(75)	37	(192)
Rural	38	(130)	32	(82)	27	(136)	38	(135)
Education								
Eight grades or less	32	(71)	17	(46)	12	(64)	29	(80)
High school	53	(141)	33	(98)	43	(120)	36	(228)
Some college or more	62	(79)	48	(56)	48	(64)	44	(135)
Subjective Social Class								
Working class	43	(148)	27	(102)	30	(135)	35	(240)
Middle class	56	(100)	42	(55)	32	(25)	36	(159)
Upper-middle class	70	(36)	45	(22)	74	(27)	50	(36)
Income								
Less than $6,000	31	(87)	11	(66)	17	(110)	30	(169)
$6,000-$9,999	52	(98)	49	(59)	51	(65)	39	(130)
$10,000 or over	69	(98)	47	(56)	56	(70)	45	(140)
Race								
White	53	(252)	37	(164)	40	(224)	38	(395)
Black	38	(38)	29	(24)	8	(24)	29	(38)
Other	0	(1)	50	(2)	–	–	18	(11)

were more likely to respond to controversial referenda, as indicated by the closeness of the electoral contest, than were other respondents.

As has been observed, respondents of lower social and economic status participated more consistently on referenda in the salient category than on referenda in the other categories. Higher participation on the part of these groups is obviously by no means a necessary indication that these referenda attracted a larger proportion of the politically alienated and disaffected to the polls. Here again, the indices of political interest and attitudes toward government and the political process employed in preceding pages are of interest. As would be expected, individuals who expressed higher levels of interest in public affairs were more likely to vote on referenda than were those who expressed lower interest (see Table 8). A larger proportion of respondents who indicated only limited interest in public affairs voted on referenda in the salient and special categories than on other referenda, suggesting perhaps that the greater controversiality of these issues attracted the attention of a larger segment of the marginally uninterested. Even so, the association between level of expressed interest in public affairs and voting on referenda is impressively consistent across all categories of referenda.

The indices of attitudes toward government and the political process present a more varied and less consistent picture (see Table 8). On the whole, respondents who expressed medium or high levels of trust in government and confidence in the responsiveness of government and in their own capacity to effect the governmental process were most likely to vote on referenda; those with low scores on these indices were least likely to vote. The distribution of respondents on the index of personal political efficacy is particularly consistent. Higher levels of confidence in personal political efficacy were consistently and strongly associated with higher turnout on referenda in all four categories. There were, however, variations in the distribution of respondents on these indices from one referenda category to the other. Participation rates for respondents at all levels

TABLE 8
PARTICIPATION RATES OF SELECTED ATTITUDINAL GROUPS ON REFERENDA IN SALIENCE CATEGORIES

	Salient		Popular		Special		Routine	
	%	n	%	n	%	n	%	n
Follow Public Affairs								
Hardly at all	21	(43)	4	(22)	24	(50)	13	(83)
Only now and then	40	(52)	26	(31)	27	(48)	28	(86)
Some of the time	50	(91)	32	(68)	38	(73)	41	(127)
Most of the time	68	(105)	54	(67)	49	(77)	53	(148)
Political Efficacy								
Low	29	(58)	12	(34)	20	(64)	22	(88)
Medium	47	(107)	34	(88)	36	(80)	43	(170)
High	67	(117)	53	(62)	46	(93)	41	(162)
Government Responsiveness								
Low	48	(68)	32	(41)	32	(65)	29	(109)
Medium	55	(116)	38	(71)	36	(95)	47	(165)
High	57	(81)	45	(49)	43	(55)	40	(114)
Trust in Government								
Low	42	(41)	39	(31)	31	(58)	33	(73)
Medium	54	(143)	34	(130)	41	(117)	42	(236)
High	54	(56)	38	(29)	34	(35)	31	(62)

of expressed confidence on the three indices were highest on referenda in the salient category. Thus referenda in this category were marked by higher levels of participation on the part of individuals who were mistrustful of government and who doubted their own capacity to affect the political process.

But despite these variations in participation levels, such individuals constituted only a small proportion of the electorate for referenda in the salient category. Only 14% of the respondents who reported that they had voted on referenda expressed a low level of trust in government as compared with 62% who scored at the medium level and 24% who scored at the high level. On the index of government responsiveness, 23% of the referenda voters indicated low levels of trust as compared with 45% and 32% who indicated medium and high levels of confidence. The distribution of the salient referenda electorate in the index of personal political efficacy was even more one-sided: only 12% of the respondents who voted on referenda in this category expressed a low level of confidence in their personal political efficacy; 34% scored at the medium level; and 54% indicated a high level of confidence in their own capacity to affect the political process.

Taken in total, our efforts to control on issue salience, although they were probably not perfectly effective, do not suggest that the characteristics of referenda voters observed in the preceding section were related to the controversiality and salience of the issues at stake. There were variations in participation patterns from one category of referenda to the other. The participation rate of respondents marked by lower social and economic characteristics and by those who expressed limited confidence in government and the political process were higher on referenda in the salient category, but the participation rates of other groups were also higher on these referenda. It is, of course, quite possible that specific highly controversial and emotion-laden issues—open-housing and the fluoridation referenda of the past are examples of such issues—attract disproportionate numbers of the politically disaffected to the polls and provoke a highly negative response as a number of studies have indicated. The data employed in this study do not permit fully effective investigation of this phenomenon. In more general terms, however, and at least where less controversial and emotion-laden propositions are at issue, there is little reason to believe on the basis of the data employed here that referenda are decided by an electorate that is disproportionately alienated, disaffected, or negative in its political inclinations.

RESIDENTIAL CHARACTERISTICS

As observed above, residents of the suburbs and smaller cities constituted the largest proportion of the total group of respondents who reported that they had voted on referenda. Residents of rural and outlying areas provided the next largest proportion, and the smallest group resided in the central cities of major urban areas (Table 2). The problems and the population characteristics of the major cities, the relative isolation of rural areas, and the presumably greater affluence of the suburbs and smaller cities, make it reasonable to expect that the response on referenda and the characteristics of referenda voters would vary from one of these residential categories to the other. Indeed, it is not unreasonable to assume that the prevalence and impact of the politically disaffected and alienated would also vary from one residential category to the other.

To explore these and related possibilities we have grouped respondents in three categories in terms of the characteristics of their places of residence. As described elsewhere, residents of the central cities of large urban areas have been grouped in one category; residents of smaller cities surrounding larger cities and of smaller and more remote urban areas have been grouped in a second category; and residents of rural and outlying areas have been classified in a third. Here again, our classification procedure is less than adequate. In particular, we have no way of distinguishing consistently between smaller cities and urban areas that can appropriately be termed suburbs of larger cities and those which cannot appropriately be so classified. Similarly, and as a more trivial matter, our classification does not discriminate, on the one hand, between rural dwellers who commute daily to urban areas and who are oriented toward those areas, and, on the other hand, those rural dwellers, whose orientation is more narrowly limited to their immediate pastoral surroundings. Our difficulty, of course, is primarily a product of the sociological and psychological complexities of the conceptual categories urban, rural, and suburban. But despite these difficulties, we doubt that the shortcomings of our classification procedures seriously distort the findings reported below.

As would be expected, the three subsamples differed significantly in social and economic characteristics (Table 9). Only about three out of four respondents in the urban category were white, while the proportion of whites in the other two residential categories was greater than nine out of ten. The proportion of older respondents was greatest in the rural category, the urban subsample was the poorest and the least well-educated, and respondents from smaller cities and suburbs the wealthiest and the best-educated. Finally, the urban category included the largest proportion of respondents who

TABLE 9

RESIDENTIAL AND SOCIOECONOMIC CHARACTERISTICS OF REFERENDA AND NON-REFERENDA VOTERS

(in percentages)

	Urban		Smaller Cities and Suburbs		Rural	
	Voted For:					
	Offices Only	Referenda	Offices Only	Referenda	Offices Only	Referenda
Race						
White	69	79	92	96	92	97
Black	28	18	6	4	8	3
Other	3	3	2	0	0	0
n=	(101)	(115)	(123)	(190)	(185)	(164)
Age						
30 years or less	17	25	18	19	16	13
31-45 years	32	25	35	32	28	32
46 or more years	51	50	47	49	56	55
n=	(101)	(115)	(123)	(189)	(184)	(164)
Income						
Less than $6,000	41	28	27	18	51	24
$6,000-$9,999	31	34	27	33	28	39
$10,000 or more	27	38	46	49	21	37
n=	(99)	(112)	(121)	(189)	(179)	(161)
Education						
Eight grades or less	17	21	14	9	31	13
High school	50	51	50	54	51	49
Some college or more	33	28	36	37	18	38
n=	(101)	(115)	(123)	(190)	(185)	(164)
Subjective Social Class						
Working class	57	54	41	43	54	37
Middle class	30	38	50	35	39	48
Upper-middle class	13	8	9	22	7	16
n=	(98)	(107)	(120)	(184)	(182)	(158)

perceived themselves as members of the working class while the smaller city and suburban category included the largest proportion that identified themselves as upper-middle class.

In view of the social and economic characteristics of respondents in the three residential categories, it is not surprising that urban referenda voters included a larger proportion of blacks, of individuals of low income and education, and of those who classified themselves as members of the working class, than did either of the other residential categories. On the other hand, variations in the characteristics of the referenda electorate as compared with the partisan electorate and in turnout patterns on referenda are observable from one category to the other that do not conform fully to our prior expectations.

As we would expect, the suburbs and the smaller cities were marked by the highest electoral participation rates. Approximately 79% of the respondents in these areas reported voting for at least one office and 48% indicated that they had voted on referenda. The comparable figures for urban areas are 76% and 41% and 72% and 33% for rural areas. Although in the total sample rural voters constituted a larger proportion of referenda voters than urban respondents, the turnout rate in both referenda and partisan contests was higher among urban respondents than among those from rural areas.

As Table 9 indicates, referenda voters in the suburbs and smaller cities and in rural areas included a larger proportion of whites, of the more affluent and the better-educated, and of individuals of higher subjective social class than did respondents who indicated that they had not voted on referenda. Middle-class respondents constituted a smaller proportion of the suburban referenda electorate than of the corresponding partisan electorate, and the proportion of suburban working-class respondents among referenda voters was slightly larger than among those who voted only for partisan offices. On the other hand, suburban residents who classified themselves as upper-middle class were significantly better represented in the referenda electorate than in the partisan electorate.

The characteristics of the urban refernda electorate do not conform fully to the general patterns observed in the suburban and rural residential categories (Table 9). Poorly educated respondents, those with an eighth-grade education or less, made up a slightly larger proportion of the urban referenda electorate than of the partisan electorate in these areas. Conversely, respondents who had attended college and those who classified themselves as upper-middle class constituted a significantly smaller segment of the referenda electorate than of the partisan electorate in urban areas. Taken in total, moreover, the data presented in Table 9 suggest that respondents marked by lower social and economic characteristics voted more consistently on referenda in urban areas than in the other residential areas.

Comparison of the participation rates of these social and economic groups from one residential category to the other provides partial confirmation of this impression. As Table 10 indicates, participation rates on referenda among respondents with an eighth-grade education or less were higher in the urban areas than among the poorly educated in the suburbs and rural areas. Blacks voted on referenda at least as consistently in urban areas as in the suburbs and significantly more consistently than in rural areas. The lowest-income group voted less consistently on referenda in urban areas than in the suburbs, but in this case as well, participation rates were higher in urban areas

TABLE 10
PARTICIPATION LEVELS OF SOCIAL, ECONOMIC, AND RESIDENTIAL GROUPINGS ON REFERENDA

	Urban		Smaller Cities and Suburbs		Rural	
	%	n	%	n	%	n
Race						
White	44	(206)	49	(367)	35	(456)
Black	31	(67)	30	(27)	18	(29)
Other	33	(9)	0	(5)	–	–
Age						
30 years or less	40	(72)	42	(92)	28	(79)
31-45 years	36	(80)	48	(128)	38	(139)
46 or more years	44	(130)	50	(179)	34	(264)
Income						
Less than $6,000	28	(111)	33	(99)	18	(217)
$6,000-$9,999	45	(84)	50	(127)	45	(139)
$10,000 or more	53	(81)	55	(170)	53	(111)
Education						
Eight grades or less	41	(59)	28	(60)	15	(141)
High school	42	(142)	49	(209)	36	(221)
Some college or more	40	(81)	55	(130)	53	(119)
Subjective Social Class						
Working class	38	(152)	43	(184)	27	(285)
Middle class	45	(91)	44	(149)	38	(145)
Upper-middle class	33	(24)	73	(55)	59	(42)

than among the comparable group in rural areas. In more general terms, turnout on referenda among respondents of lower social and economic status was consistently lower in rural areas than in the other residential categories. Finally, and rather surprisingly, participation levels among respondents who had attended college and those who classified themselves as upper-middle class were markedly lower in the urban areas than in the other residential categories. In view of these variations in turnout patterns, it appears that on the whole referenda voters in urban areas were more representative of the total population of those areas than was the case in either of the other two residential categories.

We have not presented data bearing upon the partisan characteristics of referenda voters in the several residential categories. The proportion of Democrats among referenda voters was significantly greater in the urban areas than in the other two residential groupings. In all three categories, however, the large majority of referenda voters identified with one or the other of the parties, and only a small proportion classified themselves as independents. In all three categories, moreover, disproportionate numbers of referenda voters indicated high levels of interest in public affairs, and the distributions did not differ sharply from one residential category to the other (Table 11).

Attitudes toward government and the political process as measured by the indices of political efficacy, government responsiveness, and trust in government, present a more confusing picture. Here again, the characteristics of the three total subsamples are of some interest. The proportion of rural respondents with low scores on these indices was greater than the corresponding proportion of respondents in the other categories, although the differences were not always great. Perhaps surprisingly, the levels of confidence in personal political efficacy and in government expressed by urban respondents, including both referenda voters and respondents who did not vote on referenda, were similar to those expressed by respondents from suburbs and smaller cities and were higher than those indicated by rural respondents.

In general, referenda voters in the three residential categories were at least as confident in government and in their capacity to affect government as were the corresponding partisan electorates (Table 11). Urban referenda voters were more mistrustful of government than either the urban partisan electorate or suburban referenda voters, but they were at least as confident in government as referenda

TABLE 11

ATTITUDES TOWARD THE POLITICAL SYSTEM AND PLACE OF RESIDENCE OF REFERENDA AND NON-REFERENDA VOTERS

(in percentages)

	Urban		Smaller Cities and Suburbs		Rural	
	Voted For:					
	Offices Only	Referenda	Offices Only	Referenda	Offices Only	Referenda
Follow Public Affairs						
Hardly at all	11	18	16	8	20	6
Only now and then	19	17	24	14	21	12
Some of the time	43	27	32	33	29	33
Most of the time	27	48	28	45	30	49
n=	(100)	(115)	(123)	(190)	(185)	(164)
Political Efficacy						
Low	19	10	16	15	26	9
Medium	41	42	38	34	45	45
High	40	48	46	51	29	46
n=	(93)	(111)	(116)	(181)	(184)	(163)
Government Responsiveness						
Low	29	16	28	22	28	28
Medium	43	54	40	42	43	45
High	28	30	32	36	29	27
n=	(86)	(106)	(107)	(176)	(171)	(158)
Trust in Government						
Low	16	20	15	14	27	20
Medium	60	61	65	68	54	67
High	24	18	20	18	19	14
n=	(74)	(99)	(101)	(160)	(164)	(147)

voters in rural areas. The rural referenda electorate included a larger proportion of respondents who expressed low levels of confidence in the responsiveness of government than was the case in either of the other two residential categories. Finally, referenda voting in all three residential areas was consistently and strongly associated with confidence in personal capacity to affect government and the political process.

As we have seen, the characteristics of referenda voters varied from one residential category to the other. These variations were not entirely consistent, however, and were often rather small. It is worth observing that rates of participation on referenda in urban areas among lower social and economic groups were at least comparable to participation rates among the equivalent groups in the suburbs and were consistently higher than in rural areas. Thus differences were not accompanied by lower levels of confidence in government and personal political power, and only a small proportion of referenda voters in the three residential groupings expressed low levels of trust in government and their own ability to affect the political process.

CONCLUSIONS

Numerous studies of electoral behavior have found that those of higher education and social and economic status are most likely to interest themselves in political affairs and participate regularly in the electoral process. Those who participate in the electoral process, it has also been found, are more likely to identify with one or the other of the parties, to be well informed about politics, and to be more confident in government and their own capacity to influence governmental processes than those who participate less regularly. There is no reason to believe that referenda voting in the 1968 general election constituted an exception to these generalizations. The referenda electorate was small, but virtually all referenda voters also voted in partisan races and were thus highly consistent in electoral participation. Referenda voting was even more closely associated with income, education, and subjective social status than was voting in partisan races. The referenda electorate was more predominantly white, affluent, better-educated, and of higher subjective social class than was the partisan electorate. Few referenda voters considered themselves political independents; high levels of political interest and information and of confidence in government and personal capacity to influence the political process were further characteristics of the referenda electorate. Thus referenda voters appear as a small, well-informed, well-politicized, and, in some respects, elite segment of the total electorate.

In broad terms, at least, there seems little reason to believe that the politically alienated and disaffected or those marked by a sense of political powerlessness exerted a major influence upon the collective outcome of referenda elections in 1968. The concepts of political alienation, disaffection, and powerlessness are cloudy ones. Even so, the characteristics of referenda voters, as we have suggested at several points, were not of the sort that we would normally associate with alienation, disaffection, or a sense of powerlessness.

Furthermore, referenda voters did not include a disproportionate number of individuals who doubted their own capacity to influence government and the political process, who strongly doubted the responsiveness of government, or who were highly mistrustful of government and government officials. Referenda voters were at least as confident in these terms as were individuals who voted only in partisan elections.

This is not to argue, of course, that alienated and disaffected voters played no role in the 1968 referenda elections. Some referenda voters were poorly educated, and some apparently had little confidence in government and in their own capacity to influence the governmental process. Taken alone, however, these characteristics are not necessary, or sufficient, evidence of political alienation and disaffection. Referenda voters who were marked by characteristics such as these constituted only a minority of the referenda electorate, and, as we have seen, the national response to referenda was not overwhelmingly negative but overwhelmingly positive.

It is also possible that the response of alienated and disaffected voters was a major factor influencing the outcome of particular referenda elections or shaping the pattern of the vote in specific and limited areas. The nature of certain issues and of the publicity given them may have provoked in some instances a disproportionate number of the politically alienated and disaffected to come to the polls and vote in the negative. As we have noted at several points, our data do not permit investigation of the responses to specific referenda or referenda voting in limited areas.

We are forced to conclude, however, that such occurrences, if they took place, were the exception rather than the rule. Again, the national response to referenda was overwhelmingly positive. The association between higher participation and negative voting in terms of the statewide aggregate vote was weak or nonexistent. Our efforts to control on issue salience were less than satisfactory, but they confirm the absence of strong association between participation levels and negative voting found in the aggregate vote. The referenda electorate in states in which salient referenda–defined as those characterized by high relative turnout and closely contested outcomes–were at issue, did include a larger proportion of voters marked by characteristics that are sometimes considered indications of political alienations and disaffection. But the proportion of referenda voters without these characteristics was also larger, and even in these states voters with characteristics that might be associated with alienation and disaffection constituted only a small proportion of the referenda electorate. Similarly, we did not find that a higher proportion of referenda voters marked by such characteristics were concentrated in either urban or rural areas or in the suburbs and smaller cities, although here again our efforts to control on these residential characteristics were not entirely satisfactory. Individuals of lower social and economic status did vote on referenda somewhat more consistently in urban areas than in the suburbs and smaller cities or in rural areas. The urban referenda electorate, however, did not include a larger proportion of individuals who expressed low levels of confidence in government and their own capacity to affect the political process.

In this context we are particularly impressed by the finding that virtually all referenda voters included in our sample voted in at least one partisan contest while only a slightly smaller proportion voted for both President and representative. It is clear that only a very small number of alienated and disaffected voters could have come to the polls solely to express their resentment by voting on a particular referendum. On these grounds, moreover, and on the basis of the expressed partisan identifications of referenda voters, it is difficult to conclude that the referenda electorate included a larger proportion of alienated individuals who rejected partisan politics and the mechanics of representative government. Rather, the referenda electorate appears as an integral, although select, segment of the more general electorate.

Our findings suggest, then, that the concepts of political alienation and disaffection do not provide anything approaching an adequate or general theoretical framework for interpreting and explaining referenda voting. It may be that the response of the alienated and disaffected is sometimes a critical factor in limited areas or when particular issues are at stake. It seems obvious, however, that general and major responsibility for the outcome of referenda cannot be placed upon the alienated and the disaffected. Rather, primary responsibility for the failures, and successes, of referenda must be assigned to the nature of the issues themselves and to the decisions of a small but interested, well-informed, and politically sophisticated segment of the larger electorate.

NOTES

1. The data employed in this study are available from the Inter-university Consortium for Political Research, Box 1248, Ann Arbor, Michigan 48106. Obviously, the Consortium is not responsible for either the findings or interpretations presented here.

2. In more specific terms, the procedure which we employed involved dividing the highest vote on a referendum on a state's ballot by the lowest vote as a measure of relative dropoff. Those states in which the ratio was 1.15 or greater–in which, in other words, more than one-sixth of the electorate was lost between the high and low turnout referenda–were classified as having high relative turnout. Those having a ratio of 1.01 to 1.14 were classified as having low relative turnout. In six states there was only one referendum on the ballot. By comparing the vote on these referenda with the vote for United States representative and by investigating the nature of the issues, two states (New Mexico and Pennsylvania) were classified as having low relative turnout, and four (Indiana, Kansas, Mississippi, and Oklahoma) were classified as having high relative turnout.

As a second step, the referenda in each of the two groups were ranked according to the absolute value of the margin of victory. Those in which the margin of victory was ten percentage points or less were called contested; those in which the margin was greater were called uncontested. The new fourfold categorization of referenda became salient (high turnout, contested), consensual (high turnout, uncontested), routine (low turnout, uncontested), and special (low turnout, contested). The states falling in each category for which respondents were included in our sample were:

Salient: Alabama, Indiana, Louisiana, Maryland, Michigan, Missouri, Oklahoma, South Dakota, Tennessee, and Washington.

Consensual: Arizona, Georgia, Illinois, Mississippi, Nebraska, Oregon, and Texas.

Routine: California, Colorado, Maine, Massachusetts, Minnesota, New Jersey, Pennsylvania, South Carolina, Utah, and Virginia.

Special: Arkansas, Florida, Iowa, North Carolina, and Ohio.

REFERENCES

ABERBACH, J. (1969) "Alienation and political behavior," Amer. Pol. Sci. Rev. 63 (March): 86-99.

BONE, H. A. (1968) "Easier to change." National Civic Review 57 (March): 125-131.

BOSKOFF, A. and H. ZEIGLER (1964) Voting Patterns in a Local Election. Philadelphia: Lippincott.

CAMPBELL, A., P. E. CONVERSE, W. E. MILLER, and D. E. STOKES (1960) The American Voter. New York: John Wiley.

CAMPBELL, A., G. GURIN, and W. E. MILLER (1954) The Voter Decides. Evanston: Row, Peterson.

CARTER, R. F., and W. G. SAVARD (1961) Influence of Voter Turnout on School Bond and Tax Elections. Washington, D.C.: Government Printing Office.

GAMSON, W. A. (1961) "The fluoridation dialogue: is it an ideological conflict." Public Opinion Q. 25 (Winter): 526-537.

HAHN, H. (1970) "Ethos and social class: referenda in Canadian cities," Polity 2: 295-313.

--- (1968a) "Northern referenda on fair housing: the response of white voters." Western Pol. Q. 21 (September): 483-495.

--- (1968b) "Voting in Canadian communities: a taxonomy of referendum issues." Canadian J. of Pol. Sci. 1 (December): 462-469.

HAMILTON, H. D. (1970) "Direct legislation: some implications of open housing referenda." Amer. Pol. Sci. Rev. 64 (March): 124-137.

HORTON, J. E. and W. E. THOMPSON (1962) "Powerlessness and political negativism: a study of defeated local referendums." Amer. J. of Sociology 67 (March): 485-493.

McDILL, E. L. and J. C. RIDLEY (1962) "Status, anomia, political alienation, and political participation." Amer. J. of Sociology 68 (September): 205-213.

National Education Association of the United States (1962) Financing the Public Schools: 1960-1970. Washington, D.C.

PLAUT, T. F. A. (1959) "Analysis of voting behavior on a fluoridation referendum." Public Opinion Q. 23 (Summer): 213-222.

ROBINSON, J. P., J. G. RUSK, and K. B. HEAD (1968) Measures of Political Attitudes. Ann Arbor: Institute for Social Research.

STONE, C. N. (1965) "Local referendums: an alternative to the alienated-voter model." Public Opinion Q. 29 (Summer): 213-222.

U.S. Bureau of the Census (1968) Voter Participation in November 1968 (Advance Statistics). Washington, D.C.: Government Printing Office.

WATSON, R. A. and J. H. ROMANI (1961) "Metropolitan government for metropolitan Cleveland: an analysis of the voting record." Midwest J. of Pol. Sci. 5 (August): 365-390.

WILSON, J. Q. and E. C. BANFIELD (1964) "Public-regardingness as a value premise in voting behavior." Amer. Pol. Sci. Rev. 58 (December): 876-887.

WOLFINGER, R. E. and F. I. GREENSTEIN (1968) "The repeal of fair housing in California: an analysis of referendum voting." Amer. Pol. Sci. Rev. 62 (September): 753-769.

Part II

INFLUENCING PUBLIC OFFICIALS

INFLUENCING PUBLIC OFFICIALS

In addition to individual efforts to impress popular views on public officials, attempts to communicate with government leaders may be launched by organized groups with similar political interests. In fact, pressures exerted by interest groups frequently are regarded as one of the most effective forms of political influence available to members of the public. The four chapters in this section, therefore, examine the activities of organized interests in urban politics.

Both the internal and the external problems that affect the ability of community organizations to exert political influence are analyzed by Lipsky and Levi. Their conclusions suggest that the capacity of relatively poor and powerless segments of the urban population to achieve political objectives through organized activities may be severely limited.

Despite such problems, however, the activities of organized interests permeate many aspects of urban politics. The role of interest groups in educational decision-making is examined in a study by Jennings and Zeigler of school boards throughout the country. Their investigation of both the sources and the consequences of interest group activity seems to indicate the important function that those groups may perform in the formulation of educational policy.

In addition, the effects of interest groups upon the policy-making process are explored by Zisk in her survey of city councilmen in 81 communities in the San Francisco Bay area. The results of this investigation, which indicate a relationship between political activity and policy outcomes when the processes of political conversion and translation are considered, also appear to imply that organized interests may be a critical factor in urban decision-making.

Since many policies affecting urban areas are enacted by state and national rather than by municipal governments, the actions of interest groups in city politics are not confined to local communities. Organized interests not only seek to influence city government, but cities also may find it necessary to organize themselves as interest groups. The chapter by Browne and Salisbury, therefore, discusses the activities of those organizations of urban lobbyists.

7

Community Organization as a Political Resource

MICHAEL LIPSKY
MARGARET LEVI

□ ONE WAY TO CONCEIVE THE POLITICAL PROBLEM of relatively poor people in America is to consider politics a bargaining arena from which they are excluded because they have nothing to trade.[1] Groups which fail to command numbers, money, skills, and status[2] have minimal standing in the political marketplace. To overcome such conditions relatively powerless groups–groups which, relatively speaking, lack the conventional political resources–may engage in various strategies to improve their bargaining position. Protest is one such strategy (see, e.g., Lipsky, 1970). Through the manipulation and projection of threats and appeals, more powerful groups may be induced to enter a conflict and enhance the probabilities of success. In this paper we analyze another such strategy–that of *community organization.*

Community organization is a mode of political action undertaken to increase group cohesion and the potential for collective action.

AUTHORS' NOTE: *This article is based upon the authors' independent studies of housing-oriented community organizations in New York City, Boston, Chicago, Baltimore, and Philadelphia, as well as extensive examination of written accounts of community organizations of low-income groups. We have tried to develop a persuasive and useful framework for analyzing the dynamics of community organization. We recognize the limitations of the case data on which*

Community organizations, as we understand the term, are organizations, in that they are deliberately established and designed collectivities of individuals with relatively patterned, stable relationships and modes of behavior (see, e.g., Blau, 1968: 297-298). And they are organizations which draw membership from communities of shared interests and perspectives often but not exclusively determined by geographic considerations (i.e., neighborhoods).[3] Community organization helps relatively powerless groups improve their political bargaining position by increasing the stability, persistence, and standing of the group.

Community organizations are like other organizations in that they seek to increase and stabilize membership, consolidate loyalty, and maintain themselves. Although community organizations are not defined by the class and income of their members, for two reasons we choose to focus in this paper on community organizations comprised of relatively poor and powerless people. First, "community organization" refers in general usage to poor peoples' organizations.[4] Second, the organizational and strategic problems of community organizations of relatively powerless groups differ considerably from those of organizations comprised of people of higher status.[5] Our discussion concentrates on the constraints on community organizations deriving from the scarcity of resources among relatively powerless groups, and the need to rely on outside agencies in developing incentives for members and in attaining goals. We conclude that the dilemma for community organizations of the poor is that often they must trade their autonomy in action and in goal creation for the capacity to initiate and maintain the organization.

---

we have relied in refining our thoughts, and anticipate modification of assertions and hypotheses on the basis of further empirical research. An earlier version of this paper, "Community Organization as a Political Resource: The Case of Housing," was delivered at the Annual Meeting of the APSA, Los Angeles, California, September 8-12, 1970. We appreciate the research and excellent analysis of Chicago housing organizations conducted by James Barron, the help of Margo Conk in preparing the text, and of Jeffrey Steingarten, William Pastreich, and Thomas Glynn in commenting on an earlier draft, and the financial support of the Massachusetts Institute of Technology in the writing of this paper. Mr. Lipsky would like to express appreciation to the Institute for Research on Poverty of the University of Wisconsin for sustained support which facilitated much of this research. Miss Levi would like to acknowledge indebtedness to members of the South End Tenants' Council of Boston.

The analysis of community organization as a political strategy of the poor clearly reflects on prevailing conceptions of American politics. According to one of the most sophisticated of the pluralist analyses of American politics, the "normal American political process" is "one in which there is a high probability that an active and legitimate group in the population can make itself heard effectively at some crucial stage in the process of decision" (Dahl, 1956: 145-146). Although clearly this "does not mean that every group has equal control over the outcome," we may inquire into the patterns of inequality that nonetheless may exist.

These propositions are ambiguous at precisely those places where clarity is most urgently needed. In what ways may groups be "heard?" Do different groups receive different kinds of "hearings?" Perhaps most important for our present purposes, what are the probabilities that various groups will become active? The pluralist model posits that organized and legitimate groups will be heard. But it ignores the fact that the extent to which groups become organized is itself a political process, and is itself affected by any prevailing biases in the system.

The study of community organization may address these concerns by the following questions. To what extent are relatively powerless groups comprised of individuals who are prepared to join politically oriented organizations? To what extent do relatively powerless groups have access to the resources on which concerted action in part depends? To what extent are relatively powerless groups, once organized, able to pursue group goals with independence? The answers to these questions should contribute to the continuing enterprise of describing the structure of inequality in the American political system.

CREATING ORGANIZATION

The creation of any politically oriented organization is likely to be a function of at least three interrelated factors: the resources initially commanded by a specific organization, the incentives it offers for membership, and the attractiveness to potential members of political activity generally. Potential community organizations, whose constituents are likely to be cynical toward political effort in the first place, also have fewer resources or incentives than organizations which recruit from other strata. For community organization of the poor, the requirements of organization often must be developed for interaction with other participants in the political process.

In attracting members, community organizations, like other organizations, must provide incentives compelling enough both to erode initial resistance to joining and to overcome the personal costs of political activity.[6] Chester Barnard (1938: 139), in his classic work, noted:

> The contributions of personal efforts which constitute the energies of organizations are yielded by individuals because of incentives. The egotistical motives of self-preservation and of self-satisfaction are dominating forces; on the whole, organizations can exist only when consistent with the satisfaction of these motives, unless, alternatively, they can change these motives.

Other theorists have also emphasized that organizational dynamics can largely be explained by examining the inducements offered to attract and retain members (see, e.g., Olson, 1968; March and Simon, 1958; Simon, 1945; Clark and Wilson, 1961; Kelley et al., 1967). But such explanations account only for part of the difficulties experienced by groups whose members tend to be organizational nonparticipants. Where organizational theorists focus on variations in the nature of incentives, we focus as well on the constraints in obtaining those incentives. Where students of social movements start with the assumption that followers are attracted because of a congruence between their sentiments and organizational goals, we focus as well on the problems of developing such congruence in an environment characterized by cynicism toward organizational potential (see, e.g., Zald and Ash, 1970). Where theorists (Zald and Ash, 1970) observe the routinization and accommodation of "movement organizations" over time as leadership becomes entrenched, we focus as well on accommodations that occur prior to the full development of community organizations.

By looking at what aspiring community organizations must do to attract members and allies and by focusing on the need for resources, incentives, and the attractiveness of political activity, we hope in the following discussion to identify the problems of organizing the relatively powerless, to isolate variables critical in the creation of community organization, and to show the dependence of community organizations on outside interactions.

(1) *To attract members and allies,* new organizations must demonstrate a potential for obtaining their objectives. The command of organizational resources—such as money, skilled leaders, active membership, and technical assistants—might serve as evidence of ability to achieve goals. But few community organizations start with sufficient resources to convince constituents of their future. Furthermore, the universe of potentially active members may be sharply reduced by the nature of constituents' employment and the scarcity of compensations for participation. Organizational activity is difficult to sustain after an exhausting work day. Widespread and reliable membership is unlikely to develop when labor is physical, workers hold two jobs, when both meetings and jobs are scheduled at night, or when time off for organizational business cannot be arranged.

Community organizations sometimes seek to demonstrate potential by accepting sponsors. Agencies of the federal government, the Ford Foundation, and local settlement houses may even start community organization efforts themselves. Although aid of this sort usually implies limitations on community organizations' activities or orientations, community organizations may sometimes accept these constraints in return for the accompanying status and resources.

(2) *To obtain membership commitment,* organizations must demonstrate that participants will have more than a marginal effect on anticipated rewards, and that rewards can be obtained only through participation. Constituents may be cynical that the organization itself will make a difference. Or they may feel that if any improvements are forthcoming, they will be general ones, affecting nonmembers as well as members.

The marginal contributor thus has little incentive to participate. As Robert Dahl (1960: 38-39) has shown in regard to voting:

> Nearly every adult in an American community has at least one resource at his disposal: his vote. Yet, for any particular individual the argument is logically unassailable that except in the most unusual circumstances where his preferred candidate ties for first place or loses by one vote, *his* vote won't count and thus his private decision not to go to the polls will not, if he keeps the decision to himself, influence the outcome.

Mancur Olson (1968) points out that often it is not rational for individuals to join organizations dedicated to securing collective goods. The costs of participation can be considerable and are always greater than zero, while the collective goods, such as a park or lower taxes, if secured, will become available to the potential member whether or not he joins.

Thus a tenant may refuse to join a rent strike because he wants to avoid dues, meetings, and the threat of eviction, and because he knows he will benefit anyway from improvements in building conditions the landlord is forced to make. Thus a public housing occupant may refuse to join a tenants' organization both because he does not want to be branded a troublemaker, and because he knows that he will be able to resort to the grievance machinery established by the tenants' organization if it is successful.

In an attempt to overcome this difficulty, community organizations may locate "selective" incentives "so that those who do not join the organization . . . or in other ways contribute to the attainment of the group's interest can be treated differently from those who do" (Olson, 1968: 51). To some extent this was the case with rent strikes in New York City, in a jurisdiction where rent reductions were available to individual tenants through legitimate official channels (see Lipsky, 1970). Similarly, the National Welfare Rights Organization avoided the problem by teaching its members how to obtain individual supplementary benefits (see, e.g., Cloward and Piven, 1968: 332-334; Wiley, 1970: 61-62).

Modest payments and peer-group approval of active participants, or the negative inducements of social ostracism, may also act as selective incentives. Community organizations may try to promote the functional equivalent of the civic feeling which seems to motivate so much electoral participation. In this regard, black power and ethnic-oriented movements may contribute to community organization efforts.

(3) *To attract members,* organizations must overcome generalized disinterest in political activity. Probability of future rewards, and confirmation of the importance of individual participation may not prevail over constituent cynicism about community organizations. At times relatively powerless groups appear intractably opposed to organizational efforts. Among the reasons for this, three seem particularly important.

First, politically oriented organizations are difficult to initiate among all income and social groups. Not only do relatively few people belong to voluntary associations generally (see Bloomberg, 1966: 385), but those organizations that do exist "probably function more as entertainment and leisure time activities than as serious mechanisms for attaining one's central life goals" (Rainwater, 1968: 31). The inherent difficulty in obtaining significant reform, a history of past failure, and experiences of exploitation by leaders primarily interested in self-aggrandizement present additional obstacles to organizations.

Second, relatively small variations in the costs and benefits of political activity may be insufficient to overcome the spectre of sanctions to be imposed by antagonistic forces. This may explain, for example, reluctance to test "rights" newly proclaimed by civil rights legislation. The murder of civil rights workers in the South may be an extreme instance of discouraging sanctions; violent police repression of black militants is a current, highly visible illustration of the possible dangers of an insurgent stance. While members of most housing action groups and tenants' councils may not fear physical harm, they often must live in fear of losing their homes. Examples abound of evictions and rent raises in reaction to demands for improvements. The fact that legislation and the courts serve the interests of landlords to a greater extent than tenants adds support to the tenant's fear of sanctions for which he has little protection or recourse (Lipsky and Neumann, 1969; Levi, 1969).[7]

Third, the disinclination of the relatively powerless toward politics may have cultural roots.

> The most pervasive fact about lower-class people as organizational participants is that they are not socialized either within the family or in their outside lives to work towards the solution of their problems on the basis of organization [Rainwater, 1968: 31; also see Haggstrom, 1968; Lewis, 1968].

Occasionally the political quiescence of the lower classes has been related to the same kind of cost and benefit analysis as we have suggested above.[8] The cumulative impact of a history of unbalanced costs in a calculus of political involvement may be conducive to the development of cultural norms which mitigate against political involvement.

To alter the cynicism of the relatively powerless probably requires more than the resources and incentives sufficient to inspire confidence in a particular community organization. Positive elite responses, and favorable changes in the political climate, may significantly affect perceptions of the possibilities of change.

MAINTAINING ORGANIZATION

The initial difficulties deriving from widespread cynicism about organization, and from scarcity of resources and incentives, do not disappear as community organizations try to maintain themselves. The survival of organizations generally depends on a capacity to sustain and enhance membership and status. But community organizations, even more than middle-class organizations, must find or extract resources and incentives to members from outside sources. They must find contributors, technical assistants, and others to help inspire membership loyalty and effectively attack antagonists. Yet reliance on outside actors as well as the inexperience of members may lead community organizations to select tactics and goals they might not have favored otherwise. In choosing strategies to gain resources, community organizations may also choose strategies which particularly tend to compromise organizational independence. The following paragraphs will focus on three requirements of community organizations as they seek to sustain membership loyalties, and how these requirements affect their strategies and orientations.

The Need to Demonstrate Strength

Community organizations must continually demonstrate that they are alive and well. They have to show that, even without the resources and incentives with which many middle-class organizations begin, they command membership loyalty and are able to act effectively. The dilemma of a relatively weak organization required to display itself often and publicly is increased by the need to appear efficacious. Otherwise, community organizations may prove they exist, but at the same time display weaknesses which raise doubts about their future.

The demonstration of strength requires evidence of a large and representative membership. To answer charges that they are unrepresentative or do not, in fact, command the loyalties they claim may require consummate skill. "Threaten to march, but never march" is the advice of one prominent community organizer well-seasoned in obscuring the true strength of his following.

The need to demonstrate constantly organizational strength may induce an organization to picket and march to exhaustion, because in the short run it is a relatively easy way to provide evidence of organizational activity. But when specific actions are necessary, such as renewal of mass rent withholdings in order to enforce a contract negotiated between building agent and tenant union, the organization may be poorly equipped to deliver the necessary membership commitment.

On the other hand, the capacity to evoke displays of membership commitment over time can enhance an organization's strength. Although the Milwaukee NAACP Youth Council cannot claim significant advances in the area of open housing in that city, the regular marches for open housing, sustained over a six-month period, were impressive testimonials to membership dedication and strength. Significantly the marches also served to increase solidarity.

Demonstrations of strength also require occasional displays of influence in the bargaining arena. Thus, community organizations are often led to select goals which are tangible and easily obtained. But there are costs attached to such a strategy. One of the first commandments of most community organizers is to demonstrate the organization's capacity to succeed by identifying a vulnerable target, pursuing a strategy designed to obtain a relatively quick and impressive victory, and securing tangible rewards. In theory this demonstration will not tax the long-term capacity of the organization to sustain itself and may succeed in overcoming long-standing doubts that political activity can pay off.

However, there is a danger that organizational goals will come to be defined in terms of the probabilities of short-term successes, with a corresponding distortion of long-term aims. Even when goals continue to be recognized as important and targets as appropriate, the compromises involved in winning often make the victory trivial. When a housing group pays deference to the high price of some repairs in the hopes that the reasonableness of the "tailored" demands will result in compliance, compliance may come but with little change in the condition of the structure. Similarly, recognition of the "expertise" of a public housing manager in one area so that he will be encouraged to give up some of his authority in another limits the protean tenants' council to far less than a full voice in management policy. An initially strong organization might be able to resist early, ultimately defeating compromise of this sort.

One way community organizations may demonstrate strength with tangible and quickly won objectives is by concentrating tactics on vetoing public policies. Even for organizations which have succeeded in gaining some legitimacy and recognition, it is easier to veto than to attempt to influence policy positively. However, retarding one plan without offering another may leave the organization without an issue and without rewards. Inexperienced and lacking necessary skills, few community organizations are able to follow up successful opposition with detailed alternatives, and they may be left with fewer possibilities of action than before.

These problems may help explain some aspects of organizational goal transformation. When the search for tangible rewards is thwarted, when victories are not forthcoming, or when concessions are gained in unimportant areas, community organizations may disguise their difficulties by projecting additional goals. Objectives such as heightened self-respect, black consciousness, or restructuring the economic system then become more attractive as the probabilities of attaining meaningful tangible rewards seem lower. In some cases, organizers may have initially withheld expression of such aims until constituents had been attracted to the organization. In other cases, the additional aims may serve to rationalize how little has been accomplished or justify the continued existence of a community organization whose original aims are satisfied.

The Need for Tactics Acceptable to Constituents

Community organizations not only must find tactics which have some chance of succeeding, but they must also find tactics in which constituents will engage. Community organization members may not wish to participate in militant high-risk actions, even when the possible gains are significant.

It is often alleged that successful community organization must involve issues in which people have a stake. But the very salience of an issue may also account for the unwillingness of constituents to jeopardize what they already have. Thus, the National Welfare Rights Organization, in its attempt to dramatize the inadequacies of present welfare allotments, had to abandon its efforts to persuade welfare recipients to spend their rent checks on other necessities. In this campaign it became clear that however dedicated to the organization and its broad goals, welfare clients would not risk eviction (see Cloward and Piven, 1968).[9]

Significantly, some of the most successful community organizing tactics in recent years have focused upon assisting members to seek specific, material benefits using opportunities already legally available in such a way as to appear to be protesting against specific policy areas. The most successful welfare rights tactics have been those which have inundated the bureaus with thousands of requests for supplementary benefits. This tactic permitted the organization to oppose and inconvenience the system, and at the same time not jeopardize the welfare or housing status of participants. By subsequently eliminating the supplementary grant program, and distributing these budget funds among all welfare recipients as "flat grants," the welfare administrations of New York City, Massachusetts, and other jurisdictions significantly reduced the capacity of the welfare rights movement to organize against the system while working within it (Cloward and Piven, 1968: 1970).[10]

Similarly, to the extent that it was successful, the rent strike movement in New York prospered because (1) people were willing to become involved in organizations directed toward the quality of housing, yet (2) the tactics in which they were asked to engage consisted of using currently available judicial and administrative routines to harass landlords and press their claims with the city. People did not want to risk eviction (Lipsky, 1970: 189).

Cloward and Piven (1968) correctly point out that orienting organizational tactics toward bureaucratic, currently acceptable channels, may submerge the organization in a morass of details from which it cannot extricate itself. Yet community organizations are not led blindly to such tactics. They emerge from organizational needs to orient tactics toward those activities in which constituents will participate. This is the cost of overcoming resistance to participation, and of planning organizational participation which will not be priced out of the market in the constituent's political accounting.

Community organizations with feeble membership rolls and uncertain direction may try to locate allies already respected for their ideology or strength. Joining with other organizations in what seems like common cause (for some groups joining "The Movement") may seem to rationalize tactics, attract supporters, and convey organizational legitimacy. However, such a strategy has its costs and pressures. Constituents who see themselves as middle class may feel that alliances with poor peoples' organizations diminish their status. Or they may object to a new political style implicit in the alliance.

Although a major reason for the coalition is to seek resources not otherwise available to the organization, the alliance may soon divide over tactics or goals. When the tangible objectives of allies shift direction, the community organization may feel compelled to go along even when the benefits are unclear. For example, a tenants' council in an urban renewal fight may be told that negotiations with the redevelopment authority are inappropriate,[11] or a service-oriented community organization may be told to work for political power by political confrontation.[12]

Tensions also result when recognized celebrities in "The Movement" appear on the scene. Although meant to support and assist local efforts, this may undermine them. In 1966 when the Congress of Racial Equality made Baltimore its target city, its attack on segregated bars and taverns diverted attention and leadership energy from the issues of housing and welfare stressed by local groups (Bachrach and Baratz, 1970: 71-72, 76-77). Even when the issues are the same, however, the result can be diversionary. This was the case when the Southern Christian Leadership Conference, led by Martin Luther King, came to Chicago in 1966 to "end slums." Although local organizations supported Dr. King as an individual, they lost momentum in their own immediate attacks on landlords and efforts to extend membership as both organizers and constituents became involved with the more broadly directed King offensive.

The Need to Command Skills

Community organizations must command the technical and leadership skills which will enable them to act effectively as issues arise. Members of relatively powerless organizations, almost by definition, initially lack such skills; hence the reliance on community organizers and technical assistants. But dependence on exogenous skilled personnel creates additional problems. Housing-oriented community organizations often cannot operate without lawyers to help steer them through the judicial and administrative legal processes, and planners and architects to assist them in developing plans with sufficient quality to convince a skeptical redevelopment authority or federal agency. But high-level assistance is difficult or expensive to obtain, and may be made available only for short periods of time and without dedication and continuity.

Similar problems arise with the organizers. They may be committed to organizational activities for only relatively short periods (as has been the case with most student organizers [Rothstein, 1969]), may have career ambitions which can only be fulfilled by changing jobs, or may contract battle fatigue and "burn themselves out" (Haber and Haber, 1969; Appleby, 1969). Some of the factors which may account for the "burning out" of community organizers are endemic to the job. It is difficult and exhausting to be the political man at all times. Yet community organizers must be on call "around the clock" if they are to continue to receive community support. This particularly may be the case where organizers must substitute energy and time to accomplish tasks that other groups can accomplish with money, status, or other available resources. When the organization does not fulfill an individual's career ambitions and when leaders are regarded suspiciously for their very success (Clark, 1965: 155-168, 182-198; Alinsky, 1969: 89), the strain is particularly heavy. The model of the urban political machine, commended for its apparent contributions to the welfare of relatively low-income urban residents, may also be commended for its systematic dispensation of rewards to ward bosses, inducing ongoing commitment through monetary, status, and power benefits such as few community organizations are able to mobilize.

Further difficulties arise when organizers, technical assistants, and constituents come into conflict over political style, goals, and values. Pressures exist to capitalize on proffered resources and, therefore, to accept the orientation of advisers even at the expense of directions promoted by indigenous personnel. These tensions are exacerbated when advisers are white, and the community organization which they have been assisting is nonwhite. Although community organizations have been known to reject such aid to gain a measure of psychological independence, it is more likely that assistants will gain a measure of influence, however much they may attempt to defer to the organizations' leaders.

The contradiction between deference to indigenous leadership and dependence on technical assistance has never really been resolved by theorists of advocate professionalism (Davidoff, 1969; Peattie, 1969; Piven, 1970). In fact, several advocate professionals, dismayed by what they regard as the conservative tendencies of some community organizations, have recently asserted an interest in influencing the ideology of community organizations for which they work (Hartman, 1970; Rosen, 1970). Saul Alinsky resolves the dilemma with a militant disinterest in goals per se and a militant concern for developing community power so that the group can pursue its own goals. Critics have split with him over this approach, since it appears to condone and strengthen the racism that may exist in some communities.[13]

To reject such assistance is often inappropriate for relatively powerless groups whose inability to command technical and professional help comparable to that available to other organizations places them at considerable disadvantage. Organizations which seek to develop leadership from the ranks and to avoid overdependence on a few individuals must be prepared to endure the inefficiencies and mistakes which will accompany such efforts. Considerable intra-organizational tensions can develop when unskilled people accomplish relatively poorly and over a longer period of time what a skilled organizer might accomplish successfully and with greater dispatch.

COMMUNITY ORGANIZATION AND POLITICAL ACTION

Community organizations must frequently obtain resources and establish attractiveness to members in interaction with other, more powerful, organizations and agencies. Indeed, the fewer the resources commanded by the organization, the more it is dependent on outside agencies. Most community organizations of the poor must develop ways of manipulating outside agencies to secure membership benefits or other incentives. While in theory they may be able to galvanize a large number of people, in fact the problems of organization rest on securing or extracting resources over which the organization originally has little control. Moreover, outside agencies and organizations,[14] pursuing their own maintenance and enhancement needs (Barnard, 1938; Banfield, 1961: 263), are able to manipulate the utilization of available resources in order to tame, deflect, or reduce community organization influence. The need to seek resources from outside suggests the fundamental constancy of power inequalities for community organizations.

At least three major strategies appear available to outside agencies and organizations which may result in blunting community organization efforts.

1. *Outside agencies and organizations can make concessions which effectively reduce community organizations' autonomy and limit their aims.* Community organizations may pay a high (and often not unintended) cost for prevailing in a conflict. Targets of community organization demands may, unintentionally or by design, cripple organizational initiatives by conferring victories on organizations ill-prepared to accept the reciprocal obligations accompanying "success." Ironically, for the sake of what turn out to be only symbolic or relatively minor tangible gains, community organizations may acquire major responsibilities which effectively limit their independence. Such responsibilities may overwhelm the young group or tie up its few resources in an impossible venture.

For example, a tenant union may gain exclusive bargaining status for all tenants in buildings managed by the capitulating agency. It may be forced to attempt organizing previously uninvolved tenants in order to consolidate gains. A tenants' council presented an opportunity to manage the buildings about which it complained may find it difficult to reject the offer and must endure tasks of apartment house administration for which it was not originally established.[15] Such activities may blunt the capacity of the organization's leaders to expand the constituency beyond its narrow base. But those who are already active may want to seize the opportunity to take responsibilities and consolidate gains. The gains themselves thus turn out to be more cost than benefits. A young tenants' council has neither the money nor the trained manpower to make the necessary building improvements, and may even become the "slumlord" in the eyes of tenants.

A variation on the theme of the liabilities of success is provided by the recent questionable victories obtained by the rent strikers in St. Louis public housing.[16] Here the victory may have been largely symbolic. With maintenance costs increasing, repair costs accelerating in older buildings, federal public housing subsidies frozen, and vacancy rates rising, the Public Housing Authority in St. Louis vested authority with a tenants' board. Yet it is uncertain whether the economics of public housing permit successful operation at current inadequate levels of government subsidy, particularly where deterioration of projects has been permitted to advance markedly.

The dilemma of the tenants' organization was undoubtedly a troubling one: accept responsibility for the operation of the houses or continue demanding reforms under the old regime. At this point no one can say with certainty if the tenants were wise in accepting responsibility for buildings and assuming the inheritance of neglect represented by public housing in St. Louis. We can, however, point out that victory for community organizations may prove hollow when symbolic or substantive victories are not accompanied by sufficient resources to permit realization of success over time. Saul Alinsky (1969) urges community organizations to engage in continuous struggle. At the same time he insists that they must demonstrate an ongoing capacity to succeed. This analysis, however, points to the costs that organizations may incur in accommodating these twin demands.

Governmental agencies can also take advantage of the relatively powerless status of community organizations by granting them program responsibility, or attempting to induce leaders to accept jobs with other organizations. A community organization may find it difficult to refuse designation as a Headstart recipient, or as a representative body for a neighborhood health center, since those appointments may be accompanied by higher community status and jobs for members. Community organization leaders may be induced to accept employment in governmental programs dedicated to the same general goals which have been championed by the organization.[17]

These developments share the following characteristics when they occur. They offer inducements which significantly contribute to the enhancement and maintenance of the community organization or the enrichment of individual leaders. But they also require substantial reciprocal obligations of individuals and organizations. Such obligations are incompatible with the single-minded goal orientation upon which organizational militance in part depends.

The realities of independence and subordination are, of course, apparent to community organizers. In many cases community organizations have resisted subordination by refusing the perquisites offered. The merits of accepting reciprocal responsibilities will be judged on the basis of their costs and benefits. The weaker the organization, the less it will be able to generate resistance to cooperative offers from stronger organizations. Some community organizations may be fortunate enough to acquire government funds for work which they were doing anyway, without incurring reciprocal obligations of any weight. But even in such a situation, the organization's independence may be compromised by a rapid growth that is impossible to sustain when government program expenditures, for reasons possibly extraneous to the organization, are reduced, and public support is no longer forthcoming.[18]

2. *Outside agencies and organizations can avoid the fixing of responsibility.* Community organizations often try to impose on members a view of reality which identifies individuals and agencies responsible for given policies or conditions. Constituents will be readier to accept strategies aimed at correcting conditions if they think they know who is to blame than if it seems impossible to determine the responsible parties. Where there is uncertainty over culpability, organizers will try to disseminate perspectives which fix responsibility firmly, whether or not such views violate available evidence and alternative perspectives.

Correspondingly, public officials and directors of private organizations will try to avoid responsibility for specific conditions or policies, and will contribute to the structuring of institutions so that it is difficult to locate responsibility in organizational units or roles. The exceptions to this generalization occur when public officials can enhance their own position by appearing to sympathize with community organization goals and tactics. But except in rare instances, this only occurs if community organizations have "tailored" their goals or "sanitized" their tactics to conform to relatively restricted definitions of goal and tactic acceptability.

Outside agencies and organizations can avoid identification as responsible parties under the following circumstances, among others.

(A) Responsibility is elusive and diffuse. Where different governmental units are responsible for aspects of a given problem, the vulnerability of any one unit is reduced. For example, at one point five different agencies in New York City had major responsibility for housing violations concerning problems with water systems (Lipsky, 1970: 105). "Coordination" of such agencies, without real integration, often means merely adding another unit to diffuse responsibility.

Not only does diffusion of responsibility, like "red tape," serve to ration the demand for services by making requests costly, it also contributes to the continued disorientation of community groups which sometimes require governmental responsiveness in order to develop the capacity to make demands. Government agencies are fundamentally ambivalent about coordinating government responsiveness. On the one hand, coordination enhances governmental ability to deliver services, enhancing agency reputation. On the other hand, coordination constrains governmental ability to disclaim responsibility, and, when unaccompanied by agency reordering of priorities, may lead to what are considered unmanageable increases in demands. Increased capacity to deliver services also enhances community organization ability to mobilize constituents by making the efficacy of tactics more believable, thus challenging government hegemony in areas where community organizations are active.

(B) Responsibility rests with plural entities. The more parties responsible, the greater the difficulties in securing rewards, benefits, or concessions from them, and the more discouraging to potential constituents are related tactics. This generalization may be illustrated by the difficulties experienced by community organizations in developing tactics directed against landlords. The multiplicity of landlords in low-income areas has forced some organizations to direct tactics toward official city policies, rather than to attempt mobilization against the legally responsible owners (Lipsky, 1970: 179). Other groups discover that they must attempt to devise tactics in which they can identify a consolidated group of responsible landlords. For example, Chicago tenant unions, to avoid confrontations with the city, had to identify targets with control over enough properties to be vulnerable as a common "antagonist."[19] Thus they focused on managing agents with major ghetto holdings, and housing developments belonging to a single owner. The tactics of selecting an offender as the target of organizational mobilization must overcome the "why me" defense of targets who can identify an army of equally responsible parties.

A major tenet of the pluralist description of American politics is that there exist multiple points of access which groups can utilize to gain rewards from the system or redress of grievances. To some degree this may help community organizations in efforts to veto public policies with relatively narrow potential impact. For relatively powerless groups, however, we may turn the proposition on its head. Where there are multiple points of access there are also multiple points of responsibility. For groups which depend upon clarity of target, and identification of responsibility in order to consolidate group membership and standing, the existence of multiple responsibility points make it correspondingly more difficult to organize and sustain membership. Moreover, this difficulty increases as public policies toward relatively poor people more and more involve questions of federal legislation and local implementation. William Riker (1964) has argued that federalism in the United States is a system which primarily enhances local parochialisms such as southern segregation and exclusion of blacks. This analysis may be supplemented by suggesting that it is also a system in which the elusiveness of responsibility contributes to the continued fragmentation of insurgent groups.

(C) Responsibility remains privatized. Community organizations will attract constituents to the extent that responsibility for a given issue generally rests with the target of organizational efforts. Demonstrations of deviation from norms of responsibility make the possible target more (but by no means absolutely) vulnerable to organizational offensives. However, in a number of areas, norms concerning the responsibility of government or private institutions are nonexistent. Thus, on this single dimension, one would anticipate greater ease in mobilizing constituents around issues of public education and police protection, which are conceded to be public responsibility, than around housing and employment issues, where only the most modest of public norms of responsibility prevail. These are the implications for community organization of much that has been written concerning American social welfare ideology (e.g., Steiner, 1966). If poor people are treated as if, and come to believe, that they are responsible for their own condition, community organizations will be less able to mobilize them for concerted action. Hence, if poor people are treated as if, and come to believe, that they are responsible for their own housing conditions (state of repair, neighborhood location, quality of upkeep, ability to move), community organizations will be correspondingly less able to mobilize them for concerted action.

To avoid the problems suggested above, community organizations may direct efforts at identifiable but not ultimately responsible parties. Clearly this is the framework of Alinsky's prescription (1969: ch. 8) to attack visible targets (see also Alinsky, 1970: 70-79). Just as government officials will try to disclaim responsibility over certain policies, so community organizations will try to fix elusive agency responsibility and force them to utilize influence and resources to consolidate whatever ambiguity exists in governmental responsibility in a given controversy.

In some instances community organizers have tried to capitalize on some of these problems by drawing constituents' attention to the difficulty of fixing responsibility for the most basic of citizen needs. Organizers may be attracted to such issues precisely because they raise fundamental issues about the performance of the political economic system. To inquire into housing politics, for example, involves questions about the inability of the American political and economic system to provide for the minimal needs of its citizens, and about the consequences of making public policy for the benefit of those who either profit from the current housing industry or draw comfort from the consolidation of a status quo characterized by economic and racial segregation (National Advisory Commission on Civil Disorders, 1968: 467-474). However, the mobilizing appeal of such positions is likely to be more than offset by the remoteness of this approach and the discouragement likely to ensue from an appreciation of the difficulties involved in affecting the direction of policies with such broad scope.

(D) Responsibility for policy implementation is often unrelated to responsibility for decision-making. Community organizations may extract promises of reform, only to discover later that the agency with which the bargain was reached was unable to deliver. This may occur because (1) the agency knowingly did not have authority to implement the agreement, as when a hospital agrees to changes in a community health center when agreements also must be reached with the federal agency financing the center. Or (2) the agency was insufficiently organized itself to make the arrangements over which it did have authority, as when a state agency agrees to provide rent supplement funds for families in need of housing, but lacks the staff necessary to process the individual applications and is unwilling to shift resources to this end. Or (3) the agency directors lack control over subordinates, as when a police precinct captain agrees to institute reforms which patrolmen will not consent to implement in practice.

Additionally, in social policy matters we generally lack an ability to monitor policy changes effectively. While to policy reformers this is a general problem of efficiency and cost accounting, it is vital to community organizations' capacities to effect public policy on a continuous basis. Ability to monitor programs is requisite to forcing compliance with an agreement. Perhaps more important, community organizations do not have the resources to monitor actions on a continuous basis, and must depend upon the uneven reliability of public agencies to monitor their own compliance or noncompliance, or upon the monitoring efforts of academics or middle-class civic organizations whose objectives may differ substantially from their own.

3. *Outside agencies can reorganize to deprive community organizations of tactics on which they have come to depend.* The response of state welfare agencies to the National Welfare Rights Organization provides a clear illustration. To attract members the NWRO depended upon the claim that it could help constituents obtain individual benefits by petitioning for items available under "special needs" and "emergency" categories. The submission of applications for these benefits, combined with a militant campaign by the organization, for some time did result in higher benefits and growing membership in the organization (see previous discussion of the importance of individualized, selective benefits in organizing). By moving to a "flat grant" system, in which every recipient received a small sum of money (not just welfare rights members) this organizational tactic was no longer available to the organization. A somewhat similar situation occurred in New York City when a variety of changes in city housing maintenance policies resulted in the obsolescence of rent reduction applications as a useful rent strike organizing tool.

Community groups frequently focus concern on the procedures by which decisions are made, since the substance of policy seems otherwise impervious to influence. Yet public agencies have a wide variety of options to exercise which would appear responsive to demands for changes in procedures, without substantially affecting policy matters. Not only do citizen participation mechanisms function to coopt individuals and organizations by including them in some of the ceremonies of decision-making; these procedural innovations have also served to appear responsive to critical community organization demands, while requiring organizations which have pressed such demands to develop new issues or spend scarce resources to try to dominate the decision process which they have just joined under disadvantageous circumstances.

CONCLUSION

This essay reflects an ongoing concern with the place of relatively powerless groups in American politics and the strategies available to such groups to alter their politically impoverished status. Previous examination of protest politics in American cities led to the conclusion that protest strategies, for a number of reasons, were inherently unstable. Community organization strategies, designed to increase group cohesion, offer considerably greater opportunities for the development of stable political resources. But, we conclude, community organizations of relatively powerless groups remain constrained by their need to secure resources and develop incentives to attract constituents. For the most part, those may be acquired only through interactions requiring compromises of group independence and goals. Although compromises are perhaps inevitable in any quasi-bargaining situation, the compromises forced upon organizations of the relatively powerless are inherent in becoming organizations.

The organizations on which we have focused are precisely those which pluralist theory describes as potentially developing to represent the poor and minority groups. But the emergent picture in our analysis is one in which groups, while more or less free to compete without sanctions, find themselves in a bargaining arena severely biased against the relatively powerless. They may be "invited" to attend the competition, but the cost of entry is high, and may be paid only by assuming identities which severely limit their competitive effectiveness.

NOTES

1. This framework is suggested in Wilson (1961).
2. Relatively powerless groups may be defined as those groups which, relatively speaking, are lacking in conventional political resources. For a useful list of such resources,

see Dahl (1960: 32). This paper may be read as a continuation of research on and analysis of the political strategies available to such groups initiated by Lipsky (1970). For further related discussion of relatively powerless groups, see chs. 1, 6, and 7.

3. This definition facilitates comparisons of community organizing with labor organizing to the extent that both may organize broad communities of shared interests and perspectives.

4. In common usage community organization also refers to organizations of the relatively poor and powerless for the purpose of engaging in at least some degree of interaction and conflict with the agencies of government. Although local political clubs or associations of white, middle-class homeowners who live in the same neighborhood could be considered community organizations under the definition, few politically oriented community organizers would mean such activities as their work. For a history of the term in social work theory, compare, e.g., Zald (1967); "Introduction," in Kramer and Specht (1969: 12-19); Brager and Specht (1967: 136-141).

5. There are, of course, also similarities in organizational strategic problems, but we will not explore them here.

6. For this perspective we are clearly indebted to Downs (1966).

7. Both the East Garfield Park Tenants' Union of Chicago and the South End Tenants' Council in Boston found it necessary to write protective clauses on eviction and rent hikes into their arbitration contracts with local slumlords. In reaction to demands by public housing tenants in Baltimore, the public housing authority unsuccessfully attempted to transfer an outspoken tenant leader to a project far from her organization (see Bachrach, 1970: 179-181; for further discussion, see, e.g., Lipsky and Neumann, 1969).

8. Although Edward Banfield (1970: 216-218) ascribes the dearth of lower-class efforts directed toward ameliorating problems to a class culture, he acknowledges that the situation may make investment in the future "impossible or unprofitable."

9. See also the *New York Times,* September 24, October 26, 1968.

10. See also the *New York Times,* August 1, September 24, 1968.

11. In Boston in 1969 the South End Tenants' Council was near the end of several months of negotiations with the redevelopment authority for a package including management rehabilitation funds and eventual ownership of over 100 buildings. However, allied groups asked the SETC to join the moratorium on discussions with the redevelopment authority imposed by local groups fighting the rules of the urban committee elections. The SETC reluctantly complied.

12. In order to maintain its independence, the West Side Organization was wary of too strong a coalition with the Chicago Students for a Democratic Society project and with the Woodlawn Organization. The groups disagree over the role of services and of anti-city actions (Ellis, 1969: 123-124, ch. 6).

13. For Alinsky's views, see Alinsky (1969: chs. 9, 10) and Alinsky (1970: 61-62). For those of his critics, see, e.g., Rothstein (1969: 274-275).

14. "Outside agencies and organizations" refers to any public or private organization, or any governmental unit, with which community organizations seek to interact in order to secure resources, gain concessions, or effect actions to increase their attractiveness to members. Examples will be drawn from community organization interaction with government agencies, but the generalizations are intended to apply over a broader range.

15. After months of complaining, successful arbitration proceedings and continued landlord neglect, Boston's South End Tenants' Council through arbitration was awarded management of seven buildings. However, because the buildings needed major rehabilitation, the tenants' council soon discovered that the rents did not cover maintenance, repairs, and mortgage, let alone payment of administrative personnel. Rather it found itself in the position of doing all the landlord's dirty work, including rent collection, but unable to improve upon the landlord's repair record.

16. See, e.g., the *New York Times,* November 2, 1969, p. 52; and *Newsweek* (August 25, 1969), pp. 49-50.

17. We have avoided using the familiar term "cooptation" because of the conceptual and normative difficulties which surround the concept. It is impossible empirically to know what direction an organization would have taken if the alleged cooptation did not take place. Moreover, the concept of cooptation is almost always used to denote a development displeasing to the observer. In this paper we are not concerned with judging community organizations. Rather, we seek to demonstrate the relationship between the needs of established civic groups and government agencies whose interests may lie in subordinating insurgent groups, and the needs of such groups to participate in their own subordination by developing relationships which will provide them with the incentives necessary to attract and retain members.

18. See Bachrach and Baratz (1970: 81-85) for a discussion of the effect of funding problems on the Baltimore Community Action Agency.

19. In the process of doing this, however, one major landlord was found to be relatively invulnerable to pressure. Even though he controlled a large number of ghetto properties, most of his holdings were outside of the ghetto so that he felt little significant pressure when attacked there.

REFERENCES

ALINSKY, S. (1970) The Professional Radical. New York: Harper & Row.

--- (1969) Reveille for Radicals. New York: Vintage.

APPLEBY, M. (1969) "Revolutionary change and the urban environment," pp. 216-232 in P. Long (ed.) The New Left. Boston: Porter Sargent.

BACHRACH, P. (1970) "A power analysis: the shaping of anti-poverty policy in Baltimore." Public Policy 18 (Winter): 179-201.

--- and M. BARATZ (1970) Power and Poverty. New York: Oxford.

BANFIELD, E. C. (1970) The Unheavenly City. Boston: Little, Brown.

--- (1961) Political Influence. New York: Free Press.

--- (1958) The Moral Basis of a Backward Society. New York: Free Press.

BARNARD, C. (1938) The Functions of an Executive. Cambridge: Harvard Univ. Press.

BLAU, P. (1968) "Organizations theories," pp. 297-305 in International Encyclopedia of the Social Sciences 11. New York: Macmillan.

BLOOMBERG, W. (1966) "Community organization," pp. 317-358 in H. S. Becker (ed.) Social Problems: A Modern Approach. New York: John Wiley.

BRAGER, G. and H. SPECHT (1967) "Social action by the poor: prospects, problems and strategies," pp. 136-141 in G. Brager and F. Purcell (eds.) Community Action Against Poverty. New Haven: College and Univ. Press.

CLARK, K. (1965) Dark Ghetto. New York: Harper Torchbooks.

CLARK, P. and J. Q. WILSON (1961) "Incentive systems: a theory of organization." Administrative Science Q. 6: 129-166.

CLOWARD, R. and F. P. PIVEN (1968) "Finessing the poor." Nation 207 (October): 332-334.

DAHL, R. (1960) "The analysis of influence in local communities," pp. 25-42 in C. Adrian (ed.) Social Science and Community Action. East Lansing: Michigan State University.

--- (1956) A Preface to Democratic Theory. Chicago: Univ. of Chicago Press.

DAVIDOFF, P. (1969) "The planner as advocate," pp. 544-555 in E. C. Banfield (ed.) Urban Government. New York: Free Press.

DAVIS, R. (1967) "The war on poverty: notes on an insurgent response," pp. 159-174 in M. Cohen and D. Hale (eds.) The New Student Left. Boston: Beacon.
DOWNS, A. (1966) An Economic Theory of Democracy. New York: Harper & Row.
ELLIS, W. (1969) White Ethics and Black Power. Chicago: Aldine.
FRUCHTER, N. and R. KRAMER (1966) "An approach to community organizing projects." Studies on the Left 6 (March/April): 31-61.
GITLIN, T. (1967) "The battlefields and the war," pp. 125-136 in M. Cohen and D. Hale (eds.) New Student Left. Boston: Beacon.
HABER, B. and A. HABER (1969) "Getting by with a little help from our friends," pp. 289-309 in P. Long (ed.) New Left. Boston: Porter Sargent.
HAGGSTROM, W. C. (1968) "The power of the poor," pp. 457-475 in L. Ferman, J. Kornbluh, and A. Haber (eds.) Poverty in America. Ann Arbor: Univ. of Michigan Press.
HARTMAN, C. (1970) "The advocate planner: from 'hired gun' to political partisan." Social Policy 1 (July/August): 37-38.
KELLEY, S. et al. (1967) "Registration and voting: putting first things first." Amer. Pol. Sci. Rev. 61 (June): 359-379.
KRAMER, R. and H. SPECHT, [eds.] (1969) Readings in Community Organization Practice. Englewood-Cliffs, N.J.: Prentice-Hall.
LEVI, B. (1969) "Baltimore's rent court is the landlord's tool." Baltimore Sunday Sun (November 16): Section C, 2.
LEWIS, O. (1968) "The culture of poverty," pp. 405-415 in L. Ferman, J. Kornbluh, and A. Haber (eds.) Poverty in America. Ann Arbor: Univ. of Michigan Press.
LIPSKY, M. (1970) Protest in City Politics: Rent Strikes, Housing and the Power of the Poor. Chicago: Rand McNally.
--- and C. NEUMANN (1969) "Landlord-tenant law in the United States and West Germany–a comparison of legal approaches." Tulane Law Rev. 44 (December): 36-66.
MARCH, J. and H. SIMON (1958) Organizations. New York: John Wiley.
MARRIS, P. and M. REIN (1969) Dilemmas of Social Reform. New York: Atherton.
National Advisory Commission on Civil Disorders (1968) Report. New York: Bantam.
OLSON, M. (1968) The Logic of Collective Action. New York: Schocken.
PEATTIE, L. (1969) "Reflections of an advocate planner," pp. 556-567 in E. Banfield (ed.) Urban Government. New York: Free Press.
PIVEN, F. F. (1970) "Whome does the advocate planner serve?" Social Policy 1 (May/June): 32-37.
--- and R. CLOWARD (1967) "Rent strike: disrupting the slum system." New Republic 157 (December 2): 11-15.
RAINWATER, L. (1968) "Neighborhood action and lower-class life styles," in J. Turner (ed.) Neighborhood Organization for Community Action. New York: National Association of Social Workers.
RIKER, W. (1964) Federalism. Boston: Little, Brown.
ROSEN, S. (1970) "Comments." Social Policy 1 (May/June): 36.
ROTHSTEIN, R. (1969) "Evolution of the ERAP organizers," pp. 272-288 in P. Long (ed.) New Left. Boston: Porter Sargent.
SIMON, H. (1945) Administrative Behavior. New York: Macmillan.
STEINER, G. (1966) Social Insecurity. Chicago: Rand McNally.
WILEY, G. (1970) "Welfare rights as organizational weapon." Social Policy 1 (July/August): 291-303.
WILSON, J. Q. (1961) "The strategy of protest: problems of Negro civic action." J. of Conflict Resolution (September): 291-303.
ZALD, M. N. [ed.] (1967) Organizing for Community Welfare. Chicago: Quadrangle.
--- and R. ASH (1970) "Social movement organizations: growth, decay and change," pp. 516-537 in J. Gusfield (ed.) Protest, Reform and Revolt. New York: John Wiley.

8

Interest Representation in School Governance

M. KENT JENNINGS
HARMON ZEIGLER

□ EMERGING FROM THE BEHAVIORAL REVOLUTION of the years following World War II, after years of neglect, interest groups once threatened to assume the role of a "first cause" of public policy. A veritable flood of case studies, dealing with either a single group or a single issue, has appeared, all paying either tacit or overt homage to the patron saints of the "group approach," Bentley (1949) and Truman (1951). The difficulty with such studies is that most of them *began* with the assumption that interest groups were powerful (otherwise why would we study them). Further, no matter how laudable the case study method may be, it is extraordinarily difficult to reach valid generalizations from studies of single issues or single groups.

AUTHORS' NOTE: *This paper was prepared for presentation at the 1970 meeting of the American Political Science Association, September 11, 1970, Los Angeles, California. The authors wish to acknowledge the support of the Center for the Advanced Study of Educational Administration during a portion of the time they devoted to the preparation of this paper. CASEA is a national research and development center which was established under the Cooperative Research Program of the U.S. Office of Education. The research reported in this paper was conducted as part of the research and development program of the center.*

Whatever the validity of case studies, they were soon challenged by a new group of research efforts, relying more on comparative and (within the limits of measurement) systematic observations (Milbrath, 1963; Bauer, Pool and Dexter, 1963; Zeigler and Baer, 1969). While no useful purpose would be served by an enumeration of the specific findings of this research, there was a theme common to it: interest groups are far less influential than the case studies would lead us to suspect. Thus, we have come full circle. Where interest groups were once thought to be the basic catalyst for the formation of public policy, they are now described as only one of a number of competitors for power, and frequently the least effective combatants.

SOME PROBLEMS IN THE STUDY OF INTEREST GROUPS

In spite of the vast amount of ink spilled on the subject of interest groups, we have not really made very much progress. The fault lies not so much with theory as with data. It is very difficult to measure the contribution that interest groups make in the formation of public policy and the resolution of policy disputes. Other political variables lend themselves much more readily to quantification; e.g., financial resources, malapportionment, party competition, and so forth. It is quite significant that the major efforts in developing systematic, empirical descriptions of the formation of public policy at the state level make absolutely no mention of interest groups (Dye, 1966). Their exclusion is clearly the result of the fact that nobody has developed an inexpensive and reliable method of measuring interest group strength. For instance, both Zeigler (1965) and Froman (1966) used the assessment of political scientists as an indication of interest group strength, hardly the sort of measure in which much confidence would be placed. In fact, the only effort—that of Francis (1967)—to develop a measure of the activity of interest groups in state politics (group competition rather than group effectiveness) is at odds with the conclusions reached by Zeigler.

SCHOOL DISTRICTS AS FOCI OF INQUIRY

This paper, while not pretending to solve the problems of measurement to the satisfaction of all, does offer the opportunity for systematic analysis using a relatively large number of units of analysis: local school districts. The use of school districts as units of analysis can be justified on grounds quite independent of their methodological advantages. Both in terms of formal governmental organizations and in terms of governmental office holders, the school districts supply an inordinate proportion of all such organizations and officials in the United States. Further, as the recent brouhaha over sex education, dress codes, decentralization, tax support, and student revolutions demonstrates, people are apt to become quite concerned about educational policy. If the interest group theorists can make an argument in support of their assertions, school districts provide an ideal setting. There are, indeed, plentiful examples of interest groups having profound effects on local school politics (Gross, 1958).[1] On the other hand it is not very difficult to find examples of interest groups accomplishing very little in educational politics. The major point is that the evidence is sporadic and incomplete.

Obviously, the time has come to make a stab at something more systematic. The questions which such an inquiry ought to address are really quite simple. We need to specify both the antecedents and the consequences of interest group activity.

The antecedents of interest groups, or more specifically the conditions leading to their formation, have recently been subjected to some critical assessment. The traditional position, as enunciated by Truman, is described by Salisbury (1969) as the "proliferation" hypothesis. Briefly stated, the argument is that social differentiation leads to specialization. Specialization, especially economic specialization, leads to a diversity of values and—under some not clearly specified conditions—formal organizations. To specify the conditions under which specialization results in the formation of formal organizations, Truman suggests that the distribution of an established equilibrium by disruptive factors (e.g., changes in the business cycle, technological innovation) leads disadvantaged groups to seek a restoration of balance by political activity.

Recently, both Olson (1965) and Salisbury (1969) have challenged Truman's assertions by use of an exchange theory of the origin of groups. They argue that entrepreneurs offer benefits (only some of which are political) to potential members in exchange for membership. Entrepreneurial activity is the first visible evidence of group formation. In essence, Olson and Salisbury look at formal organizations as business enterprises, and focus upon the key role of the organizer. In so doing they have added an important dimension to Truman's offerings, for it is apparent from the case material they present that individual entrepreneurs play a significant role in group formation.

Yet they have not rejected either the proliferation or the disturbed equilibrium hypothesis. It is clear, even in the Olson-Salisbury argument, that groups originate in response to unsatisfied demands on the part of potential group members. Although unsatisfied demands may be insufficient to stimulate group activity, they are functions of environmental change (proliferation) and unresponsive political systems (inability to restore equilibrium). Demands lie at the heart of interest group formation even though groups ordinarily need an individual leader (entrepreneur) to channel unsatisfied demands.

We assert, then, that there is still merit in "traditional" group theory. One purpose of the paper is to see what use can be made of such theories, if we do not rely (as Truman, Olson, and Salisbury have done) on case histories of particular kinds of organizations.

With regard to the consequences of group activity, less serious theorizing has been done. As noted previously, most of the debate has centered around the empirical question of how much influence a particular group is able to achieve. We wish to address ourselves both to this kind of question and also to the more fundamental problem of the effect of formal organization upon other components of the political system. The question is one of uncovering both the influential groups in educational decision-making and assessing the overall impact of group activity upon the decision-making process.

DATA SOURCE

To answer these questions we draw on interviews conducted in 1968 with board members and superintendents in 83 school districts throughout the continental United States. Because of our desire to link the school board study to a 1965 nationwide investigation of high school seniors, their parents, social studies teachers, and principals, a decision was made to study those boards having jurisdiction over the public secondary schools covered in the earlier inquiry. It should be stressed that this is not a representative sample of all school boards; rather it represents boards in proportion to the number of secondary students covered. Since most districts are rather small, a straight probability sample of all boards would have yielded a preponderance of small districts. Thus the sample may be strictly defined as those public school boards having jurisdiction over a national probability sample of high school seniors in 1965. Although changes in school district boundaries and population in the 1965-1968 interim affect the representativeness of the sample for 1968 purposes, these changes were judged to be slight enough to permit the extraordinary utility of linking the school board project with the earlier study. The resulting sample consists of 490 individuals (weighted n = 638) serving on 83 boards (weighted n = 106).[2]

ASSUMPTIONS ABOUT MEASURING INTEREST GROUP ACTIVITY

In constructing a measure of interest group activity, we are engaging in a methodological and theoretical shift from earlier work on interest groups. Surveys of elites designed to elicit their response to interest groups typically report findings based upon the responses of individual state legislators, city councilmen, or other elected public officials. We are less interested in individual responses than assessing interest group activity as part of a *total* political system. In addressing ourselves to the twin questions of the antecedents and consequences of group activity, we deem it advisable to conceptualize school board behavior as collective action. We are less interested in how individual board members behave vis-à-vis interest groups than with the impact of groups upon the policy outputs in a district. At the methodological level, we are moved to taking as our units of analysis the decision-making bodies rather than the individuals comprising these bodies. Hence the decision to seek the complete saturation of boards rather than selecting individuals from a potentially larger number of boards lends itself well to our research strategy.

The basic variable with which we will deal is interest group intensity. By intensity we mean the extent to which interest groups come to the attention of a governing body, in this case school boards. Intensity is purely an assessment of the quantity of interaction. By itself, it does not measure the technique or success of an interaction. What we are interested in measuring is the degree to which organizations play a significant role in the informational, cue-taking system of school boards. For this reason it is appropriate to rely upon the responses of board members. While it might have been desirable to have interviewed interest group leaders, these data would not have captured the same phenomena. It has been found that group leaders tend to exaggerate the number of interactions with governmental officials; and since we are interested in the world of the board members, their perceptions are more directly relevant.

Our measure of interest group intensity is constructed from eight open-ended questions in the interview schedule dealing with the activities of organized groups. These are the items in abbreviated form:

	Maximum Coded Responses
Organizations most interested in the board	3
Organizations from which the board seeks support	2
Organizations working for passage of financial referenda	3
Organizations working for defeat of financial referenda	2
Organizations critical of the board	2
Organizations attempting to influence teacher behavior	3
Organizations which defend teachers when attacked	2
Organizations which attack teachers	2
Total maximum possible	19

The measure of intensity is constructed by summing the number of valid responses for all respondents by board. That sum is divided by the number of respondents from each board. Thus, intensity is the mean number of organizations specifically mentioned by the members of a given board.

The range for this measure is .20 to 11.14; the mean is 3.92 and the standard deviation is 2.74. Table 1 shows how intensity is distributed over the boards in the sample.[3] What is perhaps most striking about the distribution is the skewness toward the lower end of the range, despite the presence of some boards in the very high ranges. If one is to judge by these figures, there are large numbers of districts with relatively impotent *formal* spokesmen for interest groups. We stress formal because it seems highly probable that organized interests are sometimes represented in informal ways and that boards do not necessarily perceive such action as interest group activity. It is also true that the maximum possible total of 19 points

TABLE 1

DISTRIBUTION OF ORGANIZATIONAL INTENSITY BY BOARDS

Range of Mentions	Number of Boards	Percentage of Boards
<1	7	6.3
1-2	23	21.7
2-3	20	18.7
3-4	15	13.6
4-5	11	9.9
5-6	8	7.9
6-7	7	6.8
7-8	2	1.6
8-9	7	6.3
9-10	3	2.6
10-11	4	3.3
>11	1	1.2
Total	108[a]	99.9

a. This n exceeds the actual weighted n of 106 due to rounding of noninteger weighted ns to whole numbers.

is a very difficult one to achieve for two reasons. First, even though two and three responses were coded per question, the majority of respondents give only the one or two most salient organizations in response to a question. Second, some say no organizations meet the criteria of the question. They may, for example, say that no organizations have worked to defeat financial referenda because no referenda have been held or because no groups worked against referenda which were held. Third, since board means are being computed, any taciturn or noncooperative respondents would lower the board's overall average.

All in all, then, our measure may well understate the intensity of organizational activity vis-à-vis the school board. Even being generous, however, one would conclude that a sizable proportion of districts are not boiling cauldrons of interest group activity. To the contrary, they seem to be functioning with a minimum of formal group life.

THE DISTRIBUTION OF GROUP ACTIVITY

At this point in the development of the measure of group intensity, no categorization by kind of group was attempted. However, it is instructive to note the major categories of groups which come to the attention of boards; later in the essay we will return to this classification in order to ascertain whether or not particular kinds of groups are associated with particular kinds of antecedent societal conditions and policy outcomes. The most frequently mentioned groups are those most intimately concerned with education, PTAs and teachers (Table 2). However, it is somewhat surprising that PTAs so decidedly outrank teachers organizations, whose members have a more immediate interest in board policy—salaries, for example. Yet we should recall that teachers organizations have been less than militant in most areas. Although some of the larger cities contain quite active teachers organizations, in general they have not assumed a very political role. From the point of view of school boards, PTAs are more of a force to contend with. Not only do they consist largely of parents, they are also often laced with and frequently dominated by key administrators and teachers.

Of the remaining groups, the more ideologically oriented ones rank rather well compared with those of an "establishment" tinge. Left-wing, right-wing, and taxpayers groups are those which assail the board from an ideological perspective. Of these, left-wing organizations (such as ACLU and NAACP), do much more lobbying than right-wing (such as veterans and John Birch Society), and the

TABLE 2
ORGANIZATIONS MENTIONED BY BOARD MEMBERS (in percentages)

Type of Organization	Mentions by Board Members[a]	Range by Board[b]
PTA	60	0-100
Teachers	32	0-100
Left-wing, civil rights	29	0-100
Service clubs	21	0-100
Business and professional	17	0-100
Taxpayers	16	0-100
Right-wing	13	0-100
League of Women Voters	14	0-100
Religious	11	0-60
Citizens Advisory Committee	11	0-80
Political	5	0-50
Neighborhood	5	0-57
Labor	3	0-40

a. In constructing this measure scores were assigned to each board reflecting the percentage of individual board members who mentioned each respective type of organization. The mean for all boards was then computed. Thus, the mean percentage of members per board who mentioned PTA was 60.

b. E.g., the mean range of PTA mentions varies from 0%—where no member of a board mentioned PTA in response to any of the questions—to 100%, where all members of the board mentioned PTA at least once.

heralded taxpayer organizations, leaders of the now famous taxpayer revolt. Still, if we take the less militant organizations such as service clubs (e.g., Kiwanis and school boosters, League of Women Voters, Citizens Advisory Committees, and the like), their impact (combined with the dominance of PTAs) tends to create the impression that the organizational climate in which school board decisions occur is more oriented toward the status quo. The extent to which broader questions of educational policy are raised is probably dependent upon the activities of the left-wing, civil rights, right-wing, and taxpayer organizations, whose relatively high ranking might tend to balance the numerical dominance of the status quo organizations.

SOCIAL COMPLEXITY AND ORGANIZATIONAL INTENSITY

In searching for the antecedents of organizational intensity in educational politics, one can fall back comfortably upon theories of group activity and look first at some indicators of social complexity. Literature on the rise of organizations is replete with references to the displacement of primary by secondary groups as the society becomes more complex and heterogeneous (Truman, 1951: 51). Wirth's classic essay (1938: 20) on the urban mode of life sums up such assumptions: "Being reduced to a stage of virtual impotence as an individual, the urbanite is bound to exert himself by joining with others of similar interests into organized groups to obtain his ends." Virtually all the empirical research available supports at least the portion of Wirth's conclusion about associational activity, if not his assertion as to its cause. We should be on fairly safe ground in beginning our inquiry with the assertion that the complexity of urban life should produce more "groupness" and hence greater organizational intensity. Here again, school districts prove to be a fertile ground, for as consolidation of small districts into larger ones continues, greater geographical and population heterogeneity has come even to the smaller districts.

In measuring social complexity, one would ideally compile a composite measure built with indicators corresponding to a model of what makes a geopolitical unit more or less complex. However, such measures are either virtually impossible to obtain or too costly. Sociologists and demographers have demonstrated that the larger and more urbanized the area, the more complex the set of social institutions and patterns therein. Similarly, political scientists have shown, indirectly at least, that the set of political institutions and processes also varies in complexity with size and urbanism.

Two different but, to some extent, interrelated measures are employed to describe the social and cultural complexity extant in the school districts' population. Metropolitanism is a dichotomized variable which divides the school districts between those not located in Standard Metropolitan Statistical Areas and those located within one. District population is the total adult population within the districts' boundaries.

The bearing of these complexity measures on organizational intensity is shown in Table 3, which contains two measures of association: simple correlations and beta weights (standardized regression coefficients), as well as R^2, the total amount of variance accounted for by the predictor variables. This table presents, as will most other tables in this paper, the predictor variables arranged according to the strength of their beta weights.

The association between social complexity and organizational intensity is clearly in the direction one would have predicted. Both metropolitanism and district population continue to make substantial contributions to the association with the other variable in the equation controlled, as demonstrated by the sizable beta weights. Taken together, nearly half the variance in organizational intensity is accounted for by these two indicators of complexity. Clearly dominant, however, is metropolitanism; i.e., whether the school district is within or outside an SMSA. In later portions of the paper, when complexity is entered into the equation with other categories of variables, metropolitanism will serve as the single indicator. A virtue of the census bureau's classification scheme is that SMSAs include not only central cities of 50,000 or more but also the remaining part of the county plus contiguous counties adjudged to be socially and economically integrated with the central city. For

TABLE 3
RELATIONSHIP BETWEEN SOCIAL COMPLEXITY AND ORGANIZATIONAL INTENSITY

Complexity Indicators	Simple r	Beta
Metropolitanism	.60	.44
District population	.55	.35
R^2	.47	

school districts this means that even some relatively small districts are within the orbit of the metropolis. They absorb and are affected by the modes of group life found in the larger environment.

MASS SUPPORT, COMPLEXITY, AND INTEREST GROUPS

Almost on a par with social complexity as an explanation of group activity is the notion of pressure groups originating to cope with alterations in the social and economic status of people with "shared attitudes" (Truman, 1951: 29). This is the idea of disturbances in equilibrium as outlined in the introductory section. Urbanization, contributing to the creation of discontinuity of established patterns of interaction, is an example of the conditions which lead to such alterations in status. Examples of pressure groups originating to cope with some immediate problem and then persisting to deal with new matters are too numerous to elaborate upon at length. Consider, for one example, the organization of independent drug stores to cope with the problem of chain stores in the 1930s (Palamountain, 1955). Once the initial problem had been resolved, the National Association of Retail Druggists continued its lobbying efforts, even though the tensions which led to its initial creation were somewhat abated. Once the organization has been given its start, other inducements must replace the initial tension.

It is certainly true, as Salisbury has argued, that there is a goodly amount of entrepreneurial skill required on the part of organizationally active individuals. Olson has persuasively argued that political motivations are insufficient to cement group loyalties. Nevertheless, stressful times should provide an initial stimulus for group activity. Groups are less active when there is less to fight about. Here again, school districts are particularly suitable for inquiry, for the late 1960s were undoubtedly times of high drama in school district politics. Still, one would expect to find a considerable range of tension across school districts. Some districts experience recurring crises while others are blessed with rather long periods of calm. Following the dictates of group theory, we expect that those districts with the highest tension will experience more pronounced group activity. Conversely, those with the lowest tension levels should be relatively free of organized demands.

To assess the level of tension we employ a measure of the relationships between citizens and boards known as mass support.

The measure looks at the public support rendered the school board, and is constructed in the following manner: a cumulative index score for each board was built from responses to three questions, one dealing with the degree to which the board takes unpopular stands, a second indicative of the prevalence of critics of the board, and a third describing the amount of congruency between the board's ideas of appropriate board behavior versus the board's perception of the public's ideas. The range of the index is .20 to 2.80; the mean is 1.71 and the standard deviation is .66.

Mass support is inversely related to organizational intensity (r = –.75). As the population becomes more supportive of the policies of the board, organized group activity diminishes. Groups clearly thrive in an atmosphere of conflict between the governed and the governors. It is probable that, once the level of public support has deteriorated to a level sufficient enough to generate fairly intense group activity, organizations exacerbate the loss of confidence in the board.

Declining support and organizational activity undoubtedly feed off each other; trying to establish a definite causal sequence is, therefore, difficult. One such effort is suggestive, however, There is a third variable linked logically and statistically to both mass support and organizational intensity. This is district consensus, the degree to which the district populace is perceived to be free of divisive inner conflicts. By adding district consensus to the equation we can perform an indirect test of causality. If one of the hypothesized relationships is affected adversely by the introduction of a third, related variable, then we may infer that the other relationship is the more probable causal pattern. Controlling for consensus, the effect of mass support on organizational intensity is considerably higher than that of intensity on support (betas = .70 versus .45). Similar results were obtained using other pertinent variables. It is possible to argue, then, that the relationship is reciprocal but that (given the magnitude of the regression coefficients), the stronger "causal" link is from mass support to organization intensity.

To return to the argument of disturbance in equilibrium, it appears that as public confidence in board policy declines, the decline in confidence is articulated and given explicit focus by interest groups. They pinpoint, according to their own objectives and interests, the specific aspects of discontent to which they will address their efforts. It is probable that declining mass support becomes more subject-specific as group intensity increases. The interactive

TABLE 4
EFFECTS OF MASS SUPPORT AND METROPOLITANISM ON ORGANIZATIONAL INTENSITY

	Simple r	Beta
Mass support	−.75	−.58
Metropolitanism	.60	.30
R^2	.63	

effect, then, is for organizational intensity to force and direct the generalized discontent toward the board, and also back to the publics which they serve.

A natural suspicion is that the strong association between mass support and organizational intensity is a function of social complexity. We could certainly assume that complexity and support vary inversely, given the strong, opposite association of both variables with organizational intensity and commonsense description of urban life. Using metropolitanism as the measure of complexity, we find that both complexity and mass support retain their strong association with organizational intensity (Table 4). Of the two, mass support emerges as the best predictor, suffering less of a loss between the simple correlation and multiple regression. It is clear, however, that each makes a unique contribution to organizational intensity.

THE ROLE OF ELECTORAL FACTORS

Up to this point we have seen striking evidence in support of the importance of broad range sociopolitical factors upon group intensity. Heeding the advice of those who assess the impact of political institutions upon public policy, we were led to examine the relation between interest group activity and characteristics of the electoral system. The next portion of the analysis looks at characteristics which are part of the more immediate political environment in which interest groups operate: the structure of electoral competition in the district and the legal parameters of elections. One of the most frequently asserted dicta of group research, for instance, bears upon the link between interest groups and political parties. It is often claimed that interest groups thrive in political systems with weak political parties. Not necessarily at odds with this proposition is one which says that a greater range of interest groups thrives when electoral competition is keen.

The results of our analysis do not support much of the conventional wisdom regarding the interplay between competition factors and interest group activity. True, most of the relationships are in the hypothesized direction, with competition and weak parties being positively associated with group activity.[4] Overall, however, these variables proved to be weak predictors, accounting for only 10% of the variance in organizational intensity scores. Moreover, adding metropolitanism and mass support to the regression analysis reduces even this modest contribution. While the structure of competition undoubtedly impinges on local school politics in a variety of ways, it is patent that no direct relationship exists between them and the vitality of group life.

Legal parameters constitute another set of electoral factors conceivably affecting organized groups. Students of interest group behavior have argued that the formal structure of government, rather than simply being a neutral framework, is a key element with which interest groups must reckon. For instance, the American federal system gave Negro organizations an opportunity to circumvent the hostility of southern legislatures by turning to Congress and the federal courts. Three legal parameters differentiating school districts are whether ward or at-large elections are held; whether coterminous referenda are held with school board elections; and length of office term.

Considering these three measures together the direction of the findings are again ostensibly reasonable. For example, the presence of coterminous referenda heightens organizational intensity (beta = .25). It is likely that when all school related elections are held simultaneously the climate of the elections is more heated than when elections are spread over a longer period of time. Interest groups may converge on both candidates and issues in this setting, the double stimulus accelerating somewhat the normal tempo of group activity. Similarly, there is a suggestion that a wider range of group activities is inspired by the more geographically (and probably socially) heterogeneous representation found on boards with ward elections. Despite these affirmative findings, the legal parameters taken alone account for only 10% of the variance in organizational intensity. And when social complexity and mass support are introduced into the analysis, the meager contribution of legal parameters diminishes—although never disappearing completely.

Granting the overarching importance of complexity and mass support, it is possible that competition levels and legal parameters serve as mediating devices in the linkage between complexity and support on the one hand, and group activity on the other. That is, even though the electoral variables have little independent effect on group life, they may condition the association between other independent forces and group life. To test this notion we separate out the subcategories of electoral variables, and examine the overarching relationships within each category of the electoral variables.

In general the findings support the view that electoral properties do condition the broader relationships in a systematic fashion. Thus the overall negative relationship between mass support and organizational intensity tends to jump among the more competitive school districts and where the legal parameters facilitate segmentation and differentiation. Illustratively, compared with the grand simple correlation of −.68, there is the −.45 where no electoral opposition was present in the preceding election and −.67 where opposition was present. Similarly the correlation is −.64 in districts with at-large elections versus −.83 in ward-type districts.

Although the differences in the basic correlations introduced by holding physically constant the electoral factors are not massive, they point toward a reconceptualization of how electoral matters figure in school district governance. A simple model of association between overtly political variables and organizational intensity is not adequate, for it is apparent that the electoral variables work in conjunction with mass support and metropolitanism to produce distinct patterns of interest group activity. This analysis suggests that the appropriate treatment of these characteristics is not to stop after they have lost out in the competition with broader, more societally based variables; rather we should consider them as interacting with such broader variables in a systematic fashion. When placed in this perspective, the traditional propositions about the interest group-electoral process nexus take on more substance.

SPECIFIC TYPES OF ORGANIZATIONAL INTENSITY

Organizational intensity has thus far been treated without regard for particular varieties of groups. Now we will examine the corollaries of specific organizational types. It is reasonable to

TABLE 5
INTERCORRELATIONS OF ORGANIZATIONAL INTENSITY AMONG SPECIFIC TYPES OF ORGANIZATIONS[a]

Type of Organization	Left-wing, Civil Rights	Right-wing	Taxpayers	Teachers	League of Women Voters	PTA	Political Party	Business and Professional	Citizens Advisory Committee	Labor	Religious	Neighborhood	Service Clubs
Left-wing, civil rights	—												
Right-wing	.51												
Taxpayers	.37	.50											
Teachers	.42	.41	.28										
League of Women Voters	.43	.21	.22	.22									
PTA	.25	.26	.23	.32	.31								
Political party	.26	.24	.39	.14	.27	.10							
Business and professional	.34	.27	.40	.30	.26	.10	.12						
Citizens Advisory Committee	.17	.16	.10	.04	.04	.12	–.00	.03					
Labor	.19	.12	.03	.08	.12	.11	.04	.18	.16				
Religious	.16	.21	.04	–.00	.06	.05	–.02	.02	.11	.05			
Neighborhood	.30	.15	.03	.10	.07	–.06	–.01	.26	–.10	.26	.00		
Service clubs	–.39	–.27	–.19	–.10	.06	.16	–.14	–.08	–.24	–.01	–.23	–.24	—

a. Entries are product-moment correlations.

assume—given the different ideologies and membership of the various organizations—that the two major determinants (metropolitanism and mass support) will have markedly different impacts upon distinctive organizations. It is true, of course, that organization tends to produce counterorganization so that where the right-wing flourishes we might expect a countermovement. We should not necessarily expect, however, that the existence of one kind of organization will automatically be associated with the existence of another, especially when such organizations do not have competing goals.

Table 5 consists of a correlation matrix for the organizations under analysis. It can be seen that there is considerable variation in the tendency of organizations to cluster together. Nor is this variation simply a function of the relative salience of the organization. Although PTAs and teachers are the two most frequently cited groups, their average intercorrelations with other groups are lower than those for the less frequently mentioned taxpayers and right-wing organizations. It should also be noted that not all the associations are positive, thereby suggesting that the existence of some types of groups is actually discouraged by the presence of other types.

The place of the manifest ideological component of various organizations in determining the intercorrelation pattern is unclear at first glance. For instance, the most pervasive organization, PTA, is most often found alongside five groups (teachers, League of Women Voters, left-wing, taxpayers, and right-wing), none of which bears any consistent ideological relation to the PTA. Yet the highest correlations exist between ideologically opposed groups, the right- and left-wing organizations. Equally high are the correlations between right-wing and tax groups. Indeed the highest positive relationships are reserved primarily for the groups with the strongest ideological commitment, whatever the nature of that commitment.

By way of buttressing this idea, examine the paired correlations involving service clubs, which consist of Kiwanis, Rotary, and so forth, and also the various local "boosters" clubs. In this case, all but two of the associations are negative. The service clubs are essentially nonideological, even more so than most of the groups which come to the attention of the school board. Religious organizations, neighborhood groups, and labor organizations also exist in relative isolation. The less manifest the ideology of the group, the less likely will its intensity level resemble that of other groups.

TABLE 6

SOURCES OF ORGANIZATIONAL INTENSITY: METROPOLITANISM AND MASS SUPPORT

Type of Organization	Simple r	Beta[a]	R^2
Left-wing, civil rights			
Metropolitanism	.63	.46	
Mass support	–.58	–.34	.48
Teachers			
Metropolitanism	.51	.34	
Mass support	–.50	–.32	.34
Business and professional			
Metropolitanism	.38	.30	
Mass support	.31	–.15	.16
League of Women Voters			
Metropolitanism	.29	.29	
Mass support	–.15	–[b]	.08
PTA			
Metropolitanism	.23	.20	
Mass support	–.16	–.05	.06
Right-wing			
Mass support	–.52	–.45	
Metropolitanism	.38	.14	.29
Taxpayers			
Mass support	–.45	–.42	
Metropolitanism	.28	.05	.20
Service clubs			
Mass support	.33	.23	
Metropolitanism	–.31	–.19	.13
Neighborhood			
Mass support	–.32	–.25	
Metropolitanism	.27	.14	.12
Political			
Mass support	–.33	–.33	
Metropolitanism[b]	.18	–[b]	.10
Citizens Advisory Committee			
Mass support	–.25	–.19	
Metropolitanism	.22	.12	.07
Religious			
Mass support	–.17	–.11	
Metropolitanism	.17	.11	.04
Labor			
Mass support	–.18	–.19	
Metropolitanism	.08	–.02	.03

a. Betas for "Metropolitanism" are with "Mass support" controlled, and vice versa.

b. If the tolerance level is very small, the second variable is virtually a combination of the variable(s) already in the equation. Inclusion of such a variable is very often the result of random error in measurement and is difficult to interpret in a meaningful way. Stepwise regression does not allow such variables to enter the equation.

Despite the apparent tendency of organizational activity to inspire counterorganizational activity, different kinds of organizations thrive in different kinds of environments. Table 6 relates metropolitanism and mass support to the intensity levels of specific organizations. Let us examine the table from the point of view of determining whether metropolitanism or mass support is the more determinative. Organizations listed above the dotted line are more affected by metropolitanism, those below more by mass support. Turning to metropolitanism, we see that this variable is the best predictor of the activities of left-wing and civil rights groups but does little toward predicting the activities of right-wing groups. Mass support, in contrast, does more toward predicting the activities of right-wing groups, with metropolitanism taking a markedly inferior position. Thus left groups are less dependent upon widespread hostility whereas the complexity of the environment is of little value in helping us to understand right-wing organizations. In a similar vein, mass support has a strong negative association with taxpayers associations, another conservative group, and metropolitanism has a trivial impact. It would seem that right-wing organizations are quicker to seize the initiative in periods of unrest, and do so whether or not the complexity of the community is generally conducive to organizational activity.

It should be noted that the mass support regressions for all ideological groups—irrespective of the direction of the ideology—are stronger than those for most other kinds of organizations, with political organizations also showing a strong pattern. The more political or ideological the organization, the greater the response to a decline in mass support. There is also a rather healthy relation between mass support and teacher organizations, but it is probable that in this case the organizational response is more defensive. As mass support declines, public acquiescence in school board policy declines, stimulating teachers organizations into a more active role.

Mass support appears to be least determinative when the organizations are supportive; e.g., PTA and League of Women Voters. In fact, when we examine the role of service organizations we notice that in this case the association between mass support and intensity is positive, the single exception to the rule. The more supportive the public, the greater the activity of service organizations. As we noted, service organization activity is negatively associated with the existence of other organizations. Here we find they are also a clear exception. Their role is one of local booster and they thrive when the public is in a mood to boost. In a sense, the positive relationship between service organizations and mass support is an extension of the low negative association for PTAs. Both organizations differ fundamentally from the ideologically and politically combative organizations.

A similar point can be made by examining the R^2 values for the various kinds of organizations. R^2 for ideologically oriented groups tends to be higher than for such groups as PTAs, League of Women Voters, and the like. The more a group tends to be an agent for the maintenance of things as they are, the less we can explain about its salience. In fact, mass support and complexity are woefully weak predictors for a number of organizations. Localistic and entrepeneurial factors probably account for the prominence of these specific interest groups as one moves across the landscape of school districts.

It is also instructive that R^2 for any single type of organization is considerably lower than the .63 found for the overall measure of organizational intensity, using the same two predictor variables. This suggests a strong threshold effect. The likelihood of a district generating intensity across a variety of interest groups increases greatly according to both social complexity and mass support whereas the likelihood of intensity for a given type of organization is customarily only modestly affected. Thus while two districts of varying complexity and mass support may both have the same level of intensity by one type of group, they are unlikely to have the same level of intensity when various types of organizations are considered. There is a more clearly defined threshold of circumstances and preconditions for the overall level of group life.

SOME CONSEQUENCES OF ORGANIZATIONAL INTENSITY: ISSUE AROUSAL AND DISPOSAL

If we view interest groups as bargaining agents in the allocation of public resources then we need to know what difference they make in the way school districts conduct their business. In the previous section of the paper we made the point that interest groups thrive when mass support for the school board is low. Does it therefore follow that group activity contributes to the heightened tension which accompanies a decline in public confidence? To put the question into another perspective, imagine a school district suffering a decline in public support. Even though interest groups will probably become active in this district, does their activity translate the loss of confidence into observable phenomenon? If not, it would make little difference to the school board if mass support is low, since the board would have little evidence of the state of public opinion.

The interview schedule contains some questions designed to tap the degree of "issue arousal" and "issue disposal" within the district. Some districts operate with little difficulty while others are plunged into perpetual crisis. The 1960s were years in which the school districts faced unprecedented demands, but even in such a heated climate, there were some districts which enjoyed relatively smooth sailing. Our question deals with the extent to which interest groups contribute to a tense atmosphere within the district, in contrast to the main question of the previous sections which considered the conditions which contribute to organizational intensity.

We have selected the following items as measures of issue arousal and disposal.

Financial defeats: Whether a district has seen a bond issue or tax referendum go down to defeat in the last three years.

Racial problems: The percentage of board members who say the district faces racial problems.

Financial problems: The percentage of board members who say the district has trouble in achieving an adequate level of financing. There is only a moderate (.36) correlation between this item and financial defeats, suggesting that close but successful financial elections still make board members think in terms of troublesome situations.

Teacher criticism: The percentage of board members who indicate that teachers' classroom performance has come under attack.

Firing of teachers: The percentage of board members who are aware of tenured teachers being dismissed because of their classroom behavior.

Superintendent turnover: Given turnover in the past three years, whether the superintendent had left involuntarily.

The items give us a fairly broad range of issues, dealing with finances, teacher behavior, racial tension, and school management. Some (e.g., financial defeats) are clearly "outputs"; that is, they provide tangible evidence of the state of dissatisfaction in the district. Others refer more to a general level of conflict. Even though the issues have a somewhat different portent, we developed a composite index of issue arrousal-disposal in order to sketch in first the general role of interest groups. This index ranges from 0 to 100 with a mean of 40.19 and a standard deviation of 23.0.[5] The composite index reflects rather well the essence of the individual items, as indicated by the following item-index correlations:

Racial problems	.69	Teacher criticism	.56
Financial problems	.68	Superintendent turnover	.52
Financial defeats	.65	Teacher firings	.46

The index score becomes the dependent variable in our attempt to isolate the effects of organizational intensity. Our procedure is to enter organizational intensity as an independent variable into the regression equation, along with mass support and metropolitanism. The last two variables were selected because of their powerful performance in predicting organizational intensity. If intensity is to make an independent contribution to issue arousal, it should have to do so under conditions which put it to a severe test. Table 7 shows that both mass support and organizational intensity make an appreciable impact upon issue arousal. As mass support declines and organizational intensity increases, the issue climate of the school district becomes heated. Surprisingly, the complexity of the environment has virtually no impact upon issue arousal once the contribution of the other two factors is taken into account.

We have previously mentioned the affinity between mass support and organizational intensity; we now add that, as a consequence of their interaction, a school district is likely to find itself immersed in a climate of hostility *whether or not the environment is socially complex.* The latter point is especially significant in view of the earlier discussion of the link between complexity and organizational intensity. In terms of the *consequences* of these patterns, we are far better off knowing the level of organizational intensity than knowing the degree of complexity.

TABLE 7
CONDITIONS ASSOCIATED WITH ISSUE AROUSAL-DISPOSAL

	r	Beta
Mass support	−.66	−.39
Organizational intensity	.66	.38
Metropolitanism	.41	−.03
R^2	.50	

We should be careful to disclaim any clearly established causal chain in these events. Since issue arousal does not necessarily occur after group activity, we might argue that the heating up of the debate over issues leads to organizational intensity rather than the reverse. However, when we construct the same regression equation with organizational intensity instead of issue arousal as the dependent variable, the beta coefficient is .22. That is, issue arousal predicts organizational intensity less well than intensity predicts arousal (.22 versus .38). Still, we do not wish to make too much of the argument simply on the basis of regression coefficients.

If one were to examine the "real world" of educational politics in local districts it would still be very difficult to construct a causal chain. A series of intensive comparative inquiries would be most helpful in this regard. Short of that it is perhaps more fruitful to think in terms of interaction between issue arousal, decline of mass support, and organizational intensity, each contributing to the other.

THE CONTRIBUTIONS OF SPECIFIC ORGANIZATIONS

It is certainly true that some groups are more active—and effective—than others. Given the inherently divisive nature of public issues, we might expect that the more ideologically oriented groups would have more of an impact. Since the goals of left- or right-wing groups differ fundamentally from the goals of, say, service organizations, does it necessarily follow that their effects will differ? To get at this question, we perform a regression using mass support, metropolitanism, and each organizational type individually. When this is done, the only groups whose strength is not diminished to the point of triviality are left-wing (beta = .32), teachers (.18), and right-wing (.13). These diminished coefficients demonstrate two principles. First is the importance of multiple versus single types of organizations in generating issue arousal. What an intense single group might not do, a combination of them will. Second is the fact—revealed below—that the effects of some groups are issue-specific. These specific effects are, to some extent, lost in the shuffle of a composite index of issues.

Clearly the most significant type of group insofar as overall issue arousal is concerned is the left, far more so than the right. As expected, the dominant groups are ideological; but this does not tell us why the left is so much more associated with issue arousal than

TABLE 8

CONTRIBUTIONS OF SPECIFIC ORGANIZATIONS IN SPECIFIC ISSUES[a]

Type of Organization	Race Problems	Financial Defeats	Financial Problems	Teacher Criticism	Fire Teachers	Superintendent Turnover
Left-wing, civil rights	.55 (1)	.11 (11)	.49 (1)	.40 (1)	.12 (7)	.03 (12)
Teachers	.18 (2)	.19 (8)	.12 (6)	.23 (4)	.05 (9)	.11 (6)
Neighborhood	.13 (3)	.47 (1)	——[b]	.11 (9)	–.09 (8)	.12 (5)
Religious	.12 (4)	.21 (6)	–.06 (8)	–.04 (11)	.17 (4)	.22 (2)
PTA	–.11 (5)	–.18 (9)	–.19 (3)	–.06 (10)	.04 (11)	–.21 (3)
Service clubs	.11 (6)	.35 (3)	.19 (4)	.24 (3)	–.05 (10)	.03 (11)
Labor	–.10 (7)	–.29 (5)	–.03 (9)	.02 (12)	–.14 (6)	.09 (7)
Citizens Advisory Committee	.08 (8)	.41 (2)	.14 (5)	.12 (8)	.18 (3)	–.18 (4)
Taxpayers	.07 (9)	.30 (4)	.20 (2)	–.13 (7)	.20 (1)	——[b]
Business and professional	.07 (10)	.05 (12)	.02 (10)	——[b]	–.15 (5)	–.03 (10)
League of Women Voters	.06 (11)	.21 (7)	–.11 (7)	–.34 (2)	–.03 (12)	.03 (9)
Right-wing	.05 (12)	–.15 (10)	.02 (11)	.20 (5)	.19 (2)	.09 (8)
Political party	.02 (13)	——[b]	–.02 (12)	.15 (6)	——[b]	.32 (1)
R^2	.46	.46	.34	.32	.32	.23

NOTE: Entries are beta weights.

a. Controlling for the effects of the other organizations. Number in parentheses indicates the rank order of the beta weights.

b. See note b, Table 6.

the right. It is useful here to recall that left-wing groups are more *active* (according to the perceptions of board members) than right-wing groups; perhaps sheer activity at least partially explains their greater impact. Left-wing groups are *not* more active than teachers organizations, for instance, and yet they apparently have greater impact. Indeed, if we perform a regression using only organizational types as independent variables (with the index of issue arousal as the dependent variable), left-wing groups rank first, and right-wing groups last! The strength of left-wing groups is roughly five times that of right-wing groups (.46 versus .09).

One possible explanation is that the left-wing groups cast a wider net than the right-wing groups—hence are effective on more issues. Let us turn, therefore, to an examination of each issue area, considering the effect of each group upon each separate issue area. It is unnecessary to require that all groups compete in an equation with mass support and metropolitanism if we want to make a point about the distribution of organizational issue response or of presumed influence. For instance, even though in a given issue area a group has less impact than mass support or social complexity, it still might be the most determinative group in competition with other groups.[6]

If we enter all groups into a stepwise regression equation for each issue area, this will give us a better notion of differential organizational impact. In Table 8, the entry of the group into the equation is listed next to the regression coefficient. Concentrating our attention on the first three groups for each issue area, we observe relatively little overlap between issue areas. Only four types of organizations appear in the top three more than once. They are: religious (2), Citizens Advisory Committees (2), left-wing (3), and teachers (3). Further, if we examine those groups which rank first, only left-wing groups repeat. Compared with right-wing groups the average beta for left-wing groups is demonstrably higher—28.3 vs. 11.0. These results help explain why left-wing groups have a greater overall strength than right-wing groups: they have impact upon more individual issue areas.

We should not necessarily argue that emergence as a strong predictor is equated with influence in a planned direction. Organizational intensity may often be a consequence of an inflamed issue area. It is apparent, for example, that not all of these groups take the same position on financial issues. Presumably, the League of Women Voters and teachers organizations are most likely to be in favor of passage of the issue while taxpayers associations are not. Yet the direction of their influence is the same irrespective of the ideology of the group. What seems to be happening is that such groups heat up the election environment and, possibly, stimulate a higher turnout. Generally, higher turnouts spell doom for school financial elections since they draw a disproportionate turnout from negatively inclined lower-status people.

Ironically, then, the result of activity by proschool forces and antischool forces is the same. The effect, if one can call it that, seems to be quite independent of the goals of the organization. A similar sort of effect can be observed with regard to racial problems, where left-wing and civil rights groups have the greatest impact. What probably happens in this case is that the racial problem has been there all along, and group activity simply brings it to the surface. Hence the school board defines it as a problem because interest groups have made the problem salient. Therefore, it is more accurate to say that left-wing groups, in seeking more integration, hiring of black teachers, and the teaching of black history, for example, crystallize an issue which may lead to its partial resolution.

There are instances where the relationships are less ambiguous, with the goals of the organization and the result of its activity being compatible. The positive association between right-wing group and teacher firing is a case in point. In this case, right-wing groups have harassed school boards to get rid of various kinds of allegedly subversive teachers; and the table indicates that they have done a good job. Similarly consider these negative associations between activity and issue areas: PTA efforts with superintendent turnover; labor organization intensity with teacher firings; and League of Women Voters activity with teacher criticism. These are reasonably clear examples of the intended consequences of group efforts and the actual consequences of their efforts being congruent.

One can interpret the evidence in Table 8 from another perspective. On the one hand, the fact that most organizations have impact upon a single issue area argues for a pluralistic interpretation. On the other hand, the fact that two types of organizations (left-wing and teachers) dominate half the issue areas, leaving the remaining groups to contest for influence in the others, suggests a concentration of influence. Further, it is possible that in a given district at a particular point in time, a single issue (and hence a single group) is most salient. Suppose, for example, that there is a struggle to oust the superintendent; then political organizations will appear most influential. If there is an effort to get rid of allegedly subversive teachers, then right-wing and taxpayer groups will dominate. In short, neither a concentrated power structure nor a pluralistic one receives unequivocal support.

Yet while it is true that issues and salient groups vary, two issues appear to attract a disproportionately high level of organizational interest. Judging from the multiple correlations, group effort abounds in the areas of financial defeats and racial problems. Such a finding fits well with a "commonsense" view of what is likely to arouse extraschool involvement in educational politics. Many school districts have been caught up in the "taxpayer revolt" and have, simultaneously, had to cope with federal and local pressures toward integration. Such issues not only require the mobilization of the resources of the school district, they also are likely to excite the passions of the various publics which normally may not be especially attentive to school affairs.

CONCLUSIONS

As identified in the introductory section, we are dealing with two unresolved problems in the study of interest groups. On the one hand, there is the empirical question of the origins and consequences of group activity. On the other hand, there are serious problems of measurement if we are to move toward providing at least tentative answers to the empirical questions.

The purpose of this paper is to contribute to the resolution of these kinds of problems by identifying the subsystem under investigation as the unit of analysis, rather than the individuals who comprise the subsystem. This methodological decision places the study in a somewhat different light than previous studies of interest groups. By using school boards rather than school board members as units of analysis, we were able to develop a measure of organizational intensity which could then be linked with other systemic variables. The cost of such an operation should not be ignored. We are unable to take into account the potentially significant behavior of the leaders of particular groups, and the unique behavior of individual board members. Thus we can provide no direct answer to the Salisbury-Olson theories of interest group origination. Nevertheless, we were able to account for the intensity of group activity by using indicators which made no reference to individual behavior. Perhaps the unexplained variance in our equations can be accounted for by the activities of individual entrepreneurs, as is suggested by Salisbury and Olson.

With regard to the antecedents of group activity, the "traditional" theories which Salisbury and Olson seek to modify hold up rather well. Both social complexity and the lack of mass support are quite helpful in predicting organizational intensity. It is with the *consequences* of group activity that traditional group theory leaves much to be desired. Here we have tried to go beyond the linking of a specific group with a single policy outcome; we have sought to explain the general role of interest groups in defining the climate of issue arousal in school districts. In so going, we can isolate two general categories of organizations. First, there are non-issue-specific groups, such as PTAs, League of Women Voters, and service organizations. These organizations provide support for the ongoing system, but inject little conflict into the system. They constitute a resource from which decision makers may draw in times of crisis. Then there are ideological and issue-specific groups whose role is to inject conflict into the system and to make conflict salient for decision makers. Intense activity by such organizations usually has an effect, but not necessarily the effect that such groups desire. The unanticipated consequences of such groups may be a result of the fact that they have the influence to make an issue salient by expanding the scope of conflict but apparently have less ability to control the outcome of a conflict once it has developed.

A somewhat surprising finding lies in the comparison of left-wing and right-wing groups. For one thing, the intensity of the left is much more predictable than that of the right, thereby suggesting more of a flash or idiosyncratic pattern for the latter. Of course some would argue that the views of the right—presumably residing in at least part of the so-called "silent majority"—may not need the explicit articulation of the left in order to be incorporated into school district policies. But our findings also provide a useful corrective to the popular views of social and educational critics that the right-wing is the better-organized, more spirited participant in school district politics. Issue arousal and disposal were actually much more reflective of left-wing energies than of the right-wing. The prominence of one or the other of these wings probably varies over time. During the late 1960s the left was the more prominent, if our data are a guide.

Finally, there is the notion of the feedback or interaction between issues and organizations. Once an issue is raised, partially as a consequence of group activity, other organizations enter the arena. In using survey research to describe this phenomenon, we are using what amounts to a stop-action camera. We get a picture of the situation at a single point in time; therefore, we can do no more than speculate upon the question of issues engendering groups or groups generating issues. We have suggested a pattern of development in which interest groups play the leading role in raising issues, but what is clearly needed is the development of dynamic methods of observation in order to resolve this impasse.

NOTES

1. While the literature on politics and education is growing at an impressive rate, systematic studies of interest groups have yet to appear in generally available form. Most of such studies are of the traditional case-descriptive variety, confining themselves to a single issue, a single group, or a single school district. Some of the more intriguing of these studies are: Rogers (1968), Gittell (1969), Iannaconne and Lutz (1970). Two recent comparative studies, each dealing with a single issue are Crain (1968) and Rosenthal (1969). Two projects deal directly with interest representation. They are: Smoley (1965) and Christie (1966). For an excellent effort to fit the various pieces together see Charters (forthcoming).

2. Of 84 potential boards from the original sample, 82 are represented. For one board that refused, another was substituted (making 83). For the second board that refused, a board in the original sample was double-weighted. Both these decisions were made on the basis of sampling criteria used in the 1965 study.

3. These figures are based upon all mentions, including those which could not be placed within a specific category.

4. The measures of competition for school board positions include de facto partisan versus nonpartisan elections; the proportion of present board members who were either appointed to office, were encouraged to run by members of the previous board, or both; the absence or presence of contested seats in the last primary or general school board election; and forced turnover—the proportion of incumbents defeated in immediately previous elections. A subjective party strength measure was constructed from individual responses in which each party was rated according to its perceived strength in the area.

5. The index of issue arousal-disposal was constructed in the following manner. For each board, the percentages of members answering in the affirmative on the teacher criticism, firing of teachers, financial problems, and racial problems questions were summed. To this total was added 100% if one or more financial referenda were defeated, and 100% if the superintendent's departure had been voluntary. These sums were divided by four if no budget referenda were held and there had been no superintendent turnover, by five if all but one of the above conditions held, and by six otherwise.

6. Separate regressions were also run in which each group was included in an equation with mass support and complexity as the other independent variables and each issue area or the dependent variable. Although the magnitude of the coefficients for each group ordinarily turned out to be lower than under the procedure described in the text, the patterns were quite similar.

REFERENCES

BAUER, R. A., I. POOL, and L. A. DEXTER (1963) American Business and Public Policy. New York: Atherton Press.

BENTLEY, A. A. (1949) The Process of Government. San Antonio, Texas: Principia Press of Trinity University.

CHARTERS, W. W., Jr. (forthcoming) School Board Research Revisited. Eugene, Oregon: Center for the Advanced Study of Educational Administration, University of Oregon.

CHRISTIE, S. G. (1966) Political Pressures on the School Board Member. M.S. thesis. San Diego State College.

CRAIN, R. L. (1968) The Politics of School Desegregation. Chicago: Aldine.

DYE, T. R. (1966) Politics, Economics, and the Public: Policy Outcomes in the American States. Chicago: Rand McNally.

FRANCIS, W. (1967) Legislative Issues in the Fifty States: A Comparative Analysis. Chicago: Rand McNally.

FROMAN, L. A., Jr. (1966) "Some effects of interest group strength in state politics." Amer. Pol. Sci. Rev. 60 (December): 952-962.

GITTELL, M. (1969) Confrontation at Ocean Hill-Brownsville. New York: Praeger.

GROSS, N. (1958) Who Runs Our Schools. New York: John Wiley.

IANNACONNE, L. and F. W. LUTZ (1970) Politics, Power and Policy: The Governing of Local School Districts. Columbus, Ohio: Charles E. Merrill Co.

MILBRATH, L. (1963) The Washington Lobbyists. Chicago: Rand McNally.

OLSON, M. L., Jr. (1965) The Logic of Collective Action. Cambridge: Harvard Univ. Press.

PALAMOUNTAIN, J. C. (1955) The Politics of Distribution. Cambridge: Harvard Univ. Press.

ROGERS, D. (1968) 110 Livingston Street. New York: Random House.

ROSENTHAL, A. (1969) Pedagogues and Power. Syracuse: Syracuse Univ. Press.

SALISBURY, R. H. (1969) "An exchange theory of interest groups." Midwest J. of Pol. Sci. 13 (February): 1-32.

SMOLEY, E. R., Jr. (1965) Community Participation in Urban School Government. Cooperative Research Project No. S-029. U.S. Office of Education.

TRUMAN, D. (1951) The Governmental Process. New York: Alfred A. Knopf.

WIRTH, L. (1938) "Urbanism as a way of life." Amer. J. of Sociology 44 (July): 20.

ZEIGLER, H. (1965) "Interest groups in the states," pp. 101-147 in H. Jacob and K. L. Vines (eds.) Politics in the American States. Boston: Little, Brown.

--- and M. A. BAER (1969) Lobbying. Belmont, Calif.: Wadsworth.

9

Local Interest Politics and Municipal Outputs

BETTY H. ZISK

□ ONE STRIKING RECENT DEVELOPMENT in the study of urban and state politics is the effort of many scholars to link the content of public policies to the economic and political context in which those policies are made. In the past decade, a burgeoning number of studies have demonstrated, with near unanimity, the intimate tie between *socioeconomic* variables such as urbanization, industrialization, and population density, and the amount of public expenditures for welfare, education, and general amenities. The allied effort to relate *political* variables such as party competition and voter participation rates to the level of government expenditures has met with considerably less success. There seems to be only a slight link between party competition and public expenditures, for example, when socioeconomic factors are controlled (see, e.g., Dawson and Robinson, 1963; Dye, 1968; Hofferbert, 1966, 1968; Sharkansky, 1968).

AUTHOR'S NOTE: *This study was made possible by a grant to the Stanford City Council Research Project, of which it is a part, by the National Science Foundation under Contract GS-496. An elaboration of the material presented here will appear in Betty H. Zisk,* Local Interest Groups: A One-Way Street *(Bobbs-Merrill, forthcoming). A preliminary version of the study was presented, under the same title as this paper, as a paper to the American Society for Public Administration, in Philadelphia, April 1970.*

There appears to be some hesitation about accepting these findings as final; the search for linkage variables continues, as do criticisms of the approach. There are, however, at least three reasons for the persistence of the attempt to establish political "linkage":

(1) the refusal of those who specialize in political analysis to believe that political processes or institutions are in themselves irrelevant to a central policy outcome—e.g., public expenditures;

(2) the criticism, typified by Jacob and Lipsky's (1968) review article, that the most frequently used measures for economic inputs and for policy outputs have not been adequately thought out;

(3) some dissatisfaction with what seems to be an ad hoc approach to questions of political linkage, and a growing realization that more effort must be devoted to theoretical questions if meaningful results are to be obtained (Hofferbert, 1970; Schaefer and Rakoff, 1970).

Jacob and Lipsky are primarily concerned with the question of validity. In what sense are measures of income or urbanization "inputs" into the political or economic system? They argue (1970: 514-515), in fact, that

> income, industrialization and education are not in themselves inputs. The measures have little relationship to the phenomena they are supposed to represent. We might conceive of them as environmental factors which might lead to the articulation of demands and supports and their communication to political authorities.

None of these studies has, in fact, attempted to relate these "environmental factors" to measures of interest articulation or political communication. It was simply assumed that a high level of education or wealth in a given political community would automatically lead to a large volume of political demands. The work on this question, over several decades, has buttressed the assumption by demonstrating a correlation between both education and urbanization and adult membership in voluntary associations, levels of political information, political efficacy and the like (see, e.g., Wright and Hyman, 1958; Almond and Verba, 1963; Babchuck and Booth, 1969). But the link between potential for political participation and the *act* of formulating and communicating individual or organizational demands to public officials has yet to be demonstrated.

A second conceptual problem concerns the adequacy of commonly used indicators for government services. Measures of per capita expenditures, teacher-pupil ratios, and the like, tell us very little about the distribution of benefits to different groupings of citizens in cities and states. Differentials in such distribution may be far more important in their effect on those involved than the absolute or per capita amounts of those benefits, and may be intimately tied to the demands made on the political system (Jacob and Lipsky, 1968).

With regard to theoretical problems, the recent studies *imply* that party competition and citizen turnout are important for state expenditure policy, but in none of these studies is there an explicit attempt to imbed the idea of political competition or participation into a more general theory of politics. Competition and participation are important—somehow—but the precise fashion in which they relate to interest articulation or political communications, or, more important, to the way in which policy makers operate, remains unspecified.

Two studies which depart from this pattern deserve some attention. These are analyses by Prewitt and Eulau of the relationship between city size, community support, political recruitment patterns, and representational response style, and Eyestone's examination of policy outputs in relation to city characteristics and councilmen's policy preferences.

City councils were found to be more responsive to interest groups and other attentive publics in large cities than in small communities in the San Francisco Bay area. The size of cities (used as an indicator of social pluralism) is strongly related to the choice by councilmen of a representational style ranging from one that is highly responsive to attentive publics (in large cities) to one which is largely self-defined (in small cities; Prewitt and Eulau, 1969). Population density seems to be the single most important economic or social characteristic of cities, for predicting municipal expenditure patterns. In addition, however, decisions on amenities and planning expenditures are systematically related not only to demography, but also to the policy preferences of city councilmen (Eyestone, 1971).

Prewitt, Eulau, and Eyestone have shown an important connection between some socioeconomic characteristics of the urban environment, the style and preferences of elected policy makers, and the pattern of government expenditures. They have not, however, dealt with a related set of questions concerning what might be called "the demand environment" and the making of public policy. What difference—if any—does the quality of interest group *activity* make for the councilman's policy preferences? Does the councilman's concrete relations with local interest groups affect the policy choices he makes on public expenditures?

The answers to these questions have obvious implications for democratic theory. Just as the degree of party competition might be assumed to have some effect on the behavior and decisions of state policy makers, the interest group life in a city may affect the behavior of that city's legislators. If we find that government services in states that are dominated by one party fail to differ significantly from those where party competition is strong, we begin to question the relevance of party politics for differing policy outputs in the states. Similarly, if we find that the demand environment is irrelevant to major urban policy decisions, we ask whether citizen efforts to influence their representatives through organizational activities are important to anyone but the involved citizens themselves. The present study is directed toward a partial answer to this question.

THEORETICAL CONSIDERATIONS AND RESEARCH INDICES

We are concerned with the manner in which environmental conditions such as industrialization and city size give rise to political demands, and how these demands are converted by policy makers into policy outputs. The policy-making process in American cities (and in other political settings) can be described as a relationship among five sets of variables: environmental factors; the political translation process; the political conversion process; policy outputs; and feedback. Our main interest here is in the processes of translation and conversion: the way in which actors in a political setting come to make demands on public policy makers, the choice of activities undertaken by organized interest groups, and the attitudes and behavior of policy makers toward such organized groups.

A description of these sets of variables, together with the indices used in this study, may be helpful.

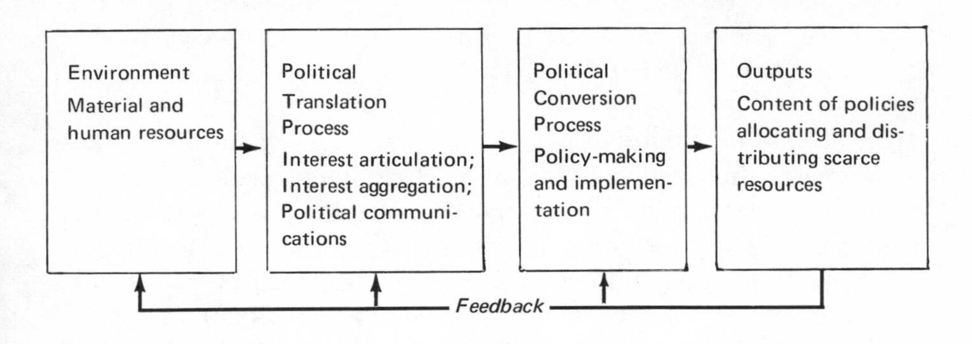

Figure 1

(1) *Environment.* Policy-making takes place within a relatively enduring material and human setting which may be viewed (depending on the research problem at hand) as either self-contained or permeable to inputs from outside that environment.[1] Cities, for example, may be described in terms of climate, geographical location, size, population density, land use, and the like. They may also be discussed in terms of human potential: the educational and occupational skills, the wealth, the heterogeneity of the actors within that environment.

From the range of census indicators at hand, four environmental characteristics were chosen: city size, industrialization, education, and family income. The measures for these variables are the city population (July 1965); the percentage of urbanized land in commercial and industrial use (July 1965); percent of high school graduates among the adult population (1960); and median family income (1960). Size and industrialization should provide a strong indication of the political and social complexity of the cities in question (potential for demand articulation), while the other two indicators serve as measures of socioeconomic status (potential for participation.)

These indicators are in common use, and only a few explanatory comments are needed. Size and industrialization seem appropriate measures of what can be called potential for conflict over the disposition of municipal resources: the larger and more diverse the socioeconomic environment, the more likelihood of argument over the use of public goods. Similarly, high economic and educational status are both *potential* political resources. Education is probably most directly related to the possession of the political skills, knowledge, and sense of efficacy of citizens that precedes the decision to participate in urban affairs.[2]

(2) *The political translation process.* How do actors formulate and communicate their demands and supports to policy makers? How, if at all, are private needs, views, and supports translated into public questions? The vehicles for such a translation are likely to differ in a city from those in a nation or a state. Two obvious differences exist: (1) the relatively greater intimacy (and limited scope) of urban government, in comparison with the sovereign and usually more complex units of state and national government, and (2) the nonpartisan legal structure of most local governments. These features of urban government imply a somewhat fragmented style of demand articulation and political communications. The fact of legal nonpartisanship leads us to look for linkage structures other than the political party, although as noted elsewhere, the phenomenon of "latent" partisanship alerts us to the continuing relevance of party identification and party activities for political campaigns in the local community (Eulau et al., 1965).

Information provided by city councilmen in 81 cities of the San Francisco Bay area was used to develop measures of local interest group activity.[3] Three questions were asked:

(1) We would appreciate your helping us rate the participation of their members in the activities of some organizations in your city. Please check the appropriate box. If no such organization exists in your city, check "not applicable."

The nine categories listed were home owners groups and neighborhood associations; general civic affairs groups such as the League of Women Voters and civic leagues; Chamber of Commerce or Jaycees; reform or protest groups; trade unions; service clubs such as Kiwanis or Rotary; merchants associations; political party clubs or organizations; garden clubs, trail clubs, and library associations. The checklist (part of a self-administered questionnaire) offered a choice between "high participation," "moderate participation," and "low participation."

(2) Do any of the community groups or organizations ever contact you personally to seek your support?

(3) In your last campaign for the council, were there any community groups or organizations which supported you? What kinds of things did they do?

Four different measures of group activity in each community were developed by aggregating the responses of all councilmen in a given city.[4] These measures are:

(a) *Number of active groups:* the mean number of groups rated as to members' participation (regardless of rating) out of the nine possible categories (53% of the council means were 7-9 groups; 47% fell below 7.0 groups).

(b) *Intensity of group members' activity:* average ranking of members' participation, computed by dividing total number of organizations whose members were "very active" or "active" by total that were ranked.

(c) *Political communications of groups to policy makers:* computed by percentage of councilmen who reported group contacts or requests for support.

(d) *Group activity in elections:* computed by percentage of councilmen who reported *formal* group support in their last campaign for office.

(3) *The political conversion process.* Almond and Powell (1968) refer to their six functions (interest articulation, interest aggregation, communications, legislation, administration, and adjudication) as "The *conversion processes* of the political system—the processes which transform the flow of *demands* and *supports* into the political system into a flow of *extraction, regulation, distribution*" (italics in original). The term is used here in a more limited sense. It refers to the orientations and concrete behavior of policy makers active in transforming (or in some cases, ignoring or refusing to transform) political demands and communications into policies.

A complete description of this process would include the orientations and behavior of policy makers toward *all* "significant others": fellow policy makers, the general public, and the large number of specialized organizations which participate in the prior process of political translation. We shall concentrate only on city councilmen's orientations and behavior toward community groups. Such a focus cannot give us a complete picture of the political conversion process, but groups seem an appropriate starting point because of their key role in the communication of community demands and supports.

We use two measures for this purpose: the extent to which councilmen seek group support on matters before the council, and the councilman's interest group role orientation. The full typology of group role orientations has been described in detail elsewhere (Zisk et al., 1965). Of interest here is the Pluralist: a councilman who values group activity, who perceives relatively many influential groups, and who is relatively sophisticated in regard to the bases of group influence. The aggregate measure is simply the percentage of Pluralists on a given council, ranging from the few councils with four out of five Pluralists through the large number with only one Pluralist or none.

Our second measure is based on responses to the question:

> Before a council decision is made do you ever actively seek support from any of the groups you have mentioned?

Once again, the council measure is the percentage of councilmen who report seeking group support.

(4) *Policy outputs* are the result of decisions and nondecisions concerning the allocation of public resources. Resources may be material (e.g., public funds and facilities), symbolic (e.g., public offices of an honorific nature), or, as is frequently the case, they may combine material and symbolic elements (see Edelman, 1964, for a discussion of the symbolic uses of politics). Demands for the distribution and use of public resources usually exceed their availability, hence their relative scarcity. The most important decisions involving local resources are probably those concerned with public expenditures and land use, and those which affect the structure of the policy-making process itself, such as voting requirements and provisions for recall elections.

Three measures of policy outputs, all of which concern public expenditures, shall be used in this study: (1) per capita expenditures, 1966-1967; (2) allocations to "amenities" as a percent of operating expenses, 1966-1967; and (3) planning expenditures as a percent of "general government" expenses, 1966-1967.

Amenities are defined as *optional* city services beyond such basic services as fire and police protection, sewerage and drainage, water, and other public utilities. They include parks and recreation facilities, libraries, and some health programs.

Planning expenditures are funds spent by planning commissions in the 81 cities. Per capita expenditures were computed by dividing total expenditures by total city population.

We would have preferred to supplement these measures with several other indicators for outputs, such as expenditures on education and welfare, or the frequency of amendments to the Master Plan, zoning variances and the like; we are limited, however, to indices which are readily available for all communities. School districts for California, for example, are not contiguous with city boundaries; thus school expenditure figures are not suitable for our purposes. We will argue, however, that while the overall level of municipal expenditures is largely dependent on the total resources of a city (e.g., taxable property), the way in which these expenditures are apportioned among differing services hinges in part on the relation between councilmen and those constituents who are making demands. Thus the share of public funds devoted to amenities spending may be a very important clue to the quality of organized group life in a city, and the relationship between city councilmen and such groups.

A summary statement of the theoretical linkage concepts discussed above, as well as the indicators used in this study, is given in Figure 2.

HYPOTHESES AND RESULTS

RELATION BETWEEN ENVIRONMENTAL FACTORS AND THE POLITICAL TRANSLATION PROCESS:

Large cities and highly industrial cities have a greater potential for citizen competition over the allocation of scarce resources than do small or primarily residential communities. Size is probably the more important of the two characteristics: it implies a social, racial, religious, and economic heterogeneity which is likely to give rise to an active and varied organizational life and a high volume of political communications from such groups (Prewitt and Eulau, 1969). Groups are also likely to perceive a high stake in the electoral process

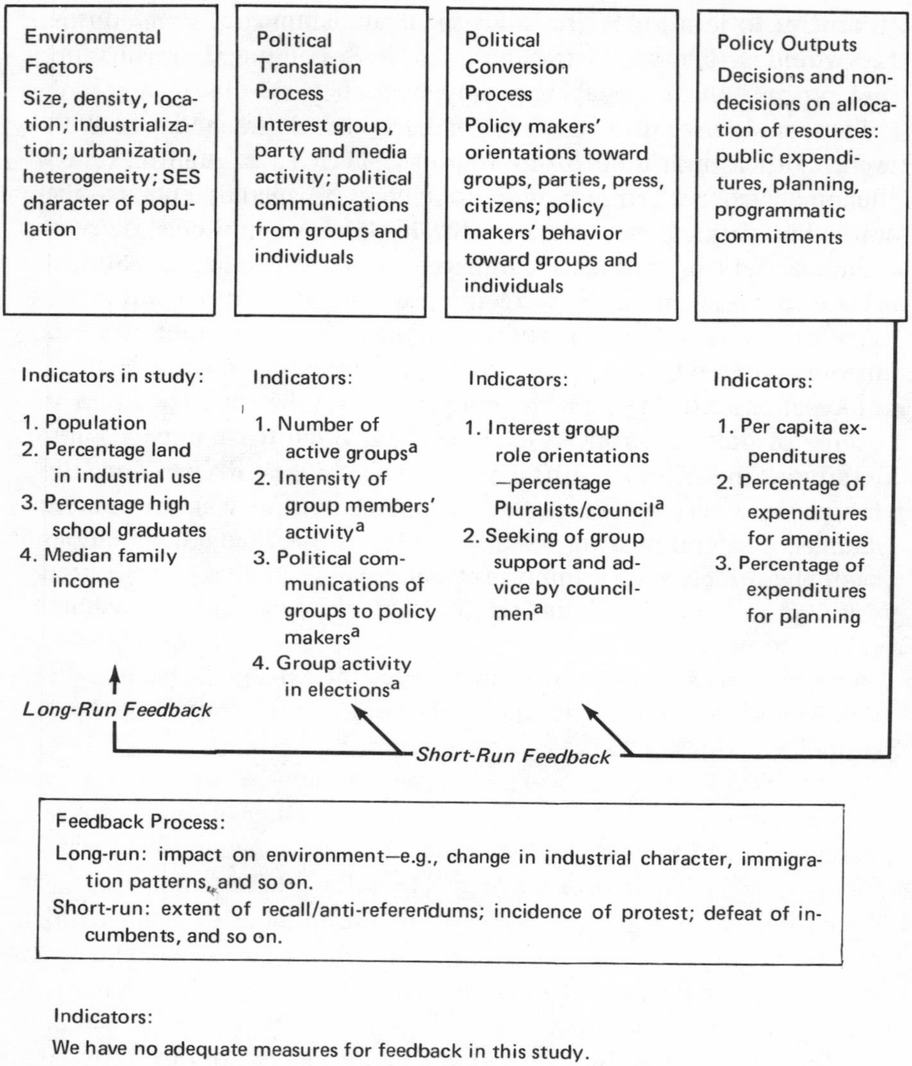

Environmental Factors	Political Translation Process	Political Conversion Process	Policy Outputs
Size, density, location; industrialization; urbanization, heterogeneity; SES character of population	Interest group, party and media activity; political communications from groups and individuals	Policy makers' orientations toward groups, parties, press, citizens; policy makers' behavior toward groups and individuals	Decisions and nondecisions on allocation of resources: public expenditures, planning, programmatic commitments
Indicators in study:	Indicators:	Indicators:	Indicators:
1. Population 2. Percentage land in industrial use 3. Percentage high school graduates 4. Median family income	1. Number of active groups[a] 2. Intensity of group members' activity[a] 3. Political communications of groups to policy makers[a] 4. Group activity in elections[a]	1. Interest group role orientations —percentage Pluralists/council[a] 2. Seeking of group support and advice by councilmen[a]	1. Per capita expenditures 2. Percentage of expenditures for amenities 3. Percentage of expenditures for planning

Long-Run Feedback

Short-Run Feedback

Feedback Process:

Long-run: impact on environment—e.g., change in industrial character, immigration patterns, and so on.

Short-run: extent of recall/anti-referendums; incidence of protest; defeat of incumbents, and so on.

Indicators:

We have no adequate measures for feedback in this study.

a. These data are derived from interviews with city councilmen in the 81 cities studied. All other data are compiled from census documents.

Figure 2: THE URBAN POLITICAL PROCESS: A SCHEMATIC PRESENTATION WITH BRIEF DESCRIPTION OF INDICATORS USED IN THIS STUDY

in large and heterogeneous communities, and thus become active in council elections. Indeed, the individual citizen may have a greater incentive to join a politically active organization in a large city than he would in a small community, simply because such organizations are probably taken more seriously by policy makers in a complex urban environment. In an environment where the policy maker is subject to a multitude of conflicting and often mutually exclusive demands from a variety of claimants, *group* claims may be given preferential treatment over *individual* claims simply because of the communications overload endured by the councilman. Politically active groups, in such a setting, may play a role in interest aggregation as well as interest articulation, from the point of view of the local policy maker.

An educated and wealthy population probably possesses a greater degree of political skill and efficacy—i.e., potential for participation in group (and other) political activities—than does a poorly educated or low-income population. We would therefore anticipate, *ceteris paribus,* a strong relation between the population characteristics and both the intensity of group members' participation in organizational activity and group utilization of local opportunities for political communication. We do not see any logical connection, however, between socioeconomic status (and educational skills) and the sheer number of groups that are active in politics or group participation in council elections.

The information in Table 1 provides strong support for our expectations concerning city size and industrialization. There is a strong correlation between these characteristics of the urban environment and all four measures of group activity. As anticipated, size appears to be a major determinant of the number of groups which are politically active. There also appears to be an unexpectedly strong relation between size and industrialization and the intensity of members' participation in organizational activities. One possible explanation for this finding is that a high stake in political outcomes in large, complex communities induces a heightened membership interest in group activity, just as it leads the organization itself to participate in local political campaigns.

There is very little relationship between income, education, and the political translation process. In the two cases where we expected to find strong links, only one—the tie between socioeconomic indicators and group communications—appeared, and even here the relationship is modest. In short, the potential for competition

TABLE 1
RELATION BETWEEN ENVIRONMENTAL FACTORS AND THE POLITICAL TRANSLATION PROCESS

Variable	Variable	Correlation (Kendall's Tau B)	Significance
Population	Number of active groups	.52	.001
	Intensity of group members' activity	.27	.001
	Political communications of groups to policy makers	.27	.001
	Group activity in elections	.19	.01
Percentage land in industrial use	Number of active groups	.20	.01
	Intensity of group members' activity	.28	.001
	Political communications of groups to policy makers	.18	.05
	Group activity in elections	.25	.01
Percentage high school graduates	Number of active groups	.03	(NS)
	Intensity of group members' activity	−.02	(NS)
	Political communications of groups to policy makers	.17	.05
	Group activity in elections	.04	(NS)
Median family income	Number of active groups	.05	(NS)
	Intensity of group members' activity	−.08	(NS)
	Political communications of groups to policy makers	.19	.01
	Group activity in elections	.05	(NS)
Internal Consistency of Environmental Factors			
Relation between			
Population and land in industrial use		.36	.001
Population and percentage high school graduates		.09	(NS)
Population and median family income		.09	(NS)
Land in industrial use and percentage high school graduates		−.23	.01
Land in industrial use and median family income		−.06	(NS)
Percentage high school graduates and median family income		.72	.001

(implied by size and industrialization) appears to be markedly more important for the group life of these communities than is the potential for participation (implied by educational and income levels of the population). Income may have other important effects on the local political process but it does not appear to be significant for organizational activity, except in the case of political communication.[5]

Because of the very high correlation between median income and education levels, and the strong relation between city size and industrialization in these cities (see bottom of Table 1), partial correlations between the environmental and translation variables were computed, controlling for the remaining environmental factors. The relationship between city size and the translation variables remained strong throughout. Industrialization had almost no independent effect (apart from size) on the number of active groups and on the political communication process, but remained important for both the intensity measure and the measure for group activity in elections. Median family income is markedly more important than our education measure, in relation to the political communication process: the relationship between political communications and level of education all but disappears when income level is controlled. The full data on partial correlations for the variables in Table 1 and subsequent tables are given in the Appendix.

To summarize: city size (and, to a lesser degree, industrialization) are positively related to four aspects of the group life in these 81 cities. The income level of the population, furthermore, is a good predictor for one of those aspects—political communications—of group activity. The potential for competition, and to some extent, the potential for participation, as measured by three commonly used environmental factors, are related to the political translation process.[6] We have found, in short, that important links do exist between the first two stages in our partial model of the political process.

RELATION BETWEEN THE POLITICAL TRANSLATION PROCESS AND THE POLITICAL CONVERSION PROCESS:

We have been warned, again and again, not to assume that policy makers' orientations and behavior will invariably mirror the human environment in which they operate (see Prewitt and Eulau, 1969; Pitkin, 1967). To assume, for example, that all elected representatives in a business-dominant community will share and act upon the values of the business community is to accept a kind of of "representational determinism" which is simply not supported by empirical evidence. Indeed, this lack of a perfect match between values of the representative and those he represents has been a common complaint of minority groups which are affronted at their seeming impotence.

At the same time, we suspect that the existence or lack of a strong and diverse set of politically active organizations in a community will have some effect on the behavior of policy-making bodies. We do not expect a one-to-one relation—i.e., that all councilmen will be highly responsive in a city where local groups are well organized and active on political issues, or that no councilmen will heed group requests in a community where organizational activity is rare, sporadic, and poorly expressed. Nor do we expect councils to be pro-labor in a working-class community and pro-business in a high-SES community. We do, however, anticipate that policy makers will be more responsive to group requests and more likely to form alliances with local organizations in communities where groups are highly organized than will be the case in those cities where groups are relatively quiescent about politics.

We thus anticipate a high correlation between the political translation process and the conversion process: there will be more Pluralists and more councilmen who seek group help in cities where there are a large number of groups; many groups with intensely involved members; and where groups are active in both the political

TABLE 2
RELATION BETWEEN POLITICAL TRANSLATION PROCESS AND POLITICAL CONVERSION PROCESS

Variable	Variable	Correlation (Kendall's Tau B)	Significance
Number of active groups	Extent of Pluralist role-orientation	.16	.05
	Seeking of group support by council	.25	.001
Intensity of group members' activity activity	Extent of Pluralist role-orientation	.21	.01
	Seeking of group support by council	.18	.01
Political communications of groups to policy makers	Extent of Pluralist role-orientation	.31	.001
	Seeking of group support by council	.23	.001
Group activity in elections	Extent of Pluralist role-orientation	.30	.001
	Seeking of group support by council	.25	.001
Internal Consistency of Variables in Translation Process			
Relation between			
Number of active groups and intensity of group members' activity		.18	.01
Number of active groups and political communications to policy makers		.24	.001
Number of active groups and group activities in elections		.15	.05
Intensity of group members activity and political communications		.12	(NS)
Intensity of group members activity and group activities in elections		.36	.001
Political communications to policy makers and group activities in elections		.22	.01
Internal Consistency of Variables in Conversion Process			
Relation between extent of Pluralist role-orientation and seeking of group support by council		.37	.001

communications process and local political campaigns than there will be in cities where groups are less active.

Table 2 confirms these expectations. All of the relationships expected are statistically significant, and with one exception (the correlation between the number of active groups and the extent of the Pluralist orientation), they are all strong. Councilmen in cities where groups are politically active are more likely to approve of group activity and to call upon groups for help on their own projects.

Partial correlations were computed (controlling for population) to test the possibility that a council's favorable stance toward groups might be attributable to the social complexity of the city rather than to the character of group life in that city. The relationship between the conversion and translation measures was weakened slightly as a result of this control, but remained significant in all but one case. This exception (see Appendix) is the correlation between number of active groups and the extent of the Pluralist orientation: when population is controlled, the figure drops from .16 to .05. It is thus clear that the sheer number of organized groups has little independent effect on councilmen's outlooks on group activity. What a group does carries far more weight than the simple fact of its existence.

Indeed, one of the most interesting aspects of Table 2 is the contrast between those variables which measure group *activity* and those which measure *potential* influence in relation to the political conversion process. Campaign activity and group communications are more effective, per se, than sheer numbers or even membership commitment. This makes considerable sense if we believe that the attitudes and activities of policy makers are in part a result of group actions rather than the reverse.

RELATION BETWEEN THE POLITICAL CONVERSION PROCESS AND POLICY OUTPUTS:

Three general hypotheses concerning the relationship between the conversion process and policy outputs are proposed:

(1) Levels of expenditure will probably vary with council orientations and behavior toward local groups, but depend primarily on the objective needs of a city and council orientations toward spending.

(2) Spending on amenities should rise with positive orientations toward groups, regardless of other environmental or organizational characteristics of the community.

(3) Planning expenditures have no direct relation to group activities: these expenditures depend primarily on a city's developmental stage.

These expectations stem from our interpretation of the literature on political development in American communities, and more specifically, from Eyestone's work (1971). General expenditures depend on a variety of factors, of which the ability of citizens to pay for new services and the objective needs of a city, at a given point in time, are probably the most important. Thus for a core city, most of whose high-status citizens have long since left for the suburbs, we expect high demands for expenditures coupled with a low resource base. In contrast, for a fringe community which has not yet undergone the pangs of rapid growth and even more traumatic demands for new services, there is a moderate resource base and low demands. An intermediate case might be a developing suburb which is experiencing rapid industrial and residential growth. This sort of community can meet expanding needs for new services with a potentially fecund source of taxation, in terms of both residential wealth and taxable industrial property.

Per capita expenditure levels, in short, do not depend primarily on political demands. They depend on what Eyestone calls "resource capabilities," or the relationship between objective resources and the willingness of policy makers to spend available revenues, within the context of the developmental stage of the community in question. The willingness of policy makers to spend community resources may depend, in turn, on their stance toward community organizations, but we cannot assume this is the case, in the face of little prior empirical work on the subject. The political translation process—at least in terms of group activity—is probably not relevant to the final level of expenditures for a given community, when we take into account other factors.

The same line of reasoning applies to planning expenditures: here the objective facts for each community are likely to be much more important for the level of spending than the amount of pressure which organized groups can bring to bear on policy makers. The group life in a city probably has little impact on the level of spending for planning; furthermore, the direction in which group demands are made is likely to depend on which groups are most active in a given city. In some communities pro-planning voices such as neighborhood associations may be loudest; in others, anti-planners such as land speculators may have the upper hand.

It is only in the case of amenities spending that we anticipate a close link between the political translation process and policy outputs. We offer two reasons for this expectation. First, most of the organized groups in these cities are likely to be advocates of some amenities spending: the most frequently named influential groups are chambers of commerce, neighborhood associations, and service clubs, most of which favor beautification and improvement projects such as parks, recreation facilities, and civic centers. There are, to be sure, a fair number of taxpayers' associations and various ad hoc groups dedicated to a reduced level of spending, but these are far outnumbered by the organizations named above. In addition, there are a large number of special-purpose groups such as youth organizations, riding associations, golden age clubs, and art, library, and music associations which are likely to request special facilities such as the construction of buildings, bridle or bicycle paths, or the purchase and setting aside of conservation land. Their claims, of course, are sometimes mutually exclusive, and they will need to compete, as well, with spokesmen for increased basic services such as health, fire and police protection. But a second reason for expecting a close relation between the degree to which councils favor and depend on local groups and the degree to which they are willing to spend for amenities hinges on the councilman's freedom of movement in this area, in contrast to the basic services area. There is relatively little argument, except in some fringe cities, about the need for sewers or well-equipped fire departments; the need for large basic service expenditures is probably the subject of a general consensus in most cities. The councilman who wishes to meet the requests and demands of his constituents, insofar as possible, probably has most freedom of choice in these optional spending areas dealing with public amenities. The push for increased expenditures—and the group struggle on the subject—are thus more likely to center on optional "frills" than on basic services.[7]

At first glance, Table 3 supports all three hypotheses. There is no relationship between level of general per capita expenditures and either translation variable; for planning expenditures, the relation is negative. Amenities spending, by contrast, is strongly related to council-orientations and behavior toward community groups. It will be recalled, however, that in the large number of studies where

TABLE 3
RELATION BETWEEN POLITICAL CONVERSION PROCESS AND POLICY OUTPUTS

Variable	Variable	Correlation (Kendall's Tau B)	Significance
Extent of Pluralist role-orientation	Per capita expenditures	.07	(NS)
	Amenities as percentage of expenditures	.19	.01
	Planning as percentage of general operating expenses	–.13	.05
Seeking of group support by council	Per capita expenditures	.10	(NS)
	Amenities as percentage of expenditures	.31	.001
	Planning as percentage of general operating expenses	–.23	.001
Alternative Explanation I: Relation Between Translation Process and Outcomes			
Number of active groups	Expenditures	.06	(NS)
	Amenities	.36	.001
	Planning	.11	(NS)
Intensity of group members activity	Expenditures	.21	.01
	Amenities	.26	.01
	Planning	–.03	(NS)
Alternative Explanation II: Relation Between Environment and Outputs			
Population	Expenditures	.03	(NS)
	Amenities	.50	.001
	Planning	.21	.01
Percentage land in industrial use	Expenditures	.29	.001
	Amenities	.33	.001
	Planning	–.01	(NS)
Percentage high school graduates	Expenditures	–.20	.01
	Amenities	.33	.001
	Planning	–.01	(NS)
Median family income	Expenditures	–.12	(NS)
	Amenities	.08	(NS)
	Planning	.13	.05

political variables were found correlated with various measures of policy outputs, such a relation disappeared when environmental factors were controlled (Dawson and Robinson, 1963; Dye, 1968; Hofferbert, 1968, 1966; Sharkansky, 1968). Examination of the third part of the table (the relation between environment and outputs) indicates that this might be the case in the present study. Partial correlations for the crucial variables are as follows:

Correlation	*Control*	*Partial Correlation*
Extent of Pluralist Role-Orientation/Amenities	Population	.09
Extent of Pluralist Role-Orientation/Amenities	Industrialization	.15
Seeking of Group Support by Council/Amenities	Population	.23
Seeking of Group Support by Council/Amenities	Industrialization	.26

Note the contrast between the explanatory strength of the two variables (Pluralist role-orientation and seeking of group support) when population and industrialization are controlled. Clearly council *orientations* toward groups have little independent effect on amenities spending, but council *behavior* toward such groups remains important, regardless of the size or industrial character of the city concerned. Once again a measure of potential lacks the explanatory power of a measure of behavior. The political translation process is important for at least one type of policy output—amenities expenditures—but chiefly when words (or attitudes) are matched by deeds.

CONCLUSION

Our study is one of the few in which concrete evidence for the relevance of *political* activity to policy outputs has been demonstrated. We have found, as have others before us, that policy outcomes are heavily circumscribed by the environment in which political decisions are made. In spite of this stricture, however, the political conversion and translation processes have an independent effect on some public policies. The complex linkage described above can be restated in simple terms. Interest groups and city councils interact within widely different socioeconomic environments, and the way and degree to which they interact varies with these environments. In large and complex cities, groups take a more active role in the political process itself. Where groups are more active, councils are more responsive. Finally, in cities where interaction between councils and groups is high (and friendly), public spending for amenities is also likely to be high. A large part of the contrast between high-amenities-spending cities and low-amenities-spending cities can be explained by the differences in the urban environment: policy makers spend more for amenities in large, industrial cities than in small residential communities. But even after such environmental factors are taken into account, the impact of organized group activity and, more important, the relations between group spokesmen and official policy makers is important for explaining outcomes in at least one policy area, the level of amenities expenditures.

We suspect that the set of political conversion and translation variables relevant to policy outcomes is markedly larger than that explored by those who have concentrated on malapportionment and party competition. If adequate measures could be developed, for example, for general citizen participation and interest in urban affairs (beyond the voting act itself), or for the quality and adequacy of press coverage of local politics, perhaps further explanation of policy outcomes via political variables would be possible. Similarly, much more work must be done to develop measures of political outcomes which go beyond the rank ordering of expenditure levels. The *impact* of expenditures on different socioeconomic groupings is an obvious candidate for such an agenda; the speed and efficiency with which policy decisions are made and implemented is another.

The finding that the political activities of interest groups does make a difference, not only for the attitudes and activities of local policy makers but for the content of public policy decisions, is an important addition to contemporary theorizing about democracy. Perhaps the local group is the major functional equivalent, in the local nonpartisan setting, to the political party or to (hidden) "power elites." Without comparable studies of the impact of the unorganized citizenry or the local press, we cannot, of course, answer this question. Obviously, all political actors—whether group spokesmen, media leaders, unorganized citizens, or elected representatives—continue to operate in an arena limited by their environment. The basic question concerns their freedom of movement within that set of constraints.

One striking conclusion which emerges from this analysis concerns the difference between what we have called the political potential and the political behavior of both citizens and policy makers. In two crucial instances, we have found that potential (words, attitudes, a paper organization) carries very little weight in comparison with concrete activity (group participation in elections; council reliance on group aid in selling their ideas to the community). The mere number of organized groups is not as important as their participation in the political articulation and communication processes; the favorable orientation of councilmen toward local groups explains very little of their decision-making behavior, in comparison with the degree to which they make allies of those groups. The lesson is clear for both scholars and activists: the scholar must move from a study of political attitudes and orientations to a study of political behavior, if he wishes to understand the policy-making process, and the local activist must match his words on behalf of his causes with concrete deeds, if he is to be effective.

NOTES

1. For purposes of the present study we have ignored exogenous factors. We believe this is justified for the problems chosen.

2. We have found, as have others, a strong relationship between the two pairs of variables: .36 for population and land in industrial use, and .72 for high school graduates and median family income (see Table 1).

3. This information, obtained in connection with the Stanford City Council Research Project, is limited to 81 out of the total of 89 cities in the six counties of the San Francisco Bay area because of response failure on the questions dealing with interest group activity. All cities where less than a majority of councilmen were interviewed in regard to interest groups were excluded from this analysis.

4. See Prewitt and Eulau (1969) for a justification of aggregating councilmen's responses as measures of whole-city tendencies. We believe that this aggregation can be justified on yet another ground: the effect of several councilmen who are responsive to groups is likely to extend to the whole council, in terms of the degree to which group claims are considered, or groups are brought into the decision-making process. Councilmen do not act as isolated individuals, and the fact that they act with regard to groups or others is likely to spill over into their colleagues' behavior.

5. It should also be noted that size and industrialization are genuine properties of the cities themselves, whereas educational and income measures are aggregated attributes of the citizens of these cities.

6. It should be recalled again, however, that our measures for the political translation process are the existence of groups, and the activities of groups, as perceived by city councilmen.

7. This reasoning may apply only at a time of general economic prosperity. It will be recalled that our study was undertaken in the mid-1960s, in an area undergoing rapid growth; a replication in less prosperous times or places might yield a high correlation between group activities and lower levels of spending on these "optional" amenities. In such a time and place we might also anticipate a different constellation of active groups–i.e., anti-spending interests may be both more vocal and more influential.

APPENDIX
PARTIAL CORRELATIONS FOR VARIABLES SHOWN IN TABLES 1-3[a]

Variable A	Variable B	Correlation	Control Variable	Partial Correlation
Population	Number of active groups	.52	Percentage land in industrial use	.49
	Intensity of group members activity	.27	Percentage land in industrial use	.19
	Political communications of groups	.27	Percentage land in industrial use	.22
			Percentage high school grads.	.26
			Median family income	.26
	Group election activity	.19	Percentage land in industrial use	.11
			Percentage high school graduates	.19
			Median family income	.19
Percentage land in industrial use	Number of active groups	.20	Population	
			Population	.02
	Intensity of group members activity	.28	Population	.20
			Median family income	.28
	Political communications of groups	.18	Population	.09
	Group election activity	.25	Population	.20
Percentage high school graduates	Number of active groups	.03	Population	–.02
	Intensity of group members activity	–.02	Population	–.05
	Political communications of groups	.17	Population	.15
			Median family income	.05
	Group election activity	.04	Population	.02
Median family income	Number of active groups	.05	Population	.00
	Intensity of group members activity	–.08	Population	–.11
	Political communications of groups	.19	Population	.17
			Percentage high school graduates	.10
	Group election activity	.05	Population	.03
Number of active groups	Extent of Pluralist role-orientation	.16	Population	.05
	Seeking of group support by council	.25	Population	.17
Intensity of group members activity	Extent of Pluralist role-orientation	.21	Population	.16
	Seeking of group support by council	.18	Population	.13

APPENDIX (Continued)

Variable A	Variable B	Correlation	Control Variable	Partial Correlation
Political communications of groups	Extent of Pluralist role-orientation	.31	Population	.27
	Seeking of group support by council	.23	Population	.18
Group activity in elections	Extent of Pluralist role-orientation	.30	Population	.27
	Seeking of group support by council	.25	Population	.22
Extent of Pluralist role-orientation	Amenities as percentage of expenditures	.19	Population	.09
			Percentage land in industrial use	.15
			Number of active groups	.14
			Intensity of group activity	.14
Seeking of group support by council	Amenities	.31	Population	.23
			Percentage land in industrial use	.26
			Number of active groups	.24
			Intensity of group members activity	.28
Number of active groups	Amenities	.36	Population	.15
			Extent of Pluralist orientation	.34
			Seeking of group support by council	.31
Intensity of group members activity	Amenities	.26	Population	.22
			Extent of Pluralist orientation	.23
			Seeking of group support by council	.22
Population	Amenities	.50	Number of active groups	.39
			Intensity of group members activity	.46
			Extent of Pluralist orientation	.48
			Seeking of group support by council	.46

a. Our selection of "control variables" was guided by the strength of the relationship between those variables and variables A and B in the table; we have not presented all possible controls, since many are clearly not important. For example, we have controlled for population in almost every instance, because population is strongly related to most variables; we have not controlled for median family income, or percentage of high school graduates, in contrast, because the initial relations were unimpressive for these variables.

REFERENCES

ALMOND, G. A. and G. B. POWELL (1968) Comparative Politics: A Developmental Approach. Boston: Little, Brown.

ALMOND, G. A. and S. VERBA (1963) The Civic Culture: Political Attitudes and Democracy in Five Nations. Princeton: Princeton Univ. Press.

BABCHUCK, N. and A. BOOTH (1969) "Voluntary association membership: a longitudinal analysis." Amer. Soc. Rev. 34: 31-45.

DAWSON, R. E. and J. A. ROBINSON (1963) "Interparty competition, economic variables and welfare policies in the American states." J. of Politics 25: 265-289.

DYE, T. (1968) Politics, Economics and the Public: Policy Outcomes in the American States. Chicago: Rand McNally.

EDELMAN, M. (1964) The Symbolic Uses of Politics. Urbana: Univ. of Illinois.

EULAU, H., B. H. ZISK, and K. PREWITT (1965) "Latent partisanship in nonpartisan elections: effects of political milieu and mobilization," in H. Zeigler and M. K. Jennings (eds.) The Electoral Process. Englewood Cliffs: Prentice-Hall.

EYESTONE, R. (1971) The Threads of Public Policy: A Study in Policy Leadership. Indianapolis: Bobbs-Merrill.

--- (1968) "Socioeconomic dimensions of the American states: 1890-1960." Midwest J. of Politics 12: 401-418.

--- (1966) "The relation between public policy and some structural and environmental variables in the American states." Amer. Pol. Sci. Rev. 60: 73-78.

HOFFERBERT, R. (1970) "Elite influence in state policy formation: a model for comparative inquiry." Polity 2: 316-344.

JACOB, H. and M. LIPSKY (1968) "Outputs, structure and process: an assessment of changes in the study of state and local politics." J. of Politics 30: 510-533.

PITKIN, H. F. (1967) The Concept of Representation. Berkeley: Univ. of California Press.

PREWITT, K. and H. EULAU (1969) "Political matrix and political representation." Amer. Pol. Sci. Rev. 63: 427-441.

SCHAEFER, G. F. and S. H. RAKOFF (1970) "Politics, policy and political science: theoretical alternatives." Politics and Society 1: 51-77.

SHARKANSKY, I. (1968) Spending in the American States. Chicago: Rand McNally.

WRIGHT, C. R. and H. H. HYMAN (1958) "Voluntary association membership of American adults: evidence from national survey samples." Amer. Soc. Rev. 23: 284-294.

ZISK, B. H., H. EULAU and K. PREWITT (1965) "City councilmen and the group struggle: a typology of role orientations." J. of Politics 27: 618-646.

10

Organized Spokesmen for Cities: Urban Interest Groups

WILLIAM P. BROWNE
ROBERT H. SALISBURY

□ A FAMILIAR LEITMOTIV OF AMERICAN political analysis develops the key words "pressure group," "lobbyist," "access," and "special interest" into a fabric of explanation of the policy-making process. One may choose the benign views of a Herring (1940) or a Truman (1951), or prefer the critical posture of a Schattschneider (1961) or a McConnell (1966), but one can hardly escape giving central attention to the organized groups at work to influence public policy outcomes. Although disputes about concepts and methodology have not infrequently substituted for the empirical treatment of group activities, there exists a sizable body of literature treating various aspects of interest group activity concerning major sectors of policy. However, despite the recent emergence of concentrated academic attention on matters involving intergovernmental relations between cities and states as well as cities and the federal government, surprisingly little attention has been given to the associations of municipal officials that often serve as intermediaries in the negotiations and discussions that precede the programs that develop. It is the aim of this essay to add to the existing literature on interest groups by discussing some of these major organizations of municipal officials, and at the same time to consider further some generic questions of interest group theory.

Interest groups have generally been regarded as a topic of pertinent concern to political scientists because of their efforts to affect policy outcomes; in short, as lobbyists. Moreover, the traditional theoretical formulation of the origins and growth of interest groups stresses their importance as mechanisms for the expression of discontent. Groups are seen mainly as instruments for affecting policy, making claims upon other groups through the medium of favorable governmental decisions. This analytic emphasis has generally been accompanied by the assumption that although organized groups might be subject to the operation of Michels' Iron Law of Oligarchy, this made no great difference to the group's lobbying efforts since in basic values the members of the group could be assumed to be pretty much in agreement.

Recently, this orientation toward interest groups has been called into question (Olson, 1965; Salisbury, 1969). It has been shown that organized groups are often not, in fact, cohesive on value and policy questions. It has been shown further that lobbying is often not the main incentive for joining such groups nor the central activity that sustains them. Organized associations are undoubtedly units of importance in American political life, but the nature of that importance needs further amplification if we are adequately to understand how groups operate.

In the urban field the group universe is large and diverse, and considering the conventional wisdom about the importance of groups, it makes it all the more bewildering that there is not more literature examining municipal interest groups. Standard works on American interest groups do not mention the National League of Cities or the National Municipal League, and urban politics texts make, at most, only passing reference to these organizations.

The discovery by professional political scientists of the organizations composed of municipal officials is quite recent and their treatment is largely restricted to the policy lobbying focus referred to earlier. Suzanne Farkas' recent and welcome analysis of the U.S. Conference of Mayors (1971) gives little attention to the internal life of that organization, nor did Connery and Leach (1960) in their brief discussion of the former American Municipal Association and USCM. Clement Vose (1966) has described some of the specialized "lobbying" of such groups as the National Institute of Municipal Law Officers, and there have been several accounts of the rather differently motivated National Municipal League (for example, Willoughby, 1969).

None of these studies adequately recognizes two quite central facts about the contemporary processes of urban political life. First, there has developed an elaborate and variegated network of formal associations of city officials which far exceeds in scope and importance anything Louis Brownlow (1955) may once have envisioned for his "1313" group of organizations. These early organizations of government officials were originally drawn together to provide a variety of technical services. But today a large share of the nation's cities are included in these organizations and officials of these cities depend upon them both for lobbying representation in Washington on behalf of urban interests and for a variety of services and benefits of direct utility to their respective communities. At the top, this network has been integrated organizationally by means of close cooperation. The largest, best staffed and most lobby-oriented of these groups, the National League of Cities and the U.S. Conference of Mayors, have largely combined their staffs and recently formed a third organization, NLC-USCM, Inc. Other organizations such as the National Association of Housing and Redevelopment Officials and the International City Management Association also work closely with the merged staffs so that despite the existence of many separate organizations of city officials there is an extraordinary degree of coordination of lobbying policy and a closeness of communication linkages among them. NLC-USCM has gone so far as to establish a State, County, City Service Center to coordinate not only the municipal associations but other interest groups as well, including the National Association of Counties, National Governors Conference, and Council of State Governments.

The second fact of importance about these organizations concerns their impact, not on public policy, but on those who belong. Perhaps the most important effect of the growth of urban interest groups is the cultivation and encouragement by the group leadership of professionalism in their jobs on the part of members, whether they be mayors or other kinds of public officials. The delineation of official role expectations has been fostered as the organizations work with members and bring them into contact with one another. Mayors, city planners, municipal law officers and other groups of officials have achieved self-conscious identities through participation in their respective organizations. Moreover, through training programs and information services, the organizations have enhanced a level of performance in the roles which could reach closer to the expectations.

Part of our discussion of urban group development will be put in historical terms so that the emergence of different types of groups and different activities and relationships can be related to the historical context in which they first appeared. One implication of this approach should be acknowledged here. It is this. The groups we will consider have not been notable for changing their activities or adapting their programs and goals as circumstances altered. Rather, as organizations they have tended to maintain traditional roles and identities. When new conditions or needs arose, new organizations have been formed, often by individuals who belonged to and remained in the old ones. Functional differentiation is thus fostered among organizations rather than adaptation by one organization to new tasks.

Our examination will focus mainly on the "peak" association of urban officials, the National League of Cities and the U.S. Conference of Mayors. These are the largest and most prominent of the many urban interest groups and they have the greatest and most far-reaching impact on the formation of national urban policies.

TABLE 1
SOME URBAN INTEREST GROUPS

Associations of Units of Government: American Association of Port Authorities, Council of State Governments, National Association of Counties, National Governors Conference, National League of Cities, National School Boards Association, National Service to Regional Councils, U.S. Conference of Mayors

Professional Based Associations: Airport Operators Council International, American Association of School Administrators, American Association of State Highway Officials, American Institute of Planners, American Public Power Association, American Public Welfare Association, American Public Works Association, American Society of Planning Officials, American Society for Public Administration, American Transit Association, American Water Works Association, Inc., Building Officials and Code Administrators International, Inc., Institute of Traffic Engineers, International Association of Auditorium Managers, Inc., International Association of Assessing Officials, International Association of Chiefs of Police, Inc., International Association of Fire Chiefs, International Bridge, Tunnel, and Turnpike Association, Inc., International City Management Association, International Institute of Municipal Clerks, Municipal Finance Offices Association, National Association for Community Development, National Association of Housing and Redevelopment Officials, National Association of State Mental Health Program Directors, National Association of Tax Administrators, National Institute of Governmental Purchasing, Inc., National Institute of Municipal Law Officers, National Recreation and Park Association, Public Personnel Association, Water Pollution Control Federation.

Independent Associations: National Municipal League, Public Administration Service, Urban America, Urban Coalition, Urban Institute.

Much of our discussion is based on extensive interviews with officials and members of these organizations and on intensive observation of a state municipal league and a local county league of municipalities. Quotations not otherwise attributed are drawn from personal interviews.

SOCIAL MOVEMENT VERSUS ROLE CULTIVATION

We will not attempt a treatment of the organized municipal reform movement, but we should take notice of the important distinction between the expressive values of reform which led to the formation of the National Municipal League and the professional improvement objectives of the contemporaneous League of American Municipalities and most of the successor groups of city officials. At the beginning, in the late 1890s, the latter were anxious to avoid the political stigma of reformism. Reform appealed variously to the business community or the anti-vice crusaders or some other section of "the better elements" of the community, but it was not an appeal that moved the hearts of most urban officials themselves. They were more often the targets of reformers, not the protagonists. If anything, they wanted technical advice and assistance from a national organization along with a forum for mutual discussion. The LAM sought to provide precisely this kind of instrumental benefit and so did the state leagues of municipalities which began to be formed after 1898.

The LAM and the NML were never hostile to one another. Their programs had a good deal in common, and both organizations were surely manifestations of the "Progressive Ethos" that dominated urban middle-class conceptions of America during the period (Hofstadter, 1955). Their memberships were based on different principles, however, and this distinction is crucial. The NML sought the support of anyone interested in the cause of municipal reform. The LAM was designed to serve the interests of municipal officials, and it was they, or rather the municipality as such, which constituted the unit of membership. As an early state municipal league director who supported LAM put it, "We are not reform groups! We are like a service organization for municipal officials." Essentially it is this concept of membership which has been utilized in the formation of the major organizations active today. Either the municipality itself as a legal entity or the holder of official municipal office qua offices are the members. Sympathizers, former office holders, or other supporters of the policy goals of the organizations have no place as dues-paying members.

Eventually the League of American Municipalities failed as an organization because its staff and leadership were unable to provide enough services that officials found truly useful. The reforming NML concentrated its efforts on "model" governmental *structures* and sought to convert "the people" to their ideal formulations. The LAM went beyond such generalized objectives and developed a series of model *programs* in such areas as finance or licensing which LAM officials would attempt to adapt to the special circumstances of a given community in an effort to increase the efficiency of its governmental performance. In addition, LAM collected data and served as a clearinghouse of information concerning municipal practices, and to further these objectives, LAM encouraged the formation of state leagues of municipalities and specialized associations of officials. However, as the latter began to provide organizational meeting grounds and proved more responsive to the felt needs of the members, LAM had little to offer except statistics and model reform programs. As John Stutz of the League of Kansas Municipalities observed, LAM was an organization "looking for something to do" (National League of Cities, 1964). On the other hand the state leagues were successful because they "offered municipal officials something, something that they felt they had a need for and could apply [to their own duties]." The organization went out of business in 1912, its staff and its functions taken over by organizations it had helped to create, but the example of a national organization of cities remained.

THE PROCESS OF PROFESSIONALIZATION

In a sense, what both the NML and the LAM sought to do was to improve cities. The state leagues of municipalities, which achieved a more enduring organizational identity in the early years of the twentieth century, concentrated on raising the competence of public *officials*. The goal was professionalism. Its criteria of accomplishment were largely those of job efficiency and in this spirit the league efforts joined the forces of Taylorism that were making comparable inroads in facilitating the role differentiation of school superintendents among others (Callahan, 1962). The idea of a professionalism produced by efficient performance had important implications for the role of the leagues also. By assisting officials–holding conferences, establishing short courses, disseminating information, and giving on-the-spot advice–a league could professionalize the man whose only actual training was on the job. Experience and example were taken as the sources of wisdom and professional officials were those who were wisely efficient.

The leagues thus appealed to those who were in fact the potential members, city officials without formal training for their jobs. The approach fitted exactly the organizational imperatives of associations seeking the dues-paying support of municipalities, which meant in practice the office holder. At the same time, the leagues generally eschewed programmatic promotion. Such controversial proposals as city manager government were left to such groups as the NML. The leagues assisted managers but avoided advocacy, for advocacy might alienate, and the leagues needed, first of all, the support of as many cities as possible. Controversy might mean the loss of membership and organizational bankruptcy. Failure was no abstract threat; about one-third of the state leagues have gone out of business at one time or another.

The leagues were not without their own research orientation. But their research was of the most practical nature directed at providing information of direct assistance to the members. Information of a comparative nature about municipal costs and wages and various regulatory codes was most often the content of their research projects. Projects were reflective of a particular need of the time, including such items as "Municipal Licensing Practices, 1928," "Regulation of Hogs within Municipal Boundaries" and "The Communication of Glanders (an infectious mouth disease among horses) in Municipal Watering Tanks." Research conducted by groups within the Chicago "1313" complex was often somewhat broader in scope than that of state leagues, but on the whole it was comparable both in its intent and its orientation. The Chicago groups, in fact, regularly collected league reports and frequently attempted to draw them together for their own purposes. After Brownlow was able to attract the organization of state leagues, the American Municipal Association, into the complex, communication and the exchange of information were greatly increased.

Professionalism through self-improvement appealed initially to the heartland of progressivism. California, Iowa, Kansas, Wisconsin, Minnesota and Michigan were leaders in early successful league creation, and within those states it was mainly the smaller cities which responded to league appeals. In the large urban centers the immediately relevant professionalism was that of party politics, and the leagues had little election advice to offer the politicians. The industrial states from Ohio east were, with four exceptions, the last to develop state leagues of municipalities. In many other states, too, however, the effort was not without its perils. Some 18 state leagues failed in the years before the Depression and in many states only a handful of cities found the concept of professionalization sufficiently attractive to pay their dues. An ironic competition developed between the leagues and the state universities. Initially many leagues sought assistance from the universities for their executive staff, and they generally encouraged the universities to cultivate bureaus of government service to provide research and advice to local officials. To a very substantial extent, however, these service bureaus offered the same services as the leagues and they were supported by taxes instead of dues. This necessitated that the leagues offer new services in order to preserve their distinctive appeal.

In time the leagues developed a more diverse service potential. During the Depression, for example, state leagues often provided legal assistance to individual cities which enabled the community to save the money it would normally have spent on attorney fees, especially in the preparation of new municipal ordinances. Comparable help was offered on problems of municipal finance and accounting as well as applying for and utilizing federal programs such as those sponsored by the Public Works Administration. This trade-off of league dues for direct services was so attractive that between 1934 and 1937 nine new state leagues were able to successfully organize.

THE EMERGENCE OF LOBBYING

State leagues of municipalities gradually developed a fairly extensive array of advisory and technical services for city officials which supplemented their educational functions and added to the benefits of information exchange afforded by newsletters and periodic meetings. These services did not, however, include lobbying. It was relatively rare for a state league official to be seen testifying before legislature committees in the pre-Depression years, and only the New Jersey director went so far as to define himself, in 1931, as a "lobbyist, politician, counsellor." What efforts the leagues did make to influence public policy dealt mainly with the legal authority of municipalities and were couched in legalistic terms rather than as issues of who in the community ought to get what. When, in the early 1930s, American cities turned to the federal government for financial help, they found, rather to their dismay, that the American Municipal Association, which since 1924 had been the national federation of state leagues of municipalities, was not disposed to represent city claims for money in Washington. There was some rather complicated maneuvering among various leaders of the AMA and some of the leading American mayors, but the result was the formation, in February 1933, of the U.S. Conference of Mayors. The new organization was designed from the beginning to be a national lobbying organization, concentrating on getting money from Congress and from administrative agencies, which money would go as directly as possible to the nation's cities. USCM membership was originally limited to cities over 50,000 population, and for many years its effective constituency were the big cities, while AMA generally represented the sentiments of smaller cities and towns which provided the bulk of its members.

USCM was created with the support of AMA and its first executive director, Paul Betters, doubled as AMA executive director. AMA opened a Washington research office and for a time the USCM relied on the older group for its research. By late 1935, however, serious tension had divided the two organizations. There were policy differences, especially over how federal aid programs should be allocated among different-sized cities, and there were administrative troubles over how organizational funds should be allocated. Finally, Betters resigned from his AMA post, and the two groups went their separate ways. The AMA returned to Chicago and "1313" to continue its nonpartisan, efficiency-oriented research and service work. Betters and USCM settled down in Washington to look for money.

The future was in Washington. In 1948 the AMA began to issue formal pronouncements on proposed national policies affecting cities. In 1954 their headquarters was moved to Washington and in 1963 the Chicago research office was closed. In 1964 AMA became the National League of Cities and through the 1960s a series of steps led to the present close reintegration of the two organizations into what amounts to a single coordinated lobbying and service enterprise. The organizations share the same set of offices. Issue positions are compromised in advance of public statements and since 1969 there has been a single person serving as Director of Congressional Relations who supervises the combined staffs of the two organizations.

SOPHISTICATED SERVICES

The main basis of organizational strength and stability for the leagues of municipalities at every level of their operation have been the services they provided their members. As we have seen, these services have evolved from the initial program assistance of the old LAM, through a period of self-help professionalization and then an emphasis on the collection and dissemination of municipal data. At every one of these stages, a primary concern of the organization was whether or not member officials would continue to pay their dues. As the organizations grew they had more resources to devote to expanding and improving their offerings. The increasing number of involved officials also began to demand more tangible returns for their investment in the group. To a large extent this meant money, and as the scope of city budgets grew and the range of state and federal programs carrying financial aid for cities also increased, the most salient service to city officials was helping them get more federal and state money. As we have noted, this was the original purpose of USCM and AMA/NLC eventually came to share it.

This raises the question of whether their lobbying efforts warrant the continued membership support of both the major organizations; NLC and USCM, by the same set of municipalities. This is an important theoretical issue as well as one of practical concern to city officials. Mancur Olson (1965) has argued that no rational person would join an organization in order simply to lobby for policy support which, if the lobbying were successful, would bring him its benefits regardless of whether he had belonged to the lobbying group. Funds for urban renewal or poverty programs are not restricted to group members. We have contended elsewhere, however, that under many conditions the group strength required to lobby successfully is a matter of great uncertainty (Salisbury, 1969). Hence, within quite broad limits rational men may join a group to give it, as their collective spokesman, the political weight necessary to be persuasive. This would appear to cover the case of the early days of USCM. In addition, however, it is not always true that public policy benefits for cities go to them equally or according to some impersonal formula. Some get more than others. It is not impossible that cities whose leaders were most active in USCM received a somewhat disproportionate share of the federal funds available. And this very possibility might well persuade otherwise reluctant officials to participate in the organization.

But Olson is surely right in stressing the importance of selective benefits, available only to members, in keeping an organization going. And urban groups have all sought ways to provide useful services to their members. In recent years such older forms of service as self-improvement have lost much of their appeal. A much larger proportion of today's urban officials have received formal training for their roles. Professionalization through self-help has largely given way to more academic processes, and as noted earlier the universities also provide many of the advisory technical services pioneered by the leagues. On the other hand, the expansion of urban government services has, in a sense, left the old-style leagues with too big a task. Although a professional league staff can contribute at the margins of contemporary efforts to cope with urban crises, to many cities, despairing over both their problems and their budgets, traditional conceptions of league-type, research-based, assistance may well seem trivial indeed. "I can't afford the time for any organization," said the mayor of one of Missouri's largest cities, "that keeps telling me things I already know and does nothing but feed me beer and sandwiches. What I need is someone to help me get to the money I need, not just tell me where it potentially is at."

And they may appear just as trivial to the leaders of these organizations themselves! A well-respected and veteran state league director offered the following comment. "I'm faced with a problem of demonstrating to my constituents [member officials] that something is being done with the money they give me. They expect to see something for everything I do but that takes a great deal of time. I don't like the term 'urban crisis' but that is what we are faced with and that is what we had better work on." The most dramatic organizational result has been the creation of the National League of Cities-U.S. Conference of Mayors, Incorporated. NLC-USCM, Inc., is an independent corporation formed primarily to contract with federal agencies in order to assist in the development and implementation of government programs. To this point, the corporation has contracted with such agencies as the Departments of Labor; Health, Education and Welfare; and Housing and Urban Development. Projects have involved the organization in operational manpower programs, "State Plan Programs" for HEW and an evaluation of the Model Cities program. Another proposed contract with the Law Enforcement Assistance Administration would establish NLC-USCM as an agency to evaluate the utilization of resources funded by the Safe Streets Act and lead to a series of recommendations on the improvement of the program.

The new organization also provides all the traditional services of advice and technical assistance to member cities. The organization's inquiry service handled approximately 1,000 requests for information on various facets of municipal government in 1970. Staff members make field visits upon request for direct consultation with municipal officials. In addition, a wide variety of technical, advisory, and general information reports are issued from NLC-USCM headquarters. In fact, the staff attempts to release pertinent information on as many of their projects as possible. These activities are on the whole very reminiscent of those undertaken by the state municipal leagues.

It is too early to see exactly how the contracting role will bear upon the member-service activities, but since the contracts concern programs of presumed benefit to cities the payoff is expected to be significant. As a senior staff member said, "Every person that is employed by this organization is contributing to the dissemination of information to our members. If we employ someone because of a federal grant or Ford money, it still means that the individual is doing work for us." These contracts provide budgetary support for the organizations; in 1970 only about one-fourth of the NLC and USCM total budget (which was nearly $4 million) came from membership dues. Contract work has underwritten very important staff expansion with spillover benefits for member cities. Finally, the organizations have gained further legitimacy in the eyes of their members and of official Washington as well from enlarging the scope of their activities and adding the expertise derived from the administration of urban program contracts. In the size of the organizational budget and in conception of the benefits to be provided through the association, NLC-USCM, Inc., is a far cry from the days of "1313."

WHAT DO THE MEMBERS PAY–WHAT DO THEY GET?

Membership in the U.S. Conference of Mayors, the National League of Cities and in state and regional leagues of municipalities has a somewhat unusual character compared to most voluntary organizations. In USCM it is the mayoral office of a city which really belongs, not the man who holds the office. In the leagues it is the city itself. Individuals participate by virtue of their official positions with their respective cities. Moreover, membership dues and at least most, if not all, the expense of attending meetings comes from the municipal budget. The only costs of membership falling directly on the individual members are those of time and energy. Individuals make the decision to join or not, of course, and despite the fact that membership costs them nothing, many officials in the past felt the organizations were not worth even a tiny fraction of their municipal budgets. USCM was heavily subsidized beyond the regular dues structure by a few of the "more involved" cities during its formative years. It was felt that a population-based dues structure which was high enough to maintain the group's endeavors would prohibit many cities from lending their support to the organization at a time when their lobbying strength was important. Similarly, AMA was able to raise a budget of only $165 from ten state municipal leagues for its initial operation in 1924.

As we noted earlier, the rather direct, tangible benefits generated by many state leagues and by USCM during the 1930s overcame much of this resistance. It is probably fair to say that today, even in fiscal crisis, few American mayors would be likely to conclude that the dues for these organizations constitute good items to cut to save the city money. In a survey of St. Louis County municipal officials, we found support to be widespread and firm. They felt that the various municipal associations had been instrumental in obtaining increased revenue from the states and federal government. Not only that, they were able to secure information and assistance from the organizations which enabled them to cut corners in their own governmental operations. Ordinances were obtained and copied without a legal fee for the city attorney, and a variety of cheaper and more efficient procedures were picked up and instituted on topics from clerical filing systems to street repair.

An important aspect of financing is that these organizations are essentially low-cost groups. Given the great size of municipal budgets, the total dues paid by a city are most often inconsequential. New York, which has the highest dues of all cities, pays NLC $7,000 per year and USCM $5,000. Combined with their dues to all other municipal associations to which New York and its officials belong, the total disbursement is less than is spent on their individual lobbying expenses in Washington, D.C., alone. Or to put it in different terms, these dues are roughly equivalent to the cost of employing three beginning maintenance employees. This implies that, given the organization's potential for increasing city revenues, there is really very little risk of any financial sort involved in the decision of whether a municipal government affiliates with these groups.

Among the other dimensions of membership costs are those derived from the positions taken by the organization on policy questions. Mayors are, after all, elected officials. If they maintain membership in an organization which takes a stand that offends large numbers of their own constituents, they may encounter political troubles. In general, all of the organizations we have examined try to avoid taking a position on issues they expect will divide their members. NLC, for example, avoided making pronouncements on the Vietnam War on this ground, and in 1970 the leadership used this argument to persuade sponsors to withdraw a motion condemning the SST. Even revenue-sharing, a prime legislative goal of USCM, threatens the perceived interests of enough cities which comprise the membership of NLC and its affiliate state leagues to lead that organization to soft-pedal the issue.

None of these organizations is primarily concerned with expressive goals, however (Salisbury, 1969: 16). They seek more tangible benefits for their members, and the members, for their part, seek mainly material services in return for their membership dues. USCM and NLC make many public pronouncements on policy issues, of course, but the main task of broad agitation for urban goals is left to such groups as Urban America and the Urban Coalition. In terms of their primary organizational concerns, therefore, USCM and NLC can afford to remain quiet on potentially divisive issues.

Active participation by members in these organizations presents a number of interesting aspects. As in nearly every organizational setting activism falls short of involving everyone. Some mayors are indifferent; some are inexperienced; some are both and likely to avoid organizational settings they do not understand or care about. No mayor defends for reelection on his activity in these groups, and few have much time on their hands for peripheral functions. Some mayors have short political lives and hardly learn the names of the organizations before leaving office. In a sense, it is surprising that attendance at meetings is as heavy and service on committees as broadly distributed as it is. Part of this activism results spontaneously from the benefits which activity confers on those who participate, and part of it is the result of conscious efforts by the leadership to gain broader involvement.

Perhaps the most important single incentive to attract the active involvement of city officials in organizations like USCM or NLC is the prestige gained from recognition by one's peers. Mayors, like other people, are gratified by the recognition of their colleagues. They are flattered by the attention. And as they become acquainted with one another at annual convocations their interest in the affairs of the organization is quickened and the value they place on success within it is raised. As a staff member said: "Officials certainly don't join our organization to come to our conventions, but at the same time these meetings are helpful hunting grounds for us to recruit active participants. We know those people that come around regularly are the ones that are interested; they come to be quite enthused about the worth of the group through these informal contacts."

Most mayors, even those of large metropolitan centers, do not have much prospect of moving on to higher political office though sometimes a state municipal league may be a useful forum in which to campaign for statewide office. Consequently, a prime source of what national recognition is available to holders of mayoral office is high position within one of the major national organizations of cities. Not only recognition from peers is involved. Attention from the national press is considerable. Network television appearances and official statements to congressional committees grow out of USCM or NLC office. On a lesser scale there is comparable prestige and attention along with peer recognition in state and metropolitan area leagues to encourage at least some officials to be active. Unlike some organizations, neither the rules nor the norms are restrictive regarding activism. Participation is welcome and open to all who will take the trouble. A former officer in county, state, and national associations of both municipalities and city attorneys said, "This is really a personal thing and people get involved largely for their own egos as well as some prestige that they see associated here. If you stand up and try at all you can be anything you want in organizations like these."

If there is oligarchy, it is mainly because so many members do not want a share of the active roles. The executive staff, which plays a crucial part, tries assiduously to attract new members into committee and officer roles. But they depend nevertheless on the active leadership of a group of veterans. Many mayors and other city officials do not serve more than a term or two. In order for USCM and NLC to retain a core of members with some degree of knowledge about the workings and the affairs of the organization, they must rely on what is in many cases an unrepresentative group of officials in leadership roles, but a group which is self-selected according to the criterion of interest and willingness to work. "A Lindsay or a Yorty may be actively sought after soon upon his election," said a ranking staff man, "but for the most part we want officials who have demonstrated over time that they understand the goals of the organization."

Both NLC and especially USCM need active contributions from the mayors of the largest ten or fifteen cities in the country in order to give legitimacy to the organization's claims to speak for the nation's cities. If, for example, the mayors of Chicago, Detroit, and Pittsburgh failed to take any part in the affairs of USCM, it would seriously undermine the organization's standing. A staff member once told a group of over twenty California city executives that faced with a choice between them and Los Angeles, USCM would have to choose Los Angeles. Generally big-city mayors themselves have more experience and better staff work to bring to national organization affairs, and their statements carry more news value and command more political attention than would the same statements coming from the mayor of one of the Springfields. The executive staff tries to involve some of the smaller-city people too, of course, in order to give them a sense of identification with the organization. But the main stress is on getting the big-city people and keeping them sufficiently interested to give their time.

This strategy, however, must be kept in careful balance when it comes to the pursuit of policy goals. Although the organizations are most dependent on the larger cities, the legislators with whom they must work, especially those on appropriations subcommittees, often have little common grounds for understanding with big-city officials. Moreoever, NLC-USCM has recently been subjected to a good deal of criticism because the feelings of the small-city officials are not in evidence as frequently as many federal officials would like to see. In response, the organizations have been making concentrated efforts to bring such officials in to testify before Congress. Accordingly, this means that these officials too must be convinced that they should devote the time to become involved in these organizations.

One incentive of special relevance for this purpose is the professional recognition of peers already noted. In the same manner that value is placed on being a senator's senator in conforming to the folkways and gaining peer recognition (Matthews, 1960), there is a premium attached to gaining the respect of other mayors through appropriate participation. This is evidenced by the comments of a small-city mayor: "To be frank, I felt that by testifying in Washington, D.C., that people would understand and see that you don't have to be from a metropolis to be a good mayor. It only takes hard work." For the large-city mayor who constantly interacts with other mayors, this is an even more important feeling. For a mayor, however popular or powerful in his own city, to be acknowledged by his colleagues as a good man is to gain a real reward in these days when mayors have all too few successes and too many crises.

More tangibly, the usefulness of the NLC/USCM lobby in securing federal programs of aid to cities, assisting in winning the money to fund those programs, recruiting and advising the personnel administering them and generally bringing help to beleaguered municipal officials has been great and acknowledged to be so by most of the big-city mayors. Indeed, their efforts through these organizations may do more for their city budgets than most other things they might spend their time on.

One other incentive to activism may be mentioned. The meetings of the national associations, like many other such groups, are occasions of social pleasure as well as instrumental benefit. The St. Louis County League of Municipalities gets its biggest attendance at its annual barbecue. The USCM host city gives its guests the most generous hospitality, and the convention can be a most pleasant semi-vacation to which many members bring their wives. For newly elected officials and those from smaller cities the chance to meet their peers, share their problems, and gain a national perspective are often highly valued. Many members indeed, especially if they are less involved in the national lobbying work, would regard these "solidary" benefits as the principal personal value derived from membership and rank these values alongside the more tangible services to their administration and their community. In the words of a mayor from a St. Louis County municipality, "My wife and I have gone to NLC's Congress of Cities for years. We wouldn't know where else to go on our vacation. What do I like best? Obviously the variety of people I've met there." Membership in USCM and NLC involves both the individual official and his city and both receive benefits from belonging.

URBAN GROUP LEADERSHIP

Governing organizations like NLC and USCM presents several special problems. We have already referred to the importance of the big cities in giving the organizations necessary political weight. At the same time, however, in most of the state and national organizations the voting rule is: one city–one vote. Smaller cities might, therefore, exercise decisive power within the organizations. But they do not. The big cities, in fact, play the predominant role, especially in the national organizations. It is they who control the bulk of both the formal offices and the informal influences. Big-city hegemony is partly derived from the way decisions are made. The rules make it easy for city officials not to bother very much with affairs of the organization unless they are really interested.

In both groups decisions are made by those who choose to get involved in the organization. And, since the decision-making involvement of members is limited to matters of general policy and the selection of officers, it behooves larger-city representatives to take an active part in order to affect the more specific elements of the organization programs. They are, after all, the primary recipients of federal programs, and their work in NLC and USCM has proportionately greater payoffs. On the other hand, small-cities officials are less likely to get involved as long as they believe that they are being effectively represented and that their colleagues from cities of similar size are included in the operational structure. The major difference between the two organizations in this respect is that state municipal league directors take an active role in leadership positions of NLC since they are still affiliates of that group.

Official policy for both groups is formally adopted on the floor at the annual conventions. However, there is but a limited amount of discussion and only a few minor changes have ever been made in the course of floor debate. This activity is mostly a ratification of decisions that have been made earlier in the various standing committees and their subcommittees. These specialized committees have much autonomy to develop policy and strategy in their respective areas of concern. USCM has only four such committees and they are rather small. NLC has eight and they are large (150 members each) with very active subcommittees that do most of the actual work.

The formal device of most importance in centralizing organizational control is the executive committee composed of the officers of each organization. The committees are each headed by the organization's annually elected president and vice-president. The other 21 members of NLC's executive body are elected for two-year staggered terms. This contrasts with that committee of USCM which includes the past presidents of the group and 8 elected trustees all of whom serve for the remainder of their time in office. Their authority to control lobbying tactics and strategy and to develop the details of policy positions is very broad since all major policy decisions are considered by these committees. Both organizations also have advisory boards which serve partly as mechanisms for screening prospective executive committee members and partly as loci for cooptation and socialization of the more vocal dissidents among the membership. The selection of these committee and board members, like those of the substantive committees, is formally left to the current organization president. But here, as in so much else, it is actually the executive staff which exercises most of the real initiatives.

THE ROLE OF THE STAFF

We need not review all the advantages and sources of power available to full-time staff personnel of large organizations (Truman, 1951). In the case of NLC and USCM the key factors are these:

(1) the executive directors are vastly more experienced in the affairs of organizations of city officials than any members;

(2) the executive directors have more staff, more contacts and more information regarding what is going on in Washington and in the other cities of the nation than has any member;

(3) the executive directors are in a position to suggest promising nominees for association office, and to provide program and policy guidance which, with very rare exceptions, the members are happy to adopt;

(4) the executive directors have over the years provided services and done favors for many of the member officials which cumulatively strengthen the directors' positions.

With regard to experience, the present incumbent at USMC, John Gunther, has been with the organization since 1958 and has held the position of executive director since 1961. Patrick Healy has headed NLC since 1953 and was director of the North Carolina state league as early as 1934. Both organizations have drawn their principal staff people heavily from state leagues and from complementary functional organizations such as the National Association of Housing and Redevelopment Officials (NAHRO). Indeed, there is considerable movement of staff personnel among the many urban organizations, both along a local-state-national axis and among the different national organizations, and this has greatly facilitated the interorganization coordination of lobbying in recent years.

At the top the directors of these organizations receive rewards commensurate with their organizational success. There are a few salaries above $40,000 per year. NLC-USCM, Inc., employed 130 staff members at the beginning of 1971 and ten state leagues had ten or more employees. In the larger organizations, moreover, nearly half the personnel are professionals of one kind or another with appropriate formal academic training. In the small-state leagues, by contrast, the staff director generally suffers a syndrome of lower salaries, little or no assistance, and the reputation among their peers in other states of being "unprofessional."

The contemporary picture of the executive staff of the urban lobby organizations contrasts sharply with the situation of a generation ago. For the first decade of its existence, from 1924 to 1930, the National League of Cities (neé American Municipal Association) had its headquarters in the office of the Kansas state league. Then the executive secretary, founding father and general entrepreneur of the AMA, was John Stutz, who continued also as director of the Kansas league. In 1932 Paul Betters was hired to succeed Stutz as AMA director, and, as we have noted, Betters also became the main moving force behind the formation of the USCM. When Betters came under fire in his dual role, it was Stutz who was particularly critical of Betters' emphasis on Washington lobbying. Stutz's position in AMA was still formidable and Betters resigned to go full-time with USCM, where he remained until his death in 1956.

Stutz never accepted the more expressly political role which USCM undertook nor was he oriented toward Washington. Neither was his entire generation of state league organizers and catalysts. Several came from state universities and returned to academic life once their respective leagues were well under way. AMA gave money to help organizing efforts, and Stutz sought to identify promising organizers and place them in state leagues which were in organizational trouble. Through this process and the complementary energies of Louis Brownlow and his "1313" array of research and improvement organizations in Chicago, a generation of urban group leaders (and of political scientists interested in city affairs!) was given shape. Until the post-World War II era a large share of the state groups depended heavily for their success on the skill of their particular director. For example, the highly successful Georgia league collapsed when its director, John Eagan, died in 1934, and in New Jersey the death of Sedly Phinney had a substantial negative impact on that vigorous state organization. The postwar years, however, have seen a steady growth in staff and budget of nearly every urban group, and among the state leagues and the two principal national organizations, survival would no longer seem to depend quite so much on the personal qualities of the executive directors. Their guidance in the substantive business of the association, however, is probably more significant than ever. The old-style leagues with few members and small budgets had little choice but to stress research and self-improvement. Today the urban lobby groups are large, vigorous, and comparatively well-financed, and this gives their leaders many more options and the opportunity for far more extensive autonomous initiative than their predecessors could enjoy.

URBAN GROUPS AND GROUP THEORY

Let us conclude this discussion by noting some of the implications for interest group theory which seem contained in the phenomena of organizations of city officials. One already mentioned is the fact that contrary to Olson's argument, organizations of rational men can under some circumstances be formed in order to lobby for collective policy goods. It seems clear that USCM was so formed and that the 101 mayors who joined in 1933 had reason to believe that their joint effort would be decisive in obtaining federal aid. Secondly, however, and in partial contradiction to the above, membership in USCM and in the other groups of interest to us here is not personally costly. It is the city which pays, and therefore the rational choice component of membership is placed in a different context. If membership is essentially free to a mayor, then once the group is going and he is part of it, inertia will operate to keep him in rather than force him to reexamine his position. Today, the groups are firmly established and inertia, even apathy, may work to their advantage rather than constituting an obstacle as it does with groups whose members must weigh the benefits against their direct costs in the form of annual dues.[1]

The urban lobby groups provide their members first of all with material benefits. The tangible assistance of information, advice and guidance, especially regarding the programs and processes of the federal government, is joined with impressive lobbying on behalf of additional financial and service aid for cities. In part, given the nature of NLC-USCM membership, this lobbying itself is a form of direct service. As important political executives in their own right, mayors would generally seek, and be expected to have something to say about, programs affecting cities. The organizations and their staff help to organize and coordinate these statements, facilitating the performance of a task most big-city mayors at least would have to do anyway but with greater difficulty and less effect. Material benefits are supplemented for some members by the solidary benefits of interaction with peers and the resultant recognition and camaraderie. Expressive benefits, on the other hand, are seldom of much importance. More often members find themselves frustrated because the organizations will not speak on some question for fear of alienating a segment of the group. Seldom are NLC, USCM or the state leagues found taking a public position which is subject to much disagreement among the members. The members are a very select group, of course, and where, as is often the case, there is widespread agreement among urban officials concerning what policy ought to be adopted, the organizations speak vigorously. But public expression is not the most important way by which these groups seek policy results. In order to optimize their political effectiveness and, at the same time, avoid offending members who might not share the leadership policy preferences, key staff and activist members do much of their work quietly. One staff veteran put it this way: "If many people understood all that we were trying to do, we'd be in trouble trying to promote the legislation we feel is essential."

The other side of this coin, as we have noted, is the tendency to form new organizations to express new value concerns rather than adapt or enlarge old organizations in ways that might too severely alter the established ways of either the executive staff or the members. In the urban field this has resulted in a large and complex network of organizations which are, for much the greatest part, complementary to one another in political impact rather than competitive. And through the interchange and interaction of staff, but not very much through their very considerable overlapping membership, these organizations have achieved a well-coordinated lobbying effort in Washington. They have also greatly enlarged the communication among urban officials themselves. At one time a central contribution of urban groups was the upgrading of competence and professionalization of city officials. This still goes on, but the more important contribution perhaps comes from the greatly intensified interaction among city officials. All big-city mayors, for example, come very quickly to be familiar with the conditions, problems, and efforts at solution of every other city. They all operate as part of a communications network which under NLC-USCM, Inc., auspices has been sufficiently active to make the urban crisis of America truly a national crisis and not simply a collection of diverse local concerns.

NOTE

1. We are talking only about holding the organization together, not mobilizing its political strength for lobbying purposes. In the latter situation membership inertia is the bane of the executive staff.

REFERENCES

BROWNLOW, L. (1955) A Passion for Politics. Chicago: Univ. of Chicago Press.

CALLAHAN, R. (1962) Education and the Cult of Efficiency. Chicago: Univ. of Chicago Press.

CONNERY, R. H. and R. H. LEACH (1960) The Federal Government and Metropolitan Areas. Cambridge: Harvard Univ. Press.

FARKAS, S. (1971) Urban Lobbying: Mayors in the Federal Arena. New York: New York Univ. Press.

HERRING, E. P. (1940) The Politics of Democracy. New York: W. W. Norton.

HOFSTADTER, R. (1955) The Age of Reform: From Bryan to F. D. R. New York: Alfred A. Knopf.

McCONNELL, G. (1966) Private Power and American Democracy. New York: Alfred A. Knopf.

MATTHEWS, D. R. (1960) U.S. Senators and Their World. Chapel Hill: Univ. of North Carolina Press.

National League of Cities (1964) "When AMA was young." Nation's Cities 2 (July): 14, 26.

OLSON, M. (1965) The Logic of Collective Action: Public Goods and the Theory of Groups. Cambridge: Harvard Univ. Press.

SALISBURY, R. H. (1969) "An exchange theory of interest groups." Midwest J. of Pol. Sci. 13 (February): 1-32.

SCHATTSCHNEIDER, E. E. (1961) The Semi-Semi Sovereign People. New York: Holt, Rinehart and Winston.

TRUMAN, D. B. (1951) The Governmental Process. New York: Alfred A. Knopf.

VOSE, C. E. (1966) "Interest groups, judicial review, and local government." Western Pol. Q. 19 (March): 85-100.

WILLOUGHBY, A. (1969) "The involved citizen: a short history of the National Municipal League." National Civic Rev. 63 (December).

Part III

LOCAL RESPONSES TO PUBLIC DEMANDS

LOCAL RESPONSES TO PUBLIC DEMANDS

Perhaps the ultimate objective of both individuals and organized interests involved in local politics is to secure a favorable reaction from government decision makers. Unless there is some probability that such efforts might be successful, neither groups nor individuals would have any incentive to participate in urban politics. Yet, politics does not consist of a simple translation of popular desires into public policy. The needs and aspirations of the people may be mediated or overridden by a complex set of forces that tend to mold the outcome of the decision-making process. Research included in this section, therefore, focuses on the influences that may affect governmental responses to public demands.

Clark, for example, not only presents a careful review of research on both power structures and decision-making structures, but he also proceeds to offer some practical suggestions for persons interested in affecting policies at the local level. Based on his own research in 51 American cities, Clark indicates that the ability to shape policy outcomes may be determined by such factors as the centralization or decentralization of political authority, the type of local leadership available, and the nature of the issue, as well as by the socioeconomic characteristics of the population.

The response of political leaders to public needs, of course, also may depend upon the availability of economic resources and upon the means by which those funds are allocated. In their investigation of budgetary decision-making, Downes and Friedman suggest a series of hypotheses concerning shifts in local expenditures that may assist in the development of generalizations about methods of achieving policy changes.

Fowler and Lineberry report the results of a comparative investigation of policy expenditures both within one country and between different nations. Despite the difficulties entailed in this type of cross-national research, they find interesting similarities in the correlates of city expenditures. In both countries, fiscal outcomes seem to be more closely related to stimulative impact of intergovernmental aid and to environmental factors than to political participation.

In addition to the investigation of municipal expenditures or policy outcomes, the adequacy of city services also may be studied by examining the attitudes of local residents. In their survey of public opinion concerning schools, parks, police protection, and garbage collection in fifteen major cities, Schuman and Gruenberg discover that major discrepancies in the evaluation of those services not only are related to city and racial differences, but they are also associated with the characteristics of the neighborhoods in which people resided.

11

The Structure of Community Influence

TERRY N. CLARK

BASIC CONCEPTS: POWER STRUCTURES, INFLUENCE, AND DECISION-MAKING

□ Over the past two decades, local communities have attracted observers and commentators of differing orientations and intents. Some have been detached scholars; others have not. Occasionally research on the local community has been pressed into service for one or another ideological concern. And where ideological concerns are strong, the very selection and designation of concepts is a portentous act. A careful reading of this literature, however, reveals far more smoke than fire. Notwithstanding some very real disagreements over values, the amount of consensus on basic results is impressive. Many problems, it seems, could have been avoided simply by using more precise concepts.

AUTHOR'S NOTE: *This is research paper no. 32 of the Comparative Study of Community Decision-Making, supported by the National Science Foundation (GS-1904, GS-3162) and the Barra Foundation. Research assistance was skillfully provided by Kristi Andersen, Peggy Cuciti, and Wayne Hoffman. The current paper presents a straightforward review for the nonspecialist of several recent studies of community power and influence. However, many results in the last half of the paper appear here for the first time.*

A fundamental area of vagueness concerns the power-influence distinction. After Floyd Hunter (1953) used *Community Power Structure* as the title of his Ph.D. thesis, the term "power structure" became popular. Referents ranged from the Nazi party to progressive American parents, and from the mere existence of an imbalance of resources to specific social acts. Most subsequent writers went no further than Hunter in perceiving the need to separate power and influence. The distinction is nonetheless fundamental.

Power is usefully conceived as the potential ability of an actor or actors to select, to change, and to attain the goals of a social system. A key word here is potential; actors may have power but choose never to exercise it. *Influence,* on the other hand, is the exercise of power that brings about change in a social system. Resources (money, technical skills, personal charm, and so forth) constitute the bases of power; it is their application in specific decisions that comprises the exercise of influence. It then follows logically that a *power structure* is a patterned distribution of power in a social system, in contrast to a *decision-making structure,* or patterned distribution of influence in a social system (see Clark, 1968a: ch. 3 for further discussion). If these distinctions are carefully observed, most apparent theoretical disagreements and many presumed differences in empirical findings may be resolved.

THREE APPROACHES TO COMMUNITY POWER AND DECISION-MAKING

Hunter was far from clear or consistent on conceptual issues, but he made a major contribution in what has come to be called the reputational approach or method. It consisted basically of asking a panel of judges in Atlanta to rank individuals "who in your opinion are the most influential persons . . . influential from the point of view of *ability* to lead others" (Hunter, 1953: 258. Italics added). The persons named were then interviewed in turn and asked to rank names on the same basic list, adding new names where appropriate. Persons consistently chosen–generally businessmen and certain professionals–were considered members of the Atlanta power structure.

The simplicity and apparent precision of Hunter's approach led it to be imitated many times during the 1950s. It nevertheless contained two fundamental weaknesses. First, Hunter confused power and influence. The results of his method isolated powerful but not necessarily influential individuals; yet in his loose discussion of his findings, Hunter often implied that the powerful could and did have their way in most matters. Still, how and to what degree their power was converted into influence, Hunter never made clear. He presented long lists and sociometric stars, but virtually no discussion of actual decisions or processes by which decisions were reached. Second, Hunter did not specify the scope of power for different leaders. Without addressing the matter explicitly, he implied that leaders were powerful across the entire range of community issues. His methodology did not lead him to inquire to what degree different leaders were powerful within different issue areas.

In 1961 two studies (Dahl, Banfield) were published which avoided these weaknesses of Hunter. Although Dahl and Banfield discussed the resources available to different community actors, and in this sense portrayed the power structures of New Haven and Chicago, they focused primarily on influence, on the decision-making structures, of their respective communities. They were less formal in their methods of data collection than Hunter, eclectically using newspapers, documents, interviews, and participant observation. But as these various sources were studied with the goal of reconstructing specific decisions, their methodology may be characterized as a decisional approach. The analysis of concrete decisions in this manner leads naturally to comparisons of the degree to which leaders active in one issue area are also active in others. In comparing influence patterns in mayoral elections, urban renewal, hospital construction, and similar issue areas, Dahl and Banfield generally concluded that few actors overlapped from one issue area to the next. The major exception in both New Haven and Chicago were the mayor and his staff, who tended to play an important role in most issue areas. Dahl in particular argued that the monolithic portrayal of community power implied by Hunter and his followers was grossly distorted. More appropriate was a pluralistic view where different actors compete with one another for influence in discrete issue areas. The fragmentation that resulted was such that efforts by different actors to achieve concerted influence were often unsuccessful. This could even be true for the mayor, unless he was skillful enough to develop the informal attachments and obligations which permitted him to exercise more extensive influence than specified in his more narrowly circumscribed legal powers. In Chicago, of course, the Democratic Party served as a political machine to help the mayor implement his goals. The important point was that without the machine the fragmentation among actors would likely be closer to that of New York, where, without any strong political machine, the mayor was frequently unable to muster the support necessary to achieve his (or anyone's) goals.

The approach followed by Dahl and Banfield, however, also contained certain shortcomings. In focusing as they did on the exercise of influence in specific decisions, they necessarily had to be selective; it is impossible to study all decisions, even at one point in time. Generalization from a narrow range of decisions may be misleading, especially if the decisions are not very carefully selected, and neither Dahl nor Banfield sampled their decisions in a theoretically explicit manner.

A second weakness is a specific instance of the first: studying decisions made at approximately one point in time ignores past decisions which may have fundamentally changed the distribution of resources. Such past decisions could of course be analyzed using a decisional approach, ignoring for the moment problems of data availability, but they seldom have been. On the other hand, the reputational approach summarizes, if only very crudely, something of the legacy of these past events. Decisions at one point in time can lead to the institutionalization of new relationships among actors, to redefining norms and values which set the context for future decisions. This obviously does not imply that the power structure is isomorphic with the decision-making structure, but it does suggest that an ever-present element of influence is the structure of values and roles, and the corresponding ability of persons in concrete decisions to anticipate reactions of key actors and thereby to constrain the range of suggested actions accordingly.

A third weakness of Dahl and Banfield's studies was shared with that of Hunter: the focus on a single community. No matter how adequately their respective methodologies may have portrayed the power or decision-making structures in Atlanta, New Haven, or Chicago, there was no evidence that analogous results would emerge from studies elsewhere. Atlanta was in the reputedly authoritarian South, New Haven was a college town, Chicago had one of the last political machines in the United States. Given the vast diversity in citizen composition, economic base, and cultural outlook which seem to characterize American cities, it is simply foolhardy to atttempt to generalize from any single community to the whole. There can be no "average community" in a highly diversified national system.

It was such considerations as these which led certain researchers in the mid-1960s to undertake comparative studies of two or more communities, and to utilize a variety of methods for collecting data about power and decision-making. Presthus (1963) studied two towns in upstate New York, and Adrian and Williams (1963) and Agger et al. (1964) each investigated four cities. Larger comparative studies were initiated. With this shift from the case study to comparative work, the types of questions posed were also altered. Whether the focus had been on power or influence, the case studies had essentially asked: Who governs? The comparative studies began with who governs, but also asked where, when, and with what effects? Not only were the patterns of leadership a matter of concern, so were the community characteristics which gave rise to differing leadership patterns, and which in turn help generate different policy outputs. Neither a monolithic power structure nor a pluralistic decision-making structure was posited; in lieu of these value-laden terms, which implied a simple dichotomy, the relativistic concept of centralization was introduced. A set of propositions was formulated about the conditions under which centralization of

	Researchers		
	Hunter and Followers	**Dahl and Banfield**	**Comparative Researchers**
Method	Reputational	Decisional	Combined methods
Focus	Power structure	Decision-making structure	Power and decision-making structures
Centralization	Monolithic	Pluralistic	Variation along a centralization continuum
Leadership	Generalized leadership by businessmen and certain professionals	Issue-specific leadership: political leaders more generally active	Variation by community
Guiding questions	Who governs?	Who governs?	Who governs, where, when and with what effects?

Figure 1 CONTRASTING APPROACHES TO COMMUNITY RESEARCH

decision-making would emerge, along with variations in leadership, and how these factors would in turn affect policy outputs (Clark, 1968a). The basic types of relationships posited among variables are depicted in Figure 2. It is convenient to examine some of the results emerging from comparative work in terms of this framework. We consider first power and decision-making structures, then focus on leadership patterns, and finally on policy outputs.

Before turning to these results, however, a few remarks about the nature of the comparative studies are in order. We have already cited certain studies of two to four communities. These early comparative studies suggested insights about why patterns of leadership and decision-making differ across communities. They also helped contribute to a body of propositions of a specifically comparative orientation. But for purposes of testing such propositions, larger studies were essential. Consider, for example, two hypotheses: first, the larger the community population, the more decentralized the decision-making structure; second, the more economically diversified the community, the more decentralized the decision-making structure. To test such hypotheses it is essential to compare communities which vary simultaneously on all three variables: population size, economic diversification, and centralization of decision-making. And to test ten such propositions, all of which are hypothesized as likely influences on a single dependent variable, the number of cases must

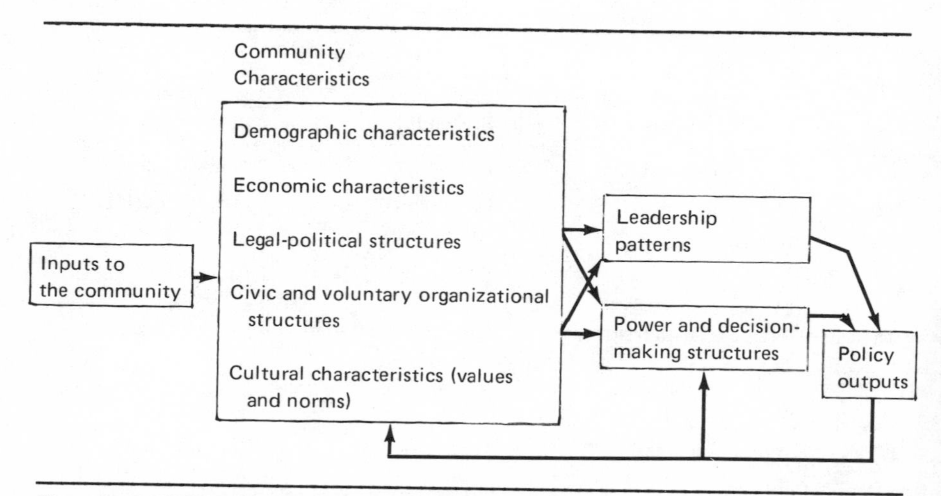

Figure 2. BASIC VARIABLES IN A COMPARATIVE APPROACH TO COMMUNITY POWER AND DECISION-MAKING

be considerably increased. For this reason, serious efforts were made in the late 1960s to accumulate comparable information for large numbers of communities. One procedure was to use a standardized coding scheme to classify results from case studies of communities. In this way data for as many as 166 different communities were compiled in a standardized framework (in Clark, 1968a: 139-156; Clark et al., 1968; Aiken, 1970, 1969; Walton, 1970). Certain findings emerged from this approach, but the vagueness of certain case studies and their corresponding noncomparability placed limits on the fruitfulness of such comparison of case studies.

Another approach is to collect comparable information from a large number of communities in a unified research program. This was the guiding idea behind the Permanent Community Sample: comparable data were to be collected from the same national sample of communities year after year to permit both cross-sectional and time series analyses (Rossi and Crain, 1968). Much of the empirical analysis reported here derives from a first study using this sampling frame. Representatives from the National Opinion Research Center at the University of Chicago interviewed 11 informants in each of 51 communities ranging in population size from 50,000 to 750,000. The same informants were interviewed in each community,[1] using the same basic interview schedule. Data were also assembled from the U.S. Census and derivative publications, *The Municipal Yearbook,* urban renewal handbooks, and similar sources. Some 800 variables have been computed for each community.[2]

POWER AND DECISION-MAKING STRUCTURES

As indicated above, the major dimension of power and decision-making structures around which debate and analysis has centered has been a continuum ranging from centralized to decentralized. The more useful theoretical approach to centralization seems to be in terms of the concept of social differentiation. Specifically, a number of middle-level propositions and empirical findings may be subsumed under the following formulation:

The greater the horizontal differentiation in a social system, the greater the differentiation between potential elites, the more decentralized the decision-making structure, which without the establishment of integrative mechanisms leads to less coordination between sectors and a lower level of outputs.

Let us consider the degree to which this formulation helps us understand the findings from four studies dealing with centralization of power and decision-making. The first three are those of Gilbert, Walton, and Aiken, based on coding results from case studies. In coding the case studies, classifications characteristically ranged from "pyramidal" to "coalitional" to "amorphous," which, due to the ambiguities of the original reports, often were applied to either power or decision-making structures, or to both.[3]

The fourth study is that of the 51 American communities. Most work to date on the 51 communities study concerning centralization of the decision-making structure has employed the following approach to operationalization. Seven of the eleven informants were posed questions concerning decisions in four issue areas: the last mayoral election, urban renewal, federal antipoverty programs, and air pollution control activities. For each issue area, a series of questions was devised to elicit information about the history of the major decisions. Although the specific wording varied across issue areas, they generally asked:

(1) Who initiated action on the issue?

(2) Who supported this action?

(3) Who opposed this action?

(4) What was the nature of the bargaining process? Who negotiated with whom?

(5) What was the outcome? Which groups tended to achieve their stated goals?

Actors mentioned in response to these questions were classified according to an inductively created code of 73 categories. The Index of Decentralization was based on counting the number of different statuses from the total list of 73 which appeared in each community.[4]

The broad concept of structural differentiation subsumes a number of more specific institutional arrangements. One demographic characteristic which tends to be associated with structural differentiation is population size: *the larger the number of inhabi-*

tants in a community, the greater the structural differentiation is a useful proposition (see Clark, 1968a: 96 ff.).

Ignoring momentarily the intervening linkages, we may examine the degree of association reported between population size and decentralization.[5] Gilbert found a definite positive relationship; neither Walton nor Aiken found any consistent relationship; in the 51 community study, we found a clear positive zero-order relationship (see Figure 3 for a summary of the findings).

Logically and probably temporally, however, population size seems to precede structural differentiation. Differentiation is a process affecting a broad range of institutions, but some of the most fundamental are those associated with the economy. As population increases, new economic activities tend to develop; correspondingly,

	Gilbert[a]	Walton[b]	Aiken[c]	Clark[d]
Population size	+	0	0	+
Economic diversification	0	0	0	+
Absentee ownership		+	+	—
Industrialization	0	0	0	0
Military installations				—
Index of Reform Government				—
Nonpartisan elections	—			
City manager	0	0	—	
Direct election of mayor			+	
Competing political parties		+		
Civic voluntary activity				±
Educational level of citizens	0		0	±

+ = positive relationship
— = negative relationship
0 = no significant relationship
± = mixed findings depending on model specification
blank = relationship not reported

a. Gilbert (1968). Data are for 166 communities, coded from earlier case studies. Findings are zero-order dichotomous relationships based on Fisher's exact test of probability, .10 or higher in significance with discipline and methodology of the reseacher controlled.

b. Walton (1970). Data are for 61 communities from 39 studies. Findings are zero-order relationships, with significance based on Fisher's exact test, computed from Q statistics and gammas. Control introduced for method of data collection.

c. Aiken (1970). Data are for 31 communities from previous case studies, supplemented from other sources. Findings are significant at at least the .10 level using partial correlation coefficient with methodology and discipline of the researcher controlled.

d. Clark (1968b, 1972). Data are for 51 communities, collected by NORC interviewers and from other sources. Findings are zero-order correlation coefficients and regression coefficients, significant at at least the .10 level. (The basic regression model is presented in note 6.)

Figure 3 RELATIONSHIPS BETWEEN COMMUNITY CHARACTERISTICS AND DECENTRALIZATION

distinct sets of interests, sources of conflict, and bases of power tend to emerge.

Or so many theorists have tended to argue, from Durkheim through Parsons and Eisenstadt. Gilbert, Walton, and Aiken found no relationship between economic diversification and centralization, however, although the relationship may have been weakened by the absence of consistent definitions of economic diversification. Still, in the 51 communities study, we found a definite strong relationship between economic diversification and decentralization. And consistent with our theory, when population size was included in a multiple regression equation along with economic diversification and other differentiation variables, the relationship between population size and centralization disappeared. Population increases seem to help generate differentiation, but it is such differentiation in turn which provides the more immediate support for decentralization of decision-making.

Considered in terms of its relationships with other local firms, when a firm is purchased by an extracommunity (most often national) corporation, this extralocal relationship is likely to differentiate it from the others. In this sense absentee ownership may contribute to economic diversification. A second process, however, is also likely to occur with absentee ownership. As the channels for promotion of executives are generally outside the local community, and as part of a national corporation they tend to be rotated after a few years, they tend to withdraw from local community activities, especially the more partisan, instrumental, and potentially conflictual activities bordering on the political arena. As the bulk of resources in most communities belong to private businesses, if some actors from the business sector withdraw from local affairs, the result is likely to be a more decentralized pattern of decision-making. After a certain point, of course, continued withdrawal of businessmen may lead to centralization of influence around other actors (e.g. political parties and unions). Nevertheless, the results from the two of our four studies which included data on absentee ownership showed a positive linear relationship between absentee ownership and decentralization.

The next economic variable, industrialization, was hypothesized as being positively related to decentralization. But as the relationship, even more than that of population size, is surely not a direct and immediate one, it is not surprising that there was no support for it in the four studies. It is the indirect effects of industrialization—wealth, education, more time for leisure and civic activities—which are likely to contribute to decentralization. Given the common ecological separation in American society between industrial establishments and the places of residence of their owners, it is perhaps surprising that a negative relationship did not emerge between various measures of industrial activity and decentralization. The general proposition would still seem to be plausible, but clearly only when broad cross-national or historical comparisons are made, and when the appropriate ecological units are compared.

A final economic variable was discovered through a factor-analytic approach to model construction (see Clark, 1972): the presence of military installations. Our immediate reaction to the finding was that it was most likely an aspect of economic diversification, or small city size, or a possible reflection of the authoritarian culture of small southern towns. But when controls for each of these alternatives were introduced, the relationship remained strongly negative, as it did in a regression equation with seven other variables.[6] Whether towns with sizable military installations were centralized before the installation arrived, or attract different types of citizens once present, or help diffuse a nonparticipatory outlook, we cannot tell from these specific findings. But we can say that towns with sizable military installations do have more centralized patterns of decision-making, and that the relationship stands firm when obvious controls are introduced.

Turning to governmental and legal variables, our general formulation again suggests the importance of structural differentiation. The basic variation in American cities in legal structure is in terms of the presence of "reform" characteristics. Conceived as part of a unified reform program, nonpartisan elections, at-large constituencies, and the professional city manager have been adopted by many American cities since the early twentieth century. Design to "professionalize" government and to make it operate on a "business-like" basis, these reforms helped concentrate authority in the hands of a city manager and his staff, and correspondingly to weaken the small ward and machine-oriented mayor and city council (see, e.g., Banfield and Wilson, 1963: parts II and III). Each of these three reform characteristics exerts some influence toward centralization, but when they are combined in a single Index of Reform Government, the Index is more important than any other item in an eight-variable regression model. The Gilbert, Walton, and Aiken studies did not compute such an index, but Gilbert found a similar relationship for nonpartisanship, and Aiken found that decentralization was correlated negatively with the presence of a city manager and positively with direct elections for mayor. All of these significant relationships support our differentiation hypothesis.

Competing political parties are another manifestation of structural differentiation; the one study (Walton) that included this variable found a definite positive association with decentralization.

Active voluntary associations were also hypothesized as contributing to decentralization, although the one study by Clark which included a measure of civic voluntary activity found a weakly negative simple correlation. But this relationship must be interpreted in conjunction with that of education.

More highly educated persons, it was hypothesized, would have more fully internalized the civic participatory norms taught in most of the American educational system. Higher education also implies more wealth and more free time for women in particular to support various civic organizations. But as cities with more highly educated citizens have disproportionately adopted reform institutions, and these in turn strongly contribute to centralization, the reform institutions seem to cancel out much of the effects which individual participation or civic voluntary activity could contribute to decentralization. Although the Gilbert and Aiken studies included measures for education, they did not analyze the separate impacts of education and reform governmental institutions using multivariate analysis.

In general, then, the diverse empirical findings from the four studies for the most part are consistent with our general formulation. Structural differentiation, as examined in a variety of community institutions, promotes more decentralized patterns of decision-making.

LEADERSHIP PATTERNS

Leadership, as indicated in Figure 1, refers to the relative importance of different types of actors—businessmen, political leaders, and the like. Leadership may indicate the importance of actors either in terms of the power structure or the decision-making structure. What do we have in the way of theory to explain differential patterns of leadership? Much less than it might appear. Writers from Aristotle through Marx and Weber have suggested that the socioeconomic structure of a community tends to influence the patterns of leadership which emerge. Recent work on resources suggests a new approach to leadership (Clark, 1968a: ch. 3; Coleman, 1971). Another line of theorizing focuses on maximizing the values of the median voter (e.g. Davis et al., 1970). These theoretical perspectives might be combined in the following sort of formulation: *The larger the number of resources and the higher the exchange value of the resources available to a particular sector, the greater the involvement of the sector in decision-making and the more policy outputs reflect the sector's values.*[7] Few of our findings, unfortunately, seem to contribute in clear and unambiguous fashion to these leadership theories. Doubtless more intensive analysis of these and other data are called for, but at this stage if our empirical findings are interesting it is not so much because they fit neatly into any single theoretical perspective.

But let us turn to the findings. As we pointed out, a basic weakness of Hunter's operationalization of the power structure was the difficulty in specifying the scope of an actor's power. In the 51 communities study, this was remedied by using an "issue-specific reputational method." Questions were posed about the degree to which given actors were powerful, but they were always posed with reference to a specific issue area. Two formats were used, one open-ended and another closed-ended. Due in part to the structure of the open-ended questions, only one or two actors tended to be mentioned in responses.[8] These were remarkably consistent across all five issue areas: the mayor and the newspaper. The mayor was the leading actor in urban renewal, the municipal bond issue, and air pollution; the newspaper was more important for the school board election and the mayoral election. In all issue areas, however, if the mayor was first, the newspaper was second in importance, and vice versa.

Clearly, however, mayors and newspapers are not the sole actors in American communities; they act in conjunction with, and as representatives of, broader sets of actors. The power scores of the 15 actors on the closed-end format[9] provide a more balanced portrayal of community actors than the open-ended format (see Table 1). Again the newspaper ranks at the top. Although the mayor was not included as an option, the Democratic Party ranks high. What the closed-end measures clearly add to the picture is the overall importance of the business community—the chamber of commerce and the four business actors.

Examining the results in this aggregate manner, however, does not distinguish leadership types across communities. What distinctive leadership patterns emerge and how are they related to various community characteristics? To answer these questions it is burdensome to report findings concerning the importance of 15 different actors in five different issue areas. We thus sought out leadership configurations by factor analyzing the intercorrelation matrix of the 15 actors. Two quite clear factors emerged. The first was an essentially business factor, with high loadings for the chamber of commerce, newspaper, industrial leaders, retail merchants, bankers, and other businessmen[10] (see Table 2). The second factor represented predominantly political leaders: the Democratic Party, Republican Party, labor unions and heads of local government agencies. The factors were very consistent across all five issue areas and for the five combined.[11] Because of this clarity of the two factors, we used them in a subsequent stage of the analysis: we correlated the score for each city on the two factors with other community characteristics.[12] Two reasonably distinct patterns seem to emerge (see Table 3).

The business cities tend to be in the West, the political cities in the Midwest (and specifically Democratic cities in the East). The business

TABLE 1

Power Sources for 15 Community Actors, Means for the 51 Cities

Actor	Mean Score
Democratic Party	1.96
Republican Party	1.80
Chamber of Commerce	2.04
Church leaders	1.68
Newspapers	2.36
Bar association	1.41
Labor unions	1.87
Ethnic groups	1.64
Neighborhood groups	1.80
Heads of local government agencies	1.83
City and County employees	1.57
Industrial leaders	1.99
Retail merchants	1.77
Bankers and executives of financial institutions	1.88
Other businessmen	1.79

TABLE 2
Factor Loadings of the 15 Actors on Business and Political (Varimax Rotated) Factors

Actor	Business Factor	Political Factor
Democratic Party	−.1165	.7860
Republican Party	.1305	.6054
Chamber of Commerce	.7321	.0325
Church leaders	.2305	.1263
Newspapers	.5791	.1121
Bar association	.2304	.1337
Labor unions	.2674	.5390
Ethnic groups	−.0397	.3278
Neighborhood groups	.1829	−.2021
Heads of local government agencies	.5204	.6081
City and County employees	.3429	.1877
Industrial leaders	.8407	.2109
Retail merchants	.8601	−.1484
Bankers and executives of financial institutions	.7363	.2499
Other businessmen	.8834	−.1032

cities tend to be smaller, less densely populated, and include more homeowners; political cities are larger in population size. There is less population turnover in the business cities. The political cities have more factories, but there are few differences in terms of economic diversification. In legal-political structure, the business cities include more reform characteristics, the political cities the opposite. The patterns of decision-making are more decentralized in political cities. The mayor is more likely to be a Republican in the business cities, and the citizens to vote Republican in presidential elections.

As we move on to cultural characteristics consider some of the findings in terms of the above formulation about number and exchange value of resources. Although money and manpower are two fundamental resources which tend to be disproportionately available to private business, votes remain the basic currency of political markets. Thus the simple population size of a community sector is normally among the most central factors influencing community leadership. Our business towns seem to include more highly educated persons, but not to deviate much from an average profile in terms of income distribution. The political cities, on the other hand, are disproportionately middle-income cities, and include few upper-income residents. The business cities tend to include more Protestants, the political cities (especially the Democratic cities) Roman Catholics. The business cities have a smaller number of citizens who

TABLE 3
Correlations of Four Leadership Measures with Community Characteristics

	Business Factor	Chamber of Commerce	Political Factor	Democratic Party
Northeast	–.360	–.515	.124	.334
Midwest	–.016	.094	.418	.367
South	.100	.098	–.146	–.170
West	.280	.319	–.418	–.548
SMSA population	–.208	–	–.365	–
City population	–.147	–.257	.201	.071
Density	–.308	–.339	.019	.133
Percentage population change, 1950-1960	–.307	–	–.113	–
Owner-occupied housing	.289	.374	–.034	–.044
Industrial activity (percentage in manufacturing establishments with 20+ employees)	–.098	–.185	.275	.349
Economic diversification	.109	.007	–.023	–.122
Index of Reform Government	.217	.303	–.527	–.597
Index of Decentralization	.031	–.127	.319	.279
Last Mayor Democrat	–.440	–	.009	–
Percentage Democratic vote in 1960 presidential elections, county	–.315	–	.350	.337
Civic voluntary activity	.012	.030	–.047	–.035
Median education	.297	.450	–.310	–.334
Percentage low income (under $3,000 annual family income)	.204	–.086	–.072	–.156
Percentage middle income ($3,000 to $10,000 annual family income)	–.137	–.186	.366	.442
Percentage upper income (over $15,000 annual family income)	.111	.204	–.320	–.314
Percentage Protestants	.310	–	–.062	–
Percentage Jews	–.201	–	.077	
Percentage Roman Catholics	–.427	–.490	.207	.415
Percentage Irish	–.283	–	.147	–
Percentage Germans (square root)	–.166	–	.109	–
Percentage Polish	–.409	–	.209	–
Percentage Mexican (log)	.175	–	–.214	–
Percentage Italian (log)	–.208	–	.134	–
Percentage Northern European (U.K., Ireland, Norway, Sweden, Denmark, Netherlands, Switzerland, France)	.045	–	.055	–
Percentage Central European (Germany, Poland, Czechoslovakia, Austria, Hungary, Yugoslavia)	–.259	–	.229	–
Percentage Southern European (Greece, Italy, Yugoslavia)	–.201	–	.263	–
Percentage rural and farmers in state	.059	.075	.296	.209

NOTE: If these simple correlations are over .230, they are significant at the .05 level; over .320 they are significant at the .01 level.

Most variables come from the 1960 U.S. Census, many as summarized in U.S. Bureau of the Census (1967). Region is defined as in the Census. Economic diversification is a dummy variable described further in Clark (1971a). The Index of Reform Government is a score based on summing the number of reform characteristics present in a city: city manager, non-partisan elections, at-large electoral constituencies, as reported in **The Municipal Yearbook 1966** (1966). The Index of Decentralization is discussed in the text. Civic voluntary activity is the number of members of the League of Women Voters per capita in the city, data supplied by the Washington headquarters of the League. The religious variables are estimates for the county, generally in 1952, reported in National Council of the Churches of Christ in the United States (1956).

are immigrants or one of whose parents immigrated to the United States. The political cities have more immigrants from Central and Southern Europe.

These findings seem to emphasize most the importance of the social composition of the community, in terms of the size of different sectors, as influencing leadership patterns. On the other hand, without more precise and adequate data concerning the nature and distribution of political resources on the one hand, or specific preferences of citizens on the other, it is difficult to feel confident about support for any specific theory. Still, our results are generally consistent with a long line of empirical studies of American city politics which have documented types which seem to bear at least a general resemblance to our business and political cities.[13] Although formulated as ideal types rather than as descriptions of empirical cities, the public and private regarding types of Banfield and Wilson (1963, passim) did synthesize rather neatly many years of case studies, and they do not appear overly inconsistent with our business and political city types. Clearly, however, we need much more in the way of precise, refutable hypotheses about leadership patterns as well as empirical work designed to articulate with such alternative hypotheses.

We have reported some basic findings for leadership in terms of our measures of the power structure. Influence scores were computed for 73 actors in each community in an effort to characterize the decision-making structure.[14] But in contrast to the power measures, the influence measures for different actors did not form any clear or consistent patterns. There was great diversity across communities, and across issue areas within individual communities. Contrary to our expectations, businessmen tended to be more involved in a wide variety of decisions than were political or civic leaders, more involved in the sense that, taken in aggregate, they scored higher on influence measures than they had on the power measures. Some critics of Floyd Hunter might have predicted the opposite. Limitations of space, unfortunately, preclude more extensive discussion of the influence findings.

POLICY OUTPUTS

There has been considerable debate in recent years, especially among political scientists, concerning the manner in which policy outputs should be conceptualized and studied (compare Downes, 1971, and Lineberry and Sharkansky, 1971, for recent reviews). Without denying the occasional utility of alternative approaches, we suggest that at present the two most illuminating classifications of policy outputs are in terms of benefits and public goods. We will discuss benefits here briefly and take up the public goods dimension below.

A concern to answer "who benefits" from public policies has been present for some time (Wirt, 1971: 201ff.), but few have gone beyond horatory assertions to demonstrate how particular policy outputs benefit particular sectors of a community. There seem to be three possible approaches. First is the assessment of the individual observer (who may also happen to be a social scientist), in terms of his personal ideological outlook, possibly linking this outlook with that of others in the society. Perhaps we have less faith in ideology than those who favor this approach. Second is some type of theory which postulates the "true" interests of various sectors in a community. The leading candidate under this heading is Marxism, which can invoke "false consciousness" when the presumed beneficiaries of a policy do not respond to it as expected. Apart from the validity of the theory, the range of policies in most political systems at one point in time is generally too narrow for Marxism to distinguish them with precision. We thus favor a third approach over these first two: consideration of the degree to which a policy is consistent with the values of particular sectors in a community. With the increase in would-be spokesmen for various sectors in American communities during the 1960s, it is much easier to demonstrate empirically just how acceptable or unacceptable different ideological positions (especially those of the radical left) are to most sectors than was the case a few years earlier. This approach emphasizes the importance of ascertaining with as much precision as possible the value configurations of citizens and articulating them with distinct policy outputs.[15] Although the approach is obviously applicable to many dimensions of policy output, to simplify the analysis we here focus mainly on one basic policy dimension: the degree to which the government is more or less active in general and in specific areas. Long-term debates between conservatives and liberals about the scope of government testify to the importance of this dimension of policy. For purposes of comparative empirical analysis, it is also a useful dimension to consider as it is captured reasonably well in the level of expenditures in the city budget. It is these expenditures (per capita) which constitute the basic dependent variable in the present section.

This approach to the assessment of benefits of public policy articulates well with our general formulation considered above: *the larger the number of resources and the higher the exchange value of the resources available to a particular sector, the greater the involvement of the sector in decision-making and the more policy outputs reflect the sector's values.*

One basic value configuration for public policy is the degree to which a sector favors corporative support for a given service. By corporative support we refer to the tendency for individuals to prefer some form of collective activity in an area instead of individual activity to the same end. Certain ethnic and religious groups seem to value corporative support more than others: Catholics more than Protestants, ethnic groups with extensive kinship ties (e.g., the Irish) compared to others, and so on. The converse of the corporative orientation is that of the individually autonomous European peasant or American pioneer farmer or rancher. Although the number of actual farmers or ranchers in the American population is small, the cultural style which they symbolize is more widely shared. The presumed individualism, the reliance upon hard work and only immediate family and neighbors to solve most problems, the need for few direct services from governmental agencies, and the corresponding distrust of strong government may have been elements of an ideology once shared by many farmers and ranchers. Certainly subsequent political groups, not all as conservative as Barry Goldwater, have sought to emphasize and perpetuate these and related themes. Several studies indicate the farmers and rural persons generally prefer less spending by federal and local governments (e.g., Key, 1963: esp. ch. 5; Boskoff and Zeigler, 1964: 42ff.), and that persons in the West and Southwest of the United States prefer less governmental activity, even when educational and income differences across regions are controlled (Patterson, 1968). In the 51 cities study, we wanted to capture this cultural orientation. Because we hypothesized that regions would tend to share similar outlooks, we used the percentage of farmers and rural nonfarm residents in the state where each community was located. This variable showed a definite negative relationship with city spending (see Table 4).[16]

Of course values tend to interpenetrate occupational and ethnic activities, and to be reinforced by benefits accruing from these activities. A significant example of this phenomenon seems to be the following: *the more a given sector tends to supply manpower for governmental agencies, the more that sector will favor governmental expansion.* Although the proportion of the city population employed by the government may be small, city employees are among the most politically active of any local group. Given the frequently low levels of turnout for such matters as bond referenda and primary elections, and the frequent readiness of city employees (spurred on by their unions and political organizations) to help mobilize support in door-to-door campaigns among more apathetic citizens (compare with Banfield and Wilson, 1963: ch. 15), they can exercise a disproportionate influence in city politics.

This proposition about governmental manpower ties in with the corporative support idea in especially striking fashion in the case of

TABLE 4
Simple Correlations of Individual Variables with Policy Outputs

	Expenditures Per Capita for Nine Common Functions
Percentage rural and farmers in state	–.378
Percentage Catholics	.261
Percentage Irish (log)	.380
Upper income (% over $15,000) (log)	.357
(Upper income)(lower income) (log)	.194
Middle income (% $3,000-$10,000)	–.329
(Middle income)(lower income)	–.124
Lower income (% under $3,000)	–.075
Population size	.292
Industrial activity	–.360
Economic diversification	.085
Index of Reform Government	.156
Index of Decentralization	.155
Business leadership factor	–.190
Chamber of Commerce leadership factor	–.160
Political leadership factor	–.313
Democratic Party leadership factor	–.296

NOTE: The nine functions are those common to the 51 cities: highways, police, fire, sewerage, sanitation, parks and recreation, financial administration, general control, and general building. Excluded therefore are functions which in some cities pass through the city budget but in others are in the jurisdiction of the county of a special district. The source for the common functions is U.S. Bureau of the Census (1969). Unless stated otherwise, other data are from the U.S. Census of 1960, mainly as summarized in U.S. Bureau of the Census (1967). Other sources: for percentage Catholic, the data are estimates for the county, in National Council of Churches of Christ in the USA (1956); percentage Irish refers to persons born in Ireland or persons with one or more parents born in Ireland; industrial activity refers to the percentage manufacturing establishments in the community with more than 20 employees; the Index of Reform Government and Decentralization were discussed above as were the four leadership variables.

Irish Catholics. The Irish have been disproportionately involved as political party activists, as elected officials (Litt, 1970), and as governmental employees. The corporative support element seems to have been reinforced by the Roman Catholic church and the political machine. The Irish arrived in the United States at a fortuitous moment when the new immigrant groups were becoming numerous enough to displace the old Yankees, but none could rival the Irish for their knowledge of the English language and the basics of the Anglo-American legal system. The considerable trust maintained by the kinship ties of the Irish also distinguished them from potential rivals such as most Polish, Italian, and subsequently Negro groups. Once this pattern of political involvement was established near the end of the nineteenth century, subsequent generations of Irish have helped maintain it. Clearly the political machine and traditional ethnic politics are disappearing rapidly, but the strength of our correlations involving Irish and Catholics suggests their continuing importance in explaining variations in policy outputs in American cities[17] (see Table 4).

Ethnic and religious groups are important in defining sectoral cleavages in some instances; income on the other hand is generally distributed more along a continuum, but nonetheless provides a fundamental basis for differentiation of value orientations concerning public policy. To simplify discussion, it is useful to consider three income levels: upper, middle, and lower. As the marginal utility of a dollar is normally less to a more affluent person, he should be less concerned about a given tax fee than someone less affluent. And accustomed to higher levels of consumption in their personal lives, upper-income persons should prefer superior education, streets, parks, and other city services as well.[18] But preferences such as these are not translated directly into public policies; they are mediated by tax structures, which in most local communities means primarily the property tax. As most American communities also provide services for their residents on a more egalitarian basis than in proportion to taxes paid, the more affluent thus have an incentive to lower their revealed preferences for a given level of public services as the proportion of less affluent citizens increases. This same tendency to resist supporting the lower-income sector would seem to apply to the middle-income sector, with the difference that the marginal utility of a single dollar is higher. Correspondingly middle-income persons should favor lower levels of governmental services than upper-income residents. Finally, lower-income persons should favor higher service levels as they pay the least for what they get, but, on the other hand, the marginal utility of a dollar is higher to them, so they should oppose increased taxes.

Our findings generally support this line of economic reasoning. Table 4 presents the simple correlations between the different income sectors and governmental expenditures. The larger the upper-income sector in a community, the greater its expenditures, although this relationship is decreased when the percentage upper income is multiplied by the reciprocal of the percentage lower income. More middle-income persons means less spending. There is little relationship between the size of the lower-income sector and spending, and the middle-income/lower-income multiplicative term is just below significance.

Although our major propositions alert us to the value configurations of different sectors concerning policy outputs, these preferences are inevitably shifted by the context of the particular community. Certain community characteristics are consistently related to policy outputs. Cities with larger numbers of inhabitants spend more, while those with more industrial plants spend less. Economic diversification has no relationship to spending. Cities with reform characteristics tend to spend more.

But let us turn from community characteristics to the decision-making structure. Several studies up to 1967 suggested that more decentralized structures of power and decision-making led to fragmentation in the political arena, and a frequent inability to create a coalition with sufficient unity to general substantial policy outputs (e.g., Hawley, 1963; Rosenthal and Crain, 1966). These studies focused on early urban renewal programs, water fluoridation, and school desegregation, and had less-than-optimal measures of centralization of power and decision-making. When adequate data were collected for a large sample of cities, the opposite relationship emerged for budgetary and urban renewal expenditures: more decentralized cities spent more (Clark, 1971a). The earlier findings appear not so much incorrect as incomplete. A crucial distinction needs to be made between public and separable goods. Public goods, following Samuelson (1954), are indivisible (they cannot be enacted partially or in stages) and their exclusion costs are high (it is difficult to refuse citizens access to them). Separable goods we designate as the opposite of public goods in these two respects. The two above sets of empirical findings then may be reconciled with the two following propositions: *For public goods, the more decentralized the decision-making structure, the lower the level of outputs. But for separable goods, the more decentralized the decision-making structure, the higher the level of outputs.*

Additional support for these propositions is found in the pattern of expenditures by function in the 51 cities. The Index of Decentralization is positively correlated with police, fire, sewerage, and sanitation expenditures, but is negatively related to financial administration and general control. These last two categories cover taxation and juridical functions which are much less easily assigned to any single neighborhood or sector of the community.

Two other recent studies provide additional support for the proposition about separable goods. In his reanalysis of results from 31 case studies, Aiken (1970) found that decentralization was positively related to participation of a city in the housing programs of the federal government and to urban renewal expenditures per capita. Although just below statistical significance, eight other output measures of public housing, urban renewal, antipoverty programs, and Model Cities participation were higher in more decentralized cities. Then in analyzing data for some 400 cities, Aiken and Alford (forthcoming) found that city size and indicators of decentralization of power and decision-making were positively correlated with output levels for public housing, urban renewal, and poverty program participation.

When we turn finally to the relationship between leadership variables and outputs, we find that all of our four measures of leadership are negatively correlated with outputs (see Table 4). The two political leadership measures are more strongly negative, however, than the two business measures. Probably these findings are best interpreted as indicative of the preferences of citizens in cities scoring high on our political and business factors. The political cities, made up of predominantly middle-income persons, and thus persons with preferences for low spending, tend to spend less than the average city. The case of the business cities is less clear; they have less distinct income characteristics than the political cities. It should be added that in a regression equation with income and income distributional characteristics of the community included, there is virtually no significant relationship between business leadership and policy outputs. The absence of a relationship here should at least alert us to the fallaciousness of a certain stereotype: business-dominated cities do not spend less than other cities. Those individual cases which are sometimes cited as examples of the opposite, our findings suggest, are more likely the result of underlying community characteristics than of the business leadership structure.

CONCLUSIONS AND POLICY IMPLICATIONS

The recent salience of the American city as a central arena for public policy has led many persons to inquire what in fact we know about how cities make decisions. Although many apparent disagreements may be detected in the literature, a common core of findings seems to emerge reasonably clearly. This paper summarizes some of the basic conclusions from the tradition of community power and influence inaugurated by Floyd Hunter, Robert Dahl, and Edward Banfield. Most specific findings reported here, however, come from a national study of 51 cities where data were collected through standardized interviews about mayoral elections, urban renewal, air pollution control activities, federal antipoverty programs, and similar issues. Doubtless further studies are called for to verify our results, but if our interpretations are correct, they suggest a number of reasonably important policy implications.

There is no substitute, of course, for the intimate knowledge which long personal acquaintance with a city can provide. Persons with such knowledge, however, are not always easy to locate, and, depending on one's auspices, they may not be ready to reveal their secrets. Further, they seldom have much perspective on how their city compares with others. Thus, for example, a federal official who must devise certain policies for more than a dozen cities in a short period of time may find that our generalizations, even when rough, are more useful than most realistic alternatives. Planners of different sorts, developers, community organizers, and aspiring candidates for public office, whatever their political persuasion, might be other potential users of these results. What can we suggest that may be of interest to such persons?

An important distinction must be made initially between the power structure of a city and its decision-making structure. The *power structure* includes actors who have access to important resources; they are potentially important actors. But they often take no part in concrete decisions. Hence a *decision-making structure,* by contrast, includes actors who are actually involved in concrete decisions. If one is concerned with influencing policies similar to those currently followed in an issue area, major attention should probably be directed to persons involved in the decision-making structure. But if significant alterations in policy are envisaged, the views of actors in the power structure of the issue area should be considered as well.

Variations of the reputational method tap the power structure, while the decisional approach is better to characterize the decision-making structure. Persons who conduct a brief study for policy purposes may want to use one or both of these methods, but in any case we recommend that "issue-specific" versions be used, subsequently aggregating if a general measure is desired.

Although original data may usefully be collected for policy purposes using such procedures, this is often impossible. In the absence of fresh empirical data on power and decision-making structures, correlations between them and a number of community characteristics which have emerged from our work may nevertheless be suggestive. If this appears to be the case, it may be worth the short amount of time necessary to gather minimal basic data from such standard sources as the *County and City Data Book* and *The Municipal Yearbook*. With such information at his disposition, the policy maker may wish to consider the implications of relationships between such data and our findings. These we summarize in terms of centralization-decentralization, leadership, and policy outputs.

CENTRALIZATION-DECENTRALIZATION

In cities and issue areas that are highly centralized, major attention should be directed to the few leading actors. Personal visits or other means of persuasion effective with such leaders may be sufficient to establish new policies in the area. In a more decentralized city or issue area, however, such an elitist strategy is likely to be disappointing. Converted leaders may not continue to agree with one another or be able to persuade their followers of the new policy. In such situations, strategies of more general appeal are called for: mailings of letters and brochures, use of the mass media, meetings with grass-roots organizations, and so forth.

One is likely to find more centralized structures in those cities which are smaller in population, more economically specialized, have predominantly local ownership of industry, a city manager, nonpartisanship at large elections, and few competing political parties and voluntary groups.

LEADERSHIP

In general, the mayor and newspaper tend to be the two most powerful actors in American cities. Their support, at some stage and in some manner, is likely to be essential for implementing major programs. Still, their support is seldom enough. Other actors nevertheless vary considerably by city type and issue area. Two broad configurations of city types and sets of actors within cities, however, may be distinguished: a business type and a political type. Business-related actors–mainly the newspaper, chamber of commerce, and various business leaders–are most powerful in cities in the West, with smaller populations, city managers, nonpartisanship, at-large elections, and more affluent citizens who tend to vote Republican. Business-type cities also tend to be more centralized. Political actors–especially the Republican and Democratic parties, labor union leaders, and heads of governmental agencies–tend to be more important in the Midwest and East, in cities with larger populations, more industry, mayor-council government, partisan and ward-based elections, middle-income residents ($3,000 to $10,000 family income in 1960), Roman Catholics, and immigrants from Central and Southern Europe. Political cities tend to be more decentralized. These generalizations about business and political leaders, however, refer to power rather than decision-making structures. The specific actors important for decision-making vary considerably across cities and from one issue area to the next.

POLICY OUTPUTS

The stereotype that upper-income persons and business-dominated cities are opposed to higher expenditures is largely incorrect. It follows from an unjustified generalization from the national to the local level, which has been incorrectly maintained even by leading sources (e.g., Banfield and Wilson, 1963: ch. 16). Our results show that upper-income individuals, and cities with more affluent residents and often a business-oriented leadership, tend to prefer to spend more per capita than less affluent individuals, cities with less affluent residents, and more politically oriented leaders. On the other hand, certain special groups, in particular Roman Catholics and persons of Irish descent, tend to prefer higher levels of governmental activity. The opposite is true of cities in states with large percentages of farmers and rural residents. If appeals that involve increasing governmental activity are thus to be made in cities which, according to these criteria, tend to spend less, an effort should be made to avoid salient increases in local taxes. The same reticence is not so important in more affluent cities.

The action strategy should also be planned as a function of the type of policy output. Reasonably public goods, those not easily divisible and quite broadly shared—e.g., an air-pollution program for an entire city or a school desegregation plan—should be sharply distinguished from more separable goods, i.e., those which may be allocated to distinct subsectors of a community, such as garbage collection, rent subsidies, or individual welfare payments. Public policies involving more public goods seem to be more easy to implement in a more centralized decision-making situation. For example, many water fluoridation programs a few years ago were often accepted in cities where the decision was made by public health officials in conjunction with a city manager or a small number of other leaders. On the other hand, in cities where fluoridation became widely discussed in the newspapers and in grass-roots organizations, and was voted on in a referendum, it was often defeated. But separable goods seem to be just the opposite. Poverty program expenditures, garbage collection, snow removal, public housing construction, and programs which in general affect mainly specific individuals or subsectors of a city tend to be carried out at higher levels in more decentralized situations. Clearly a threshold effect is present here: in particular one thinks of "decentralized" situations in New York City that have easily become chaotic. But given a minimal level of consensus and fundamental viability of the decision-making institutions, then a higher level of separable goods is likely to emerge from a more decentralized action strategy. Use of the mass media, mailings, handouts, meetings with grass-roots organizations, and similar decentralizing strategies are correspondingly in order if the goal is to increase outputs of separable goods.

If these suggestions may be useful for certain policy-related matters, it is only because they build on an extensive body of basic research. We hope that this is not overlooked.

NOTES

1. The positions interviewed were those of mayor, chairmen of the Democratic and Republican parties, president of the largest bank, editor of the newspaper with the largest circulation, president of the chamber of commerce, president of the bar association, head of the largest labor union, health commissioner, urban renewal director, and director of the last major hospital fund drive.

2. Further details of the study are reported in Clark (1968b).

3. The findings reported below for these studies (summarized in Figure 3) should be interpreted with caution as they are based on essentially zero-order relationships. Controls were introduced in each study for discipline or method of researcher, or both, but no other multivariate analysis was reported. When the data from the largest of the three studies, that of Gilbert, were reanalyzed using multiple regression analysis, the zero-order relationships were seriously attenuated (see Clark et al., 1968). No doubt the same would occur for the two smaller studies, but the number of cases hinders such multivariate analysis, not to mention the problems of statistical inference based on the very nonrandom character of the samples. On the other hand, the same study (Clark et al., 1968) led to the opposite finding from that of Walton and Aiken: neither discipline nor method of the researcher exerted a significant influence on the reported degree of centralization. This disparity was interpreted as resulting from inclusion of a larger number of cases (166 as against 61 or 31), many by persons who were not involved in the Hunter-Dahl disagreements, and who had less definite commitments to substantive or methodological positions.

4. If a single status was mentioned in more than one issue area or more than one stage of a decision, it was still counted only once in constructing the Index. The rationale was that centralization is appropriately conceived in terms of two dimensions: participation and overlap. The Index should, and does, allow for a higher score if there are more actors participating in the various decisions. But if the same actors overlap from one issue area or one decisional stage to the next, the Index score is not increased. Corrections were included for missing issue areas and missing informants.

5. The term *de*centralization is used so that numerically positive relationships between the number of actors involved in decisions and other community characteristics may be labeled as positive.

6. Reform government, economic diversification, industrial activity, community poverty (percentage under $3,000 income), education, percentage Roman Catholics, population size, civic voluntary activity (see Clark, 1972, Table 4).

7. Sector refers to any collection of actors within a community who have enough in common to work together occasionally. Possible examples: retail merchants, poor blacks.

8. The open-ended questions were of the following form: "Is there any single person whose opposition would be almost impossible to overcome or whose support would be essential if someone wanted to . . ." followed by five different endings, each relating to a specific issue area: "(run for/be appointed to) the school board in (city)?; organize a campaign for a municipal bond referendum in (city)?; get the city to undertake an urban renewal project?; . . . essential for a program for the control of air pollution in (city)?; run for mayor in (city)?" If the initial response was negative, the informant was asked for the name of the person who "comes closest to this description." A power index was created from these items in the following manner:

$$P_s = \sum_{i=1}^{7} \sum_{a=1}^{5} \frac{s_{ia}}{t_{ia}}$$

The informants, i's, were seven. The issue areas, a's, were five. The score for a status was based on all mentions for that status, s_{ia} divided by those for the total number of statuses, t_{ia}.

9. The closed-end format was as follows. The informant was presented with a list of 15 actors (listed in Table 1). The interviewer stated: "Here is a list of groups and organizations. Please tell me for each whether their support is essential for the success of a candidate for the school board, whether their support is important but not essential, or whether their support is not important." The same format was then repeated, substituting for the school board "a municipal bond referendum," "an urban renewal project," "a program of air pollution control," and "a candidate for mayor." A score was then computed for each of the 15 actors in each community simply by taking the mean of the mentions, scoring essential as 3, important but not essential as 2, and not important as 1.

10. The .52 loading of Heads of Local Government Agencies on this factor probably refers to the city manager in these predominantly reform cities. If this is the case, the business factor remains a quite consistent configuration.

11. For this reason we present only the results combining all five issue areas in Tables 1 and 2.

12. Despite the general clarity of the two factors, our own skepticism about factor analysis and problems of model muddling (compare with Clark, 1972), and possible skepticism of others, led us to include (in Table 3) the power index scores for two leading individual actors, the chamber of commerce and the Democratic Party, alongside those for the two factors. The picture which emerges for these two individual actors is quite consistent with that emerging from the factors. At an early stage we experimented with a third factor which emerged, and tentatively labeled it a civic factor because of the high loadings of church and neighborhood groups. But the simultaneous importance of several governmental actors, and the very low and inconsistent correlations between this factor and community characteristics, led us not to continue with it in subsequent analysis.

13. One of the most valuable sets of studies for our purposes has been the ambitious series undertaken in conjunction with the Harvard-MIT Joint Center for Urban Studies, summarized in Banfield (1965). The original reports deserve more attention than they seem to have received. We should also note that two leadership factors quite similar to those which we found have been reported by Morlock (1971). Strictly speaking, of course, the political and business factors, like public and private regardingness, are ideal types. As they refer to configurations of leadership, not to cities per se, identification of any single city with a given type should be considered a simplification legitimate solely for illustrative purposes.

14. These are analyzed in Clark (1969) and a volume now in preparation. Some of the flavor of the results is perhaps conveyed by the following ranking of actors in terms of their influence in probably our most important issue area: the mayoral election. The percentages are means for the 51 cities combined.

Business (unspecified)	9.7
Local newspapers	8.9
Organized labor	7.8
Negroes and civil rights groups	6.5
Democratic Party	5.9
Republican Party	5.0
General public	4.5
U.S. Chamber of Commerce	3.1

The heterogeneity of the issue area is indicated by the fact that these eight leading actors account for only 51.4% of the total mentions in the issue area.

15. Many approaches to measurement of this congruence between citizen values and policy outputs are possible–fieldwork, analysis of documents, and so forth. We are currently involved in analyzing old survey questions posed to individuals about such matters, and in designing a more adequate questionnaire format for eliciting such information in the future. Some of the results which we report build on these analyses of surveys, but clearly far more is called for in this area.

16. Zero-order Pearson correlation coefficients for most variables and expenditures per capita for nine functions common to the 51 cities are presented in Table 4. Most relationships discussed in the text which are significant in Table 4 were also significant in one or more regression equations computed with varying combinations of the variables listed in Table 4. Still, the degree of intercorrelation among some of these variables led us to use only selected items in many of the regressions. For example Irish and Catholic were strongly positively correlated (r = .812), and strongly negatively correlated with percentage rural and farmers in the state; hence only Irish was used in most regressions as it had a stronger relationship to the other two variables than did either of the other two. Only selected income distribution measures were used as several were very highly intercorrelated.

One regression equation including many of the variables discussed in this section is the following:

Dependent Variable: Common Functions

Independent Variables	Regression Coefficient	Standard Error
Irish	4.72	2.54
Business factor	–1.86	2.76
Civic factor	2.08	2.97
Party factor	–5.95	3.25
Index of decentralization	2.44	1.98
Middle income	–201.4	57.0
Lower income	–85.6	46.4
R = .741		

When the same regression equation was reestimated with transfer payments from higher levels of government and median income for inhabitants of the city included, the middle-income and lower-income variables dropped below significance. The results were broadly similar when the equations were estimated using each of nine common functions separately as the dependent variable.

17. Although we cannot present correlations for each city function, the relationships are especially strong for the Irish and Catholics for fire and police expenditures.

18. Surprising as it may seem, almost no serious empirical work has been done on the preferences of different socioeconomic groups for local governmental activity. However, some evidence concerning the relationship between income and preferences for public expenditures is available from previously unanalyzed items in national attitude surveys. We are currently in the process of analyzing several such surveys, but the results are available from one which bears on this matter. Representatives from the National Opinion Research Center posed the following questions to a national sample of 1,482 persons: "Suppose the local government wanted to raise taxes to pay for new roads, schools, and hospitals. In general, would you probably be in favor of raising taxes or would you probably be against raising the taxes?" The favorable responses, broken down by income, were as follows:

	Respondent's Total Family Income (Before Taxes)			
	Under $3,000	$3,000 to 9,999	$10,000 to 14,999	Over $15,000
In favor	47.6%	53.1%	61.1%	71.7%

Source: NORC-4050, April 1968

REFERENCES

ADRIAN, C. and O. P. WILLIAMS (1963) Four Cities. Philadelphia: Univ. of Pennsylvania Press.

AGGER, R. E., D. GOLDRICH, and B. E. SWANSON (1964) The Rulers and the Ruled. New York: John Wiley.

AIKEN, M. (1970) "The distribution of community power: structural bases and social consequences," pp. 487-525 in M. Aiken and P. E. Mott [eds.] The Structure of Community Power. New York: Random House.

--- (1969) "Community power and mobilization." Annals 385 (September): 76-88.

AIKEN, M. and R. ALFORD (forthcoming) "Community structure and innovation: public housing, urban renewal, and the war on poverty," in T. N. Clark (ed.) Comparative Community Politics.

AIKEN, M. and P. E. MOTT [eds.] (1970) The Structure of Community Power. New York: Random House.

BANFIELD, E. (1965) Big City Politics. New York: Random House.

--- (1961) Political Influence. New York: Free Press.

BANFIELD, E. and J. Q. WILSON (1963) City Politics. Cambridge: Harvard Univ. and MIT Press.

BOSKOFF, A. and H. ZEIGLER (1964) Voting Patterns in a Local Election. New York: J. B. Lippincott.

CLARK, T. N. (1972) "Urban typologies and political outputs," in B. J. L. Berry (ed.) Handbook of City Classification. New York: John Wiley.

--- (1971a) "Community structure, decision-making, budget expenditures, and urban renewal in 51 American communities," pp. 293-313 in C. M. Bonjean, T. N. Clark and R. L. Lineberry (eds.) Community Politics. New York: Free Press.

--- (1971b) "Community decisions and budgetary outputs: toward a theory of collective decision-making." Presented at NSF-MSSB Seminar on Mathematical Theory of Collective Decisions, Hilton Head Island, South Carolina, August.

--- (1969) "A comparative study of community structures and leadership." Presented at annual meeting of the American Political Science Association, September.

--- [ed.] (1968a) Community Structure and Decision-Making: Comparative Analyses. San Francisco: Chandler.

--- (1968b) "Community structure, decision-making, budget expenditures, and urban renewal in 51 American communities." Amer. Soc. Rev. 33 (August): 576-593.

CLARK, T. N. et al. (1968) "Discipline, method, community structure and decision-making: the role and limitations of the sociology of knowledge." Amer. Sociologist 3 (August): 214-217.

COLEMAN, J. S. (1971) Resources for Social Change. New York: John Wiley.

DAHL, R. A. (1961) Who Governs? New Haven: Yale Univ. Press.

DAVIS, O. A., M. J. HINICH, and P. C. ORDESHOOK (1970) "An expository development of a mathematical model of the electoral process." Amer. Pol. Sci. Rev. 64 (June): 426-448.

DOWNES, B. T. [ed.] (1971) Cities and Suburbs. Belmont: Wadsworth.

HAWLEY, A. H. (1963) "Community power and urban renewal success." Amer. J. of Sociology (January): 422-431.

HUNTER, F. (1953) Community Power Structure. Chapel Hill: Univ. of North Carolina Press.

International City Manager's Association (1966) The Municipal Yearbook 1966. Chicago: ICMA.

KEY, V. O. (1963) Public Opinion in American Democracy. New York: Alfred A. Knopf.

LINEBERRY, R. L. and I. SHARKANSKY (1971) Urban Politics and Public Policy. New York: Harper & Row.

LITT, E. (1970) Ethnic Politics in America. Glenview: Scott, Foresman.

MORLOCK, L. (1971) "Business influence, countervailing groups and the balance of influence in 92 cities." Presented at annual meeting of Committee for Community Research and Development of the Society for the Study of Social Problems, Denver, Colorado, August.

National Council of the Churches of Christ in the U.S. (1956) Churches and Church Membership in the United States. New York: NCCC in the U.S.

PATTERSON, S. C. (1968) "The political cultures of the American states," pp. 275-291 in N. R. Luttbeg (ed.) Public Opinion and Public Policy: Models of Political Linkage. Homewood, Ill.: Dorsey.

PRESTHUS, R. (1963) Men at the Top. New York: Oxford.

ROSENTHAL, D. and R. L. CRAIN (1966) "Structure and values in local political systems: the case of fluoridation decisions." J. of Politics 28 (February): 169-196.

ROSSI, P. H. and R. L. CRAIN (1968) "The NORC permanent community sample." Public Opinion Q. 32 (Summer): 261-272.

SAMUELSON, P. A. (1954) "The pure theory of public expenditures." Rev. of Economics and Statistics (November): 87-89.

U.S. Bureau of the Census (1969) City Government Finances 1967-1968. GF68-No. 4. Washington, D.C.: U.S. Government Printing Office.

--- (1967) County and City Data Book 1967. Washington, D.C.: U.S. Government Printing Office.

WALTON, J. (1970) "A systematic survey of community power research," pp. 443-463 in M. Aiken and P. E. Mott (eds.) The Structure of Community Power. New York: Random House.

WIRT, F. M. [ed.] (1971) Future Directions in Community Power Research: A Colloquium. Berkeley, Calif.: Institute of Governmental Studies, University of California.

12

Local Level Decision-Making and Public Policy Outcomes: A Theoretical Perspective

BRYAN T. DOWNES
LEWIS A. FRIEDMAN

□ IT HAS BEEN MORE THAN TEN YEARS since Richard Dawson and James Robinson (1961: 267) argued that, "The task of political science is to find and explain the independent and intervening variables that account for policy differences." Although a growing number of political scientists have taken up the gauntlet thrown down by Dawson and Robinson and have directed their research efforts toward explaining the authoritative actions (public policies) of political systems, we still appear to be some distance from that goal (Ranney, 1968; Jacob and Lipsky, 1968; Coulter, 1970; Rakoff and Schaefer, 1970).

In this article, we first intend to discuss a number of reasons why this goal has yet to be realized in the area of governmental budgeting. Although a great deal of research has been undertaken in this area, much of it exhibits serious theoretical, conceptual, and methodological shortcomings. Hence, many findings of this body of research are both contradictory and ambiguous. It is important that these

---

AUTHORS' NOTE: *This is a revised version of a paper delivered at the Annual Meeting of the American Political Science Association, Chicago (September 1971), entitled,* "Local Level Decision Making and Budgetary Outcomes: A Theoretical Perspective on Research in Fourteen Michigan Cities." *The authors gratefully acknowledge the financial support of the National Science Foundation*

shortcomings be corrected, for if they are perpetuated, not only will our understanding of the factors affecting governmental fiscal policy be distorted, but a great deal of subsequent research will be misdirected.

Once we have critically examined research on governmental budgeting, we have a second concern, and that is, to suggest an alternative and hopefully more productive research strategy.[1] We are particularly interested in developing a strategy that will facilitate more adequate explanation of why changes occur in the allocation of municipal fiscal resources. Although much of the research on governmental budgeting has been process-oriented, that is, concerned with describing the "how's" of budgeting, our research explores the impact which variation in the budgetary process has on the amount of yearly change in expenditures in fourteen Michigan cities.

A REVIEW OF PAST RESEARCH

BUDGETARY DECISION-MAKING[2]

Sweeping claims have been made for the universality of an executive-dominated, incremental model of budgetary decision-making. For example, John Crecine (1969: 221, 222, 224, 228), in what is probably the major work on municipal budgeting, asserts that "there is little reason to believe this model would not describe the budget process in most large cities in the U.S. . . . smaller municipalities and local governments . . . state and federal governments . . . and the four major remaining categories of non-market organizations." In a similar vein, Meltsner and Wildavsky (1970) claim that "budgeting in Oakland is not much different from budgeting elsewhere."

In reaching such conclusions, previous studies of budgetary decision-making have made use of two basic approaches.[3] One focuses on the patterns of interaction among the participants in this process, particularly the distribution of power and influence. The

---

and the assistance of the Urban Survey Research Unit of Michigan State University under the direction of Dr. Philip Marcus. We are also indebted for support to the Political Science Departments at Michigan State University and the University of Missouri, St. Louis.

other focuses on the cognitive and evaluative processes individual decision makers use when choosing among alternative courses of action (Braybrook and Lindblom, 1963; Smithies, 1955; Schultze, 1968; Lyden and Miller, 1968).

Beginning with a summary of the *first approach,* regarding patterns of interaction (and executive domination), four actors have been identified in the budget process (Crecine, 1969). These are department heads, the executive (either the manager or the mayor), the legislature (called the city council), and community interest groups and elites. Budget decisions are made in a three-stage sequence. First there is the department head's preparation of *requests* for the following year; then the executive reviews these requests and formulates his own budget *recommendations;* while the final stage involves legislative action and the formal *adoption* of the appropriation (budget) ordinance.

In the first stage the department head faces the problem of deciding how much to ask for, particularly how much more to request for next year above what he is currently receiving. Once these requests are formally submitted, Crecine found that department heads are essentially passive participants in budget decisions. They adhere closely to the executive's letter of instruction as it provides cues to the acceptable range of expenditure increases that will be permitted. They make few efforts to affect the decisions being made upon their budget requests by others. For example, they "do not go over the mayor's head to the council to request more funds for their programs or to restore cuts made in the mayor's office" (Crecine, 1969: 50-51).

The executive faces the problem in the second stage of balancing requests with estimates of revenues. Departmental requests almost always exceed available resources. The decision to be made is who shall be cut and by how much. This decision is not made in consultation with either department heads or the legislature.

In the third stage of this process, it has been found that the council does little more than formally adopt the executive's budget into law. Presented with a budget balanced between expenditures and revenues, they are either ill-equipped to make changes in the executive's budget during their review, or else feel constrained by the balanced budget requirement and conflicting political pressures. The result is that they rely most heavily upon the executive's recommended budget. Crecine (1969: 35) reports that the council in "virtually every case, approved the mayor's budget almost exactly as submitted . . ." The council is merely a rubber stamp and plays only a small part in making municipal budget decisions.

Throughout these three stages in the budget process it has been found that participation by community interest groups and external elites is minimal. Hence, their influence is almost nonexistent. Again Crecine (1969: 192) observes, "What our model suggests is that budgets in municipal government are reasonably abstracted documents bearing little direct relationship to specific community pressures."

It is evident, according to this body of research, that the municipal budget process is executive-centered and executive-controlled. For example, Crecine (1969: 38) concludes, "the municipal budget is the mayor's budget in which the mayor's policies dominate. . . . The council and department heads have surprisingly little to say about municipal resource allocation on a macro level." Similarly, and based on their research in Oakland, Meltsner and Wildavsky (1970: 344) conclude that the manager "is the key figure in making most of the decisions . . . the city manager reviews all the budget and, for the most part, makes the decisions. He guides the city Council in its consideration. He feels that it is his budget. And he uses it to make his influence felt throughout city government."

The *second approach* to budgetary decision-making focuses on individual cognitive and evaluative process when in a choice situation. These researchers have concluded that faced with the "limited capacities of human problem solving," the decision maker cannot follow the prescriptions of a rational deductive model. Instead, he proceeds in an incremental fashion. He simplifies the search and selection of alternative policies. Aids to calculation or rules are developed and applied that call for a noncomprehensive and nonexhaustive search for alternatives and their consequences. The result is the selection of courses of action that differ in relatively small degrees from existing ones.

This approach to making budgetary decisions involves a process in which the present level of appropriations to departments are not reviewed in their entirety each year. Instead, existing levels of funding are accepted as the legitimate base for future decisions. Next year's budget is based upon this year's, much as this year's budget was based upon last year's. The decision that now has to be made is simplified, for attention can be concentrated upon the narrow range of expenditure changes in next year's proposals. Budget decisions are then based upon an evaluation of the size of this percentage difference. For example, Wildavsky (1964: 15) observes, "The men who make the budget are concerned with relatively small increments to an existing base. Their attention is focused on a small number of items over which the budgetary battle is fought." And in another article he concludes that, "decision making on the budget takes place in terms of percentages" (David et al., 1966).

Thus the substantive content of what is being decided can be ignored, as can evaluations of program performance. "The criteria employed by financial decision makers do not reflect a primary concern with the nature of the economy, the platforms of the political parties, or articulated policy desires. . . . The criteria of financial decision makers are non-ideological and frequently non-programmatic" (Sharkansky, 1970: 13).

EXPENDITURE POLICY

Within the extensive body of empirical research that has recently been conducted on governmental expenditure policy, two separate foci can be distinguished. The first and most thoroughly researched is variation in spending levels at one point in time. The second and less extensively examined is change in levels of spending over time. Since our research focuses on this latter question, we will only discuss research on expenditure change.

The most notable research on the question of expenditure change is that of David et al. (1966) on the national level and Sharkansky (1970) on the state level. The former study examines the relationship between the President's requests for 56 non-defense agencies and congressional appropriations between 1947 and 1962. They are able to identify a regression equation that accounts for the budgetary outcomes of each agency with very little of the variance unexplained. This equation "is similar to a set of simple decision rules that are linear and temporarily stable." An empirical referent of an incremental model of decision-making has now been identified. The behavior of each actor can be represented by percentage increases or decreases in expenditure levels. The existence of a linear regression equation is taken to represent the history-dependent character of a pattern of small or incremental changes in expenditure levels over time.

Sharkansky reaches a similar conclusion based on an analysis of state expenditures from 1903 to 1965. By means of correlation coefficients, he first demonstrates that expenditure levels within this 62-year span are significantly correlated with each other. "Interstate differences in spending during 1965 bear a resemblance to the differences that prevailed 62 years earlier." He then develops a regression equation to represent the pattern of expenditure change over this time period. Here he finds a close correspondence between expenditure levels estimated on the basis of the regression equation and the actual level of expenditures. On the basis of this analysis he concludes that expenditures change in small degrees from year to year and are, therefore, incremental in nature.

LIMITATIONS OF PAST RESEARCH

A discerning and critical reading of these two bodies of research raises several conceptual and methodological issues. It also causes one to question the validity of the claim regarding the universality of an executive-centered, incremental model of budgeting and of an incremental pattern of expenditure change.

THE CONCEPTUALIZATION AND MEASUREMENT OF EXPENDITURE CHANGE

The present formulation of the concept of incremental policy change is ambiguous, offers few guides for its measurement, and is almost nonrefutable. Braybrook and Lindblom (1963) offer a definition which states that "a small or incremental change is one that within some short period, such as five years, is small or incremental, regardless of the indefinite future . . . [for] any change is nonincremental if one counts its indefinitely cumulative consequences from here to eternity." Certainly to define a "small or incremental change" as one that is "small or incremental" does little to clarify the meaning of the concept or to specify how it is to be measured. It does not provide an unambiguous means of quantitatively distinguishing between an incremental and a nonincremental change in policy. However, it is questionable whether the attempt to provide an absolute standard for the construction of mutually exclusive and jointly exhaustive categories of different size changes in expenditure policy can be achieved. Instead, it is suggested that the size of policy change should be conceptualized in terms of a continuous distribution from small to large. Within any sample of observations a known distribution exists in the size of expenditure change from one year to the next.

Although the conceptualization of incrementalism as well as its measurement can be questioned, the definition offered by Braybrook and Lindblom does point to the size of yearly change in expenditures as its central component. However, previous research, by the statistics employed, has not actually measured incrementalism in this manner. The use of correlation coefficients and regression equations can only determine the existence of a linear statistical relationship and the strength of that relationship, but not its absolute size. These statistics demonstrate that there is a consistent relationship between spending levels in one year and spending levels in another year but not whether the spending levels between years are small or large. Thus even a 50% yearly increase over five years, according to this type of analysis would be interpreted as an incremental pattern of expenditure change, when surely this would not be the case.

Therefore a different measure of expenditure change is offered—one that is based on the size of yearly expenditure changes and one that enables the researcher to quantitatively measure variation in this variable. This measure is simply the percentage change in the level of total per capita expenditures.[4] Taking last year's expenditures as the base, the proportion of next year's spending level is computed. In a later section, this as well as other measures of expenditure change will be discussed more fully.

THE CONCEPTUALIZATION AND MEASUREMENT OF BUDGETARY DECISION-MAKING

Contrary to the claims made for the universality of an executive-centered, incremental process of budgeting, a close scrutiny of past research reveals inconsistencies in several components of this model. There is evidence that the executive does not dominate budget decisions in all situations, and that decision-making rules which emphasize evaluation of program performance are used in making budget decisions.

Returning to the three-stage model of budgeting, Meltsner and Wildavsky (1970: 331) report a different pattern of relationships between the executive and department heads. Here the letter of budget instruction "simply initiates the budgetary process; it does not structure or delimit the actor's decision problem." Hence, the discretion departments have in formulating their budget requests may vary among cities and an important element of executive control may be absent. In this situation the executive is only responding to departmental initiatives, instead of influencing them from the start of the process.

The part legislatures play in budgeting also exhibits considerable variation. Thomas Anton (1964) reports in two of the three cities he studied how the council makes the decisions necessary to balance the budget. The executive merely transmits departmental requests to the council exactly as he receives them, for their review and decision. As previous research has pointed to the centrality of the balanced budget requirement, a situation in which the council makes these decisions cannot be interpreted as an executive-centered budget process.

Just as the relationship between the executive and department heads, and the executive and the legislature can vary, so too can the relationship between department heads and the council display a different pattern. Again Anton (1964: 15-16) describes a situation where extensive interaction between these two actors takes place. He writes that "the cutting process [of the council] takes place in relatively informal and generally private meetings between departmental representatives and the council Finance Committee." Solutions are "worked out in the confrontation of committee with departments" (Anton, 1964: 15-16). Where such a pattern exists, the formation of a coalition by these two actors against the executive is a possibility.

The research of David Caputo (1970: 11-12), points out how external community interests, which have been factored out of the dominant model of budgeting, may participate meaningfully in budgetary decision-making. In reporting on his four-city study he writes that community interest groups did participate in the budgetary process and "attempted, usually with some degree of success, to influence the budgetary process and to change budgetary allocation decisions."

Because the executive's influence over other budget actors has been shown to vary, the conclusion that the executive dominates the process of municipal budget-making is open to question. According to Caputo (1970: 11), "it is possible to conclude that no one participant (or group of participants) can formally or informally dominate the budget process." Additional support for the existence of a decentralized decision-making process, in which the executive does not dominate, is found in the work of two other researchers. Meltsner and Wildavsky (1970: 336) write that the budgetary process is made up of a series of bargaining arenas and opportunities: "It contains numerous levels of review and appeal." Decisions that are made at one stage can and are appealed to the next and higher stage. Finally the assistant city administrator of Ann Arbor, Michigan (Borut, 1970: 298-299) described that city's budget process as an "advocate process" where "Department heads plead their case and the city administrator argues for reductions on behalf of a balanced budget. If a department head is persuasive . . . cuts may be restored." Thus, this aspect of the process of budgeting exists as a continuous distribution from a hierarchical, executive-controlled process at one end, to a legislative-dominated process at the other end of the scale, with a decentralized bargaining structure in the middle.

Evidence can also be presented to show that rules or criteria other than incremental ones are being used in making budget choices. The development and application of a Program Planning Budgeting System (PPBS) is such an alternative and is similar to a rational deductive model of decision-making. The reports of its spread throughout different levels of government is ample evidence that program evaluation criteria are being used to make budget decisions (Mushkin, 1970, 1969).

When PPBS is being used, expenditures are organized into program categories which represent the various goals or objectives that are being sought. Budget decisions are then made on the basis of a systematic and comparative analysis of alternative spending proposals, both within and among program categories. The selection among alternative spending proposals is based on the criteria of program cost and program performance. There is no program that is justified regardless of cost, and there is no cost that is justified regardless of program performance.

The use of such criteria, although not a fully developed program planning budgeting system, is reported by Anton (1964: 11). In one of the three cities he examined, the budget "takes on the character of a financial plan, which states the dollar and cents costs of achieving specified policy goals." Department heads are "required to present detailed justifications for each request and are charged specifically with analyzing their entire departments operations with a view of effectuating every economy possible." At the same time the budget is a tool for planning future activities, as each department is once again responsible for "presenting programs and proposals for new and higher standards of service."

In this secondary analysis of past research, then, there is ample evidence to suggest that variation in the process of municipal budget-making exists. With this variation established, our next step is to formulate hypotheses regarding the relationship between municipal budgetary processes and expenditure change. But before these hypotheses are presented, other aspects of our alternative research design will be discussed.

AN ALTERNATIVE RESEARCH STRATEGY

Several of the problems uncovered in past studies of municipal budgeting can be traced, in large part, to two additional methodological limitations of this body of research. First, case studies cannot establish the representativeness of particular cases being studied; therefore, they cannot support the sweeping generalizations that have been made. Second, by narratively recreating events, these studies do not generate precise measurements of the phenomena under observation. Our research, while building upon the substantive concerns of past case studies, seeks to go beyond these two limitations of previous research designs. By examining 14 cities at one time, a number that itself is greater than the total of all previous empirical research on municipal budgeting, a comparative focus is achieved. By gathering data through structured interviews with 170 decision makers, systematic and quantitative data will be collected. This will enable us to systematically measure variation in the way local decision makers in our 14 cities go about formulating and adopting their budgets.

THE SELECTION OF CITIES

The first problem with case studies, as just stated, is their inability to support an extension of their findings beyond the circumstances of the particular cases studied. There is no way to evaluate whether the findings are the result of some unique characteristic present in a particular situation. Whether the description holds in other situations cannot be determined. Therefore, the model of an executive-centered, incremental process of budgeting cannot be verified by the study of one, two, three, or even four cases.

However, the absolute number of cases under investigation does not totally preclude making empirical generalizations, if one takes care to specify precisely the universe to which the generalizations pertain. The problem of past research has often been its insensitivity to this particular methodological question, and, consequently, all-inclusive generalizations have been made. The major work in this field, by John Crecine, makes near-universal claims for the applicability of a model of budgeting derived from the study of three cities–Detroit, Pittsburgh, and Cleveland, whose superficial characteristics, at least, make them atypical of all cities in the United States. Not taken into consideration, for example, in generalizing from these cities are differences in governmental structure among municipalities, as well as population, budget size, and budgetary procedures. It is noteworthy that many of the differences in municipal budgetary processes are found in middle-sized cities (populations between about 25,000 and 250,000) and in cities having different structural characteristics. But these two factors cannot explain different patterns of budgeting for they are but nominal categories for a number of as yet undefined concepts. Furthermore, it is not the intention of this research at the present time to account for variation identified in the process of budgeting. Instead, this discussion is presented in order to sensitize us to the nature of the generalizations that can legitimately be made about municipal budgeting from a particular "sample" of cities.

With these considerations in mind, we excluded Detroit and its suburbs, and concentrated our attention upon the middle-sized core cities of metropolitan areas that dot Michigan. The mean population of these cities is 97,057, ranging from a low of 26,144 to a high of 195,892. There are 11 manager cities and three mayor cities. Thus, the range of generality of this study does not extend to all cities, or even to all middle-sized cities, but to the class of core cities of small metropolitan areas.

DATA-COLLECTION PROCEDURES

The second limitation of case studies has to do with their data-collection procedures. By a narrative recreation of events, case studies provide a vivid and detailed description of what took place. However, this data-collection procedure often results in retelling a story. This is not to say that the description offered is inaccurate but that the data gathered are impressionistic and highly unsystematic.

That the claim for the universality of an executive-dominated, incremental model of budgetary decision-making could exist at the same time that inconsistencies and contradictions are found, testifies to the problems inherent in case studies. In the absence of any measurements, comparisons among cases are limited. All that can be reported are the interpretations of the researcher based upon his intimate, yet highly personalized knowledge of the facts. Measurement is limited to classificatory statements regarding the presence or absence of a certain phenomenon. But this dichotomization of the real world into either/or categories, while a necessary first step in making comparisons, is one that needs to be followed by more precise measurements.

Unfortunately researchers doing case studies are not able to make such measurements. A case in point is the discussion of incremental decision rules. Most often they have been presented as mutually exclusive and jointly exhaustive. However, this is a dichotomy that is being imposed upon the reality of a continuous distribution. Case studies cannot determine the position of cities along this scale, not only because too few observations are being made, but because they do not generate the quantifiable data that are necessary to do so. Consequently, variation in the process of budgeting is neither identified nor recognized.

In our research we have gathered data in a form that will enable meaningful quantitative comparisons. The first step in this direction was the administration of a structured interview schedule to 170 authoritative decision makers in the 14 cities. These included three department heads in each city (police, public works, and parks and recreation); all the chief executives and other administrative officials who participated in making budget decisions; and as many legislators as would consent to be interviewed. An attempt was also made to go beyond the nominal categories of most social research by designing questions in terms of an ordinal scale of measurement.

THE MEASUREMENT OF EXPENDITURE OUTCOMES

Within a three-stage model of budgeting, there are separate expenditure outcomes for each of the three stages. It is variation in these outcomes that we seek to explain. The conceptualization of municipal budgeting in terms of a staging model allows for a more accurate and detailed investigation of budgetary decision processes and expenditure-change policies in each city. Previous reports that expenditures change incrementally have usually been based on the analysis of spending figures from the budget formally adopted by the council. However, this single measure masks the varying relationships among budget actors that can be revealed by a more detailed analysis of the outcomes of each particular stage. An incremental pattern of change at the final stage can be the result of any one of several possible combinations of actions by budgetary actors. For example, an incremental pattern may result from each actor along the sequence of decision-making asking for only a small percentage increase above last year's spending. However, a more likely pattern is where department heads ask for substantial changes only to be reduced to an incremental pattern by the actions of the executive or the council. Even when the executive supports department head requests, these may still be reduced by the legislature. A situation may even exist where department heads do not ask for large increases, but the executive does, and it is his recommendations that are reduced by the council to conform to an incremental pattern of change.

Such a conceptualization not only reveals the complexity and multidimensionality of the budgetary process and patterns of expenditure change, but also two dimensions to the behavior of budgetary actors. One is a *vertical dimension,* measured by the percentage increase or decrease in per capita expenditures from one year to the next. The other is a *horizontal dimension* which measures the actions of the executive and the legislature as they act in response to the budget requests or recommendations they receive for review. This is represented by the percentage change in the budget made by each of these two actors in the budget of the previous actor (the changes made by the executive in department head requests, and the changes made by the council in executive recommendations).

These five measures of expenditure changes can be ordered sequentially to represent decisions made in the course of formulating and adopting the budget. First there is departmental budgeting, the outcome of which is singularly measured (see Table 1) by the percentage increase in funds requested above what is currently being received.[5] This measure shows a mean increase of 15.59% (median 15.00%) which is relatively large. The underlying assumption that

TABLE 1
PERCENTAGE PER CAPITA INCREASE IN DEPARTMENTAL REQUESTS, 1970-1971

City Number	Percentage Increase
07	30.96
05	29.18
01	28.73
09	20.06
12	18.48
14	17.72
13	15.76
02	14.24
03	13.92
10	11.21
06	9.17
04	7.26
11	3.51
08	−1.86

Mean +15.59
Median +15.00
Standard Deviation 9.65

TABLE 2
PERCENTAGE CHANGE MADE BY EXECUTIVE IN DEPARTMENTAL REQUESTS, 1971

City Number	Percentage Change
07	−14.26
01	−13.04
13	−10.06
09	− 7.43
06	− 4.86
14	− 4.86
03	− 3.62
05	− 2.72
04	− 2.05
12	− 1.92
02	− 1.12
11	− .54
08	− .13
10	− .13

Mean −4.76
Median −3.17
Standard Deviation 4.71

department heads always ask for more money is not quite supported by the data, for in one city there is an absolute decrease of –1.86% in the amount of funds requested from what was previously received. The data reveal considerable variation in the amount of departmental requests, ranging from this low value of –1.86 to a high of +30.96%. The standard deviation of 9.65 is a high value, and is more than half the mean. Clearly there is considerable variation in the size of departmental expenditure requests in our 14 cities.

The outcome of the second or executive stage in municipal budgeting is represented by two measures. The first of these (see Table 2) is the executive's response to departmental requests. The first thing noticeable about this response is that in all 14 cities the executive acts to reduce departmental budgets. The mean reduction is 4.76% (median 3.17%). Once again we see large variation in the size of this response, as this reduction ranges from a very small decrease of –.13% to the very large one of –14.26%. The variability in the executive's response to departmental requests is indicated by the size of the standard deviation, whose value of 4.71 is almost exactly the same as the mean. Once again, variation in the behavior of budgetary actors is quite evident.

The second measure of expenditure change in the executive stage of municipal budgeting focuses on executive decisions on departmental requests and is measured by the size of expenditure change from the previous year's budget that is recorded in his submitted recommendations to the council. Here we find (see Table 3) a mean increase of 8.41% (median 9.62%) and positive in all cities except one. The range of this measure extends from a reduction of 2.01% to an increase of 24.44%. The standard deviation of 6.96 is close to the mean and again indicates the wide range of differences among cities in the behavior of the chief executive and in the expenditure outcomes of the budgetary process.

The outcomes of the decision-making process involving the legislature are represented by the same two measures as in the executive stage. The first looks at the council's response to the executive's recommended budget (see Table 4). Here we observe a much lower absolute level in the size of council changes than the executive made in reaction to departmental requests. Thus the average change was a modest reduction of –.74% (median –.06%). But this hides more than it reveals, for the variability of council action is great. Three councils increased the executive's budget, three councils accepted executive recommendations, and eight councils

TABLE 3
PERCENTAGE PER CAPITA INCREASE IN EXECUTIVE RECOMMENDATIONS, 1970-1971

City Number	Percentage Increase
05	25.44
12	16.06
14	11.74
09	10.53
07	10.52
01	10.44
11	9.76
03	9.49
04	4.98
06	3.59
13	3.48
11	2.94
02	.81
08	−2.01

Mean +8.41
Median +9.62
Standard Deviation 6.96

TABLE 4
PERCENTAGE CHANGE MADE BY LEGISLATURE IN THE EXECUTIVE'S RECOMMENDATIONS, 1971

City Number	Percentage Change
01	1.64
14	.85
07	.01
05	.00
08	.00
10	.00
09	− .05
07	− .06
12	− .18
02	− .22
03	− .38
04	−2.45
13	−4.73
11	−4.84

Mean −.74
Median −.06
Standard Deviation 1.91

reduced the size of the executive's budget. These changes range from an increase of 1.64% to a decrease of 4.84%. The standard deviation of this measure is extremely large: 1.91, which is almost three times the size of the mean. The kind of decision made by the council shows the largest degree of variability among any of the measures of budgetary change.

The second (see Table 5) examines the percentage change in per capita expenditures in the budget adopted this year (1971) as compared to the budget adopted last year (1970). This is the measure employed in previous studies of expenditure change. The mean value of this change is an increase of 7.61% (median 9.40%), a modest increase that could easily be interpreted as a "small or incremental" change in policy. However, we immediately see that variability in this measure of change is quite large. The range of yearly changes in spending levels extends from 2.08% reduction to a 25.44% increase. Four cities experienced an absolute decline in spending levels and 10 cities show an increase. The standard deviation of this measure (7.95) is again larger than the mean, indicating the extreme variability of the size of expenditure change. Whether these changes are incremental is somewhat beside the point. The more interesting and the more challenging problem is to explain

TABLE 5
PERCENTAGE PER CAPITA INCREASE IN
THE LEGISLATURES ADOPTED BUDGET, 1970-1971

City Number	Percentage Increase
05	25.44
12	15.84
14	12.73
01	12.32
07	10.51
09	10.00
10	9.76
03	9.05
06	4.28
04	2.33
02	– .14
13	–1.48
08	–2.01
11	–2.08

Mean +7.61
Median +9.40
Standard Deviation 7.95

the wide variation in expenditure changes among the 14 cities, not only in this last measure, but in all expenditure change measures devised, since significant differences are found to exist among each of them in the cities studied.

BUDGETARY DECISION-MAKING AND EXPENDITURE OUTCOMES

It is not logically possible to account for variation in the absolute level of spending among cities by examining budget decisions made in any given year. Funding levels at any one point in time are the cumulative result of the community's history, social and economic conditions, and decisions made in the past. However, it is possible to explain changes made in those budget levels from one year to the next by the study of the decision-making process in that time period. Thus, the explanation of change in expenditures from fiscal year 1970 to 1971 can be made by the investigation of the budget process in 1971.

Linking the process of municipal budget-making to expenditure-change policies, however, is difficult, since up to this point research on budgeting has not systematically examined this relationship. In its emphasis upon a case study description of what occurs in the course of formulating and adopting the budget, the examination of its consequences—its impact on the actual content of the expenditures decisions being made—has been neglected.

An example is provided by Thomas Anton (1964: 14) who reports on variation among the three cities he studied in such features as the "formalization, rationalization, professionalization, length of time devoted to budgeting, amount of information gathered, existence or nonexistence of the budget concept, and presence or absence of budget review." However, at no point does he examine expenditure patterns in these three cities. He does not ask whether these cities, first of all, have substantively different expenditure policies and then whether such policy differences could be attributed to the observed variation in their budgetary process. The question of what difference it makes that cities go about budgeting differently, is left answered.

The hypotheses formulated and to be tested in this study are but a first step, and a very tentative one, toward answering this question. These hypotheses are often simplistic and are formulated in a one-variable cause-and-effect manner that obviously will need re-formulation as the data analysis proceeds. The relationships are formulated in a univariate direction to explain the behavior of each set of actors across the three stages of decision-making and to account for both the vertical and horizontal measures of expenditure change. This too may need revision as the data reveal the varying impact of these process characteristics upon the different expenditure changes of each stage. The specific hypotheses are derived from the previously described model of budgeting and the two approaches to the study of decision-making. A third variable, decision maker attitudes, is added and will be discussed later.

The direction of these hypotheses are deduced from the specific findings of past research. An "incremental" or small amount of expenditure change is reported as the policy outcome of an executive-centered, incremental process of decision-making. Therefore it is assumed that any divergence from this pattern in the budgetary process will result in a different size change in expenditures from one year to the next. A decentralized program-evaluation system of budgeting produces larger amounts of expenditure change. The research of Terry Clark (1968) on 51 cities lends partial support to this view as he reports a higher level of budgetary outputs in cities having a decentralized decision-making structure.

An analysis of the expenditure data additionally supports the formulation of the hypotheses in this particular direction. The mean percentage of expenditure change is highest at the first stage of departmental requests (15.66%). Decisions made in the next two stages serve to reduce the size of these increases. The executive cuts these requests by an average of –4.77%, and the legislature in turn adds a further modest reduction of –.74%. The result of these two decisions is the reduction of the size of expenditure increases from the previous year. At the time of the executive's recommendations, the budget is an average 8.41% above the previous year, as almost one-half the additional funds were reduced. The council, by its smaller reductions, finally adopts a budget that is an average 7.61% above the budget they adopted in the previous year. We can assume, then, that department head requests function as demands for budgetary increases which are then reduced by others. In an executive-centered, incremental process of decision-making, the executive possesses sufficient power and influence to bring about a reduction in these requests. The result is a budget more in line with current levels of spending. On the other hand, a decentralized program evaluation process is one in which the executive does not have the resources to act as an "economizer" and keep down departmental requests. In this situation, department heads are able to obtain more of what they originally asked for, and the size of the budget finally adopted displays a larger percentage increase in expenditure levels. When decisions are not made hierarchically by the executive on the basis of an incremental comparison of spending levels, expenditures will exhibit a larger rate of increase.

However, it should be noted that an alternative direction to these hypotheses and an alternative explanation for these relationships can be formulated. Some researchers have pointed to a centralized decision-making system as being most conducive to policy innovation and change. A decentralized bargaining structure, this research contends, produces stalemate and "incremental" policies (Banfield, 1961). Which of these alternative explanations will be supported by this particular research is still unknown.

It is for these reasons that it should again be emphasized that the particular hypotheses presented here are but tentative and preliminary explorations of the relationship between process characteristics and public policy.

The *pattern of executive domination* of the budget process has been broken down into five different components and five separate hypotheses.

HYPOTHESIS 1

> The more discretion each actor has in preparing his own budget, the *more* that budget will display expenditure increases from the previous year.
>
> The more discretion each actor has in preparing his own budget, the *more* changes will be made in that budget by subsequent actors.

Initially, we will examine the inputs each actor receives from other actors in the course of preparing his budget. Specifically, the discretion or freedom one has in formulating requests and answering the question of how much to ask for, can vary. As has been reported earlier in this paper, inconsistencies appear in this aspect of department head-executive relationships. In some cities, departments are constrained in making requests by executive instructions and policy guidelines. In other cities, departments appear to have substantial leeway in making their budget requests. This hypothesis directly explores the consequences of such differences for the kind of requests that actually will be made. We think the less control an executive exerts over the initial formulation of departmental budgets, the larger will be the expenditure increases department heads request.

It is also important to note that this feature of the budget process applies to the relationship between the executive and the legislature. The council may establish guidelines for the executive to follow in his dealings with department heads and in formulating his own budget recommendations. It may be that the executive is not an independent actor, but is merely implementing the policies of the council. In this situation, the executive would surely not be the dominant budgetary decision maker.

The relationship between the amount of discretion each actor has and the amount of change made in his budget by subsequent actors is more tenuous. It is derived from the assumed positive relationship between the size of one's requests and the size of reductions made in those requests. The more one asks for, the more one gets cut (Sharkansky, 1968; Sharkansky and Turnbull, 1969). Thus, if having leeway results in larger requests, then a causal connection exists between having leeway and experiencing larger reductions in those requests.

HYPOTHESIS 2

> The more each actor plays an active role in making budget decisions, the *more* that actor's budget will display expenditure increases from the previous year.
>
> The more each actor plays an active role in making budget decisions, the *more* changes will be made in the budget by subsequent actors.

Previous research has pointed to the existence of a specific role for each actor in the budget process. Department heads are supposed to be expansionary and continually seeking to increase their spending levels, while the executive and the legislature act in varying degrees as economizers to reduce these requests. Do department heads who see their role as obtaining more funds for their department actually ask for more funds as opposed to department heads who do not define their role in this particular way? The same question would pertain to the relationship between normative role expectations and behavior for the executive and the legislature.

A second aspect of the concept budget roles taps the involvement of each actor in making budget decisions. The responses of the council, for example, range from passive participation of only formally adopting the budget ordinance, to an intermediate position of reviewing the executive's recommendations and making additions or deletions to those recommendations, to playing the primary part in making budget decisions. These budget roles can be scaled along a passive-active continuum in terms of each actor's involvement in the decisions that are being made on the budget. The past findings of a passive role for the council can be investigated by testing the following hypothesis—the more active a role the council perceives for itself, the more changes they will make in the budget submitted by the executive.

HYPOTHESIS 3

> The more the interaction among actors is characterized by a bargaining relationship, the *more* that actor's budget will display expenditure increases from the previous year.
>
> The more the interaction among actors is characterized by a bargaining relationship, the *more* changes will be made in the budget of subsequent actors.

These hypotheses direct our attention to the behavioral properties of the interactions among budgetary actors and, in particular, the degree of executive centralization in the decision-making process. The alternatives range from a hierarchical, executive-controlled process of budgeting, to a decentralized bargaining system, to one in which the legislature assumes a very significant if not a dominant position in making budget decisions.

While formal meetings are held between department heads and the executive and often between department heads and the legislature, what takes place at these meetings as well as the purpose these formal contacts serve in the course of budgeting can vary. These meetings may indicate an executive-controlled process and be used merely to inform departments of previously made decisions; or department heads may have their "day in court" and the opportunity to plead their case; or budget decisions may actually be made in the meeting. A similar, if slightly different scale ordering, would pertain to the purpose of department heads meeting the council, and would also indicate variation in executive dominance. Of particular interest would be the case where department heads use this meeting as an opportunity to go around the executive and appeal to the council to restore cuts made by the executive.

Another way to gauge the extent of executive control is to analyze the way disagreements among budgetary actors are resolved. Again, as in the example of the relationship between department heads and the executive, the resolution of disagreements may take place in a way that indicates the dominance of decision-making by the executive. This would be a situation where disagreements may be aired, but are then resolved hierarchically by the executive. However, they may be settled through bargaining and negotiation among these two actors, or by department-head attempts to form an alliance with the council in order to influence the executive. Thus, an ordering of responses pertaining to how disagreements are resolved may also reveal further variation in the degree of executive centralization. The existence of bargaining and negotiation among the actors is a breach in the executive's control and domination of budgetary decisions and would be reflected in the department's ability to obtain a larger part of their requests, and by the council making more changes in the budget they receive for review.

HYPOTHESIS 4

> The greater the extent that a "non-decision-making" process takes place in the course of making budget decisions, the *less* that actor's budget will display expenditure increases from the previous year.
>
> The greater the extent that a "non-decision-making" process takes place in the course of making budget decisions, the *less* changes will be made in the budget by subsequent actors.

Previous studies of budgeting have only reported the formal, structured pattern of interactions among the participants in budgetary decision-making and have ignored the informal contacts and unexpressed calculations involved in the process. These aspects of budgeting can be characterized by using the concept "non-decision-making" as formulated by Bachrach and Baratz (1970). In order to obtain a complete picture of the distribution of power and influence within the decision-making structure, it is necessary to identify what budget requests were not made by each actor because of calculations of the anticipated opposition of relevant others. Often such considerations enter into the cognitive and evaluative process of the individual as he decides what level and kind of budget should be submitted to another actor for review. The questions that were employed to uncover this process were: what kinds of decisions were left to others to make; if requests were avoided because of the expected opposition of others; whether informal indications were received of budget items that would not be acceptable to others; and whether informal approval of specific budget items was sought before a formal request was made.

An analysis of only the budget figures that were officially recorded in the printed budget document would indicate only part of the process of municipal budget-making. It does not reveal what expenditures went formally unrequested but were informally sought and not approved, or whether the expected opposition of others was taken into consideration and was sufficient to prevent the formal voicing of a request. The lack of overt change by the council in the executive's budget can partly be explained by this non-decision-making process. The executive, in his dealings with the council throughout the entire year, knows the priorities and policy positions of the council. He does not often make a budget proposal that he knows will meet with immediate opposition. He does not generate conflict if it is avoidable. In these situations the council has a greater part to play in determining budgetary outcomes than would be revealed by simply looking at their formal changes in the budget document. The interesting question regarding the role of the council in these cases is not so much what changes they made in the budget, but what spending proposals were not made by the executive because of anticipated council opposition.

The effect of this process upon the amount of expenditure change from one year to the next and from one actor to the next is to decrease its size. When non-decision-making considerations exist, the formal budget figures are less important and should show a smaller increase, or change, than actually occurred and which would be recorded officially in those cities where this process was not in operation.

HYPOTHESIS 5

> The more that groups and individuals external to the formal structure of government participate and have influence in the budget process, the *more* expenditures will change from the previous year.
>
> The more that groups and individuals external to the formal structure of government participate and have influence in the budget process, the *less* expenditure changes will be made by each actor.

The final component of the first approach to budgetary decision-making has to do with the level of participation and influence of community interest groups and individuals outside the formal structure of government. It will be recalled that previous research has reported a decision-making process on the budget that is relatively closed and insulated from these external pressures. It is assumed that when such forces are present they act as an impetus for the expansion of municipal services and, hence, an accompanying increase in the level of municipal spending. They would function to support departmental expenditure requests for more funds. Therefore, the more they are active within the budgetary process and the more influential they are in this process, the more expenditures should increase from one year to the next and the less change would be made by each actor in the budget they receive for review.

If, on the other hand, such participation results in greater controversy and conflict, a decision about how particular fiscal resources should be allocated may be more difficult to reach. In addition, the final decision, because it must be acceptable to many actors, may fall short of the original demands made by the supporters of expenditure change. We might expect, then, that the more individuals and groups external to the city government who participate in a decision, and the more such participation engenders conflict, the less expenditures will change.

In addition, if community groups are regular participants in budgetary decision-making, and the decision process as a result is constantly engulfed in conflict, budgetary decision makers may be reluctant to consider expenditure changes which are potentially conflictual. In this case, decision-making may be confined to consideration of relatively noncontroversial increments in expenditures.

HYPOTHESIS 6

> The greater the extent that program evaluation rules are applied in the course of making budget choices, as opposed to the application of incremental rules, the *more* expenditures will change from the previous year.

The second approach to the study of budgetary decision-making focuses upon the use of different *decision-making rules* in a choice situation. Many separate questions were asked in an attempt to explore the nature of the decision rules used by various actors. We are arguing that the use of different decision rules should affect the size of yearly expenditure changes. For example, Allen Schick

(1969: 137) has pointed out that, "Because of its future orientation, system budgeting (PPBS) is likely to induce somewhat larger annual budget shifts than might derive under process rules (incrementalism)."

HYPOTHESIS 7

> The more "activists" the orientations held by decision makers about the policy-making role of local government, the more expenditures will increase from the previous year.

This hypothesis introduces a third component into our analysis of the budgetary process and one that heretofore has not been examined, and that is, the *attitudes of individual decision makers.* These attitudes, however, are only an underlying impetus to behavior, and their impact can be mitigated by either the interactions taking place among budgetary actors or decision-making rules.

Many different kinds of attitudes are relevant and could affect the behavior of budgetary actors. The ones we consider include the policy priorities of decision makers in different spending areas, general evaluations of the role of local government in the community (i.e., encouraging economic growth, increasing services, and so on), fiscal relations with higher levels of government, and local taxation policies. The underlying conceptual dimension is the willingness to use government and its resources to achieve some specific policy objective. This can be scaled along a passive-active continuum of local government policy-making responsibilities. In some cities, decision makers desire government to be innovative and aggressive in initiating and promoting new policies, while in others this is not the case. The more activist these attitudes, the more change should be seen in expenditure levels from one year to the next, since the budget is used to achieve these policy goals.

HYPOTHESIS 8

> The greater the extent a revenue constraint is perceived by decision makers, the *less* expenditures will increase from the previous year.

One last concept and hypothesis needs to be presented, and that is the impact of the *availability of revenues.* Certainly if no additional funds are available, then the budget cannot increase. The data show that while the final budget increased an average of 7.61% over the 14 cities, in four cities there was an average absolute decline of −1.43%.

However, limitations on the availability of additional revenues are less relevant than perceptions of the availability of additional funds by budgetary actors. Therefore this question is formulated in terms of attitudes and not objective measures of the ability to support additional taxes. The more these revenue constraints are felt by authoritative decision makers, the less will the total size of the budget increase.

CONCLUSION

Throughout this discussion we have kept three related goals in mind. Our first goal has been to critically examine research on budgetary decision-making. Our second goal has been to suggest how variation in the way in which budgetary decisions are made may affect the likelihood of expenditure change at the local level. Our third and somewhat implicit goal has been to further the development of research which will give rise to generalizations about how differences in decision-making processes affect the prospects for policy change. For those interested in social engineering or making government more responsive to citizen and group demands, the realization of these goals should provide important clues as to where, when, and how to intervene most effectively in the policy-making process in your local community (Downes, 1971: 1-27).

NOTES

1. Throughout our critical review of research we will focus particular attention on materials on municipal budgeting. However, despite this focus, most of the criticisms we will make are equally applicable to work on governmental budgeting at the state and national levels, as well as research in other policy areas. Particularly critical has been the failure to adequately conceptualize and measure (1) various components of the policy-making process, and (2) various types of public policy outcomes. Equally important has been the general unwillingness of researchers to (1) comparatively examine the impact which variation in the policy-making process has on various types of public policies, and (2) the interrelationship of various policy outcomes, that is, *public policy,* or the actions of government; *policy outputs,* or the service levels affected by public policy; and *policy impacts,* or the effect policy has on people's lives, the alleviation of particular problems, or citizen demands for policy change.

2. Some representative studies of budgetary decision-making include Wildavsky (1964), Crecine (1969), Anton (1966, 1964), Gerwin (1969), Fenno (1966).

3. This is an analytical distinction which has been imposed on this body of research. Most research in this area incorporates both approaches.

4. Only general-fund expenditures were employed in the calculation of this measure. This served to maximize local discretion and control. Additional measures could also be developed of expenditure changes in various broad functional or specific policy areas.

5. Specific names of cities have not been used in order to protect the confidentiality of the respondents.

REFERENCES

ANTON, T. J. (1966) The Politics of State Expenditures in Illinois. Urbana: Univ. of Illinois Press.

--- (1964) Budgeting in Three Illinois Cities. Urbana: Institute of Government and Public Affairs.

BACHRACH, P. and M. S. BARATZ (1970) Power and Poverty. New York: Oxford.

BANFIELD, E. C. (1961) Political Influence. New York: Free Press.

BORUT, D. J. (1970) "Implementing PPBS: a practitioner's viewpoint," in J. P. Crecine (ed.) Financing the Metropolis: Public Policy in Urban Economies. Beverly Hills: Sage Pubns.

BRAYBROOK, D. and C. E. LINDBLOM (1963) A Strategy of Decision. New York: Free Press.

CAPUTO, D. A. (1970) "Normative and empirical implications of budgetary processes." Delivered at the 1970 Annual Meeting of the American Political Science Association, Los Angeles.

CLARK, T. N. (1968) "Community structure, decision making, budget expenditures and urban renewal." Amer. Soc. Rev. 33 (August): 576-593.

COULTER, P. B. (1970) "Comparative community politics and public policy: problems in theory and research." Polity 3 (Fall): 22-43.

CRECINE, J. P. (1969) Governmental Problem Solving: A Computer Simulation of Municipal Budgeting. Chicago: Rand McNally.

DAVID, O., M. A. H. DEMPSTER, and A. WILDAVSKY (1966) "A theory of the budgetary process." Amer. Pol. Sci. Rev. 55 (September): 529-547.

DAWSON, R. E. and J. A. ROBINSON (1961) "Inter-party competition, economic variables, and welfare policies in the American states." J. of Politics 25 (May): 265-289.

DOWNES, B. T. [ed.] (1971) Cities and Suburbs: Selected Readings in Local Politics and Public Policy. Belmont, Calif.: Wadsworth.

FENNO, R. (1966) Power of the Purse. Boston: Little, Brown.

GERWIN, D. (1969) Budgeting Public Funds: The Decision Process in an Urban School District. Madison: Univ. of Wisconsin Press.

JACOB, H. and M. LIPSKY (1968) "Outputs, structure, and power: an assessment of changes in the study of state and local politics." J. of Politics 30 (May): 510-538.

LYDEN, F. J. and E. G. MILLER [eds.] (1968) Planning Programming Budgeting: A Systems Approach to Management. Chicago: Markham.

MELTSNER, A. J. and A. WILDAVSKY (1970) "Leave city budgeting alone!: a survey, case study, and recommendations for reform," in J. P. Crecine (ed.) Financing the Metropolis: Public Policy in Urban Economies. Beverly Hills: Sage Pubns.

MUSHKIN, S. (1970) "PPB for the cities: problems and next steps," in J. P. Crecine (ed.) Financing the Metropolis: Public Policy in Urban Economies. Beverly Hills: Sage Pubns.

--- (1969) "PPB in cities." Public Administration Rev. 29 (March/April): 167-178.

RAKOFF, S. H. and G. F. SCHAEFER (1970) "Politics, policy, and political science." Politics and Society 1 (November): 51-77.

RANNEY, A. [ed.] (1968) Political Science and Public Policy. Chicago: Markham.

SCHICK, A. (1969) "System politics and systems budgeting." Public Administration Rev. 29 (March/April): 137-151.

SCHULTZE, C. L. (1968) The Politics and Economics of Public Spending. Washington, D.C.: Brookings Institution.

SHARKANSKY, I. (1970) Spending in the American States. Chicago: Rand McNally.

--- (1968) "Agency requests, gubernatorial support, and budget success in state legislatures." Amer. Pol. Sci. Rev. 57 (December): 1220-1231.

--- and A. TURNBULL III (1969) "Budget making in Georgia and Wisconsin: a test of a model." Midwest J. of Pol. Sci. 13 (November): 631-645.

SMITHIES, A. (1955) The Budgetary Process in the United States. New York: McGraw Hill.

WILDAVSKY, A. (1964) The Politics of the Budgetary Process. Boston: Little, Brown.

13

The Comparative Analysis of Urban Policy: Canada and the United States

EDMUND P. FOWLER
ROBERT L. LINEBERRY

□ DESPITE NUMEROUS APPEALS for comparative urban research in political science, what often passes for comparative analysis is either (1) analysis and "comparison" of samples of American cities (Crecine, 1969; Clark, 1971; Lineberry and Fowler, 1967; Wilson, 1968), or (2) studies of individual non-American cities or urban systems within a single polity (e.g., Cattell, 1968). We do not wish to reflect adversely upon the quality or seminal character of these studies. But we may nonetheless conclude that the former, however conceptually rich and systematic, suffers from parochialism, while the latter, however cosmopolitan, is neither truly comparative nor always methodologically exacting.

The present study compares the role of sociopolitical variables and fiscal policies in two national systems of cities, the United States and Canada. Canadians are surely wearied by comparisons of *Canada and Her Great Neighbour* (Angus, 1938). But, as we attempt to

---

AUTHORS' NOTE: *For financial support, we are indebted to a grant from the Canada Council, though all arguments and errors are the product and property of the authors. We have also incurred debts to numerous provincial and municipal officials who made our data gathering less arduous.*

demonstrate in the first section of this paper, there are enough similarities between urban systems in the two nations to make comparative analysis possible, yet sufficient differences to make it interesting. This study is, in a sense, doubly comparative, since it compares the relationship between urban policy and environment both within and between two polities.

A COMPARATIVE OVERVIEW

An appreciation of some of the major similarities and differences between urban systems in the United States and Canada will provide a useful backdrop to the remainder of our analysis.[1]

SOCIOECONOMIC AND POLITICAL ENVIRONMENT

The familiar American pattern of industrial and demographic decentralization has been repeated in Canada's urban areas. While Canada's population was increasingly concentrated in major cities from 1901 to 1931, suburbanization was apparent by the decennial census of 1941 and more pronounced thereafter. The deconcentration, as in the United States, was more evident with persons than with production (Slater, 1968). As one Canadian political scientist observed, "Metropolitan development followed familiar North American lines. There was a steady exodus of business firms and middle-class citizens from the central city to the suburbs and a steady migration of lower-income families from elsewhere in the nation and from other countries" (Kaplan, 1967: 41). In 1945 Toronto was heavily British and middle-class; today "New Canadians" make up half the population. Most are Catholics from Eastern and Southern Europe.

Both suburbs and central cities have experienced financial difficulties attributable to this decentralization. New suburbs have relatively well-to-do citizens, but capital investment requirements may outweigh their ability to obtain credit; central cities have lost industries and residents who contributed most to tax revenues. In both nations, fiscal burdens have been higher in central cities than in suburbs and life-style differences have tended to generate resistance to metropolitan unification.

The similarity of economic environments in American and Canadian cities is not fully reproduced in their respective political cultures. Generalizing about political culture is a hazardous business in any case. But Harold Kaplan estimates that Canadian urban political culture is more deferential, less intense, and more easily led than American. There is in Canada, he contends (Kaplan, 1967: 210)

> greater respect for law and order and for persons in positions of authority; [a] lesser importance in Canada of individualism, experimentation, and a spirit of revolt; and [a] greater strength in Canada of an aristocratic or class tradition as opposed to an egalitarian tradition. . . . Canadians, it seems, hold a unitary rather than a pluralist view of the public interest. In the unitary view, informed persons in positions of authority, proceeding *in camera* and free from political pressures search out the public interest. Canadians seem less willing than Americans to accept the notion that the public interest will emerge through the open agitation of issues and the clash of opposing groups in a free political marketplace.

Horowitz (1966), attempting to explain the stronger tendency toward socialism in Canada than in the United States, suggests that in Canada there is a "Tory strain" not present in the liberal and bourgeois United States. The Tory tradition emphasizes a corporate, unitary image of the polity, and Tories are not at all afraid of "big government," a *bete noire* of nineteenth-century liberals. This tolerance for a larger scope of government is a necessary component of a political culture's willingness to accept socialism at the middle- and upper-class level as well as among the working class.

Despite its federal system, Canada resembles Great Britain in its urban decisional styles more than it does the United States. Policy initiative is more likely to come from higher levels of government and local governments are more managerial and administrative. Frank Smallwood (1966) characterizes Canadian local politics as "feedback" rather than the "game" politics which dominates American cities. In the latter, participation is an obligation and interest groups continually compete not only to have their say, but to secure the power to decide. In feedback politics, competition is less keen, and negotiations and public hearings are more important.

The administrative character of Canadian cities is reinforced by their lesser autonomy from higher levels of government. Urban governments in North America are dependent upon higher levels not only for crucial financial assistance, but for their very existence. But circumscribed as they are by state authority, American cities possess more autonomy than their northern counterparts. The harsh dictum that "the state (province) giveth and the state (province) taketh away" is followed even more rigorously in Canada than in the United States. The impact of Dillon's Rule[2] has been somewhat mitigated by home rule provisions, for which no Canadian counterpart exists. The variously named state departments of urban or community affairs possess meager power alongside provincial departments of municipal affairs. On the other hand, the Canadian federal government has all but ignored cities in comparison to the massive assistance to urban areas from Washington (Simmons and Simmons, 1969: 13). One prominent Toronto politician, a former mayor, ran for and won a seat in the federal parliament, seeking to make the federal government more sensitive to urban needs. After three years as an MP, he resigned in frustration in favor of running for the provincial legislature. He won and then became the opposition Liberal Party's shadow urban affairs minister.

Electoral politics and political cleavages are similar but by no means identical. Although the myth of "government close to the people" is more ingrained south of the border, the mean local election turnout is actually lower in the United States. About a third of American eligibles turn out while 43% of Canada's potential urban electorate votes on polling day.[3] The reform institutions of at-large elections and nonpartisanship are entrenched in both systems, but a much larger proportion of Canadian than American cities are nonpartisan, both legally and politically. Indeed, the high level of "reformism" in Canadian cities suggests that factors other than reform institutions (such as cumbersome registration systems) may be the major determinants of low turnouts in American cities. In local elections, the ethnic vote is cultivated as actively in Canada as it is in the United States, although the periodic violence so characteristic of American cities is less prevalent. Political issues are remarkably similar. The conflict over urban renewal and highway construction is as fervid in Toronto and Montreal as in Boston, San Francisco, and San Antonio. In both nations, a new breed of Jacobins[4] has arisen to oppose conventional ideas of urban progress.

GOVERNMENTAL STRUCTURE

Canadian municipal governments are normally more decentralized than those of American cities. In Britain, the centralizing effects of the council-committee system, widely used in Canada, are mitigated if not eliminated by well-organized party systems. But only scattered local parties exist in Canada.[5] Canadian cities have neither a strong mayor nor a party system and are thus characterized by "the lack of adequate means for bringing about unity in policy and effective coordination of administrative management" (Plunkett, 1968: 34). Such charges might seem applicable to many American cities, but the centralization of power in either the strong mayor or manager system blunts their force. Canadian mayors are almost always "weak mayors," and managers in Canadian cities have considerably less control over administration than have their American brethren in the International City Management Association (Young, 1969; Plunkett, 1968: 20).

A "balkanization" of government characterizes Canadian no less than American metropolitan areas (Lineberry, 1970; Plunkett, 1968: ch. 8; Kaplan, 1967). In the United States, however, only one state (Indiana) has ordered reorganization of its major metropolitan area without voter consent. As evidence of the relatively greater provincial authority over cities, Canadian provinces have been much more willing to impose forms of metropolitan integration, the most dramatic examples being Toronto and Winnipeg. But both countries remain plagued by myriads of special districts and commissions, nominally freed from meddlesome politicians (and, sometimes, voters) that cut across city and metropolitan boundaries.

FISCAL POLICIES

Fiscal patterns exhibit considerable similarity, at least in terms of traditional urban taxation and expenditure functions. The principal revenue source of both is the real property tax, which accounts for about four-fifths of American and two-thirds to three-quarters of Canadian revenues from local sources. In Canada as in the United States, educational spending and taxing is typically separated from city government responsibility, although in both nations school boards are "practice fields for budding city politicians" (Lorimer, 1970: 132). Table 1 suggests that Canadian local governments, as the

TABLE 1
MUNICIPAL EXPENDITURES IN CANADA AND THE UNITED STATES[a]

	Sample		
Category	U.S. Over 50,000	Canada Over 50,000	Ontario-B.C.
Per capita expenditures	$85.45	$127.37	$125.20
Per capita recreation expenditures	7.99	10.99	8.48
Per capita protection expenditures[b]	13.98	26.79 (14.29)	22.90

a. If we conservatively assume a 10% exchange differential between Canadian and U.S. dollars, the U.S. figures in the table would be increased to $93.99, $8.79, and $15.38, respectively, still giving Canadians a clear edge at least in total spending.

Tory acquiescence in "bigger government" might imply, outspend somewhat their American counterparts.

SOME IMPLICATIONS

There are several important differences between the two countries which have implications for policy levels, our major concern in this article. First, Canadian provincial governments have an iron grip on local governments which affects both their organization and their policies.[6] Whether this is reflected in the relationship between local sociopolitical attributes and policy levels, however, is an empirical issue which we examine below. Second, both the interest aggregation and decision-making functions at the *local* level are more unstructured in Canada than in the United States. In Canada, there are fewer organized political groups articulating interests and making policy demands upon weaker local executives. Third, Canadians, especially middle-class Canadians, are less likely to shrink from assigning a larger role to government.

RESEARCH DESIGN

SAMPLES

Ensuring adequacy and accuracy of sample and data in a study within a single nation is difficult enough; it is doubly complex in comparative analysis. Despite our best efforts, there is no way in which a single Canadian sample can be compared with a single American sample.[7] We have thus used two Canadian samples and one of American municipalities. Neither Canadian sample by itself offered the range of data available for American cities over 50,000 population. The two complement each other and each offers certain advantages. The smaller one consists of the universe of Canadian cities over 50,000 in 1966 (n = 19). Its advantages and disadvantages are

(1) It is directly comparable by population size with the most reliable American data, i.e., those cities over 50,000 population.

(2) Its data on fiscal policies are both more comparable within the sample and more exactingly collected.[8]

(3) Its data are more recent, with spending and taxing data derived from 1966 reports.

(4) Yet it is too small in size to lend itself to really satisfactory statistical manipulation.

We have also used a second Canadian sample, consisting of all cities over 25,000 population in 1966 in Ontario and British Columbia. Its advantages and disadvantages for our purposes are

(1) Its size (n = 45) permits more confidence in statistical analysis.

(2) But its representativeness is more tenuous, as Ontario and British Columbia are the richest provinces.

(3) And its data are neither so comparable to the American sample, nor so carefully collected.

The anomalies of census gathering being what they are, it is possible that no individual samples in two nations, even those as similar as Canada and the United States, will produce fully satisfactory data. We have tried—either to eliminate beforehand or acknowledge as we report our findings—the major peculiarities and incomparabilities between the American and Canadian data.

The United States sample is more straightforward, consisting of a one-third sample of American cities with 1960 populations of 50,000 or more. Fiscal data for cities of this size are more reliable than for smaller places and more confidence can be placed in the measurement of the dependent variables.

TABLE 2
VARIABLES AND SOURCES OF DATA

Variable: Short Name	Sample: U.S. Over 50,000 (n = 88)	Canada Over 50,000 (n = 19)	Ontario-B.C. Over 25,000 (n = 45)
Population	Log of population[a]	Log of population[b]	Log of population[b]
Turnout	Percentage adults voting in city election[c]	Percentage adults voting in city election[d]	Not available
Income	Median family income[a]	Median family income[e]	Median family income[b]
Owner occupancy	Percentage dwelling units owner occupied[a]	Percentage dwelling units owner occupied[b]	Percentage dwelling units owner occupied[b]
Catholic	Percentage metropolitan area Roman-Catholic[f]	Percentage city population Roman-Catholic[b]	Percentage city population Roman-Catholic[b]
White-collar	Percentage in white-collar occupations[a]	Percentage in managerial occupations[b]	Percentage in managerial occupations[b]
Grants	State and federal aid[g]	Provincial grants[h]	Provincial grants[i]
Assignment	Local proportion of state-local spending in state[j]	Local proportion of provincial-local spending in province[k]	Local proportion of provincial-local spending in province[k]
Spending	Total noneducational operating expenditures[g]	Total noneducational operating expenditures[h]	Total noneducational operating expenditures[i]
Taxation	Revenue from own sources[g]	Total taxation[h]	Total taxation for general municipal purposes[i]
Protection	Police expenditures[g]	Police and fire expenditures[h]	Police and fire expenditures[i]
Recreation	Parks and recreation expenditures[g]	Parks and recreation expenditures[h]	Parks and recreation expenditures[i]

a. U.S. Bureau of the Census, **City and County Data Book** (Washington: GPO, 1962).

b. Dominion Bureau of Statistics, **Census of Canada, 1961** (Ottawa: Dominion Bureau of Statistics, 1962).

c. These data are derived from Eugene C. Lee's survey of municipal elections in 1962, reported in his "City Elections: A Statistical Profile," **Municipal Year Book** (Chicago: International City Managers' Association, 1963). We are indebted to Professor Lee and to Mrs. Ruth Dixon for making these data available to us.

d. Canadian Federation of Mayors and Municipalities, **Voting Turnout in Municipal Elections of Canadian Cities over 100,000,** 1968. For other cities, reports from Chief Returning Officers.

e. Department of National Revenue, **1966 Report** (Ottawa: Department of National Revenue, 1967).

f. National Council of Churches, **Churches and Church Membership in the United States,** Series D, no. 1, 1957.

g. U.S. Bureau of the Census, **Compendium of City Government Finances** (Washington: GPO, 1963).

h. L. Sandford, **Report on Taxation in the City of Halifax** (Halifax, N.S.: Institute of Public Affairs, Dalhousie University, 1968).

i. Provincial auditors' reports, 1964, for Ontario, British Columbia, Nova Scotia, and New Brunswick; city auditors' reports for Alberta, Saskatchewan, Manitoba, and Quebec.

j. These data are taken, by states, from Alan Campbell and Seymour Sacks, **Metropolitan America** (New York: Free Press, 1967), Appendix A2.

k. David Perry, "Fiscal Figures: Provincial and Municipal Expenditures," **Canadian Tax Journal,** 16 (1968), 395-397.

DEPENDENT VARIABLES

There are two major dependent variables (hereinafter called *spending* and *taxation*) used in this study, although briefer analysis will also be made of two additional variables, *protection* and *recreation.* Our object is to select both independent and dependent variables of maximum possible comparability in order to examine comparatively the determinants of policy levels in the two nations. *Spending* consists, in both systems, of municipal noneducational operating expenditures. *Taxation* includes, for the American sample, the census bureau category "Revenue from Own Sources." In the Canadian samples, *taxation* includes "total taxation"[9] for the cities over 50,000 and taxation for "general municipal services" in the Ontario-British Columbia sample. Sources of information for these variables, as well as two other dependent variables which we shall briefly examine *(protection* and *recreation)* are identified in Table 2.

INDEPENDENT VARIABLES

Two principles guided the choice of independent variables in this research. First, we attempted to secure data on the major political and environmental features of cities utilized in previous public policy research in political science and economics. The objective here was *continuity* with previous research. Second, insofar as possible, we selected variables which were logically and conceptually equivalent in both American and Canadian municipalities. This was obviously a stringent test, but we believe that our equivalency standards have been met or at least approximated by each of the variables in Table 2. Our second objective, therefore, was *comparability.*

For the most part, eight of the independent variables listed in Table 2 are conventional ones used in state and local policy research in the United States (Dye, 1966; Fabricant, 1952; Schmandt and Stephens, 1963; Davis and Harris, 1966; Lineberry and Fowler, 1967; Clark, 1971), and require little explanation. Two variables however, may need some clarification. A number of studies have identified *grants* as a major predictor of municipal expenditures (Brazer, 1959; Campbell and Sacks, 1967). In the United States, however, intergovernmental transfer payments to cities come from both the states and the federal government, while in Canada almost all aid originates at the provincial level. *Assignment* is derived from

Campbell and Sacks' (1967) study of metropolitan expenditures. The *assignment* variable measures the "division of governmental labor" between local governments and state authorities within a given state, by calculating the percentage of total state and local expenditures made by local governments. In Campbell and Sacks' study, the *assignment* variable emerged as the most important single predictor of local expenditures. Since there are also important differences among provinces in provincial-local division of labor, we used a comparable measure in the Canadian analysis.

We have available, therefore, an American sample and two Canadian samples, together with substantially comparable measures of sociopolitical and public policy variables. It is thus possible for us to elaborate and replicate some of the findings of the policy analysis research from a comparative perspective, a task we approach in the next section.

PARALLEL POLICY ANALYSIS

The research on the correlates of policy levels is as complex as it is voluminous. Economists, political scientists, and even sociologists have contributed to the specification of factors assumed to cause variations in policy outputs of state and local governments. In contemporary political science, this stream of research takes it cue from Dawson and Robinson's (1963: 266) seminal paper, which argued that "public policy is the major dependent variable which political science seeks to explain. The task of political science, then, is to find and explain the independent and intervening variables which account for policy differences." Yet most of this research has been parochial, confined exclusively to the United States. Parallel analyses of comparable indicators, however, may add a cosmopolitan touch to this literature, and may also specify the degree to which well-accepted findings in the American context are idiosyncratic. This research is of great value to us because it suggests hypotheses which can be incorporated into the present study. But such research must be taken as suggestive rather than definitive because (1) it has all been conducted within the United States, and (2) a variety of different units of analysis, time periods, and measures have been used, limiting its precise comparability even within the United States.

Much of this research upon which we draw is, therefore, not merely ambiguous, but actually contradictory.

HYPOTHESES

Table 3 provides a summarized index of the major directions of relationships we hypothesize from previous research. The rationale for these hypotheses can be briefly discussed at this point.

The relationship between *population* and municipal spending was, in earlier studies, hypothesized to be positive partly because larger cities are likely to undertake a range of public policies—civic centers, transportation networks, welfare programs, and so on—unnecessary or unwanted in smaller areas. For a variety of reasons, possibly economies of scale or unavailability of resources, the most recent evidence (Rivard, 1967; Schmandt and Stephens, 1963; Dupre, n.d.) suggests that there is no significant relationship between population and fiscal levels.

One of the most controversial hypotheses stimulating previous research is the intuitively appealing Key-Lockard argument that higher levels of *turnout* lead to larger working-class participation and thus more liberal social welfare policies (Key, 1951: 298-311; Lockard, 1959: 320-340). This hypothesis predicts a positive relationship between participation levels and the scope of government. But there are several difficulties in hypothesizing such a

TABLE 3
HYPOTHESES[a]

	Hypothesized Direction	
Variable	U.S.	Canada
Population	0	0
Turnout	±	±
Income	0	+
Owner occupancy	—	—
Catholic	+	—
White-collar	—	—
Grants	+	+
Assignment	+	0

a. A + sign indicates a relationship hypothesized to be positive, while a — sign indicates a variable whose effects are hypothesized to be negative. The sign ± indicates a variable whose effects are indeterminate from previous research. A sign 0 indicates a hypothesized nil relationship.

positive relationship between *turnout* and *spending* and *taxation*. First, however consistent with common sense, evidence from state-level studies casts doubt upon the viability of the Key-Lockard hypothesis. Dye (1966), Dawson and Robinson (1963), and others have shown that the participation-policy nexus is spurious when account is taken of socioeconomic development of states.[10] Second, even if evidence could be unearthed to demonstrate a clear causal connection between participation and policy, it is not at all clear whether the effects of working class participation at the local level would be to raise or lower expenditures. The Banfield-Wilson (1963: 36) "ethos theory" holds that the lower class support high levels of policy and spending, but the "alienated voter" literature (Horton and Thompson, 1962) describes the lower class as hostile to government activity, appearing sporadically at the polls to defeat both incumbents and referenda.[11] Unraveling this tangled skein is beyond our capacity, so we assign the *turnout* variable a schizophrenic ± in our hypothesis summary in Table 3.

One of the most extensively used variables in policy analysis study is *income*, first used in the classic Fabricant (1952) study and included in many successive investigations. So much research has been done at different levels, and with slightly varying indicators, that the findings are complex and even superficially contradictory. However, once measures and units of analysis are taken into account, something like the following pattern begins to emerge in the United States:

(1) income is positively and strongly related to total state-local expenditures (Dye, 1966);

(2) it is not significantly related to state expenditures excluding a local component (Sharkansky, 1967);

(3) income is positively and strongly associated with local educational expenditures, and, in most studies, with total local expenditures (Brazer, 1959; Pidot, 1969); but

(4) it is not related to total local noneducational expenditures (Campbell and Sacks, 1967: 140).

The last of these propositions seems to parallel most closely the measures of spending used in this analysis and we thus hypothesize a nil relationship between income and fiscal levels. In Canadian cities, however, previous research (Rivard, 1967) has found that income is positively related to spending, and we so hypothesize regarding our Canadian samples.

Owner occupancy has been included as an independent variable in two American studies (Froman, 1967; Lineberry and Fowler, 1967) and each found it strongly and negatively related to expenditures and taxation. Both direct and indirect evidence suggests a negative relationship for the *white-collar* variable and fiscal levels. Banfield and Wilson (1963: 36) argue that the "middle-income group [presumably coterminous with the white-collar class] generally wants a low level of public expenditures. It consists of people who are worrying about the mortgage on their bungalow and about keeping up the payments on the new car." More directly, some earlier research (Lineberry and Fowler, 1967) has shown a negative relationship between proportions of white-collar workers and public finance levels. We expect a similar relationship in Canada.

The *Catholic* variable was included because of its presumed importance in both polities (perhaps even more powerful north than south of the border) in shaping political attitudes. Terry Clark's (1971: 308) investigation of outputs in 51 American cities found the relationship between percent Catholic and per capita local expenditures to be "phenomenally strong" and positive. Research using a surrogate of religious affiliation (private school attendance) found a strong and positive relationship between that measure and fiscal levels (Lineberry and Fowler, 1967). From such research, we hypothesize a positive association between *Catholic* and *spending* and *taxation* in the United States. In Canada, however, we suspect that the opposite is the case, i.e., that *Catholic* and outputs will be negatively associated. There are two reasons for this hypothesis. First, French Canadian Catholics[12] are dependent upon their church for many functions typically performed elsewhere by civil government. Libraries, schools, parks, and even health and welfare services are provided by the church (Rioux and Martin, 1964). Such services necessarily lessen the need for a large scope of government. Second, the church is a powerful institution, and owns many lands, especially in Quebec, which are untaxable. A recent estimate suggests that if church lands were taxed, municipalities could increase their revenues by 25%.[13] Because the church in Canada performs important governmental functions and because its immunity from taxation reduces urban revenues, we hypothesize a negative relationship between proportion *Catholic* and policy levels in Canada.

We hypothesize positive relationships between both *grants* and the *assignment* variable and policy outputs. No investigation which has incorporated intergovernmental aid as a variable has failed to find it strongly and positively associated with state, state and local, or local fiscal levels. It is in fact for this reason that Elliott Morss (1966: 97) has criticized the use of *grants* as a variable as theoretically meaningless. He argued:

> One might conclude that the only aim [of these studies] was to increase the coefficient of multiple determination (R^2) over what was recorded in earlier studies. . . . Using [state and federal] aid to explain expenditures is analogous to using taxes to explain expenditures in the sense that both aid and taxes are sources of funds. The fact that these variables turn out to have substantial explanatory power serves as little more than verification of the quite obvious fact that government receipts and expenditures are closely related.

If one uses intergovernmental aid as a theoretically meaningful variable, says Morss, why not also use taxes and reach nearly perfect correlations?

However superficially appealing Morss' argument may be, we do not agree with it. Actually, aid is not conceptually identical to taxes, because it can independently affect the local resource level. Aid may be seen as playing one of three roles in affecting local resources. In the first place, transfer payments may be *substitutive,* i.e., they may reduce the recipient unit's own efforts. This is the hallowed conservative criticism of federal assistance, verbalized as the charge that federal aid "reduces local initiative." (For a Canadian example of the substitutive function, see note 6.) Second, intergovernmental assistance may be merely *additive,* in that the recipient's efforts are affected neither positively nor negatively. And, third, aid may be *stimulative,* in that recipients receiving larger proportions of aid actually increase their own efforts. Morss' argument against the use of grants as a variable assumes that all intergovernmental aid is only additive, i.e., that, like a tax dollar, a dollar of aid is simply an addition to the resource base. Aid is thus seen as neutral in its impact on local efforts. Unfortunately for Morss' argument, strong evidence suggests that intergovernmental aid in the United States is not additive but actually stimulative. Pidot (1969), for example, found that each $1.00 increase in state aid to metropolitan areas is associated with a $1.40 expenditure increase, and every $1.00 of federal aid is associated with a $2.45 expenditure increase. We therefore suggest a positive impact of *grants* on *taxation* directly and *spending* indirectly should be observable, even when other variables are taken into account.[14]

Campbell and Sacks (1967) tell a similar story about the stimulative effect of expenditure *assignment.* Greater local responsibility is associated with disproportionately increased local tax efforts, although the effects on expenditures is minimal. We hypothesize, therefore, following Campbell and Sacks, that increases in the *assignment* variable are positively associated with *taxation* and not significantly related to *spending.*

METHODS

The conventional statistic used to describe the size and sign of the relationship between environmental and policy variables has been the correlation coefficient. Despite its familiarity, it has several drawbacks for our purposes, the principal one being that it only measures one's confidence that a relationship exists, while we are concerned with the size and direction of the relationship. We use, therefore, the beta weight, also called a coefficient of relative importance. The beta weight is similar to, although not identical with, both the partial correlation coefficient and the partial slope in a multiple regression equation. The beta weight measures how many standard deviation units the dependent variable in a multiple regression equation changes with a change of one standard deviation unit in the independent variable, controlling for other independent variables (Blalock, 1960: 344-346). Like the correlation coefficient, it is a "standardized measure." It is also likely to have a higher value by chance in smaller samples, a characteristic of many other measures of association. In the sample of large Canadian cities, for example, a coefficient would have to be .40 to be considered significant, while in the Ontario-B.C. and United States samples, the figures would be .35 and .30, respectively.

FINDINGS

We lack the space and patience to parade four dozen specific findings before the reader. We shall highlight only those which seem

TABLE 4
COEFFICIENTS OF RELATIVE IMPORTANCE (beta weights) FOR TAXING AND INDEPENDENT VARIABLES

	Sample		
Independent Variable	**U.S.**	**Canada Over 50,000**	**Ontario-B.C.**
Log population	.07	.32	−.26
Turnout	−.08	−.14	N.A.
Income	.13	.13	.25
Owner occupancy	−.20	−.86	−.39
Catholic	.31	−.41	−.47
White-collar	.05	−.37	−.12
Grants	.47	.42	.30
Assignment	−.20	.42	.37
	F = 12.1	F = 2.5	F = 3.0
	R^2 = .55	R^2 = .61	R^2 = .32
	p = $<$.001	p = $>$.05	p = $<$.05

most important to us, but the reader has Table 4 to check or dispute our readings of the evidence.

Consistent with our expectations for the United States, *population* shows no very significant relationship to fiscal levels. In Canada, however, a vexing inconsistency appears with results depending upon the sample. In the larger Canadian cities, there is a modest positive relationship between population and outputs, while in the Ontario-

TABLE 5
COEFFICIENTS OF RELATIVE IMPORTANCE (beta weights) FOR SPENDING AND INDEPENDENT VARIABLES

	Sample		
Independent Variable	**U.S.**	**Canada Over 50,000**	**Ontario-B.C.**
Log population	.13	.30	−.28
Turnout	−.08	.08	N.A.
Income	.09	.23	.29
Owner occupancy	−.27	−.69	−.46
Catholic	.05	−.68	−.40
White-collar	.10	−.33	−.06
Grants	.54	.47	.32
Assignment	.00	−.01	.17
	F = 8.7	F = 9.2	F = 2.5
	R^2 = .48	R^2 = .85	R^2 = .28
	p = $<$.001	p = .001	p = $<$.05

British Columbia sample, the relationship is negative. This may be a function of the greater generosity of *grants* to larger cities in Canada. (In the large-city Canadian sample, there is a correlation of .25 between *grants* and *population,* while the *grants-population* correlation is actually negative, –.11, in the Ontario-British Columbia sample.) Whatever the explanation, the effect of population upon spending varies not only between but within the two nations.

As we hypothesized, *turnout* is unrelated to levels of public expenditures and taxes in either nation. When the other variables are held constant, there is very little support for the Key-Lockard hypothesis associating higher participation with more generous expenditures. Although Munger (1969) and his associates have demonstrated that popular opinion may be related to nonfiscal, statutory outputs of states, few studies have successfully linked mass opinions and opinion-registering mechanisms to fiscal levels of state or local governments. Our analysis extends this bit of, by now, conventional wisdom to the urban scene in both Canada and the United States. Perhaps, indeed, the initial expectations were somewhat unrealistic. The links in the Key-Lockard chain are, after all, somewhat tenuous: a larger turnout must mean that more working people actually do vote; their preferences must be for higher spending; politicians must interpret the vote correctly and must be willing to spend more heavily; and the money must be available to the polity (Fowler, 1969: 166-169).

As we hypothesized, *income* is not significantly related to fiscal levels in the United States, but it is moderately and positively associated in Canada.

The behavior of the *owner occupancy* variable is fully consistent with our hypotheses in both Canada and the United States, though its negative effects are more discernible in the former.

Clark's (1971) finding that *Catholic* is strongly and positively associated with expenditure levels in the United States is not consistently supported by our data. But the relationship in Canada—as we hypothesized—is strong and negative. Clark sought an explanation in the attitudes and values of Catholics (bordering dangerously, perhaps, on an ecological fallacy), but it seems to us that the explanation relates more directly to the church and its functions, at least in Canada. As we noted above, vast holdings of church lands are exempt from the property tax, thus reducing local taxation, and many church functions substitute for traditional municipal ones, thus reducing municipal expenditures.

TABLE 6

COEFFICIENTS OF RELATIVE IMPORTANCE (beta weights) FOR PROTECTION AND INDEPENDENT VARIABLES

	Sample		
Independent Variable	U.S.	Canada Over 50,000	Ontario-B.C.
Log population	.39	–.19	–.15
Turnout	–.06	.11	N.A.
Income	.35	.29	.04
Owner occupancy	–.46	–.08	–.57
Catholic	.01	–.40	–.30
White-collar	.00	–.13	.16
Grants	.06	.19	.28
Assignment	–.07	–.84	.00
	F = 6.8	F = 1.7	F = 2.5
	R^2 = .41	R^2 = .52	R^2 = .28
	p = < .001	p = > .05	p = < .05

TABLE 7

COEFFICIENTS OF RELATIVE IMPORTANCE (beta weights) FOR PARKS AND RECREATION AND INDEPENDENT VARIABLES

	Sample		
Independent Variable	U.S.	Canada Over 50,000	Ontario-B.C.
Log population	.18	–.10	–.49
Turnout	.01	.32	N.A.
Income	–.05	.13	.40
Owner occupancy	–.09	–.19	–.50
Catholic	–.21	–1.08	–.39
White-collar	.28	–.54	.00
Grants	.24	.17	.22
Assignment	–.09	–.16	–.07
	F = 2.0	F = 2.5	F = 4.7
	R^2 = .17	R^2 = .61	R^2 = .43
	p = > .05	p = > .05	p = < .001

The *white-collar* variable diverges somewhat from our expectations, especially in the United States. Although we predicted a negative correlation, we find virtually none at all in the United States. In Canada, there is the expected strong and negative relationship. Perhaps differences in measurement account for differences in findings. The American *white-collar* variable measures all persons employed in such occupations, while the Canadian measure is a "managerial" measure including large numbers of small proprietors and shopkeepers.

The *grants* variable is a strong and positive predictor of expenditures and taxation in both nations. The fact that grants account for only a small proportion of total revenues (about 20-30% in both nations) suggests that intergovernmental aid plays a stimulative role in Canadian and American cities. With the possible exception of *owner occupancy, grants* seems to be the one factor most clearly associated with fiscal choices in each sample.

The effect of the *assignment* variable, however, deviates somewhat from our hypotheses, especially in the United States. There, a negative relationship exists between local responsibility and *taxation* and none at all with *spending.*[15] But in Canada, the *assignment* variable is consistent with the hypothesis, showing no significant relationship to *spending* but strongly correlated with *taxation.*

RECREATION AND PROTECTION SPENDING

Tables 6 and 7 provide the results of parallel policy analyses for two additional outputs, parks and recreation expenditures, and a variable defined as "protection." The latter consists of police expenditures in the United States and police and fire expenditures in Canada. Although there are some dramatic similarities between the findings for these specific policies and the more general analyses of *spending* and *taxation* (note, e.g., the continued strong and negative impact of the *Catholic* variable in Canada), there are also some clear differences. In the United States, for example, both wealth and size-of-place seem strongly related to protection, even though their relationship to total spending and taxing was weak or nonexistent. The data in Tables 6 and 7 suggest once again that the results of policy analysis depend upon (1) time periods used, (2) the nature of the sample, and the exact measurement of the (3) independent and (4) dependent variables.

CONCLUSIONS

There are some very significant and enduring differences between American and Canadian political cultures. One strains to imagine an American Trudeau in the presidency or a Canadian Nixon as Prime Minister. At the local level, Canadian cities are steeped in British traditions of public stewardship in the administrative city, while American cities are a melting pot of populism, bossism, reformism, and elitism. In the first section of this article, we outlined some of the divergent patterns of culture and government in Canada and the United States. Perhaps the most notable among them were a Canadian tolerance for a large public sector, a greater concentration of power there at the provincial vis-à-vis the local level, and a more deferential and less politicized style in Canada. To be sure, it is difficult to incorporate measures for such general and abstract variables as centralization or politicization. But we can measure to a degree, even across political cultures, certain features of the socioeconomic environment, political participation, and local-federal centralization. And, despite all the idiosyncracies of particular variables, in particular samples, we are consistently struck by the similarities shown by our parallel policy analyses. Some variables–*Catholic*, for example–show opposite influences in the two polities, but there are probably more parallelisms than divergences. In both nations, environmental factors explain a moderately large proportion of the variation in expenditures and taxation. (The R^2's range from .17 to .85.) In both polities, political participation, albeit a variable most imperfectly approximated, showed little relationship to fiscal outcomes. In Canada as well as in the United States, intergovernmental aid seems to have a stimulative effect on local spending. This is not to argue that variables of political culture and political style are unimportant simply because they are unmeasurable with these data. But neither should a political cultural approach underestimate the importance of socioeconomic and governmental variables in determining what governments do, or at least, where they spend their money.

We have attempted in this paper to explore comparatively the determinants of public policy in two nations, Canada and the United States. To be sure, the analysis of public expenditures is far from the alpha and omega of public policy research. As several critics have argued (Coulter, 1970; Jacob and Lipsky, 1968), public policies do not stop at expenditures. In the long run, it is probably more important to understand the impact (Dye, 1971) or distribution (Fry and Winters, 1970) of expenditures than the expenditures themselves. But comparative urban policy analysis must be built, we suppose, from the ground up, and one must begin with available data. We do hope that research into comparative public policies will find a place in the burgeoning literature of comparative urban politics (see, e.g., Fried and Rabinovitz, forthcoming; Daland, 1970).

NOTES

1. The study of Canadian local politics has been nascent until recent years. Some of the best work, upon which our discussion draws, includes Kaplan (1967); Plunkett (1968); Lorimer (1970); Feldman and Goldrick (1969); and Rowat (1969).

2. Enunciated by Justice John Dillon in *City of Clinton v. Cedar Rapids and Mo. R.R.*, 24 Iowa 455 (1868).

3. The data on turnout in American cities are derived from Alford and Lee (1968); the Canadian data sources are described in our Table 2.

4. After Jane Jacobs (1961), who has argued wittily and persuasively for less "progress," more neighborhood, and a lively, concentrated inner city. See, e.g., the neighborhood opposition to highways described in Lupo et al. (1971).

5. Literature on Canadian local parties is as rare as the phenomenon itself. The only city with a genuine party "system," ironically, is Montreal, a completely one-party city dominated by the machine of Mayor Jean Drapeau. On local parties in Canada, see Bureau of Municipal Research (1971) and Masson (forthcoming).

6. For example, in 1970, the Ontario provincial government required local school boards to reduce spending by 10-15%–it was a provincial election year–before receiving their provincial grants. Reductions in the property tax were made possible (or, actually, forced) on a large scale.

7. Newfoundland was omitted from the analysis because the province's highly centralized control of municipal finances makes it a deviant case.

8. Lawrence Sandford's (1967) thorough and careful analysis of municipal fiscal policies is the source of the expenditure data on the sample of Canadian cities over 50,000.

9. "Total taxation" includes residential, commercial, and industrial real property and personal property or occupancy tax, and, in various cities, utilities tax, poll tax, water tax, sales tax, amusement tax, special and supplementary assessments, and local improvement charges.

10. Fry and Winters (1970), however, present evidence that the distributions of state expenditures are determined in part by political variables.

11. Hahn (1969) shows that, in Canada, there is a direct relationship between turnout and approval of "agreement-oriented" issues such as changes in governmental form. But, on other issues such as fluoridation and Sunday entertainment, there were indications of the "alienated voter" pattern of higher turnout and defeat.

12. In Canada, Catholicism is almost synonymous with francophone culture, with the exception of Toronto, where there are 400,000 Italians.

13. For the estimate, see the Toronto *Globe and Mail,* May 29, 1971. The church's assessed land in Quebec alone is probably worth $6.5 billion. Rivard (1967) does not use Catholicism as a variable, but his dummy variable "Quebec" is negatively related to all his spending measures.

14. However, Dupre (n.d.) found only a slight relationship between total spending and total grants. He makes the important point that certain grants, like welfare and highways, are conditionally tied to expenditures on those items. Unconditional grants make up 20-35% of the total and are, in Ontario, unrelated to total spending.

15. We do not readily conclude that our analysis of the *assignment* variable disputes Campbell and Sacks' (1967) conclusions about its importance. The measures are similar, but not identical. They utilized the expenditure assignment variable to explain variation in metropolitan area expenditures by state; we coded a single number representing the aggregate state-local assignment by state and then assigned that value to any city in that state which fell into our sample.

REFERENCES

ALFORD, R. R. and E. C. LEE (1968) "Voting turnout in American cities." Amer. Pol. Sci. Rev. 62 (September): 796-813.

ANGUS, H. F. (1938) Canada and Her Great Neighbour. Toronto: Macmillan.

BANFIELD, E. C. and J. Q. WILSON (1963) City Politics. Cambridge: Harvard Univ. Press.

BLALOCK, H. M. (1960) Social Statistics. New York: McGraw-Hill.

BRAZER, H. E. (1959) City Expenditures in the United States. New York: National Bureau of Economic Research.

Bureau of Municipal Research (1971) Parties to Change. Toronto: Bureau of Municipal Research.

CAMPBELL, A. K. and S. SACKS (1967) Metropolitan America. New York: Free Press.

CATTELL, D. T. (1968) Leningrad: A Case Study in Soviet Urban Government. New York: Praeger.

CLARK, T. N. (1971) "Community structure, decision-making, budget expenditures, and urban renewal in 51 American communities," pp. 293-314 in C. M. Bonjean et al. (eds.) Community Politics: A Behavioral Approach. New York: Free Press.

COULTER, P. B. (1970) "Comparative community politics and public policy." Polity 3 (Fall): 22-43.

CRECINE, J. P. (1969) Governmental Problem-Solving. Chicago: Rand McNally.

DALAND, R. T. [ed.] (1970) Comparative Urban Research. Beverly Hills: Sage Pubns.

DAVIS, O. and G. H. HARRIS, Jr. (1966) "A political approach to a theory of public expenditures." National Tax J. 19 (September): 259-275.

DAWSON, R. E. and J. A. ROBINSON (1963) "Interparty competition, economic variables, and welfare policies in the American states." J. of Politics 25 (November): 265-289.

DUPRE, J. S. (n.d.) Intergovernmental Finances in Ontario: A Provincial-Local Perspective. Toronto: Queen's Printer.

DYE, T. R. [ed.] (1971) The Measurement of Policy Impact. Proceedings of a conference on the measurement of policy impact, Florida State University.

--- (1966) Politics, Economics and the Public. Chicago: Rand McNally.

FABRICANT, S. (1952) The Trend of Government Activity in the United States Since 1900. New York: National Bureau of Economic Research.

FELDMAN, L. D. and M. D. GOLDRICK [eds.] (1969) Politics and Government of Urban Canada. Toronto: Methuen.

FOWLER, E. P. (1969) "Government spending as response and initiative." Ph.D. dissertation. University of North Carolina.

FRIED, R. and F. RABINOVITZ (forthcoming) Comparative Urban Performance. Englewood Cliffs, N.J.: Prentice-Hall.

FROMAN, L. A., Jr. (1967) "An analysis of public policies in cities." J. of Politics 29 (February): 94-108.

FRY, B. and R. WINTERS (1970) "The politics of redistribution." Amer. Pol. Sci. Rev. 64 (June): 508-522.

HAHN, H. (1969) "Voting in Canadian cities: a taxonomy of referendum issues," pp. 161-169 in L. D. Feldman and M. D. Goldrick (eds.) Politics and Government of Urban Canada. Toronto: Methuen.

HOROWITZ, G. (1966) "Conservatism, liberalism, and socialism in Canada: an interpretation." Canadian J. of Economics and Pol. Sci. 32 (May): 143-171.

HORTON, J. E. and W. E. THOMPSON (1962) "Powerlessness and political negativism: a study of defeated referendums." Amer. J. of Sociology 67 (March): 485-493.

JACOB, H. and M. LIPSKY (1968) "Outputs, structure and power: an assessment of changes in the study of state and local politics," J. of Politics 30 (May): 480-509.

JACOBS, J. (1961) The Death and Life of Great American Cities. New York: Random House.

KAPLAN, H. (1967) Urban Political Systems. New York: Columbia Univ. Press.

KEY, V. O. (1951) Southern Politics. New York: Knopf.

LINEBERRY, R. L. (1970) "Reforming metropolitan governance: requiem or reality." Georgetown Law J. 58 (March-May): 675-718.

——— and E. P. FOWLER (1967) "Reformism and public policies in American cities." Amer. Pol. Sci. Rev. 61 (September): 701-716.

LOCKARD, D. (1959) New England State Politics. Princeton: Princeton Univ. Press.

LORIMER, J. (1970) The Real World of City Politics. Toronto: James Lewis and Samuel.

LUPO, A., F. COLCORD, and E. P. FOWLER (1971) Rites of Way: The Politics of Highway Construction in Boston and the American City. Boston: Little, Brown.

MASSON, J. [ed.] (forthcoming) Emerging Party Systems in Canada. Toronto: Prentice-Hall of Canada.

MORSS, E. R. (1966) "Some thoughts on the determinants of state and local expenditures." National Tax J. 14 (March): 95-103.

MUNGER, F. (1969) "Opinions, elections, parties, and policies: a cross-state analysis." Presented at the Sixty-Fifth Annual Meeting of the American Political Science Association, New York.

PIDOT, G. (1969) "A principal components analysis of the determinants of local government fiscal patterns." Rev. of Economics and Statistics 51 (May): 176-188.

PLUNKETT, T. J. (1968) Urban Canada and its Government. Toronto: Macmillan of Canada.

RIOUX, M. and Y. MARTIN (1964) French Canadian Society. Ottawa: Carlton Univ. Press.

RIVARD, J.-Y. (1967) "Determinants of city expenditures in Canada." Ph.D. dissertation. University of Michigan.

ROWAT, D. C. (1969) Essays on the Improvement of Local Government. Toronto: Methuen.

SANDFORD, L. (1967) Report on Taxation in the City of Halifax. Halifax, N.S.: Institute of Public Affairs, Dalhousie University.

SCHMANDT, H. J. and G. R. STEPHENS (1963) "Local government expenditure patterns in the United States." Land Economics 34 (November): 397-406.

SHARKANSKY, I. (1967) "Economic and political correlates of state government expenditures: general tendencies and deviant cases." Midwest J. of Pol. Sci. 11 (May): 173-192.

SIMMONS, J. and R. SIMMONS (1969) Urban Canada. Toronto: Copp Clark.

SLATER, D. (1968) "Decentralization of urban people and manufacturing in Canada." Canadian J. of Economics and Pol. Sci. 27 (February): 72-84.

SMALLWOOD, F. (1966) "'Game politics' vs. 'feedback politics,'" pp. 313-316 in R. Morlan (ed.) Capitol, Courthouse and City Hall. Boston: Houghton Mifflin.

WILSON, J. Q. [ed.] (1968) City Politics and Public Policy. New York: John Wiley.

YOUNG, D. (1969) "Canadian local governmental development: some aspects of the commission and city manager form," pp. 207-219 in L. Feldman and M. Goldrick (eds.) Politics and Government of Urban Canada. Toronto: Methuen.

14

Dissatisfaction with City Services: Is Race an Important Factor?

HOWARD SCHUMAN
BARRY GRUENBERG

□ THE ADEQUACY OF THE SERVICES a city provides its citizens cannot be judged accurately by the amount of money expended or the number of persons paid to provide the services. High levels of either may simply indicate inefficiency, excessive patronage, or some other feature of urban life irrelevant to satisfactory services. More appropriate criteria of adequacy are objective indices of performance and results: frequency of garbage collection, low crime and high arrest rates, reading levels of school children. This article examines still a third measure of civic adequacy: subjective reports by random samples of citizens about their satisfaction with four essential city services.

Subjective evaluations are of fundamental importance insofar as we regard citizen satisfaction both as the ultimate goal of city services and—in the form of dissatisfaction—as a major factor

---

AUTHORS' NOTE: *We are indebted to Angus Campbell, whose earlier writing first suggested to us the question investigated here. Elizabeth Keogh Taylor aided in the analysis of data and carried out most of our computer runs. Otis Dudley Duncan provided a number of helpful suggestions on an earlier draft, and useful comments were also made by Hubert O'Gorman. Some of the findings reported here were discussed at the Conference on Comparative Community Studies, Fort Collins, Colorado, August, 1971.*

TABLE 1

1. "First, I'd like to ask how satisfied you are with some of the main services the city is supposed to provide for your neighborhood. A. What about the quality of the public schools in this neighborhood—are you generally satisfied, somewhat dissatisfied, or very dissatisfied?" [Repeat question for B, C, and D] (in percentages)

	A. Quality of Public Schools		B. Parks and Playgrounds for Children		C. Police Protection		D. Garbage Collection	
	Black	White	Black	White	Black	White	Black	White
1. Generally satisfied	53	68	36	56	51	71	70	82
2. Somewhat dissatisfied	28	20	30	23	21	18	14	9
3. Very dissatisfied	19	13	34	21	28	10	15	8
Total	100	101	100	100	100	99	99	99
n	(2,265)	(1,837)	(2,386)	(2,226)	(2,553)	(2,468)	(2,738)	(2,492)
Mean score	1.64	1.46	1.97	1.65	1.77	1.40	1.46	1.25
Standard deviation	0.77	0.71	0.84	0.80	0.86	0.67	0.75	0.59

2. "Thinking about city services like schools, parks, and garbage collection do you think your neighborhood gets better, about the same, or worse service than most other parts of the city?" (in percentages)

	Black	White
Better	11	19
About the same	65	71
Worse	24	10
Total	100	100
n	(2,581)	(2,312)
Mean score	2.12	1.89
Standard deviation	0.57	0.52

prompting change in the delivery of services. But such reports are also, as we know from much experience, ambiguous in nature and origin since they may be influenced not only by objective reality, but also by expectations, ideology, and a host of other individual and group characteristics. We deal here with a crucial perspective on municipal functioning, but one which, like data on expenditures, cannot be taken at face value.

Four apparently straightforward questions about "satisfaction" with specific neighborhood services were posed at the beginning of a lengthy interview in 1968 with cross-sections of black (n = 2809) and white (n = 2584) citizens in fifteen American cities[1] The four services covered are: "quality of public schools," "parks and playgrounds for children," "police protection," and "garbage collection."[2] Table 1 presents the percentage of respondents of each race showing different degrees of satisfaction with each service.[3] Each item is scored on a three-point scale (where 3 = very dissatisfied), and means and standard deviations based on such scoring are presented as well. A related question on the sense of relative deprivation in these services felt by respondents is also shown in the table.

Although all the questions are rather abstract in phrasing, follow-up inquiries to a random subsample of 100 respondents indicate that most people answered the questions within the intended frame of reference.[4] For example, the following are typical explanations by two respondents about their initial answers on "police protection":

> Generally satisfied: "Whenever anything happens they are here right away. If there is an accident or anything. Nothing else happens."

> Very dissatisfied: "Police are of little use to call here. Rapes, robberies, and murders are committed. Police give no service to citizens in this area. I feel that there is no concern by the law enforcement people for those who live here."

The main finding of interest in Table 1 is that blacks are more dissatisfied with each service than whites. The differences are not pronounced, but they are clear-cut and consistent.[5] We take their explanation as the main task of this article. In the course of pursuing it, we will also account for some of the variation in levels of satisfaction within each race—that is, for urban inhabitants as such, regardless of the color of their skin.

We may note initially two quite different lines of explanation for racial differences in dissatisfactions with city services. One possibility is simply that blacks experience objectively worse services than whites. For example, blacks could live disproportionately in cities that have poorer services; or in areas *within* cities that have poorer services. The problem with such an explanation is the assumption that objective reality is perceived without distortion and that these perceptions in turn are transformed directly into levels of satisfaction. Concepts like "relative deprivation" have sensitized social scientists to the problematic nature of the link between the social world as seen by the detached observer and the same world as experienced by actors in it.[6] For example, satisfaction with a service may be based on limited past personal experience or on ideological beliefs, rather than on broad observation of the current state of things.

Such reflections raise the possibility that blacks show more dissatisfaction with city services than whites for reasons unrelated to the objective character of the services themselves. We are often told that blacks are alienated from American institutions generally, and there is some evidence to support such claims.[7] Questions about city services may simply furnish a convenient screen onto which a disillusioned racial group projects its dissatisfactions with life in America. This reasoning would lead us to look for the causes and correlates of dissatisfaction with services among ideological rather than ecological variables, and to focus on the way politics affects personality rather than on the way politics affects city streets.

The two perspectives just outlined provide some broad bearings for our search, but do not exhaust or even clearly identify specific variables. We turn to these, and to the evidence, after noting two other features of Table 1. The reader may have noticed that the ordering of dissatisfactions by service is not exactly what one would expect. The greatest dissatisfaction is not over police protection or schools, which dominate the newspapers as sources of complaint, but with "parks and playgrounds for children." We are unable to illuminate this ordering very much, since the four services are objectively incommensurable and relative "satisfaction" from one to another involves intrinsically subjective factors. The primacy of parks is real enough, however, for it holds for each race and for most of the fifteen cities (as shown in Table 2). Perhaps overall citizen satisfaction could be increased substantially by municipal emphasis on park construction and management. One thinks of the abandoned commercial areas in cities that suffered rioting in the late 1960s: reclaiming some of these central city areas for park land might be well received.

At the other end of the scale, it is no great surprise that dissatisfaction is lowest for garbage collection for both races. In the middle are schools and police, rather close together but with the ordering for blacks the reverse of that for whites. We will be able to throw some light on this reversal at a later point.

The ordering of the four services leads to the related question of how much respondents distinguish among them. In one sense the items in Table 1 could be treated simply as four indicators of an underlying construct about general dissatisfaction with city services. Indeed, the format of the questions encouraged such a "set," since the items were grouped together and asked in exactly the same ways. Yet the quite different levels of dissatisfaction across the four services indicate that considerable differentiation did in fact occur. It is noteworthy that such differentiation seems to be greater for blacks than for whites. For example, the gap between satisfaction with garbage collection and with parks is 36% for blacks and only 26% for whites.[8] This does not fit the hypothesis that blacks are responding with a more ideological set of attitudes toward all city services than are whites.

CITY DIFFERENCES IN DISSATISFACTION LEVELS

The presentation of our analysis will move from a social structural to an individual level, and from larger to smaller units. We begin with major cities as units of analysis. Then we look within cities at areas defined in terms of racial and economic proportions. Finally we shift to individuals and their characteristics, starting with demographic and socioeconomic attributes, and ending with personal attitudes and ideology.

Our total sample actually comprises 15 city samples of each race. City of residence is a particularly relevant level of analysis because the city is the governmental unit responsible for the four services. It is possible, for example, that cities differ sharply in the services they provide, and that racial differences in dissatisfaction result from the

TABLE 2
Black and White City Mean Dissatisfaction Scores for Four Services*

City	Schools			Parks			Police			Garbage		
	B	W	d	B	W	d	B	W	d	B	W	d
1. Baltimore	1.35	1.53	–.18	1.85	1.90	–.05	1.73	1.41	.23	1.99	1.18	.81
2. Boston	2.07	1.74	.33	2.30	1.96	.34	2.36	1.92	.44	1.82	1.23	.59
3. Brooklyn	1.81	1.53	.28	1.86	1.83	.03	1.92	1.53	.39	1.57	1.33	.24
4. Chicago	1.74	1.47	.27	1.97	1.52	.45	1.66	1.20	.46	1.53	1.12	.41
5. Cincinnati	1.59	1.32	.27	2.13	1.55	.58	1.69	1.24	.45	1.26	1.11	.15
6. Cleveland	1.55	1.25	.30	1.98	1.65	.33	1.73	1.44	.29	1.52	1.27	.35
7. Detroit	1.66	1.47	.19	2.09	1.71	.38	1.82	1.45	.37	1.52	1.41	.11
8. Gary	1.49	1.46	.03	2.26	1.96	.30	2.18	1.84	.34	1.57	1.28	.29
9. Milwaukee	1.85	1.16	.69	2.28	1.38	.90	1.84	1.24	.60	1.60	1.26	.34
10. Newark	2.05	1.77	.28	2.27	1.88	.39	1.99	1.48	.51	1.52	1.11	.41
11. Philadelphia	1.60	1.47	.13	1.93	1.68	.25	1.76	1.34	.42	1.40	1.16	.24
12. Pittsburgh	1.59	1.38	.21	1.88	1.93	–.05	2.06	1.56	.50	1.60	1.77	–.13
13. St. Louis	1.48	1.34	.14	1.92	1.41	.51	1.78	1.35	.43	1.53	1.32	.21
14. San Francisco	1.69	1.40	.29	1.78	1.34	.44	1.66	1.40	.26	1.32	1.17	.15
15. Washington	1.53	1.77	–.24	1.84	1.32	.52	1.67	1.32	.35	1.36	1.26	.10
A. Mean of 15 city means	1.67	1.47	.20	2.02	1.67	.35	1.86	1.45	.41	1.54	1.26	.28
B. Variance explained by city (%)	4.3	3.3		2.5	5.6		2.7	6.8		2.5	6.3	
C. Standard deviation among city means	.197	.174		.176	.228		.201	.197		.177	.159	
D. Mean variance within city	.565	.481		.678	.615		.710	.476		.563	.340	
E. Black-white r across cities	.36			.32			.90			.37		

*The higher the score, the greater the dissatisfaction. **Minimum** sample sizes on which Table 4 means are based are as follows, with blacks given first and whites second for each city: Baltimore (143, 111), Boston (91, 106), Brooklyn (127, 168), Chicago (168, 87), Cincinnati (149, 115), Cleveland (122, 128), Detroit (148, 114), Gary (226, 136), Milwaukee (185, 117), Newark (185, 159), Philadelphia (202, 178), Pittsburgh (140, 140), St. Louis (143, 108), San Francisco (101, 109), Washington (135, 61). Total sample (2,265, 1,837). These are the exact sizes for the schools question, where the D.K.s removed were highest, but understate the n for other questions. For example, the lowest city n just given is for Washington whites (61), but this rises to 97 for parks, 111 for police, and 101 for garbage.

differing distribution of the races across the 15 cities. Thus our black sample (when properly weighted) comes more from Washington than Milwaukee, while the reverse is true for our white sample. If services are poorer in Washington than Milwaukee, this would tend to produce the racial differences in dissatisfaction in Table 1.

Data for each of the 15 central cities are presented in Table 2, along with summary statistics. On the question first of whether racial differences in dissatisfaction can be reduced to city of residence, the answer is clearly no. A racial comparison is possible for each city on each of the four services–a total of 60 comparisons, not all independent, of course. With the exception of five scattered small reversals, blacks are more dissatisfied than whites on each service in each city. On the whole, when city of residence is controlled, racial differences in dissatisfaction appear larger than in the aggregated data presented earlier in Table 1.[9]

A second question answered by Table 2 is whether cities vary in level of dissatisfaction. Row B at the bottom of the table shows that "city of residence" accounts from 2.5% to 6.8% of the variance in satisfaction levels, depending on the particular race and service.[10] These are all significant proportions, and indeed somewhat greater in magnitude than the variance in satisfaction levels explained by race (see note 5). We can also be reasonably confident from previous work (Schuman and Gruenberg, 1970) that variation in dissatisfaction levels by city of residence is not merely a reflection of socioeconomic or demographic differences among city populations.[11]

The fact that city and race produce independent variation in dissatisfaction provides some perspective on the aggregate racial differences reported in Table 1. While blacks are almost always less satisfied than whites when city is held constant, blacks in one city are often more satisfied than whites in another city. To take a concrete and extreme illustration, the average black person in Boston could reduce his dissatisfaction with parks from 2.30 to 1.78 by moving to San Francisco (assuming he becomes an "average black person" in the latter city), while he could reduce it only to 1.96 by remaining in Boston but "becoming" white. We are not suggesting cross-continental busing as a way to solve the problem of parks, but the example does remind us that racial differences are only one component in dissatisfaction with city services.

Having shown that cities differ markedly in levels of satisfaction with services, can we discover what causes these differences? Row E

TABLE 3

Mean Dissatisfaction with City Service Scores in Suburbs and in Central Cities (whites only)[a] (in percentages)

	Schools	Parks	Police	Garbage
Detroit				
Central city	1.46	1.71	1.45	1.41
Suburbs	1.25	1.50	1.16	1.22
Difference	.21	.21	.29	.19
Variance explained	3.1	1.8	5.9	2.2
Cleveland				
Central city	1.25	1.65	1.45	1.27
Suburbs	1.12	1.34	1.06	1.10
Difference	.13	.31	.39	.17
Variance explained	2.0	5.3	11.3	2.8

a. The higher the score, the greater the dissatisfaction. The number of cases varies from question to question, but the minimum is 114 for the Detroit central city and 180 for the Detroit suburbs. For Cleveland, the minimums are 128 and 125, respectively.

TABLE 4

Correlations Between City Mean Dissatisfaction Scores and Objective City Characteristics (n = 15 cities)

Satisfaction Question[a]	City Service Characteristics	Product Moment Correlations: Blacks	Product Moment Correlations: Whites
Parks	Absolute number of parks and playgrounds[b]	–.37	–.20
Parks	Number of parks and playgrounds per capita[c]	–.05	–.27
Police	Number of police[d]	–.07	–.09
Police	Crime rate[e] per capita	.18	.32
Garbage	Number of garbage men[f] per capita	–.21	–.05

a. High scores indicate dissatisfaction.

b. Hawkins, **Recreation and Park Yearbook,** 1966: Table 41, Washington, D.C.

c. Calculated using [b] and total population figures.

d. **Municipal Yearbook,** 1968: Table IX, Washington, D.C.: International City Management Association.

e. Hoover, J. Edgar (1968) **Uniform Crime Reports for the United States,** Washington, D.C. (Index based on following offenses: murder, forcible rape, robbery, aggravated assault, burglary, larceny of fifty dollars or more, and auto theft.)

f. **Municipal Yearbook,** 1971: Table 1/7, Washington, D.C.: International City Management Association.

of Table 2 shows that the ordering of black and white city means (represented by correlation coefficients) is somewhat similar for three of the services and strikingly so for the fourth (police protection), indicating that a city tends to appear similar to both its black and its white inhabitants. This suggests some objective reality to overall city differences in services. We find additional evidence for interpreting satisfaction scores as representing objective reality when we compare city whites with suburban whites. Comparisons are available for Cleveland and Detroit, where suburban as well as central city white samples were obtained. Table 3 reveals greater suburban satisfaction with each of the four services, although most notably with police protection. A cross-section survey allows no firm conclusion about causal direction, but the most reasonable assumption is that services are indeed better in suburbs than in central cities, and that this fact is the source of the difference in satisfaction levels. If people move to suburbs partly because they desire improved municipal services, our data indicate that they are not disappointed.

With these two indications that satisfaction does reflect objective differences in services, we expected variations in satisfaction among the 15 central cities to be associated with differences in objective measures of the four services. Table 4 presents correlations between three of our satisfaction questions and five supposedly relevant city statistics. For example, we would expect a *negative* correlation between the number of parks in a city and the degree of dissatisfaction about parks (where dissatisfaction is scored as the high end of the scale)—which we do find in Table 4. Indeed, all the correlations in Table 4 are in the direction one would predict if satisfaction reflects the objective character of city services. However, the correlations are all modest in size and some are trivial. Apart from their lack of statistical significance, to be expected with only 15 cases, we are reluctant to make much of such small associations. The consistency over five tests in both racial samples suggests that dissatisfaction is related at least slightly to measurable aspects of city services, but the relation is either very small or it is seriously attenuated by measurement problems with the indicators used. Perhaps if we had direct observations on regularity of garbage collection, rather than an indirect measure such as number of sanitation employees, we would find stronger associations. Future research in this area could usefully obtain direct observational data on the quality of such services in order to test its correspondence with attitudes measured at the same time through survey interviewing.

We have also examined the ordering of cities in Table 4 inductively to allow for the possibility that other city characteristics might emerge as possible causes of dissatisfaction levels. Boston turns out to be the locus of greatest dissatisfaction on all four services for blacks and on three of the four services for whites. Newark and Gary also tend to be high in dissatisfaction. Since these three cities are among the smaller ones in our sample, there is a negative correlation between city size and dissatisfaction. For blacks, there is also a positive association between northern location and dissatisfaction, with the above-mentioned cities high, while Baltimore, St. Louis, and Washington (the only "southern" cities in our sample) are all relatively low. We have no interpretation of either of these results; until they are confirmed with a larger set of cities we mention them mainly as suggestions for checking in future research. It is difficult to explain why Boston is at the top of the dissatisfaction list; we are less inclined to regard it as a sign of a newly discovered dimension of urbanism than as the product of a unique constellation of factors located on the Massachusetts Bay.

AREAS WITHIN CITIES

Although cities differ considerably in levels of dissatisfaction, these variations have not helped us account for racial differences. We turn now to look *within* cities—but still with an areal emphasis. All four services can be thought of as distributed over neighborhoods: parks and schools in the most visible sense, police protection and garbage collection somewhat less so. Our inquiry concerns the extent to which racial differences can be reduced to neighborhood differences.

We begin with residential segregation, perhaps the most salient social fact of life today in most American cities. A question in our interview schedule asked each respondent to define his own neighborhood in this regard:

> In this block and the two or three blocks right around here, are all the families (Negro/white), most (Negro/white), about half Negro and half white, or are most of them (white/Negro)?

(The first term in parentheses was used when the respondent was black, the second when white.) Answers to the question allow us to characterize neighborhoods as: all-black, most-black, about half and half, most-white, or all-white.[12] Mean dissatisfaction scores for respondents in each of these types of neighborhoods are given in Table 5. The results are striking: with few exceptions, black dissatisfaction declines each step of the way from all-black to mostly white areas, while white dissatisfaction rises each step of the way from all-white to mostly black areas. There are inversions in the case of schools and garbage, but even these items continue to show the overall trends portrayed more clearly by parks, police, and the relative deprivation question. Degree of neighborhood "whiteness" can be thought of as an important source of satisfaction levels for both blacks and whites.[13]

An even more direct comparison can be made between black and white dissatisfaction levels in terms of segregation of area. The largest

TABLE 5
Mean Dissatisfaction Scores for Each Race by Degree of Residential Integration (integration areas)[a]

	Racial Mix of Residential Area					
	All-Black	Most-Black	Half and Half	Most-White	All-White	Variance Explained (%)
Schools						
Black respondents	1.75	1.61	1.66	1.51	–	0.7
White respondents	–	1.70	1.86	1.52	1.37	3.2
Parks						
Black respondents	2.00	2.00	1.87	1.72	–	0.8
White respondents	–	1.91	1.84	1.74	1.55	2.0
Police						
Black respondents	1.87	1.81	1.54	1.51	–	1.9
White respondents	–	1.69	1.62	1.46	1.32	2.3
Garbage						
Black respondents	1.52	1.50	1.30	1.28	–	1.2
White respondents	–	1.29	1.30	1.28	1.23	0.2
Neighborhood Worse Off						
Black respondents	2.21	2.16	1.96	1.70	–	5.0
White respondents	–	2.18	2.05	1.89	1.85	1.9
Minimum black sample sizes	(556)	(1,256)	(332)	(87)	–	
Minimum white sample sizes	–	(65)	(104)	(703)	(950)	

a. The higher the score, the greater the dissatisfaction.

racial difference for each question occurs when we subtract the scores of whites in all-white areas from blacks in all-black areas:

	Difference
Schools	.38
Parks	.45
Police	.55
Garbage	.29
Relative Deprivation	.36

The differences are roughly twice the size of those obtainable from Table 1 for the total black and white samples. We now ask what happens when we compare blacks and whites who live in areas with the *same* degree of integration. The most precise comparison possible is for "half and half" neighborhoods:[14]

	Difference
Schools	–.20
Parks	.03
Police	–.08
Garbage	.00
Relative Deprivation	–.09

For this comparison, with area of integration held constant, the black/white difference not only vanishes, but tends to be slightly reversed, with whites more dissatisfied than blacks. Thus we have in an important sense completely solved the problem we posed initially: blacks are more dissatisfied than whites because most of them live in largely black areas. It is not color of skin, but color of area that is associated with dissatisfaction.

Moreover, black and white differences on the four services disappear even more completely in mostly white areas, with the added advantage that absolute dissatisfaction is generally low for both races in such areas. Projecting these results in a purely hypothetical fashion, if the black population were dispersed into outlying areas so as to create a smaller proportion of blacks than whites in each area, racial differences in dissatisfaction should disappear, with only a slight rise in white dissatisfaction. Indeed, comparing white dissatisfaction in all-white suburbs from Table 3 with white levels in all-white parts of cities, we see that area continues to be important even when color is controlled. Hence both black and white levels of dissatisfaction could theoretically be "adjusted downward" still further by moving both racial groups from city to suburban towns. The end point of this exercise is to suggest—only half seriously, to be sure—that a dispersal of blacks (and whites) from central cities to suburbs, with care to keep the white middle-class proportion high everywhere, would not only eliminate racial differences in dissatisfaction with services, but considerably reduce the overall level of such dissatisfaction in the metropolitan area.[15] This would, of course, leave central cities vacant, but a creative metropolitan administration could no doubt come up with imaginative uses for so much well-located space. Central cities might become giant airports servicing surrounding areas! Or perhaps they could be turned back into agricultural production, reversing the age-old direction of flow between center and hinterland.

If we have solved our original problem in one sense, there are two important senses in which it remains. First, there is the question of giving a clear causal interpretation to the association between the racial mix of an area and its dissatisfaction with services. One interpretation that suggests itself immediately is differential treatment of neighborhoods, as was indeed indicated explicitly by at least one black respondent: "This is a partially white neighborhood and they give better service in a mixed neighborhood than they give in an all-colored neighborhood." Our findings would occur if in black neighborhoods garbage is collected less frequently, police come less quickly or act less concerned, parks are fewer or less cared for, teachers are less qualified or less interested. These interpretations all assume that the racial character of the neighborhood leads to differential treatment by the city government, which in turn leads to differences in satisfaction.

An alternate interpretation exists, however, which points to a direct causal link between the racial mix of an area and the level of dissatisfaction there, without differential treatment by city administrations being involved. A school that is "half and half" may seem less attractive to white parents than an all-white school, even though it may be more attractive to some black parents than an all-black school. Likewise, "street crime" rates (and thus a sense of lack of police protection) are probably correlated inversely with the "whiteness" of the neighborhood; hence, blacks in mixed neighborhoods may feel safer, but whites less safe, than comparable blacks and whites in completely segregated neighborhoods. More generally, respondents may be reacting directly to the effects of racial mixture, rather than to the nature of the services provided by the city.

It is difficult to locate decisive evidence bearing on these two radically different interpretations, but several considerations suggest that the first—differential treatment—has at least some importance. First, garbage collection seems clearly the responsibility of city administration, yet it shows much the same trend as the other services, at least on the black side. (However, the disappearance of a trend on the white side may well be significant). Second, the white suburbs show even greater satisfaction than the all-white areas of the city; thus where racial composition is controlled, "administration" makes a difference on each service. Finally, citizens themselves perceive the problem as one of differential treatment: the relative deprivation item at the bottom of Table 5 shows the same trends by neighborhood as the other items, if anything a bit more sharply.[16] Indeed, the item presents a noteworthy reversal, with white respondents in mixed areas especially likely to cry foul with respect to the provision of services by the city. All these considerations suggest that the relation of dissatisfaction with services to areas of the city reflects at least in part differential treatment by city administrations and not merely racial composition as such.

A second basic question involves the nature of the areal units we have been considering. The degree of integration of an area correlates with other characteristics of the area, notably its socioeconomic level. What appears to be discrimination (or composition) in terms of the racial character of the area may actually be discrimination (or composition) on the basis of social class. Some respondents spoke explicitly of economic level, rather than of race: "The city officials think this is a poor neighborhood and they don't care and think that people who live here don't care either. So they pick up garbage just whenever they get around to it."

In order to determine more specifically the nature of the areal unit, it is necessary to look at dissatisfactions with services both by areal integration of neighborhoods and by economic levels of neighborhoods, with each controlled for the other. We do so in Table 6, where mean dissatisfaction scores are presented for both "integration areas" and "income areas," with each adjusted to remove the effects of the other (as well as the effects of city of residence) by means of multiple classification analysis.[17]

TABLE 6
Mean Adjusted Dissatisfaction Scores by Integration Area and by Income Area (black sample only)[a]

	Schools	Parks	Police	Garbage	Relative Deprivation	Minimum n
Integration Areas						
All-black	1.78	2.00	1.85	1.49	2.19	(556)
Most-black	1.61	2.00	1.82	1.50	2.17	(1,256)
Half and half	1.65	1.88	1.57	1.33	1.98	(332)
Most-white	1.48	1.72	1.51	1.30	1.74	(87)
Income Areas						
Low-income	1.64	2.00	1.92	1.52	2.24	(733)
Medium-income	1.64	1.96	1.77	1.52	2.13	(758)
High-income	1.65	1.93	1.64	1.35	2.00	(740)

a. High scores indicate greater dissatisfaction. These adjusted means are obtained from a multiple classification analysis, with integration areas, income areas, and the 15 cities as predictors of dissatisfaction scores.

In general, integration areas continue to be associated with dissatisfaction scores after this adjustment. The distinction between all-black and most-black areas proves to be of little importance, which is not surprising if we assume that the latter category would be better characterized as "almost all black."[18] From the standpoint of both racial composition and the view from city hall, the distinction between the two categories is probably slight. On the other hand, the difference in means between most-black and half and half is sharply as predicted for all but one question (schools). The results are equally consistent for half and half versus most-white, despite the small number of cases in the latter category. It is also instructive to compare the all-black and most-white area means: the differences are consistent and large in each case—about as large as the differences between black and white means in Table 1.

Clear trends also appear for income areas, with level of integration controlled, although the differences are not quite as consistent. Schools no longer show any difference and parks only a slight one. But the overall finding is that both the color and the income of one's neighborhood count when it comes to perceived quality of services. We must think of a more general dimension of status, which includes race, class, and perhaps other attributes, all of which affect satisfaction with city services. So far as we can tell, these attributes are additive, so that living in a neighborhood which is high in "whiteness" and high in income leads to the greatest satisfaction.

Examination of tables where both variables are combined shows little evidence of nonadditive effects.

DIFFERENCES AT THE LEVEL OF INDIVIDUAL ATTRIBUTES

It might be thought that the areal differences first reported can be reduced further to demographic or socioeconomic attributes of individuals. This is not the case. For example, when the "income areas" discussed earlier are included in a multiple classification analysis along with respondent reports of their own family incomes, the areal measure is improved slightly as a predictor on four of the five questions, while the individual level measure shows negligible relationships to the same four dissatisfaction questions. This is also true when respondents' education is included with either income areas or integration areas. These individual level socioeconomic variables generally show no consistent association with dissatisfaction over city services.

The one exception to the preceding summary is instructive, for it involves respondent education in relation to the schools question. For both blacks and whites, though especially for the former, dissatisfaction with schools increases with *increasing* education of respondents (see Table 7). (This is not a result of area of residence, for the relationship increases for blacks and is maintained for whites

TABLE 7
Black and White Dissatisfaction Scores on School Question by Educational Level

Respondent's Own Education	Blacks		Whites		Difference Between Means
	n[a]	Mean	n[a]	Mean	
0- 8 years	(484)	1.42	(277)	1.33	.09
9-11 years	(916)	1.67	(499)	1.48	.19
12 years	(706)	1.65	(629)	1.41	.24
13-15 years	(262)	1.83	(231)	1.56	.27
16+ years	(116)	1.93	(228)	1.60	.33

a. The ns given here (unlike other tables) are in weighted form, since this indicates the contribution of each subclass mean to the overall mean for that race. Weighted and unweighted ns are not greatly different for these categories.

when "integration area" is controlled.) It appears that those whose own education makes them place a high value on good schooling are the persons most apt to be critical of the current state of the schools in their neighborhood.

Furthermore, this relationship is stronger for blacks than for whites, and thus, as Table 7 shows, the difference between black and white dissatisfaction levels increases sharply as one goes up the educational ladder. However, the overall effect of this increase tends to be canceled out because the two races are distributed differently over the educational levels. Black college graduates are particularly dissatisfied with schools, but they constitute a much smaller proportion (less than 5%) of the total black sample than do white college graduates (more than 12%) of the total white sample. We can expect black criticism of the quality of schools to increase in the coming years as black education itself increases.[19]

IDEOLOGICAL FACTORS IN DISSATISFACTION

At the beginning of this chapter we raised the question of ideological *versus* objective factors in racial differences in dissatisfaction with city services. Most of our evidence thus far is consistent with the assumption that services actually are worse in certain areas and that this causes variations in dissatisfaction. We failed to demonstrate such a relationship directly, however, since strong associations did not emerge (Table 4) between objective indicators of services and subjective dissatisfactions. This leaves a more ideological explanation as an important possibility. Perhaps blacks (and especially blacks in segregated areas) are simply more negative about government in general, and this negativism is added to "normal" dissatisfaction to produce the racial differences recorded in Table 1.

The best measures available to us of ideological disenchantment with government for both blacks and whites are questions on the city mayor and the federal government:

> How about the Mayor of [CITY NAME]? Do you think he is trying as hard as he can to solve the main problems of the city or is he not

> doing all he could to solve these problems? [IF NOT DOING ALL HE COULD] Do you think he is trying fairly hard or not hard at all?

The question on the federal government was the same except that "federal government in Washington" was substituted for "mayor," and "they" for "he."

Both items are related moderately to the four dissatisfaction items (data not shown). If the causal direction is simply that poor city services lead to criticism of government heads, then the relation should be stronger for the mayor than for the federal government, since the mayor is the person most directly responsible for the quality of services. However, the relations are about equally strong for the mayor and the federal government. This leaves as the more likely possibility the existence of an ideological propensity to be critical of both the products and the leaders of government.

Does this ideological propensity contribute somehow to the *greater* dissatisfaction of blacks? Apparently not. When ideology is held constant (for example, when we look only at those who feel the mayor is not trying hard at all), the gap between black and white dissatisfaction scores is not reduced. "Ideology" seems to elevate *equally* both black and white dissatisfaction levels, but is not a source of the racial *difference* in such levels. Moreover, there is no evidence that ideology is involved in the relationship between dissatisfaction scores and the racial mix of a neighborhood: when either the mayor or federal government items are introduced as controls, the basic associations presented in Table 5 are unchanged, namely, satisfaction tends to increase with whiteness of neighborhood.

SOME CONCLUSIONS AND PRACTICAL IMPLICATIONS

Our analysis indicates that the primary source of racial difference in dissatisfaction with city services lies in variations by neighborhood within cities. These variations are essentially along a status dimension defined in our study by race and social class. It is not one's own race or class that is more relevant, however, but that of one's neighborhood. Persons living in largely black and lower-income areas are most dissatisfied with the services they receive—regardless of their race or income; persons living in largely white and upper-income areas are most satisfied with their services—again regardless of their own race or income.

We cannot determine whether these neighborhood differences are due primarily to discriminatory treatment by city administrations or primarily to racial composition in a more direct sense. Both may well be involved. On the one hand, for example, an increase in lower-class blacks in an area is associated with a rise in certain types of crime; at the same time, police patrols may become less visible and concerned in such a neighborhood, at least from the standpoint of the average citizen. *Both* changes would lead to dissatisfaction with the "quality of police protection in this neighborhood."

Other variables were also identified as sources of dissatisfaction with city services. In particular, cities themselves differ in levels of dissatisfaction—enough so that despite the general trend of racial difference, whites in some cities are more dissatisfied than are blacks in other cities. Thus while city variation does not account at all for racial *differences,* it is an important factor in producing variation in satisfaction for *both* races. We suspect that these city differences in subjective evaluations are due to objective differences in the delivery of municipal services, but we were only slightly successful in documenting this crucial point. Future research on attitudes in this area needs also to identify and measure more clearly the objective characteristics of neighborhoods and of cities that are presumed to underlie reported dissatisfactions.

The main competing explanation of dissatisfactions with services assumes that they are simply facets of a broader alienation from government. Our study provides some evidence that general ideology is involved in dissatisfaction, but not that it is an important factor in racial differences. Whites show much the same relationship between general ideology and specific dissatisfactions as do blacks. Alienation, therefore, acts largely as a constant factor raising dissatisfaction for both races, but not greatly affecting the gap between them.

Beyond our interest in raising civic satisfaction, the measures we have focused on may have a larger importance. After the 1967 urban riots, a search went on to identify the type of city or of area within city in which riots were more likely to occur. The search was generally unsuccessful, and it was soon recognized that almost any large city could have a riot (see Spilerman, 1970). Somewhat more success was obtained by separating out the parts of cities in which riots occurred, but even here the economic differences appeared much less sharp than had been expected (National Advisory Commission on Civil Disorders, 1968: 77, 322, 348-355). Indeed, other than having a disproportionately large black population, it is not clear from the commission's tables (using 1960 census data) that so-called "disorder areas" differed socioeconomically from the rest of the city once race was held constant. But such areas *do* differ in level of dissatisfaction with city services, as shown in Table 8, for the black sample. Blacks in riot tracts are noticeably more dissatisfied than blacks in the same four cities living outside such tracts. The relationships are reduced when integration areas and income areas are controlled, as might be expected, but this does not change the descriptive fact that black citizens living in riot areas experienced greater dissatisfaction with city services than blacks living elsewhere in the same cities. Assuming that these dissatisfaction levels are not simply a result of the riots, but preceded them, we may have here one factor providing legitimacy for the kind of behavior and ideology that characterized the riots. In particular, the belief that the city administration discriminates against one's neighborhood in the provision of basic services could be used to justify actions ordinarily viewed as criminal. We shall not pursue the finding further in this paper, but it points up some possible implications of citizen dissatisfaction with municipal services.

TABLE 8
Mean Dissatisfaction Scores for Riot and Nonriot Tracts in Four Major Riot Cities[a] (black sample only)

	Minimum n	Schools	Parks	Police	Garbage	Relative Deprivation
A. Dissatisfaction means: adjusted only for city of residence						
Riot tracts	308	1.93	2.29	2.01	1.62	2.24
Nonriot tracts	359	1.62	2.07	1.73	1.40	2.07
Variance explained (%)		3.8	1.8	2.7	2.0	2.3
B. Dissatisfaction means: adjusted for city, integration areas, and income areas						
Riot tracts	303	1.89	2.25	1.92	1.59	2.16
Nonriot tracts	349	1.65	2.09	1.79	1.41	2.12
Variance explained (%)		2.4	1.0	0.6	1.4	0.1

a. High scores indicate greater dissatisfaction. The four cities in our sample that had had major riots in 1967 according to the National Advisory Commission on Civil Disorders (1968: 348-358) were Cincinnati, Detroit, Milwaukee, and Newark. Tracts in which the "disorders" occurred are identified in the commission's report.

NOTES

1. More details on the sampling design for this study are provided in Campbell and Schuman (1968: Appendix A) and Campbell (1971: Appendix B). For the present analysis, as in the two just cited, the total sample for each race has been weighted so that each city population is represented in its correct proportion in the total 15-city population. (However, n's shown in tables to indicate reliability of results are unweighted unless otherwise indicated.) Thus when we deal with the entire sample we represent the combined population of the 15 cities—a population approximating that of the major cities in the non-South United States. Black and white cross-sections, ages 16 to 69, were drawn separately in each city, and except where noted are presented separately. Suburban white cross-sections were also drawn around two cities (Cleveland, n = 141, and Detroit, n = 225), and will be introduced at a later point.

2. This is the order in which the services were presented. However, a fifth service was also included (between parks and police): "sports and recreation centers for teenagers." The nature of the latter service overlaps conceptually "parks and playgrounds for children," and as expected the two are much more highly correlated (.51 black sample, .53 white sample) than is any other pair of services (.37 is the highest of the remaining correlations, and most are considerably lower). Both to simplify our analysis and to avoid overrepresenting one particular type of service, we have omitted "sports and recreation centers" from the present article.

3. Percentages in Table 1 have been calculated after omitting "don't know" (DK) responses. The city service questions show a surprisingly wide range of DK responses, from close to zero for garbage collection to about a quarter of each total racial sample for schools. The fact that the four questions are so similar in wording suggests that "don't know" here means self-acknowledged ignorance about the service being described, rather than poor understanding of the question. The underlying dimension of ignorance is obvious: almost everyone has reason to be aware of garbage collection (3% DK, averaging the two races) but this is less true of police (6% DK) and much less true of parks (15%) and schools (24%). (DKs for the relative deprivation question are 8% for blacks and 9% for whites.) We have eliminated DK respondents from most of our analysis as the least troublesome way of dealing with an unavoidable problem. Since such responses are omitted on a question-by-question basis, our sample varies in size and exact composition over the four services; this may account for some minor differences by question in particular analyses, although we have not uncovered any obvious examples.

4. For a description of this method of follow-up inquiry, see Schuman (1966).

5. The variance accounted for by race for each service is: 1.3% (schools), 3.4% (parks), 5.3% (police), 2.3% (garbage). These are squared correlation ratios calculated with the races weighted at their estimated 1968 proportions in the total 15-city population (31% black, 69% white), but only minute changes occur if 50-50 ratios are assumed. All these associations are statistically significant at the .01 level.

6. The classic studies here are reported in Stouffer et al. (1949). A recent review and development of the relative deprivation concept appears in Gurr (1970).

7. For example, blacks have been consistently more critical than whites of the Vietnam War, as shown by Gallup data going back to the mid-sixties when the war still had widespread support. Or at a broader level, there are noteworthy racial differences in belief in a friendly and controllable social environment, with blacks on the skeptical side (Coleman et al., 1966). A more directly relevant bit of data come from our own earlier study, where

blacks are more critical than whites of mayors, governors, and the federal government (Campbell and Schuman, 1968: 41).

8. More generally, considering only the percentage "generally satisfied," the mean of the four service items for blacks is 52.5%, with a standard deviation of 12%; the corresponding mean for whites is 69.2%, with a standard deviation of 9.3%.

9. Note that the simple mean of the 15-city means (row A of Table 2) is higher in each instance than the corresponding aggregated sample mean shown in Table 1 (where cities are weighted in proportion to their populations). It is true, of course, that the racial difference is greater in some cities than in others. Sometimes this makes good sense, as with Milwaukee, where we know from other evidence that racial polarization was extreme in 1968 and this may well have affected either services or perceptions of services. However, we have not been able to locate more systematic interaction effects involving *types* of cities.

10. The relative deprivation question is omitted from this already complex table, but it also shows a small relation to "city of residence:" 1.4% variance explained for blacks, 2.0% for whites. The 15 black and white city means are: Baltimore (2.01, 1.96); Boston (2.23, 1.96); Brooklyn (2.13, 1.87); Chicago (2.13, 1.83); Cincinnati (2.19, 1.94); Cleveland (2.23, 1.90); Detroit (2.12, 1.92); Gary (2.14, 1.93); Milwaukee (2.31, 1.92); Newark (2.10, 1.98); Philadelphia (2.13, 1.89); Pittsburgh (2.16, 2.04); St. louis (2.17, 1.93); San Francisco (2.25, 1.79); Washington (1.99, 1.52).

11. City effects may seem to be related to race in still another way, namely, for three of the four services the variance accounted for by city is more than twice as great for whites as for blacks. However, it is doubtful whether substantive interpretation can be found for this difference. In row C of Table 2, the standard deviation of the city means is presented without weighting for city size; since these are not systematically greater for whites than for blacks, we cannot attribute the greater explanatory power of city among whites to greater intercity variations in mean scores. The racial difference in the explanatory power of city is the result of the greater variation *within* city among blacks (row D), a finding which does not help our understanding since this greater within city variance consists of all the unexplained variance in the system and therefore offers no theoretical direction. In fact, the greater "within variance" for blacks is probably only a reflection of the greater total variance on these items for blacks than for whites (shown earlier in Table 1 in terms of standard deviation sizes), which in turn seems to result from the fact that blacks are fairly evenly spread over the three categories of the satisfaction scales, while whites are bunched toward the satisfaction end. Thus what seems to be a difference in variance in a formal sense turns out to be a restatement of our initial problem: why are blacks more dissatisfied with city services than are whites?

12. Fields (1970-1971) reports an intraclass (within-cluster) correlation of .65 for this question for a cross-section of black respondents in Detroit. This means that 65% of the variance in responses to the question can be accounted for by the extent to which respondents live in the same neighborhood "cluster" of dwelling units. We must certainly expect some disagreement among respondents in the same neighborhood, given the vagueness of key terms such as "most," but responses seem unlikely to be seriously biased and are probably as good a measure of segregation as can be obtained in the absence of contemporaneous census information.

13. The results in Table 5 are based on the entire 15-city racial samples. Since cities differ in both satisfaction levels and degree of integration, the results could be due to the fact that some cities are especially distinctive in both. However, when the means in Table 5 are adjusted by adding city of residence to area within city as predictors of dissatisfaction in a multiple classification analysis, the reported means are only slightly changed–in no case by more than 0.03 of a point. The control for city tends to increase slightly the association between dissatisfaction and integration of area for whites and to decrease it slightly for blacks, but in neither case would conclusions be altered in any way. We have also examined

the relation between integration of area and satisfaction levels within several cities where sample sizes are large enough to make this possible–again without any change in conclusions.

14. Precisely because we can most realistically assume that both races are talking about the same neighborhood. It is more likely that differential perception enters in when blacks and whites describe a neighborhood as "most-black" (or "most-white"), since "most" can be defined to cover a wide range of proportions.

15. Of course, such a move would have other effects, no doubt undesirable from some points of view. Black political power would be reduced, at least in the sense that it now exists in cities like Newark. We have here simply carried one line of results and one line of thinking to a logical extreme, but obviously not in an entirely serious vein. From the standpoint of the average black family, however, as distinct from black leadership, these are conjectures at least worth entertaining.

16. Suburban whites also see themselves as slightly better off than do city whites on the relative deprivation item, although the difference is not great:

Cleveland city	1.92
Cleveland suburbs	1.83
	$E^2 = 1.0\%$

Detroit city	1.90
Detroit suburbs	1.82
	$E^2 = 0.8\%$

17. Multiple classification analysis is a form of multiple regression using dummy variables (see Andrews et al., 1969). "Integration areas" have already been defined and used in Table 5. "Income areas" were constructed as follows: using 1960 census data, the median family income for the *tract* in which a black respondent lived was coded. Respondents in a given city were then ordered from high to low in terms of these tract figures, the order trichotomized, and the respondent thus characterized as living in a high-, medium-, or low-income area in his city. This operation has been carried out only for black respondents and therefore the rest of this section provides data on blacks only except where noted. The variables "income areas" and "integration areas" show a correlation of .29.

18. As already noted, much of our measurement error for this question probably lies in the distinction between all-black and most-black. Two respondents living in an essentially all-black area would choose different categories if one knew of a single white family nearby that was unknown to the other.

19. A number of other individual attributes were reviewed. They are not introduced systematically here because they did not appear relevant to our main problem of accounting for racial differences. Thus for blacks:

(a) Persons with young children are slightly more apt to be dissatisfied with schools and parks (but not police or garbage) than persons without children.

(b) There are no sex differences on the four dissatisfaction questions.

(c) Owning versus renting does not affect satisfactions.

(d) Age shows little relationship to three of the services, but a small negative association with garbage (i.e., younger people are more dissatisfied) for reasons that are not clear.

(e) Recent move to the city of present residence shows no association with any dissatisfaction.

REFERENCES

ANDREWS, F., J. N. MORGAN, and J. A. SONQUIST (1969) Multiple Classification Analysis. Ann Arbor: Institute for Social Research.

CAMPBELL, A. (1971) White Attitudes Toward Black People. Ann Arbor: Institute for Social Research.

--- and H. SCHUMAN (1968) Racial Attitudes in Fifteen American Cities: Report for the National Advisory Commission on Civil Disorders. Ann Arbor: Institute for Social Research.

COLEMAN, J. et al. (1966) Equality of Educational Opportunity. Washington, D.C.: U.S. Government Printing Office.

FIELDS, J. (1970-1971) "The sample cluster: a neglected data source." Public Opinion Q. 34 (Winter): 593-603.

GURR, T. (1970) Why Men Rebel. Princeton: Princeton Univ. Press.

National Advisory Commission on Civil Disorders (1968) Report. Washington, D.C.: U.S. Government Printing Office.

SCHUMAN, H. (1966) "The random probe: a technique for evaluating the validity of closed question." Amer. Soc. Rev. 31 (April): 218-222.

--- and B. GRUENBERG (1970) "The impact of city on racial attitudes." Amer. J. of Sociology 76 (September): 213-261.

SPILERMAN, S. (1970) "The causes of racial disturbances: a comparison of alternative explanations." Amer. Soc. Rev. 35 (August): 627-649.

STOUFFER, S. et al. (1949) The American Soldier: Adjustment During Army Life. Princeton: Princeton Univ. Press.

Part IV

EMERGING ISSUES IN URBAN POLICY

EMERGING ISSUES IN URBAN POLICY

Although the relationship between the public and political leaders seems to form the core of urban politics, any examination of this subject undoubtedly would be incomplete without a careful investigation of some of the most pressing controversies facing cities. The ability of decision makers to achieve satisfactory solutions to metropolitan problems might be determined both by the adequacy of the process by which policies are formulated and by the nature of the difficulties that confront them. Consequently, this portion of the volume presents five studies concerning major issues in urban politics and public policy.

Among the questions that dominate city politics, perhaps none is of more crucial importance than the conflict over racial or ethnic relations. Moreover, many of the most basic elements in this controversy are related to the environments in which minority groups are located. In a study that may contain important implications concerning the growing trend toward racial separation, Rose presents a geographical analysis of the environmental characteristics of localities that consist wholly or primarily of black residents. Moreover, social and political behavior in predominantly black areas is discussed in an essay by Perry and Feagin, which suggests the need for a major revision of traditional concepts concerning those areas. In addition, the chapter by Barrera, Munoz, and Ornelas not only reviews prior studies of political activity in Mexican-American or Chicano communities, but it also calls for a new view of those communities as internal colonies.

In addition to the challenges posed by racial or ethnic relations, urban areas are increasingly beset by serious threats to the quality of the environment. As Loveridge observes, the problems raised by the issue of pollution may severely tax the ability of governmental institutions to resolve them.

Many of the major controversies in American cities, including ethnic or racial relations, the environment, and similar questions ultimately may be settled by Congress or federal agencies. As a result, Ripley assesses the development of four major public policies affecting economic and human resource development, and he presents a comprehensive framework for the analysis of other policy issues.

In the final chapter, Holden focuses both on the unique political problems generated by the process of urbanization and on the capacity of governments to cope with those problems. His essay traces the development of two major reform movements in the field of political science as well as the role of local, state, and national institutions in the development of policies designed to ameliorate the urban crisis.

15

The All Black Town:
Suburban Prototype or Rural Slum?

HAROLD ROSE

□ BLACK POPULATIONS IN THE UNITED STATES continue to become more urbanized as central-city populations swell, while rural and small-town populations continue to decline. By 1970, 74% of all blacks resided in metropolitan areas with the lion's share, 58%, confined to the central city. The decade of the sixties is likely to be recorded by historians as the turning point in modern American history, for it was a period of unparalleled prosperity for some, and revolutionary social change for others. From the perspective of its long-term impact, though, it is likely to be viewed as the period in which the central city began to wither, at least in terms of its functional importance and the ability to provide qualities of life which were being sought after by an ever more affluent population. Unfortunately for minorities and the poor, the urban form which had beckoned in the industrial era and had continued to serve as the anchor for postindustrial development was no longer being viewed as viable. Thus the hordes of low-income, black migrants who made the trek from the rural countryside or from the small town were arriving at a time when the city was undergoing cataclysmic economic and social change. For some, this simply represented a move from a rural slum to an urban slum environment, or a concrete reservation. The question with which we are left is: how does one facilitate the equalization of environmental opportunity with metropolitan areas, given the shifts which are currently taking place within that context?

THE EVOLUTION OF BLACK SUBURBAN COMMUNITIES

The metropolitan area has replaced the central city as the functional entity upon which economic development rests. It is generally assumed that opportunities for the good life are to be found beyond the margins of the central city, as is attested by both the movement of jobs and people from the central city to the suburbs. It is the current process of social, economic, and, subsequently, place mobility which serves as the basis for some of the stresses which are currently to be found within metropolitan America. A persistent problem, however, is the continued refusal of persons of dissimilar class or racial attributes to share a common social space. Thus the segmentation of the metropolitan area into units of common racial and social character is the outcome. Given this situation, the principal focus of this paper revolves around the ability of a set of places, previously referred to as all-Negro towns, to serve as maximizers of environmental opportunity through their role as ports of entry for black people seeking access to both the locational advantages and amenities thought to prevail in suburbia.

THE ALL-NEGRO TOWN AS A SUBURBAN COMMUNITY

It is somewhat presumptuous to assume that the set of places identified in an earlier paper as all-Negro towns could have any significant impact on black suburbanization in America (Rose, 1965: 352-364). Nevertheless, given the general intractability of the problem, both in terms of its social and perceived power attributes, an investigation of this type is justified. The original set of places identified as all-Negro towns were defined as places with populations of 1,000 or more, of whom 95%, at least, were classified as nonwhite (Rose, 1965: 352). After having extracted the pseudo-towns from the functional towns, it was found that there existed twelve places in the nation that satisfied the above-defined criteria. Ten of these were located within metropolitan areas and thus possessed the locational characteristics which could lead to their being loosely described as suburban communities.

In 1960, the total population of these ten communities numbered less than 44,000 and even by 1970 they were the place of residence of fewer than 50,000.[1] Given this very low rate of aggregate growth of these places, it might be concluded that they possess little potential for future black colonization. But given the import of the problem, nothing would be gained by ignoring a potential alternative residential environment for an already colonized segment of the population. Of course, the question left unanswered at this point is the lack of congruency between locational access and environmental access.

In 1960, the majority of these places failed to meet the criteria which are generally thought to describe suburban living in terms of environmental quality. At the same time they possessed the locational attributes which have led a number of social scientists to identify them as suburban simply because they lie within that set of communities identified as metropolitan ring communities. Needless to say, the environmental character of a place may be changed simply as a result of adding new units meeting some minimally desirable criteria. However, the major question which should not be overlooked is the fact that eight of the ten initially identified communities were politically independent and were thus seats of black power, at least in a political sense. The importance of this issue should not go unnoticed if one considers that support from the black population on behalf of the maintenance of central city ghettos, as distinctive socio-spatial residential enclaves, revolves around their being perceived as sources of political power.

COLONIZING AND GHETTOIZING SUBURBAN RING COMMUNITIES

The original study of the all-Negro town was designed to highlight the unique, a tradition which is still very much prevalent in geography. In this instance, unique qualities are of little interest in and of themselves; the potential for providing alternative residential and work environments is the basis for this investigation. Since the objectives of the present study differ greatly from the initial study, there can be some relaxation in the definitional criteria employed to identify the universal set. In the present study, attention will be

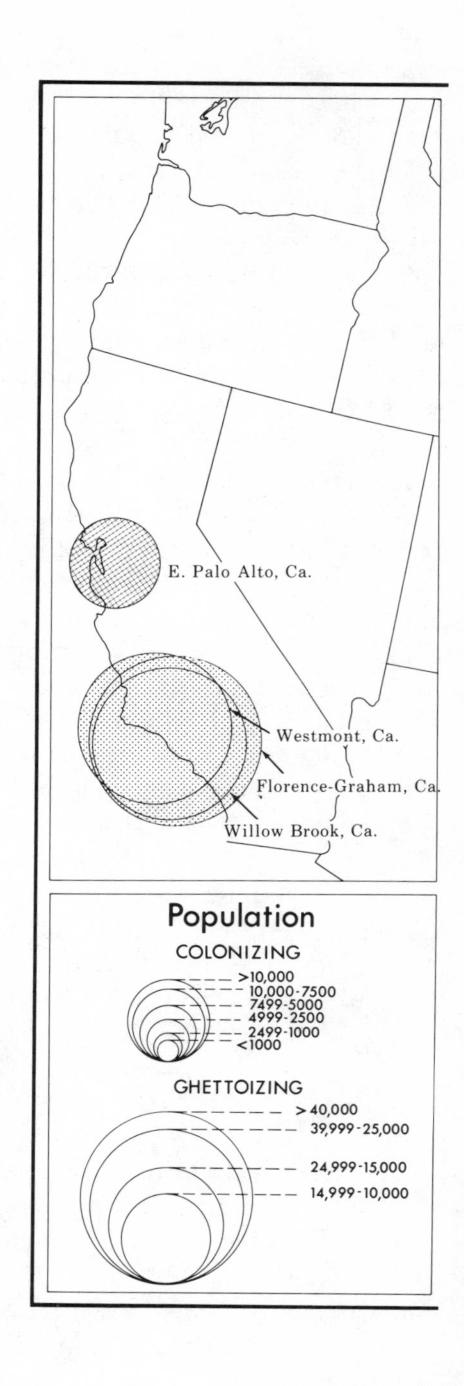
E. Palo Alto, Ca.
Westmont, Ca.
Florence-Graham, Ca.
Willow Brook, Ca.
Population
COLONIZING
>10,000
10,000-7500
7499-5000
4999-2500
2499-1000
<1000
GHETTOIZING
> 40,000
39,999-25,000
24,999-15,000
14,999-10,000

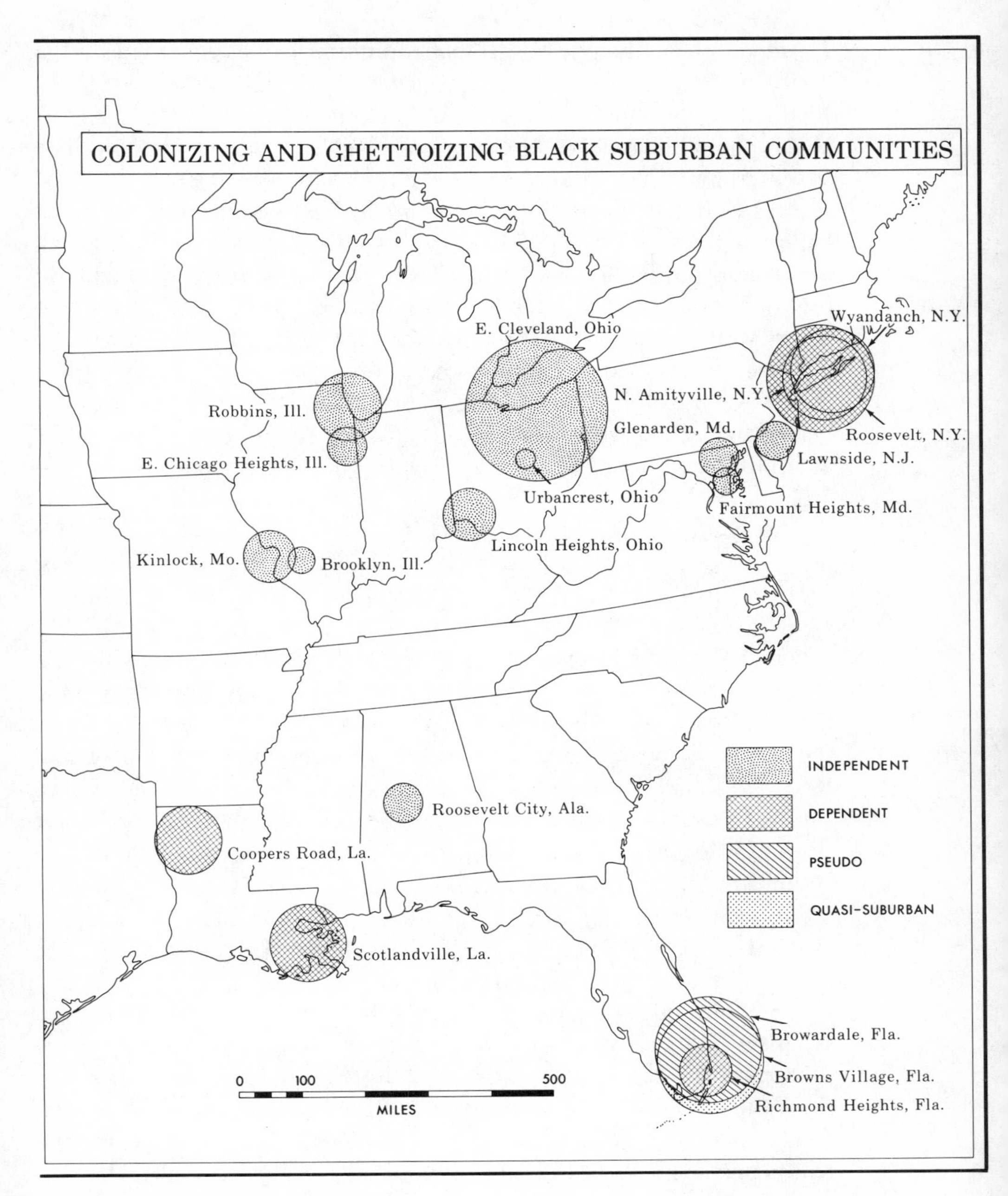
COLONIZING AND GHETTOIZING BLACK SUBURBAN COMMUNITIES
Wyandanch, N.Y.
E. Cleveland, Ohio
N. Amityville, N.Y.
Robbins, Ill.
Glenarden, Md.
Roosevelt, N.Y.
Lawnside, N.J.
E. Chicago Heights, Ill.
Urbancrest, Ohio
Fairmount Heights, Md.
Lincoln Heights, Ohio
Kinlock, Mo.
Brooklyn, Ill.
INDEPENDENT
DEPENDENT
PSEUDO
QUASI-SUBURBAN
Roosevelt City, Ala.
Coopers Road, La.
Scotlandville, La.
Browardale, Fla.
Browns Village, Fla.
Richmond Heights, Fla.
0
100
500
MILES

Figure 1.

focused on all communities within a Standard Metropolitan Statistical Area whose populations are 50% or more black (see Figure 1).[2] The decision to include additional communities was governed by the desire to more realistically assess patterns of black suburbanization. The added communities will, as a rule, represent a different genre of places than the bulk of those which made up the original set. The latter communities are more likely to be undergoing a process of ghettoization, whereas the former might best be described as black colonies. Farley (1970: 525), in his recent assessment of black suburbanization, highlights the latter form as being particularly prevalent among the older suburbs of the North and chooses to describe it as the "population replacement type." Thus, through the addition of the latter set of communities, one might compare the processes of colonization versus ghettoization in establishing black residential zones within the suburban ring. In the case of the latter communities, the base size is 10,000, whereas in the former it is 1,000.[3]

The colonizing communities which are included as a part of this investigation are Roosevelt City, Alabama; Richmond Heights, Florida; Robbins, East Chicago Heights, and Brooklyn, Illinois; Cooper Road and Scotlandville, Louisiana; Glenarden and Fairmount Heights, Maryland; Kinloch, Missouri; Lawnside, New Jersey; Lincoln Heights and Urbancrest, Ohio. The ghettoizing communities include East Palo Alto, Willowbrook, Westmont, and Florence-Graham, California; Browardal and Browns Village, Florida; North Amityville, Roosevelt, and Wyandance, New York; and East Cleveland, Ohio. The latter communities exceed in size all but one of the larger communities among the colonizing group. Unlike the colonizing communities, most of the ghettoizing communities are unincorporated and are thereby without the perceived political power which is associated with independence. An upper limit of 40,000 was placed on the ghettoizing communities as a means of eliminating from that set of communities places which might best be described as industrial satellites.[4] A key question here is: does the colonizing or ghettoizing suburban ring community provide greater access to the good life and, if so, the good life to which segment of the black population?

The trend in black town development is mixed, with the colonizing communities being characterized by both growth and decline during the previous ten years. Among the older colonizing communities, population decline was most often the rule, although Lawnside, New Jersey, the oldest of the several communities, did in fact evidence limited growth. The specific conditions leading to population decline among these communities are not well understood at this point. But it was previously noted that residential quality in these communities was generally to be found wanting in 1960. This, coupled with the fact that such communities were sometimes viewed by central-city blacks as rural backwoods places, no doubt did much to establish their image, in the absence of major landscape improvements.

Among those communities characterized by growth, a major building effort was in evidence, and subsequently these new real estate developments would impact on the communities' growth potential. Nowhere was this more evident than in Glenarden, Maryland, one of the smaller communities in 1960. Needless to say, the availability of space within the legal territory of Glenarden, coupled with a pent-up demand for single family accommodations on the part of a segment of Washington's population, led to Glenarden's rank as third most rapidly growing community in the state during the sixties; it ended the period with more than 4,500 residents. Glenarden's expansive growth can be contrasted with sizable decline in Urbancrest, Ohio. Both of these minimal threshold communities, in 1960, were characterized by growth patterns which moved in opposite directions during the most recent decade. The latter community possessed a quasi-rural landscape pockmarked by housing of vernacular form and remotely located in terms of the major direction of black expansion within the city of Columbus. A number of conjectural answers might be posited to explain the divergent pattern of growth among the colonizing communities, but at this point no definitive answers can be posited to explain the decennial changes in population numbers for each individual place.

The colonizing communities in most instances have a long history of existence with one even predating the Civil War period. On the other hand, the ghettoizing communities are either themselves relatively new or at least recent places of residence for sizable numbers of black people. Whereas the original communities most often started out as black communities, the latter communities are most often ones in which black entry has led to white abandonment, or blacks and whites occupy different zones within a single community. If the ghettoization process operates in these outlying areas in ways similar to those observed in central cities, then one can anticipate a number of these communities reaching the level of black occupancy observed in central cities. This would lead to black

TABLE 1
TRENDS IN THE PATTERN OF POPULATION GROWTH AMONG COLONIZING COMMUNITIES 1960-1970

Community	1960 Population	1970 Population	% Change
Roosevelt City, Alabama[a]	–	3,363	–
Richmond Heights, Florida	4,311	6,663	54.6
Brooklyn, Illinois	1,922	1,693	–11.9
East Chicago Heights, Illinois[b]	3,270	4,611	41.0
Robbins, Illinois	7,511	9,410	25.3
Cooper Road, Louisiana[c]	7,700	9,034	14.7
Scotlandville, Louisiana[b]	20,108	22,557	10.8
Fairmount Heights, Maryland	2,308	2,002	–13.3
Glenarden, Maryland	1,336	4,538	239.7
Kinloch, Missouri	6,500	5,629	–13.4
Lawnside, New Jersey	2,155	2,751	27.7
Lincoln Heights, Ohio	7,798	6,082	–22.0
Urbancrest, Ohio	1,029	754	–26.7

SOURCE: U.S. Bureau of the Census, 1970 Census of Population, appropriate PC (V) series.

a. Did not exist as an incorporated place in 1960.

b. Was not included among the original set of all-Negro towns because it did not satisfy the 1960 defining criteria in terms of the percentage of population nonwhite.

c. Was identified as North Shreveport, Louisiana, in 1960.

TABLE 2
THE SIZE OF THE BLACK POPULATION IN GHETTOIZING SUBURBAN RING COMMUNITIES: 1970

Community	Total Population	Black Population	Black % of Total Population
East Palo Alto, California	17,837	10,846	61.0
Florence-Graham, California	42,895	24,031	59.0
Westmont, California	29,310	23,635	80.6
Willowbrook, California	28,705	23,616	82.2
Browardale, Florida	17,404	15,751	90.7
Browns Village, Florida	23,452	21,471	91.0
North Amityville, N.Y.	11,905	7,747	65.8
Roosevelt, N.Y.	15,008	10,133	67.5
Wyandance, N.Y.	14,906	8,797	59.0
East Cleveland, Ohio	39,600	23,196	58.6

SOURCE: U.S. Bureau of the Census, 1970 Census of Population, appropriate PC (V) series.

proportions which characterize the majority of the colonizing communities. The possibility is strengthened when one considers that few of these communities had significant numbers of black residents in 1960. Nevertheless they were the places of residence of almost four times as many black residents in 1970 as were the colonizing communities.

Ghettoizing Communities and Distance from the Central City

Among the ghettoizing communities, two different patterns can be distinguished that are largely based on distance from the central city. One group of communities is contiguous to central-city black communities and is thus situated in the unincorporated territories found immediately beyond the central city. These, in some instances, are simply no man's lands which fall between the cracks of the incorporated municipalities. Thus one hesitates to accord them the same status as those communities which are more distantly located from the central city. This group includes the three communities found within the Los Angeles metropolitan area and the two Florida communities. Given the process of spatial residential development in southern cities, the unincorporated communities of Browardale and Browns Village no doubt started out as black communities and thus have always maintained a uniracial character. At this time, Browns Village is simply a physical extension of one of Miami's ghetto nucleations. Salter and Mings (1969: 85), in discussing an aspect of the 1968 Miami riot, had this to say about Browns Village:

> Field investigation indicates that the political boundaries separating the unincorporated county portion of Liberty City from the City of Miami portion of the Liberty City area are purely artificial. No physical barriers separate the two areas. Homogeneous house types, both single and multiple type dwellings, extend well beyond both sides of the city line.

Each of the other nonsouthern communities is of more recent vintage and no doubt reflects the goals of a segment of the population to free itself from the very dense central-city environments which prevail in their respective central cities. East Palo Alto is the best known of these several communities, as it received national notoriety in 1968 as the outgrowth of a referendum held to determine the community's name. A segment of the population was desirous of changing the community's name to Nairobi in recognition of its dominant ethnic makeup. The referenda item was defeated by a two-to-one margin. But nevertheless, the term *Nairobi* is being employed to identify a number of local institutions found within the community. East Palo Alto, evolving as a black community during an era of emerging black nationalism, has been strongly influenced by the prevailing philosophy of that era. Quite aside from the publicity received as an outgrowth of its strong identity advocacy, this community is located in an environment that has benefited by the presence of a number of growth industries that offer employment opportunities to its residents.

The Long Island communities have been generally overlooked, but they tend, as a rule, to represent the farthest penetration of blacks into a rapidly growing area in which the black population tend to be more heavily concentrated in the western end of the island. Wyandance, the easternmost community, is located approximately 28 miles from Manhattan.

SUBURBAN STATUS

The social scientist who devotes scholarly attention to aggregations of black people in the United States usually focuses his attention on central-city populations simply because they are the most visible, coupled with the fact that the central city is a cauldron of problems. Slightly more than a generation ago, the few social scientists engaged in research relating to the black population tended to focus on rural blacks, a logical decision at that time. Research focusing on blacks whose mobility characteristics have seen them move into yet a third social and economic milieu, the suburb, is indeed scant. The absence of investigations of the third type is related to the still limited, but growing, number of blacks whose place of residence is to be found in locations beyond the margins of the central city.

In both 1960 and 1970, approximately 5% of the suburban ring population were black. Given the rate of growth of suburban ring populations in general, this level of black presence indicates that central-city blacks, like their white counterparts, are showing signs of joining the movement to suburbia. In 1970, 3.7 million blacks were located outside of central cities, an increase of almost a million over the level prevailing ten years earlier (BLS Report No. 394, 1971: 12). Given the known growth rates which characterized the black population during the decade, a conservative estimate of the number of blacks establishing residence in suburban ring communities during this period is approximately 600,000, a small number indeed, but one which can be expected to increase significantly over time.

The Amorphous Suburban Concept

The whole notion of what constitutes suburban living is seldom defined precisely. Nevertheless, persons engaged in the process of seeking a satisfactory residential environment harbor certain internalized images which define for them a configuration which might be identified as suburbia, since this concept has been popularized as constituting the good life. Researchers who attempt to measure changes in the extent to which the population resides in one kind of urban milieu or another accept the dichotomous inside-or-outside central city as being at least functionally appropriate for distinguishing between suburban and nonsuburban residence. The looseness of such distinguishing criteria does not enable one to evaluate forthrightly what the movement of blacks from the central city to the ring means in terms of providing an alternative environment that results in enhanced quality of life. Where one is located within the suburban ring in large measure influences the kind of residential environment to which he might expect to gain access.

THE PATTERN OF BLACK SUBURBANIZATION

It appears that black suburbanization is intricately tied to the process of ghettoization, as it relates to those segments of space to which large numbers of blacks eventually gain access. Thus, as was previously demonstrated, the number of blacks in suburbia is a function of the period during which the central-city black community attained a critical threshold population (Rose, 1972). In 1970, 48.6% of all black suburban ring populations were to be found in the rings of fifteen Standard Metropolitan Statistical Areas. This would seem to indicate that much entry of blacks into the suburban ring is simply spillover from an adjacent ghetto edge which abuts the city-suburban boundary. The diagram below (see Figure 2) schematically illustrates this phenomenon. Black territorial communities tend to expand sectorally and thus those with large black populations might already have expanded to the margin of the city. In Washington, D.C., St. Louis, and Cleveland, this pattern is already obvious. Washington and St. Louis, places where the black population reached the threshold level at an early date, show a growing number of blacks in zone B (see Figure 3). In other Standard Metropolitan Statistical Areas, black outlyers were sited at an early date, independent of territorial development within the central city. Pittsburgh and Cincinnati represent examples of this pattern. In a number of southern metropolitan areas, and especially those which are overbounded, the only blacks to be found in the ring are those located in zone E, as the other zones may be present only on a limited scale. The quality of residential accommodations to be found within these various zones is partially a function of the age of the city. The nature of black suburbanization, as reflected in the quality

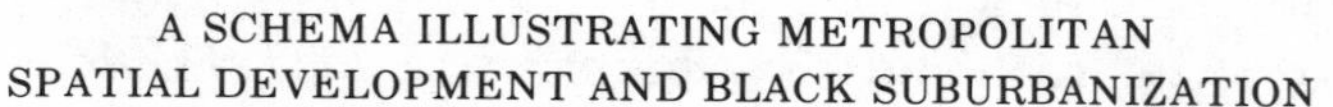

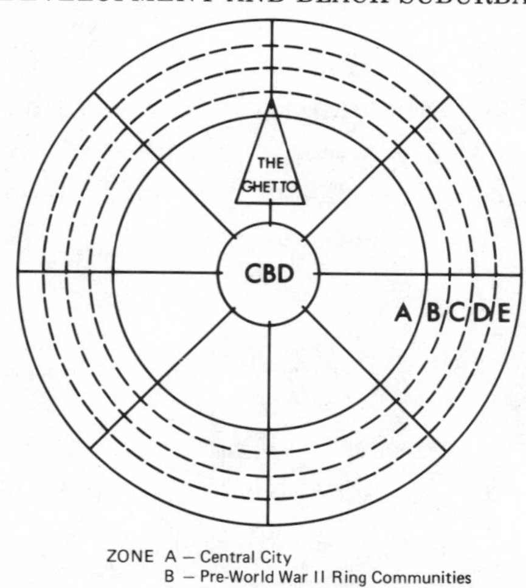

ZONE A – Central City
B – Pre-World War II Ring Communities
C – Post-World War II Ring Communities
D – Post-1960 Ring Communities
E – Rural Urban Fringe

Figure 2.

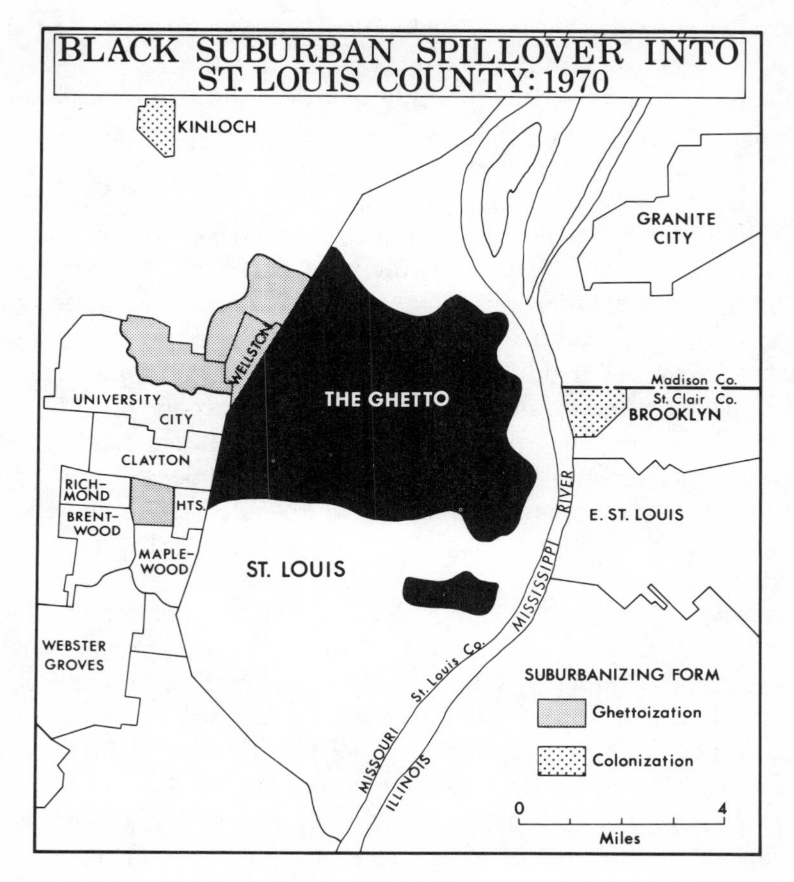

Figure 3.

of the environment available to that population, can be partially detailed by relating the location of the communities in the study set to the Figure 2 schema.

The Spatial Evolution of Colonizing Black Communities

Most of the colonizing communities came into existence when their locations were part of the rural-urban fringe. Thus, upon their degree of isolation rests their potential for future development. The demands for suburban residential space, in most instances, has been sufficiently great that other communities have grown up around the colonizing communities, thereby limiting their potential for physical expansion. This situation characterizes the plight of Lincoln Heights, Ohio, and Kinloch, Missouri, more effectively than some others.

Nevertheless, only limited development acreage is to be found in most of these communities.

Glenarden, Maryland, is the only real exception among the set, as it remained in a zone of limited real estate development prior to 1960. Its rapid expansion during the most recent period was made possible partly by its ability to annex territory during this interval, an option which was not available to most other places of this type. As a result of the nature of their development, the colonizing communities are quite varied in terms of their internal structure. Their internal appearance is basically influenced by the nature of the housing constructed during their period of maximum development. Thus, having come into existence when the site upon which they are located was a part of zone E, their current condition is related to the quality and structural design prevailing during the period of maximum development. This has led to a residential qualitative mix of varying proportions reflecting the dominant design qualities within zones B, C, and D.

Glenarden, Maryland, is the single colonizing community to be dominated by zone D residential qualities. Richmond Heights, Florida, and Cooper Road, Louisiana, are good combinations of zones C and D, but with the latter community still exhibiting many aspects of properties generally associated with rural-urban fringe locations. It is rather obvious that zones B, C, and D are not fully developed in the Shreveport metropolitan area, and therefore the schema is less explanatory in the case of Cooper Road than elsewhere. This is an attribute which might possibly characterize a number of southern metropolitan areas. A number of other small, predominantly black communities are located in the southern part of the Miami SMSA, but, because of their more-rural-than-suburban quality, they were previously distinguished from Richmond Heights by lumping them into the category of outer ring communities which were recognized as being semi-rural in character (Rose, 1964: 230-231).

Robbins, Illinois, and Lawnside, New Jersey, are both dominated by zone B landscapes, but despite the limited availability of raw land within these communities, developers have added housing in limited segments of these areas that is typical of that associated with zone D. In both instances, in-migrants from the central city have settled in these "new-look zones" who represent a nonindigenous population, a factor which has produced some minor stress in community relations. Fairmount Heights, Maryland, situated just across the district boundary, is totally zone B in appearance, while Urbancrest, Ohio, and Brooklyn, Illinois, represent a special case of zone B, in that they might be better described as semi-isolated, early twentieth-century villages that have received only limited benefits from the whole urbanization process.

The Spatial Evolution of Ghettoizing Black Communities

The ghettoizing ring communities are a much more recent phenomenon than the colonizing communities. It appears that blacks were seldom residents of these communities when they were part of the rural-urban fringe. East Cleveland, the Los Angeles metropolitan area communities, and Brownsville are spillover communities. Blacks have simply occupied residential space that was contiguous to space already occupied by the group. The California and Florida examples represent a special case of the ghettoizing communities, since they represent unincorporated areas which are physically a part of the central city with which they are associated. For this reason, this set is thought to have attracted migrants to Wyandance. The older community of the case of Browardale, it appears, more nearly conforms to what has been previously described as a pseudo-community. In addition to the identification of twelve legitimate all-Negro towns in 1960, it was also discovered that there existed seven communities which might best be described as pseudo-communities.[5] It appears that a sizable segment of the functional black community of Fort Lauderdale is not politically a part of that city, and so the term Browardale is employed to describe what is functionally a part of the city, but legally part of Broward County, a situation which has led to the disenfranchisement of as much as one-third of Fort Lauderdale's black population.

The Long Island communities and East Palo Alto, California, are leapfrog rather than spillover communities. Roosevelt, New York, the older of these communities, is of pre-World War II vintage. North Amityville is also of prewar origin, but its period of maximum development occurred during the postwar era. The more remote of these communities is Wyandance, which underwent its period of most recent growth during the sixties. The latter community is characterized by a landscape that is strong in elements of the rural-urban fringe. The low-density characteristics, coupled with a meadow-like appearance, set the black community apart from the

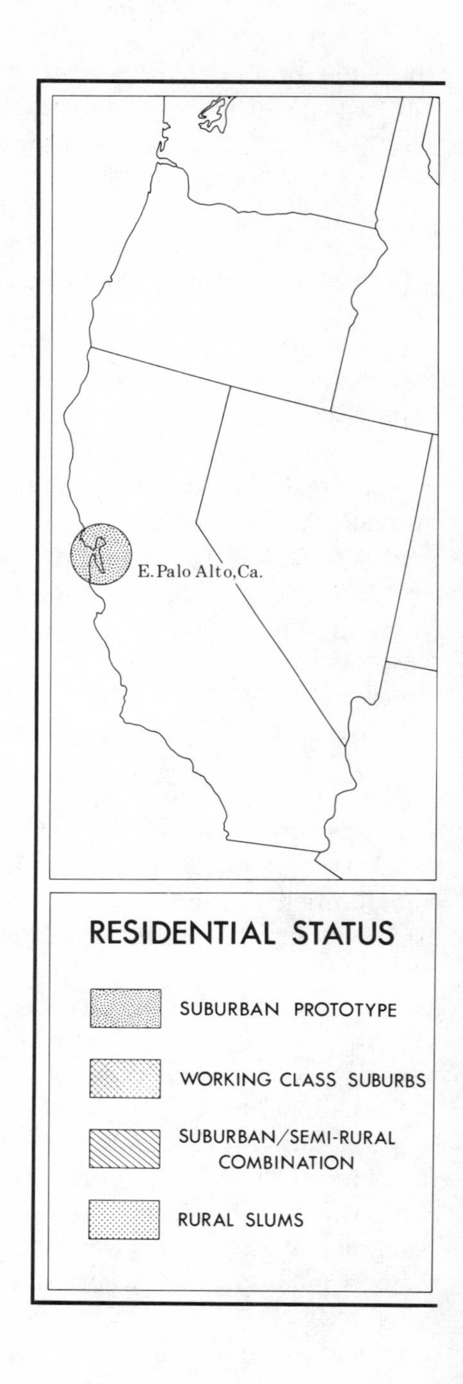
E. Palo Alto, Ca.
RESIDENTIAL STATUS
SUBURBAN PROTOTYPE
WORKING CLASS SUBURBS
SUBURBAN/SEMI-RURAL COMBINATION
RURAL SLUMS

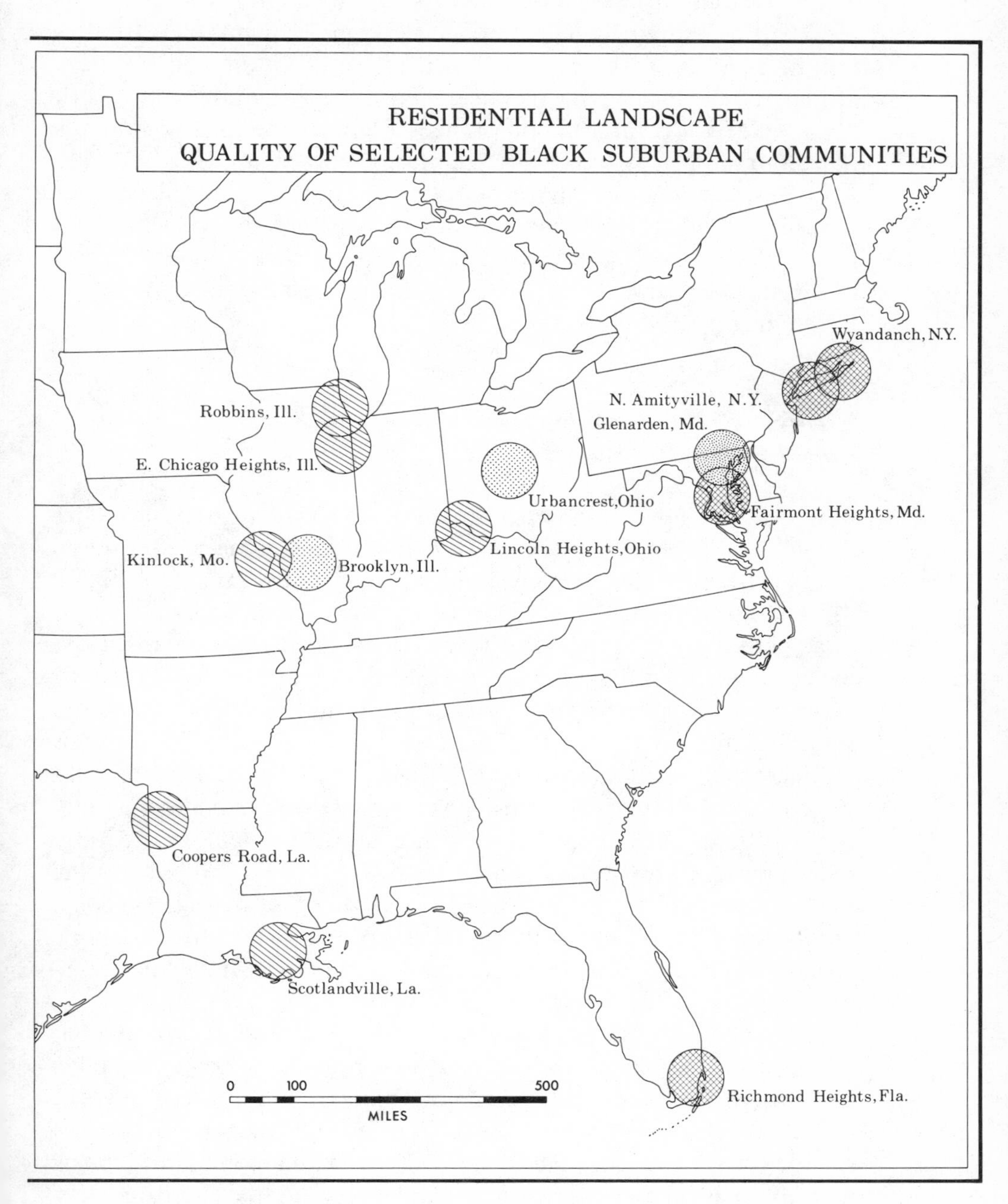
RESIDENTIAL LANDSCAPE
QUALITY OF SELECTED BLACK SUBURBAN COMMUNITIES
Wyandanch, N.Y.
N. Amityville, N.Y.
Glenarden, Md.
Robbins, Ill.
E. Chicago Heights, Ill.
Urbancrest,Ohio
Fairmont Heights, Md.
Lincoln Heights,Ohio
Kinlock, Mo.
Brooklyn, Ill.
Coopers Road, La.
Scotlandville, La.
Richmond Heights, Fla.
0
100
500
MILES

Figure 4.

white community, even though a common housing design is prevalent in both areas. The prevalence of employment opportunities for blue-collar workers in the general area is thought to have attracted migrants to Wyandance. The older community of North Amityville is less rural in appearance, although the family garden is not an unfamiliar sight. Of the four leapfrog communities, East Palo Alto projects best the image of suburbia. It is basically a postwar suburban tract development that has been essentially abandoned by its previous white occupants. The spillover communities are most often sited in zone B, while the leapfrog communities are to be found in zones C and D.

The Social Class and Landscape Characteristics of Black Suburban Communities

An assessment of landscape qualities and median housing values should enable one to place these communities along a continuum which describes them in physical and social class terms (see Figure 4). The assessment conducted here is simply preliminary, but it at least permits one to assess in a loose fashion the quality of life to which a very limited segment of the black population has gained access by shifting their place of residence from the central city to the suburban ring.

Since 1970 census income data is not yet available, it is necessary to work backwards from housing value data to an income position. Davis (1965: 241) indicated that the housing value/income ratio prevailing in 1960 was approximately 2.0.[6] Thus, by halving the median housing value, it is possible to arrive at an expected median family income which will be used to designate the economic character of the population residing within the individual communities.

From the perspective of environmental quality, it is obvious that the ghettoizing communities generally provide a more desirable physical environment than do the colonizing communities. Only one of the colonizing communities gives the impression of fitting the middle-class image of suburbia, with a second community being identified as mixed working-to-middle class in its characteristics. Glenarden, the first community, is one whose lack of drabness is essentially associated with the recency of most of its housing

TABLE 3
CLASS CHARACTERISTICS OF BLACK SUBURBAN COMMUNITIES (based on housing cost and landscape character)

Colonizing	Class Characteristics	Ghettoizing	Class Characteristics
Glenarden, Md.	Middle	East Palo Alto, Calif.	Middle
Richmond Heights, Fla.	Middle-Working	Wyandance, N.Y.	Middle
Lawnside, N.J.	Working-Lower	Roosevelt, N.Y.	Middle
Robbins, Ill.	Working-Lower	Westmont, Calif.	Middle
Fairmount Heights, Md.	Working	North Amityville, N.Y.	Middle-Working
Scotlandville, La.	Working	Willowbrook, Calif.	Working
East Chicago Heights, Ill.	Working	Florence-Graham, Calif.	Working
Lincoln Heights, Ohio	Lower-Working	East Cleveland, Ohio	Working
Kinloch, Mo.	Lower-Working	Browns Village, Fla.	Working
Urbancrest, Ohio	Lower	Browardale, Fla.	Working
Brooklyn, Ohio	Lower		
Roosevelt City, Ala.	Lower		
Cooper Road, La.	a		

a. Information not available at this time.

construction (see Figure 5). The second community, Richmond Heights, Florida, fits less well in this category, largely because of the income constraints imposed in determining position in the hierarchy. Yet, it is completely a community of single-family residences, most of which have been constructed within the last twenty years. Among the older structures in the community, there is an apparent lack of good maintenance which deters from their general attractiveness. The maintenance quality may in part be related to the relatively low income of a segment of the population. Locally, though, this community is recognized as a middle-class community when it is viewed within the context of the set of black communities found within the Miami metropolitan area. Dudas and Longbrake (1971: 165) recently remarked that, "The existence of a middle-class black suburb, Richmond Heights, has deterred integration of middle income blacks and whites in Dade County."

In those communities possessing land which could be developed for residential construction, the working- to lower-class status is appropriate. The newer housing, which constitutes only a limited percentage of the total housing stock, is not sufficient to raise the image of the community to the level that would allow it to take on the appearance of the mythical suburban community. The lack of large quantities of attractive housing restricts the entry of higher-income persons, although the image of these places has been

The above photograph depicts the quality of the residential environment in the Fox Ridge section of Glenarden, Maryland. It is this kind of housing environment that allows Glenarden to be described as a black suburban prototype.

Figure 5a.

Although Glenarden is a community wherein single-family detached units are dominant, there is no dearth of available moderate-cost apartment units. Several apartment clusters are dispersed throughout the community. The above were constructed under the FHA 221(d)3 program.

Figure 5b.

modified by the development of new tract-type homes. Both Robbins, Illinois, and Lawnside, New Jersey, fall into this category, and within these communities housing runs the gamut in quality from the brokendown shacks to the newest split-level design. In each of the lower-class communities there was a decline in population during the decade. In most instances, these were communities which were older with little remaining space that could accommodate residential development.

The ghettoizing communities tend, as a rule, to be of more recent vintage and, because they have a tendency to be located in zones where wage rates are relatively high, they tend to fare well in terms of their social class composition and their maintenance characteristics. Those communities located closest to the margins of the central city tend to fall a shade lower on the spectrum in terms of their social class composition. The housing values prevailing in these latter communities suggest that they might best be described as working class. None of the ghettoizing communities is identified as lower class, a designation which was applied to three of the colonizing group. Needless to say, the colonizing communities as a group possess one important quality which is missing from all but one of the latter type, and that is community control or political independence.

The middle-income communities can be further distinguished from those on the lower end of the spectrum in terms of selected demographic characteristics. Since suburbia is generally thought of as a family residential zone, one would assume a balanced sex ratio in such communities. Likewise, suburbia is thought to attract young homeowners, and thus would include a high proportion of persons in the prime working ages (20-44). Glenarden, Maryland, and East Palo Alto, California, both middle-income communities, possess sex ratios of 98 and include 38 and 41% of their populations, respectively, in the prime working ages. Kinloch, Missouri, and Urbancrest, Ohio, both classify as lower class and becoming depopulated, are characterized by sex ratios of 88 and 80, respectively, indicating a sex imbalance weighted toward the female side. Only 25% of the former community's population are in the prime working ages, and 20% of the latter. Characteristics of this sort tend to vary with the social class makeup of the population and their economic support base. Thus the middle-income communities normally represent growth zones that tend to attract young families in the prime working ages. Those communities lacking the potential for increased development, because of an absence of land for development or other obvious drawbacks, tend to be characterized by sex imbalances and an older age structure; a condition which leads these communities to be characterized as reserves for yesterday's people.

The Relative Deprivation of Black Suburban Communities

Generally these communities tend to reflect their condition of relative deprivation when compared with other communities in the suburban ring. Employing the median housing value prevailing in the suburban county and comparing it with that prevailing in the black community clearly illustrates the lower economic status position of these outlying communities. In only one instance was a black community found to possess a housing value median that approached the median for the county. In two instances the community medians were less than half that describing the county. In another situation a community with a very low housing value median showed up well in a relative sense, but its location in a depressed county simply highlighted the widespread condition of poverty which prevailed within the larger area. In most instances the median housing values prevailing in the SMSA's central cities also exceeded that which described outlying black communities. The question of the status of these communities posed at the outset clearly indicates that while they represent a mixed array, most often they are found to represent low-status communities that are unable to offer their residents much more than the locational and situational advantage of being removed from the central city. Given the prevailing landscape characteristics, one can categorize these communities in terms of the question initially posed.[7]

Advantages Associated with Residing in Non-Central-City Locations

Although it has been said that the housing environment in black non-central-city communities leaves much to be desired in terms of the suburban image, are there advantages to be found in the rise of communities of this sort? Given the fact that a number of these communities are growth communities, there is undoubtedly some perceived value in their presence. Can these communities aid in alleviating the plight of a segment of the central-city population, or are they simply oddities offering little more than escape from the congested and threatening environment of the central-city ghetto? On the face of things, it is readily apparent that twenty-or-so communities housing approximately 300,000 persons are not going to alter appreciably the lot of any significant number of today's ghetto residents. Nevertheless, they do possess the advantage of providing access to an alternative residential environment, when that environment is thought to possess advantages not present within the central city. It is obvious that some of these communities do not possess residential attributes that distinguish them from central-city locations. But they do possess locational advantages which increasingly enhance opportunities to secure employment in emerging workplace locations.

East Palo Alto can be listed among those communities possessing enhanced employment opportunity, as it is situated in easy proximity to more than one hundred workplace locations. There are sixteen major workplace locations found within a two-mile radius of the community (see Figure 6). A number of industrial parks exist in the area, and they house numerous firms engaged in the production of electronic components for the aerospace industry. Hewlett-Packard, Varian Associates, Control Data, Philco, Eastman-Kodak, and Lockheed Aircraft are but a few of the well-known firms serving as sources of employment in this area. Thus the acquisition of residential accommodations in East Palo Alto has led to employment opportunities that possibly would not have been available if one had chosen to remain in the central city of San Francisco or Oakland.

A similar point could be made in discussing the advantages of Wyandance growing out of its eastern Long Island location. The industrial park serving the Wyandance-Deer Park area is the site of a number of firms providing well-paid blue-collar jobs. Job vacancies in the Wyandance-Deer Park area were running higher than those in locations nearer the central city and paying wages for comparable jobs higher than those prevailing in New York City (National Committee against Discrimination in Housing, 1970: 15-19). Aerospace plants and four major hospitals are the principal employers of the older members of the labor force (National Committee against Discrimination in Housing, 1970: 15-19). It is the availability of job opportunities associated with the location of these communities that enabled them to be described as middle-income communities. Obviously, the availability of jobs does not eliminate all of the

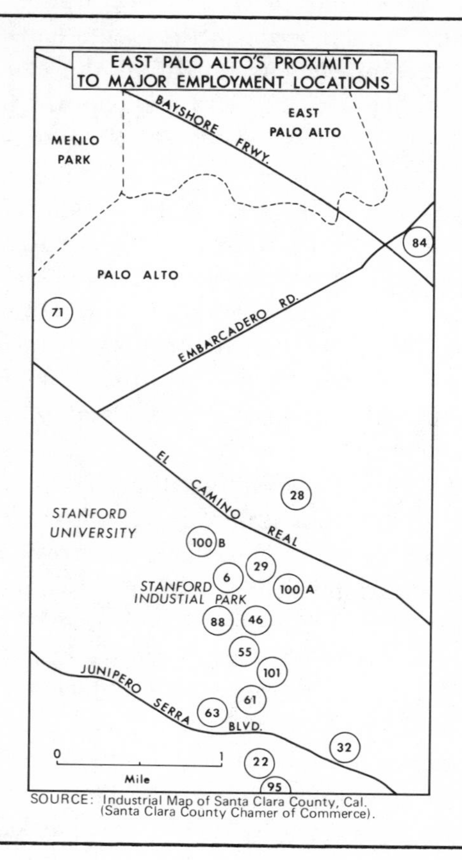

Figure 6.

problems that confront communities of this sort, but it does aid in strengthening a community in ways that would not be possible under conditions of high levels of unemployment. Thus it is possible that communities of this sort provide the potential for altering one's life chances.

Many blacks, just as their white counterparts, are desirous of escaping from the problems of the central city and strive to gain a foothold in suburbia. This is seldom an easy task given rising housing prices in the suburban housing market and the various roadblocks, financial and social, which are often present to dampen one's prospects for mobility. Unless one happens to be a member of that small army of blacks who are comfortably affluent, then passage out of the central city is likely to be difficult indeed. The availability of a suburban residential environment is seldom present outside of a suburban racial enclave. Thus, the large-scale movement of blacks to suburbia during the previous decade in a very large measure took place within the context of growing black suburban residential clusters as opposed to a condition of random dispersion constrained simply by one's income level. Farley (1970: 527) recently remarked, "we can be certain that the residential segregation patterns of the central city are reappearing within the suburbs." The patterns which he suggests are reappearing are strengthened by exclusionary zoning acts, official planning policy, and the continued intransigence of that segment of the population who sees its value system threatened by the entry of black families into a previously all-white community. To date, exclusionary zoning ordinances have generally received support from the courts (Harvard Law Review, 1971: 1649). Similarly, planning boards are able to control future development by specifying the minimum lot size that might be sold for residential development in areas that are as yet unincorporated. Recently the St. Louis County Planning Board, through the use of legal techniques, zoned an area as nonurban or not open for development at this time as a means of guiding development both in terms of its physical and economic character. Thus, on the western margins of middle- and upper-income suburbs in St. Louis County, land is being removed from active development and placed into a holding zone until orderly development can be assured (Bell, 1971: 411-418). Action of this sort effectively staves off the possibility of siting low- and moderate-income housing in some suburban locations, a mechanism that was formerly thought to aid black entry into suburbia. Thus, as Figure 3 illustrates, the suburbanization of blacks in St. Louis County is likely to be confined to those locations that are contiguous to already existing black communities, and thereby strongly supporting the spillover thesis.

In some instances, the presence of black colonies in the suburbs have served as a magnet attracting other blacks and have had the effect of promoting the development of a black suburban zone. The presence of Glenarden in Prince Georges County has no doubt abetted the development of other black residential zones in close proximity. This is partially evidenced by the spread of tract developments designed for black occupancy. But as Grier and Grier (1966: 13) have demonstrated, the Washington ghetto is basically expanding along three corridors, and two of these corridors are directed toward Prince Georges County. During the sixties, Prince Georges County's black population increased by more than 70,000, a more-than 200% increase in a ten-year period. The spillover from the central city was basically confined to these corridors, and Prince Georges County was the principal beneficiary. This has led to a change in the racial composition of a number of communities located just beyond the D. C. boundary in a very short time. Thus, Glenarden is at present on the outer margin of black suburbanization, with evidence of expansion beyond Glenarden in its embryonic stage (see Figure 7). A number of these new communities could qualify for inclusion into our universe of black colonies both in terms of minimal size and racial makeup.[7] At least three additional communities are thought to satisfy the minimal requirements for inclusion and they are Carmody Hills-Pepper Mill Village, Chapel Oaks-Cedar Heights, and Palmer Park, all of which contain

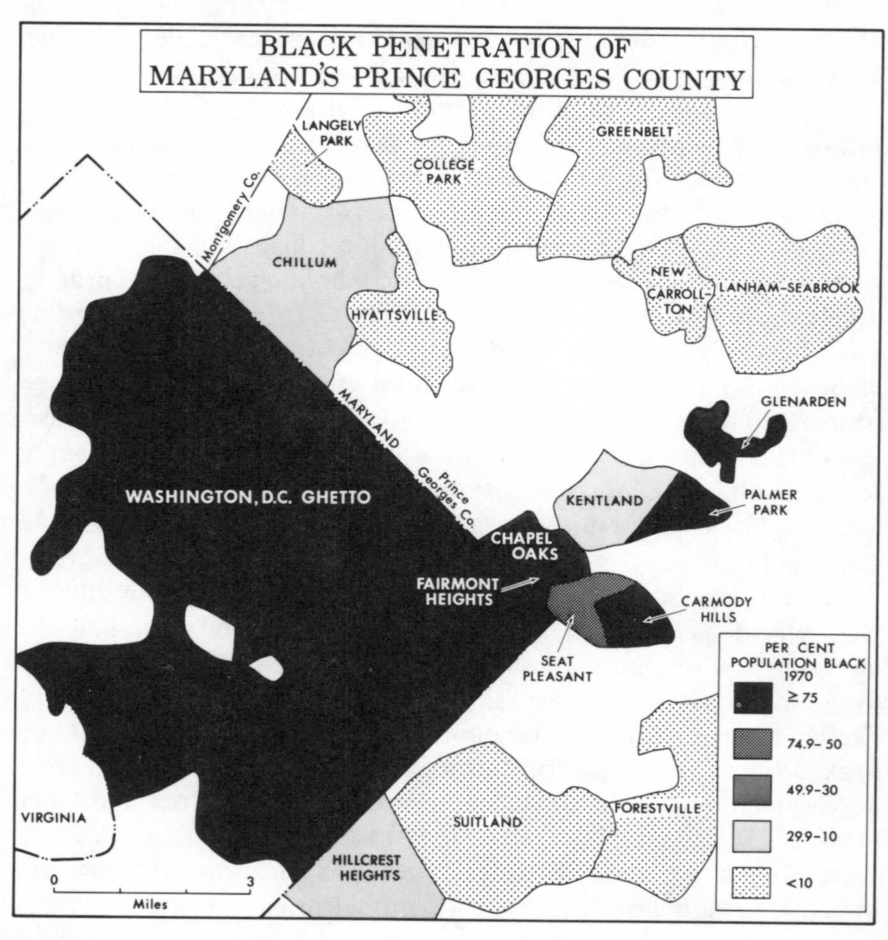

Figure 7.

populations of more than 6,000. While blacks are able to acquire access to a more desirable housing environment, Grier (1971: 48) points out that in this instance there is cause to worry, since employment opportunities will largely be associated with the movement of jobs to the northwest. Similarly, Lincoln Heights, a black suburban community which lies to the west of the Mill Creek Valley in Cincinnati, has had the same effect, but to a lesser degree. The latter community has served as a magnet attracting other blacks to the general zone and has likewise spawned a suburban tract development some few miles to the west (see Figure 8). Lincoln Heights is an older and declining community, but during the last ten years there has been a sizable increase in the size of the black population in Woodlawn, its nearest neighbor community.

Hollydale: An Incipient Suburban Prototype

Hollydale, a showplace black suburban community, which in the early sixties was described as having a rural-urban fringe location (Rose, 1965: 380), has now grown from a community of approximately 100 single-family units to a community of 180 single-family units, a condition which will shortly lead to its exhausting the land available for development within the tract. Hollydale's 1970 population of 797 was below the minimum necessary to qualify it as a black suburban colony, although its population now exceeds that of Urbancrest, Ohio.[8] For Hollydale to eventually qualify for inclusion within the group, a minimum of fifty additional units will have to be constructed to boost its population to 1,000 if the present average family size remains unchanged.

Hollydale possesses a nearly perfect suburban image, as it is totally a community of attractive single-family homes (see Figure 9). Similarly, median housing values in Hollydale exceed those describing homes occupied by blacks in all of the other Mill Creek Valley communities with sizable black populations (see Figure 8). Housing values in this community are approximately 100% higher than those prevailing in Lincoln Heights. But it should be realized that Hollydale was partially spawned by the efforts of a number of persons who had originally resided in the older community of Lincoln Heights. Its development was not without opposition from surrounding white tract developers, but that opposition was not sufficiently strong to

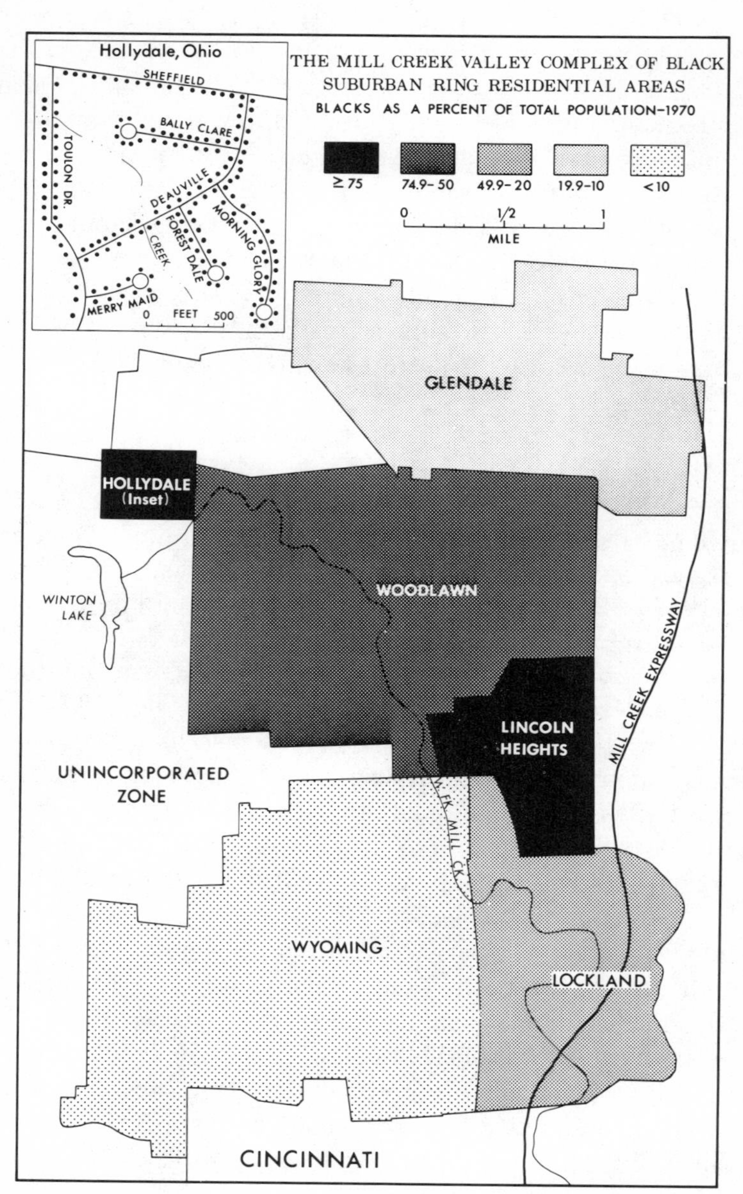
Hollydale, Ohio
SHEFFIELD
BALLY CLARE
TOULON DR.
DEAUVILLE
MORNING GLORY
FOREST DALE
CREEK
MERRY MAID
0 FEET 500
THE MILL CREEK VALLEY COMPLEX OF BLACK SUBURBAN RING RESIDENTIAL AREAS
BLACKS AS A PERCENT OF TOTAL POPULATION–1970
≥ 75
74.9– 50
49.9– 20
19.9–10
<10
0 1/2 1
MILE
GLENDALE
HOLLYDALE (Inset)
WINTON LAKE
WOODLAWN
LINCOLN HEIGHTS
MILL CREEK EXPRESSWAY
UNINCORPORATED ZONE
W. FK. MILL CK.
WYOMING
LOCKLAND
CINCINNATI

Figure 8.

Black access to the above kind of accommodations are often lacking in the older suburbs in which most blacks tend to be concentrated. The above units are among the more recently constructed in Hollydale, Ohio, an incipient black colony.

Figure 9.

prevent its coming into existence. Thus, Hollydale might logically be thought of as Lincoln Heights West.

Black Suburbanization a Function of Size of Central-City Black Population

There has not been the same magnitude of black expansion into the Mill Creek district as that which characterized Prince Georges County inner-belt communities during the recent decade. The limited expansion into this zone is partially related to the fact that Cincinnati's black population is less than one-quarter the size of Washington's black population. The rapid growth of Washington's non-central-city black population has led Schucter (1969: 362) to speculate that the suburban zone will be the place of residence of 516,000 blacks by 1980. Such an increase would result in tripling the size of the present black population found outside the central city and a continuation of the process of racial transformation of inner-belt communities; no such potential force is present within the Cincinnati SMSA.

PLANNING FOR BLACK SUBURBANIZATION

The continued rapid growth of black central-city populations aggravates the potential for black suburbanization, given the anticipated response of middle-income suburban residents, white and black. The principal purpose of this study is to assess the potentiality of communities of a given type to provide that segment of the black population, having only limited options, but the desire to reside in a non-central-city environment, the opportunity to translate that desire into reality. It has been observed that the bulk of the ghettoizing suburban communities are working-class communities, which illustrates that working-class blacks show evidence of wishing to escape the central city when a satisfactory environment in terms of both cost and cultural affinity can be identified. Class distinctions and group specific cultural dominance will continue to influence the location of sites which can be occupied by blacks within the suburban ring.

The opposition to the construction of low- and moderate-income housing beyond existing ghetto locations shows little sign of diminishing, a condition which will continue to assure that black suburbanization is likely to be confined basically to emerging suburban enclaves of the spillover variety. It is true that low- and moderate-income housing sponsored by the public sector is seldom available outside black suburban colonies. Public housing is an accomplished fact in most of the colonizing black suburban communities, either in the form of traditional public housing units or single-family units constructed under the Federal Housing Authority's 235 program, and in some locations, multiple-family units under the 221(d)3 program. The likelihood of public housing being constructed in communities other than black colonies is somewhat remote, although the plan recently devised by the Miami Valley Regional Planning Commission does provide minimal hope for the location of FHA-sponsored housing beyond the limits of the central city. The Commission has approved a plan that would result in the dispersal of low- and moderate-income housing throughout the Dayton, Ohio, metropolitan area (Bertsch and Shafer, 1971). If action of this sort is followed by other metropolitan areas, it might become possible to provide low-income blacks access to suburbia. Otherwise, low-income blacks are not likely to be able to penetrate the suburban ring beyond the possibility of securing housing within an existing black colony.

Among the ghettoizing black communities, the situation may be such that class characteristics may act to deter the development of housing for low-income populations within zones which are predominantly black. In those instances where the ghettoizing communities are located in unincorporated zones, even the opposition of local blacks may not be sufficient to alter the location of such units, as county governments may decide that these are locations that are without power and thus constitute the line of least resistance. But there is little doubt that the combination of attitudes which revolve around the desire to promote class and race homogeneity will operate to prevent low-income blacks from escaping the central-city ghetto.

Affluent blacks may be able to escape the necessity of locating within one of the community types previously discussed, and thus the demands of this group need not be taken into consideration here. Nevertheless, there is little evidence of the random dispersion of blacks within urban areas, regardless of income characteristics. There may, in fact, be developing small residential enclaves of affluent blacks in outlying areas whose pattern of spatial aggregation go undetected by researchers who rely upon census data to initially identify the location of black residential clusters. There is evidence that the more affluent black suburbanites in the Washington metropolitan area have settled in the upper-income communities of Maryland's Montgomery County rather than the lower- and moderate-income communities of Prince Georges County.

Planning will be necessary if all segments of the black population are to be given some choice in the residential selection decision. The planning of new communities is thought by some to be a technique that will aid in enhancing this choice. But, to date, it has been illustrated that new communities simply enable middle-income whites to flee the central city (Schucter, 1969: 362). Soul City, a proposed black new town, is seen as evidence of the wish to provide a planned environment for a segment of the black population within the framework of the new-town concept. With its location in rural northeast North Carolina, one might anticipate that it will attract a sizable number of low-income residents. On completion, Soul City would qualify as a free-standing, colonizing black community of the central place variety. Mound Bayou, Mississippi, presently represents a community of this type which is located in a nonmetropolitan environment. Among those who are interested in halting or at least slowing down the movement of blacks from the South, the idea of the free-standing black community is likely to have some appeal. On the other hand, those who view black cultural patterns as threatening might possibly support black new towns in suburbia, as long as they are isolated from the nonblack suburban communities.

SUMMARY AND CONCLUSIONS

An attempt has been made to assess the quality of life prevailing within a set of suburban ring communities whose populations are predominantly black. Significant differences appear among the communities depending upon where they fall into the category of colonizing or ghettoizing communities. At the present, the ghettoizing communities tend as a group to provide the superior residential accommodations, especially if one considers only the leapfrog communities as opposed to the spillover communities. This is simply because the leapfrog communities have evolved within a zone of more recent residential development. But the future of black suburban development tends to favor the spillover communities which evolve as a natural outgrowth of the spatial expansion of central city ghettos. Thus for some time to come, suburban ghettoization of the latter type is likely to be the dominant form in those metropolitan areas where the black population numbers 250,000 or more and the central city has been the place of residence of at least 25,000 black people since 1920. The leapfrog communities of the noncolony type represent a seemingly recent category of black suburban development and likewise tend to possess greater potential for providing access to both a more satisfactory residential environment and enhanced employment opportunity.

The colonizing black suburban communities, as a rule, are characterized by a mixed variety of residential accommodations.

Since these communities generally tend to be older, with many coming into existence during an era when only rudimentary shelter was available to blacks, they are often characterized by the presence of a disproportionate number of vernacular housing units. Those colonies which have emerged during the more recent era tend to project an image that is more in keeping with normative concepts of the suburban image.

The principal advantage which the colonies possess over the ghettoizing communities is that they are more often than not politically independent. To date, the only unincorporated black colonies are located in the South. The latter quality is one which might attract blacks who would normally be disposed to favor the maintenance of central-city ghettos because of the perceived advantages associated with the evolution of black territorial dominance. But it is apparent that political control per se is not sufficient to provide access to an improved physical environment. This is probably no better demonstrated than in the condition of population decline which characterized a third of the colonies during the previous decade. The colonies are frequently found to be the places of residence of low-income populations and seldom contain additional space for residential expansion. In some instances, though, it would be possible to physically redevelop segments of these black colonies in ways which would transform them from their present shacktown status to that of communities possessing an environment of pleasant residential accommodations. Some progress is being made in this direction.

It is obvious that the 23 communities cited in this investigation are likely to provide more than minimal opportunity for blacks wishing to achieve access to a suburban residental environment. But a keener understanding of the problems and potentialities of these communities should aid in the process of planning for suburban black residential development. If efforts are not made to plan for the orderly entry of blacks into suburbia, then we can expect black suburban development to take basically the form of ghetto spillover. On the other hand, information acquired through a careful assessment of black colonizing communities could lead to enhanced opportunity for the development of prototypic suburban communities.

To date the suburban prototypes are indeed limited in number, while a large number of these communities at best simply provide locational advantages as opposed to environmental advantages. In no way is this investigation advocating the development of black communities beyond the reaches of the central city to the exclusion of individuals selecting a place of residence within the environment of their choice. But one would be indeed naive to assume that a significant number of blacks are going to be able to acquire suburban residence via the latter route within the near future, given conditions of rising housing prices and local community response to a segment of the population that is viewed by some suburbanites as cultural aliens.

NOTES

1. The population of Scotlandville, Louisiana, is not included in this computation.

2. The 50% level was employed since it is frequently used by writers to identify territorial units in which the units are thought to be part of the black community.

3. The lower population limit established for the population replacement type suburban community is made necessary because of the lack of information on race for communities smaller than 10,000 at this stage of the processing of the 1970 census results.

4. Florence-Graham, California, with 42,000 people, is included among this group because its status is somewhat uncertain. It may well constitute a pseudo-community.

5. The term "pseudo-community" was used to distinguish black communities which were simply physical extensions of a larger political community which was predominantly white. In each instance, the black outlyers were politically disenfranchised. See Rose (1965: 362-365) for a more detailed discussion. Reliance on census sources as a means of identifying a universe of places can lead to the inclusion of places whose identity is questionable.

6. The economic-class characteristics were derived by designating each place a middle-income community if the derived median family income was 90% or more than the regional median. If the family income of a place was less than two-thirds of the regional median, it was described as a lower-class community. All other communities were identified as working-class communities.

7. Since much of the investigation of these communities is in a preliminary stage, field observations have not yet been conducted for the universe. Only those communities in which field observations have been made are included in Figure 3.

8. Urbancrest's population dropped below the original threshold employed in 1960, but, because it was included among the original communities, a decision was made to continue to identify it among the universe of black colonies.

REFERENCES

BELL, C. (1971) "Controlling residential development on the urban fringe: St. Louis County, Missouri." J. of Urban Law 48: 409-428.

BLS Report No. 394 (1971) The Social and Economic Status of Negroes in the United States, 1970, Current Population Reports, Series P-23, No. 38. Washington: U.S. Department of Commerce.

BERTSCH, D. and A. SHAFER (1971) "A regional housing plan: the Miami Valley Regional Planning Commission Experience." Planners Notebook (April).

DAVIS, J. T. (1965) "Middle class housing in the central city." Economic Geography (July): 238-251.

DUDAS, J. J. and D. B. LONGBRAKE (1971) "Problems and future directions of residential integration: the local application of federally funded programs in Dade County, Florida." Southeastern Geographer (November): 157-168.

FARLEY, R. (1970) "The changing distribution of Negroes within metropolitan areas: the emergence of black suburbs." Amer. J. of Sociology (January): 512-523.

GRIER, G.(1971) "Washington." City (January-February): 45-49.

--- and E. GRIER (1966) Equality and Beyond. Chicago: Quadrangle.

Harvard Law Review (1971) "Exclusionary zoning and equal protection." (May): 1645-1669.

National Committee Against Discrimination in Housing (1970) Jobs and Housing: A Study of Employment and Housing Opportunities for Racial Minorities in Suburban Areas of the Metropolitan New York Region. New York.

ROSE, H. M. (1972) The Black Ghetto: A Spatial Behavioral Perspective. New York: McGraw-Hill.

--- (1965) "The all-Negro town: its evolution and function." Geographical Rev. (July): 362-381.

--- (1964) "Metropolitan Miami's changing Negro population, 1950-1960." Economic Geography (July): 221-238.

SALTER, P. S. and R. C. MINGS (1969) "A geographic aspect of the 1968 Miami racial disturbance: a preliminary investigation." Professional Geographer (March): 79-86.

SCHUCTER, A. (1969) White Power, Black Freedom. Boston: Beacon.

16

Stereotyping in Black and White

DAVID C. PERRY
JOE R. FEAGIN

□ THE LITERATURE CONCERNED WITH RACE RELATIONS has grown in recent years so that today it is fast becoming one of the most expansive bodies of research in the social sciences. As with most areas of social science study, there are certain common conditions of agreement which, through their constant occurrence in research, are generally assumed to be important, if not essential preconditions to any accurate discussion of a social phenomenon. Such preconditions surround the study of the black ghettos in urban areas. There is a common tendency among social scientists to treat the ghetto as either a community mired in unhealthy family relations, or not viable economically because its residents cannot defer gratification, or a place of scattered and fleeting friendships of little meaning or utility, or an environment where the individual residents are isolated, anomic, and defeated–finding little meaning in life and not having a close attachment to the community in which they live. These conditions, allegedly reflecting the ghetto way of life, have been grouped in various studies under such headings as "social pathology" (Clark, 1965), "social disorganization" (Myrdal,

AUTHORS' NOTE: *The authors would like to express their appreciation to James Jones for his insightful criticisms of an earlier draft of this essay.*

1964), or the "urban jungle" (National Advisory Commission on Civil Disorders, 1968).

Yet we are not at all sure that the black ghetto is a unique and distinctive hotbed of pathology. Such selective posturing by the social sciences seems to suffer from what Gordon Allport terms the stereotyping of a culture. For Allport, a stereotype is either an erroneous, exaggerated, or oversimplified generalization which demonstrates a certain affective attitude about a group to "prevent differentiated thinking about a group" (Allport, 1958: 187). As such, stereotypes act as screening devices which reinforce selective conceptualizations and maintain "simplicity in perceptions" (Allport, 1958: 188).

In the most rudimentary sense, for something to exhibit a pathological existence, it is considered to be "abnormal . . . (that is evidencing) . . . anatomic and physiological deviations from the normal that constitute disease or characterize a particular disease" (Webster's Seventh New Collegiate Dictionary, 1965: 618). In line with this definition, the social sciences have dubbed black ghettos as *socially* pathological because they traditionally exhibit higher levels of delinquency, crime, illegitimacy, and educational dropouts than do other areas of the city. From patterns of this sort and others, such as high levels of unemployment, broken families, and even black skins, come the arguments that the socially "anatomic and physiological deviations" from normality such as social disorganization, disintegration, and atomization constitute the bases for the diseases of hopelessness, loneliness, and alienation of black ghettoites. Implicitly, many social scientists appear to be working with an uncritical sense of what is normal. That is, the normality from which ghetto life and culture may (or for that matter may *not*) deviate is the middle-class life and culture which is the background of most of these writers and researchers.

The security of such empirically and normatively based stereotyping selectively precludes approaching black communities from a positive position which would study the social organization of black ghettos. The general neglect of social organization strongly suggests the need for an intensive and comparative examination of interpersonal relations in both formal and informal settings within black communities, not just superficial comparisons with economic or domestic constellations outside the ghetto. The jungle image of the slum has been criticized by researchers, but it is still the prevailing image of the black ghetto both inside and outside academia (Gans,

1962: 80-81). To offset this undue stereotypical emphasis, this essay will investigate (1) the evolution of the pathology and ancillary disorganization hypotheses in the literature; (2) the extent to which the literature and the evidence concerning the black ghetto community are by no means consistent or conclusive in supporting existing generalizations; and (3) an alternative approach to the study of the urban black ghetto.

THE QUEST FOR PATHOLOGY AND DISORGANIZATION

While the immediate reaction of the reader might be to assume that we are in the process of making the Pollyanna argument that the black urbanite is doing quite well and that his environment is not in need of repair, this is not the sense of our argument; rather we are disturbed by the tendency of many social scientists—and those influenced by them—to view the black community as some form of radically deviant and sick society. However, such an approach is easily accepted. First, because the ghetto is commonly assumed to be representative of what is the worst in our urban society. The term itself is an example of a stereotype. Second, because of the liberal perspective of social science, there is the tendency of researchers to place the blame for social ills on the victims under study or on the victims' historical development, rather than on the larger societal framework. Thus, the life, goals, and culture of the dominant society are not culpable in the same way as they are not deviant. Third, such stereotyping by community scholars fits the more general and now several-decades-old tradition of urban scholars who have been writing obituary notices concerning the end of personal social ties in modern urban communities. Certain classical sociologists, such as Louis Wirth and George Simmel, some time ago presented a picture of urban anonymity and the absence of binding social ties, beyond the formal associational level, for urban dwellers. It seems that in spite of dissenting arguments to the contrary,[1] the ongoing development of the literature continues to adhere to the Wirthian position for ghettos (and, to a certain extent, even suburbs), namely, that the emerging urban way of life, more particularly the urban social way of life, is characterized by

> the substitution of secondary for primary contacts, the weakening of bonds of kinship, and the declining social significance of the family, the disappearance of the neighborhood, and the undermining of the traditional basis of social solidarity [Wirth, 1938: 1-24].

In a book dealing with a broad range of more contemporary studies, Maurice Stein (1964: 283-284) takes up the disorganization stereotype, when discussing the general effect of urbanization. He seems to argue that urbanization has been essentially disorganizing and that most modern urbanites are caught up in an impersonal world of intense competition and interpersonal manipulation. Depersonalizing trends have transformed personal social relationships into object relationships valued only as sources of status and material reward. As individuals have become increasingly dependent upon impersonal bureaucracies and formal organizations, personal loyalties, such as neighborhood and family ties, must, of inexorable necessity, have declined or disappeared. Alienation and anomie now predominate in our complex and impersonal urban areas.

This neo-Wirthian theme of the "eclipse of community," of isolation, impersonality, and disorganization, has also stayed with us in the more general literature bemoaning the alienation and anomie of modern urban civilization. Herbert Marcuse (1964: 246), in effect, believes that Orwell's *1984* has already arrived. Centralized authorities and impersonal manipulators, such as mass media administrators, have already replaced traditional social controls; presumably those controls once furnished by kinship, friendship, and local community ties. The critical question is: why does the rise of modern technology dictate the fall of personal worth and social organization?

A final and related question concerns the definition of terms such as pathological and disorganized; namely, to what extent are cultural patterns, family configurations, gratification cycles, and the like, detrimental to the survival and viability of the community which they characterize? While the terms pathology and disorganization have often seemed to refer, most basically, to the difference between those being studied and some real or imagined standard of the middle class or upper-middle class (although it sometimes seems that researchers see no group as normal), other definitions seem to us to be relatively more objective and fruitful for understanding the true nature of ghetto life and culture. Herbert Gans (1962: 309), for example, asked similar questions of the assumptions of the deprived condition of the Italian slum of the West End in Boston, assumptions which led to the demise of the community by the federal bulldozer. Gans suggests that, for realistic renewal purposes designed to benefit social victims, a slum should be viewed as an area that can be "proven to create problems" and pain for its local residents, not just for the larger community. What may be a pattern of behavior different from that of the larger community does not have to be pathological for all or most ghetto residents. In essence, very little attention has been paid to the black ghettos of this country as positive and normal communities, just as very little positive attention had been paid to the West End before Gans changed the perspective of analysis away from the traditional assumption that a slum is de facto a negative environment.

SLUMS AND GHETTOS: THE PROBLEM OF DEFINITION

A few analysts have noted that the term "slum" is an evaluative rather than an analytical term. "What seems to happen is that neighborhoods come to be described as slums if they are inhabited by residents who, for a variety of reasons, indulge in overt and visible behavior considered undesirable by the majority of the community" (Gans, 1962: 309). The consequence of such an assumption for government policymakers and academic analysts alike is to provide reports and studies replete with information "intended to show the prevalence of antisocial and pathological behavior in the area" (Gans, 1962: 309).

Perhaps the most influential scholar ever to write in this tradition was E. F. Frazier. Frazier's (1939) indictment of the stability of the lower-class black family stands as the basic source for the most influential recent studies of the black ghetto and its disorganized and pathological state. Not only does Frazier serve as a prime source of documentation for writers such as Glazer and Moynihan (1963), but

> an essential element of Frazier's reasoning is one that is perpetuated by later thinkers. This is a direct logical leap from social statistics, which are deviant in terms of middle-class norms, to a model of disorder and instability. Such reasoning effectively eliminates con-

> sideration of possible cultural forms that, in spite of differing from Frazier's assumed standard, might have their own order and functions [Valentine, 1968: 23].

Indeed, it can be argued that one prescriptive aim of statistics and analysis in the social sciences has been an attempt to "middle-classify" the members of black ghettos in much the same way as John Dollard described the same tendency as the dominant aim of society as a whole.[2] In line with this assumption, one of the few studies of the mobility of black ghettoites began with the assumption that middle-class blacks, when confronted with the requirement to move from an urban renewal zone in a low-income area, "would rush to embrace any opportunity to escape from their relatively segregated and declining neighborhood" (Watts et al., 1964: 8). However, it was discovered that middle-income black families living in Boston's central ghetto and with a substantial number of new homes available were highly reticent about moving. Of the 250 families interviewed and studied over a 16-month period, less than 4% chose to live among white families in predominantly white communities. Under renewal pressure there was a low rate of migration from the ghetto area; moreover, there was little effort to find a place to live outside the general ghetto area. Fewer than one in five of the middle-class families even tried to inspect a dwelling outside the ghetto.[3] Social ties (choice) and urbanity, at least as much as discrimination, appeared to shape their mobility decisions.

A second study in the same city discovered that civil rights leaders in the area, explaining these patterns of "non-mobility" of potentially mobile black urbanites, felt that while blacks should have the right to move anywhere they wished, they still should also have the right to stay where they were if they so desired. Indeed, one civil rights leader stated that this pattern of remaining in the ghetto was due to the fact that

> Negroes like where they are living. It is the individual Negro's prerogative to move. Plenty of Negroes are moving out—to places like Randolph. But people like me don't want to move anyplace. My daughter has decided to live nearby because of transportation and quality of housing. But the masses of Negroes don't want to move. *They stay where their roots are* [Feagin, 1966: 12].

While the evidence from these studies does not entirely dismiss the stereotype of the ghetto as a pathological and disorganized place to be escaped, it does provide us with enough information to question whether or not the ghetto really is a place to be escaped as soon as one can afford to flee. Is it so unlikely, as much as the literature would have us believe, that the ghetto is to an important degree a community of proximity, sharing, and growth? Proximity, either of mother and child, father and mother, or friend and friend, results more often than not in the desire for continued cohesion. Coupled with this proximity is the accumulation of shared experiences, opinions, and values. The growth of such sharing and communication increases the cohesiveness of the social force between two or more units. Thus one of the forces which keeps black Americans (who potentially could move) in the ghetto areas may well be the same as that which tends to keep many other urban areas predominantly Italian, Irish, or Polish: intimate kinship and friendship networks. Such networks have been much neglected by recent analysts of the city.

The evidence to date has been far less conclusive than the assertions concerning the social disorganization and pathology of the black ghetto would lead one to believe. Some time ago, Myrdal (1964) argued, for example, that the characteristic traits of the black community were forms of social pathology: the unstable black family, the insufficient recreational activities, the narrowness of the interests of the average black urbanite. Even the social organization which he did find, "the plethora of Negro sociable organizatons," was viewed as little more than a pathological reaction to caste pressures (Myrdal, 1964: 952-953). Coupled with Myrdal's assault upon a sense of black community or identity was Frazier who cited the combined influences of degenerative slavery and the alienating urban milieu upon the urban black as creating culture almost totally devoid of social control, creative community institutions, and stable families (Frazier, 1939: 245ff.). More recent students of the black ghetto frequently have not modified this view of the ghetto as institutionalized pathology. Glazer and Moynihan (1963) follow the lead of Frazier and Myrdal before them; they see the state of the Negro as involved more often than not in patterns of broken homes and concomitant illegitimacy which, when coupled with economic insecurity, result in poor upbringing and emotional problems—thus creating a cycle of self-perpetuating poverty. Kenneth Clark in his otherwise excellent book on Harlem furthers this view of self-perpetuating pathology and disorganization by periodically overemphasizing the disease imagery in evaluating ghetto life. In one place, for example, Clark (1965: 81) argues that:

> the dark ghetto is institutionalized pathology; it is chronic, self-perpetuating pathology; and it is a futile attempt of those with power to confine that pathology so as to prevent the spread of its contagion to the larger community. It would follow that one would find in the ghetto such symptoms of social disorganization and disease as high rates of juvenile delinquency, venereal disease among young people, narcotic addiction, illegitimacy, homicide, and suicide.

Partially influenced by this Harlem study, the widely publicized report (U.S. Department of Labor, 1965: 4-6, 47-48; cited hereafter as the Moynihan Report. Also compare Frazier, 1939: 245ff.), moves in non sequitur fashion from an analysis of the family instability of a minority of families within black ghettos to a description of the black communities as disorganized and in a state of "massive deterioration." It seems that the authors of this report, along with many others, let the alleged disorganization of a minority obscure their view of the organization of the majority. Variations in family organization may exist to a disproportionate degree in black ghettos, although the comparative evidence is far from conclusive. Even so, surely it is an extreme exaggeration to state as a presumably well-documented generalization about ghetto life that

> to those living in the heart of a ghetto, black comes to mean not just "stay back," but also membership in a community of persons who think poorly of each other, who manipulate each other, who give each other small comfort in a desperate world [Rainwater, 1966: 205].

Such views of life in a black ghetto seem to us onesided and overstated. The tend to depict the whole of ghetto life and culture as nothing more than a pathological reaction to white segregation and discrimination. They not only represent the tendency to find lack of conformity with middle-class standards as equivalent to pathology and disorganization, they are also the result of the social scientist's occasional tendency to confuse the analysis of cultural environment with surface-level social statistics of family structure, unemployment, crime, and the like. Just as factors such as educational attendance, male-headed families, single-family dwellings, and medium to high family income, do not by any means explain the suburban culture which is beginning to demonstrate high school protests (a significant number of secondary schools had recorded protest events by 1967), long-haired, "dirty" children, increasing alcoholism, increased indebtedness, and boredom; so too, it is hard for us to accept the argument that black ghettos, because they display a different set of surface-level social statistics, can be assumed to be almost entirely lacking in social organization and integrated persons. It is not inaccurate to point out that the descriptions of white urban America which point to drug addiction, divorce, and other forms of alleged social pathology and disorganization are dismissed as sporadic tendencies of a highly unrepresentative minority: such characteristics are not reflective of the social organization and psychic strength of the vast majority of middle-class white America. At the same time many analysts, public and academic, are not so quick to dismiss a minority (if they see it as minority and not majority) of juvenile delinquent blacks, a minority of structurally different female-headed black families, a minority of criminals, as being ascribed to a distinct minority of urban black America.

In fact we could raise the question about the normative stance of such analysts: are we learning more about the middle-class biases and analytical pathologies of the analysts than we are of the black ghettos they so selectively analyze? These biased distortions overlook the important organizational aspects of a black community. In admittedly somewhat overstated terms we ask: are there no unbroken homes in a ghetto? No ordinary friendships? No functioning kinship networks? Could it be that for a hundred years blacks have only been reacting? Does not ordinary social life exist, even in a black ghetto, for a majority of the residents? There has been, with only a few exceptions, a general failure to study systematically the positive social forces within black communities.

Adequate answers to these questions would be important for several reasons. In the first place such answers would help correct the "ghetto as disorganization" image commonly applied to black areas. Of the two radically different views of the slum in the literature, the "slum as jungle" stereotype has usually been applied to black ghettos (Glazer and Moynihan, 1963; Valentine, 1968; Gans, 1962) while the "slum as village" has generally not been used in regard to these same ghettos. In the second place, certain existing research evidence suggests that existing social organization within a black ghetto may be useful in explaining, in part at least, the non-migration (out of ghettos in the face of urban renewal) pattern discussed earlier.

AN ALTERNATIVE PORTRAIT OF THE BLACK GHETTO

The most important reason for viewing the ghetto from a positive stance, or at least with a minimum of middle-class bias, is that one might well discover that there is a great deal of social integration within the black community, even if we do not see very much social integration between the black and white urban communities. In fact, it appears that the well-articulated position of black-white social polarization is yet another reason the present-day social scientist continues to place this strongly felt pathology stereotype upon the black community. Such a predilection seems to emerge from the position in the field of race relations which often places social integration in a nearly synonymous position with racial integration, the bringing of blacks into positions of equal status with whites in society. This is not the sense in which we are using the term. We would prefer that social integration be used as an important term descriptive of the idea of community; it stands for the internal stability of a given community of people not necessarily the stability of relationships of that community with other communities. In other words, whatever the personal and social costs of living in a ghetto, as opposed, for example, to living in a suburb, can it be said of the residents who reside in a black ghetto that they are living in a place which does not reject them or alienate them from one another, which accepts their style of social interaction?

A perspective emphasizing the normal aspects of black communities fits well with what we know about black mobilization in cities; the growth in numbers, resources, and organization. For example, the significance of social networks is suggested in one study of the major riot in Detroit in 1967, a study that found rioters who were not involved in the initial precipitating incident had learned of the riot not primarily by means of the mass media but through social networks (Singer et al., 1970: 44). Such a tantalizing finding points to the importance of the hitherto neglected informal social networks in ghetto areas, and consequently the role of such networks in the developmental phases of ghetto rioting, and does not support urban theorizing tied closely to assumptions of anonymity, social isolation, and the breakdown of social control. Thus, an important type of integration into the urban fabric are those interpersonal links often characterized by intimacy. The importance of such primary relationships has been emphasized by Katz and Lazarsfeld (1964: 44):

> Interpersonal relations seem to be *"anchorage" points for individual opinions, attitudes, habits and values.* That is, interacting individuals seem collectively and continuously *to generate* and *to maintain* common ideas and behavior patterns which they are reluctant to surrender or to modify unilaterally.

Critical communications proceed along these informal networks; and communication need not be seen only in terms of opinions or news. Broadly conceived, communication can include the exchange of aid, money, or job information among relatives or friends, or the exchange of marriage partners among families.

The general neglect of social organization within black ghettos strongly suggests the need for an intensive examination of this organization, including interpersonal relations in both formal and informal settings. To offset the undue emphasis on social pathology and disorganization, we will now suggest an alternative investigative approach which argues most basically that the majority of urban black residents maintain strong and diverse interpersonal ties within their communities.

It is the basic argument of this essay that there is enough evidence to support the view that interpersonal relationships are of the same basic significance for blacks as they are for other urban residents, including those living in other ethnic slums. The specific indices of interpersonal relationships which will be examined here are as follows:

(1) the nature of primary family organization and structure;

(2) the extent and intensity of kinship contacts; and

(3) the extent and intensity of friendship and neighboring contacts.

Two further types of social participation will be noted in order to give more strength to our suggestions:

(1) the extent of contact with the news media, and

(2) the extent and intensity of voluntary group participation and the impact of this participation for black political organization.

FAMILY STRUCTURE RECONSIDERED

Perhaps the most important recently cited statistic in the contemporary social sciences is that put forth to demonstrate the deterioration of black society as a function of the disorganization of the black family. The Moynihan Report (U.S. Department of Labor, 1965: 6) states that 23% of ever-married black females in urban areas head families with the husbands absent or divorced. While we in no way argue that this is not a significant percentage of all urban black families, we would make the apparently obvious but almost universally overlooked point that this is a distinct minority of all families living in black ghettos. In other words, if we were to take the opposite reading of the data, 77% of all black ghetto families are similar in structure to the male-headed model used as the standard of comparison. Some see in these data the disorganization of black family structure and blame it, at least in part, upon "the effect that three centuries of exploitation" (U.S. Department of Labor, 1965: 5) has had upon the historical, as well as contemporary, development of black communities. However, it could also be argued that, in the face of the landslide of evidence which has been unearthed concerning the oppressed position of black Americans, one of the most substantial pieces of evidence of social organization and integrative strength resides in the fact that three-quarters of black families are intact and take the form of the valued or traditional family structure. In short, we are arguing that to infer the structural deterioration of the black family and hence black society in the urban communities is untenable if the major evidence is that 23% (or 25% or 30%)–a minority–of said families are female-headed.

Secondly, if we extract those whose husbands are dead, ill, away from home solely as the result of employment, the true broken-home measure would drop substantially, perhaps 18% or less. Thirdly, we have no indication of how many in this remaining percentage in urban areas no longer have children below the age of eighteen. In conjunction with this, while apparently there are no data at this time on the strictly urban female-headed families with children, there is a question raised about the significance of this figure if we turn to a study of *all* black families in America by Andrew Billingsley (1968: 19). He points out that of all single-parent black families, while the vast majority are female-headed, such families still constitute less than 6% of all black American families.

There are not only a variety of statistical machinations which are presently confusing an accurate description of the female-headed black family, but there are also a variety of subtypes of such families which bring the whole argument of the pathological and disorganized female-headed family into question. One set of subtypes includes female-headed families who are: (1) single, (2) divorced, (3) legally separated, (4) widowed, or (5) spouse absent (Billingsley, 1968: 19; U.S. Department of Labor, 1965: 6). All these subtypes imply female-headedness for different reasons, for different lengths of time, and with different results or feelings for the family members. As a result, we are hard-pressed to lump the various family subtypes together and assume a generalized set of conditions termed concurrently or separately as deteriorating, socially pathological, or culturally deprived.

In view of the above arguments, we offer the following typology (see Table 1) as a simple example of the variety of patterns which could exist as organizational alternatives for any given structural-family alignment. The problem with the Moynihan data and the Frazier and Clark assertions, for example, is that they confuse structural configurations with the organizational realities of decision-making centers within the family. They assume that the black family is disorganized, with pathological payoffs, if the family is not

TABLE 1
TYPES OF DECISION-MAKING AND FAMILY ORGANIZATION

Authority/ Decision-Making	Intact (husband-wife present)	Wife Absent (husband-headed)	Husband Absent (wife-headed)
Egalitarian[a]	+	+	+
Patriarchal[b]	+	+	–
Matriarchal[c]	+	–	+
Other[d]	–	+	+

– = Unlikely, but not entirely impossible
+ = Possible or likely

a. Husband and wife substantially share decision-making.

b. Husband is primary decision maker.

c. Wife is primary decision maker.

d. Some other relative such as a grandfather or grandmother makes a majority of the major decisions.

an intact family. In the typology we see, for example, that an intact family could logically be other than patriarchal in its decision-making pattern; it could be egalitarian or it could be matriarchal. The structure does not act as de facto evidence of the decision-making reality, as is quite often assumed in the literature. Moreover, in the cases of attenuated families, those with the husband absent or the wife absent, there is no visible guarantee from simply viewing the structural alignment, that the family will be patriarchal or matriarchal in its authority arrangements. That is, there is certainly no guarantee that an attenuated family will be any differently organized than a nearby "normal" intact family. In fact, there is good reason to expect that a variety of authority or decision-making patterns will be evident at different times in a family's life. Our point here is not to offer a comprehensive typology for study of the black family; rather we are arguing that the jury is still out as to just what structural patterns have to tell us about the reality of family decision-making in the black ghetto, or vice versa. The typology suggests that there are several different family alignments in the ghetto; but they do not indicate which are pathological and which are not.

Our argument here is not far removed from another typology-producing essay which points out that to infer family disorganization or instability from simple compositional information is very difficult. In a provocative essay, Miller and Riessman (1968: 38) point out, in addition, that variations in a composition do not necessarily have the commonly predicted pathological effect:

> as yet, it is not possible to formulate a clean-cut classification which avoids cultural biases and still is able to render a judgment about the impact of life style on individuals. For example, does the absence of a permanent male figure mean that the children are necessarily psychologically deformed by living in such a family? Assessments such as these are difficult to make because much of our knowledge and theorizing about fatherless families is based on middle-class situations.

Indeed, as a result of this assessment of the importance (or lack thereof) of family structure, Miller and Riessman (1968: 38) suggest an alternative measure of family stability. For them:

> Familial-stability patterns are characterized by families coping with their problems—the children are being fed, though not necessarily on

> a schedule, the family meets its obligations so that it is not forced to keep on the move; children are not getting into much more trouble than other children in the neighborhood.

In a study of what would apparently pass for a matriarchal black family, Stack (1970: 303) discovered that the description of the urban black family as a matrifocal complex, while it "might bring out the importance of women in family life," could not encompass the variety of domestic arrangements among black families. Rather than viewing these domestic strategies as social disorganization, Stack argues that they are evidence of organization; of the adaptiveness and coping strategies of urban blacks in regard to daily social and economic problems. She presents the fifty-year history of one large black family as it migrated from the rural South to the urban North and West to validate her arguments; she discovered fifty years of constant closeness that went beyond common nuclear family relationships or matrifocal relationships, embracing "new and expanded households and/or domestic units . . . created to care for children. The basis of these cooperative units is cogenerational sibling alignment, the domestic cooperation of close adult females, and the exchange of goods and services between the male and female relatives" (Stack, 1970: 311). Thus Stack finds that through such differently organized, but not disorganized, domestic units there is constant assurance that all children will be cared for. Finally, in a somewhat related study of black and white intact families, psychologist Delores Mack (1971: 24) discovered that there are no significant power-distribution differentials between white and black families. Thus she argues that the social scientists have had a "habit of taking cursory looks at the power relationships within black families in the United States and carrying on about 'matriarchy' and the 'unnatural superiority of the Negro woman' " (Mack, 1971: 24). Hence, we believe that through exploring alternative research hypotheses and techniques it is possible to discover that what aggregate statistics tell us are family units which differ from middle-class intact structures and appear to be matrifocal, may be *differently* organized, but not *dis*organized, and may quite easily be other than matrifocal in their decision-making structure.

THE EXTENT AND INTENSITY OF KINSHIP CONTACTS

As the Stack study indicates, one of the prime movers of kinship and primary family interaction among urban blacks is the care of the children. In a study which compared urban black ghettoites to white residents of communities of varying socioeconomic status, it was found that a significantly higher percentage of black urbanites (61%) "spent a lot of time with their children," while only 43% of upper-class whites could say the same.[4] At the same time only 21% of the black residents "felt they were not the kind of parent they would like to be," while 36% of upper-class whites felt this way. Moreover, fewer ghettoites felt inadequate as a spouse than did members of the white residential classes. Such data could be interpreted to mean that while there may be more separation and divorce among ghettoites, such disorganization of the basic family structure is more healthy in achieving a sense of adequacy, once a marriage is consummated in the ghetto.

The assumption that the ghetto is the prototype of informal alienation and social isolation becomes suspect in light of a research study of low-income blacks in New York. Most blacks are certainly not isolated from relations. In a study of New York City welfare mothers, Podell (1968: 36) found that 41% of black welfare mothers had one or more relatives living in their neighborhood and 72% had relatives elsewhere in the city. In another statewide study of welfare recipients in California it was discovered that 27% of blacks had relatives in the neighborhood and another 53% had relatives in the city of residence (Stone and Schlamp, 1966: 52-53). They also found that most black urbanites (86%) visited with relatives at least once a week, and over half of them visited their second and third relatives over forty times a year.

Corroborating evidence is provided by an in-depth study of 120 Boston ghetto residents conducted by the junior author; the sample was predominantly low-income. The dominant impression which one gets from examining our Boston survey data on primary social ties is that the black respondents interviewed there definitely do not fit the stereotype of the isolated ghetto dweller who has no concern for or contact with his kin or friends (Feagin, 1966). Overall, the sample was by no means isolated from those interpersonal ties which form important communication networks in urban subcommunities. While several American and English studies (see Lazarsfeld and Merton, 1963: 513-530; Mogey, 1956; Oeser and Hammond, 1954; Young and Willmott, 1957) have revealed the importance of kin attachments in white working-class areas, the data on relatives for the Boston ghetto sample reveal also that kin are quite important sources of close personal ties. Significantly, considering that the majority have come to Boston only in the last fifteen years or so, these black urbanites reported an average of approximately three relatives beyond their immediate family in the Boston area. They usually reported visiting these relatives frequently, one or two typically being visited (or visited with) several times a week. Moreover, about half of the respondents had received aid from relatives in the process of moving into their current residences.

In concluding their study of kinship bonds among blacks, Stone and Schlamp (1966: 62) point to the importance of the extended family networks:

> If kinship is central in the life style of low-income families then any change, even of small degree, in the network of kinship bonds can be expected to produce repercussions in other areas of the family's way of life. It may be that reduced support from kin is one factor that sets the stage for the onset of welfare dependency.

Thus it is evident that the importance of the primary and secondary network as circumscribed by the kinship complex cannot be underestimated simply because its configurations appear different from those of white nuclear families. Nor can they be construed as disorganized structures of atomized and alienated individuals. Rather, from the accumulating data briefly presented in this section, it is evident that while black-family systems evidence structures which do not always meet the ideal type of the middle-class ("normal") intact family, they are not necessarily any more disorganized, or more pathological than middle-class families.

THE EXTENT AND INTENSITY OF FRIENDSHIP AND NEIGHBORING CONTACTS

Contact with relatives and primary family is only one set of the important types of interpersonal contact available to black ghettoites. Friendship and neighbor ties are also important. A careful examination of the friendship ties of our black housewives in Boston revealed that the great majority were also not isolated from this type of intimate social interaction (Feagin, 1966). They averaged about three friends each, and typically they visited with two of these friends several times a week. Neighbor relations also were of some importance. Most of the respondents maintained speaking and visiting relationships with at least a few of their neighbors, while two-thirds had done some visiting in neighbors' homes. Thus, friendship and neighboring ties were significant components of the social lives of these black ghetto residents and could be considered objective criteria for designating their local residential area a viable neighborhood. In our estimation, the integration of these respondents into the social life of their urban community was roughly comparable to that which has been found with working-class whites (Lazarsfeld and Merton, 1963: 513-530; Mogey, 1956; Oeser and Hammond, 1954; Young and Willmott, 1957). Admittedly, the variety of indices used in the various studies of social participation makes it difficult to cumulate research in this area. Nevertheless, these black wives had informal ties generally as strong (extensive and intensive) as have been found for comparable white samples. Taken as a whole, then, such patterns definitely contravene a portrait of an alienating and uncomfortable community in which the person would experience a pathological and disorganized sense of social atomization. Furthermore, Stone and Schlamp, and Podell all found similar (in some cases more extensive) interactive patterns in California and New York. Of a sample of black families in California, no less than 82% knew approximately six families in the immediate neighborhood and 80% of the black husbands listed three or more good friends. Finally, 73% of the California respondents could enumerate one or more mutual-aid crises when they were helped by such friends and neighbors (Stone and Schlamp, 1966: 64-68). In the New York study, nearly nine out of ten welfare mothers could name a best friend and half could name one who had been a best friend for over five years. Anywhere from 61% to 75% of the black welfare mothers interviewed in this study participated in the neighborly activities of child care, aid in time of illness, and lending neighbors money (Podell, 1968: 9, 34, 37).

As a result, the black ghettoite does not appear to lack a community of social integration among his own brothers. That there is a social polarization between white and black communities within the urban complex might be true, but to infer from that phenomenon that the black community is socially disorganized when it is simply socially and culturally different and at the same time evidencing internal social integration, is an error of significant magnitude and stands as a substantial impediment to relevant research and policy change.

Nonetheless, the response of some readers to the arguments and data used so far might be that they are not really too surprising—given that we are detailing the coping patterns of an "encapsulated" people. Wellman (1971: 603) catches the flavor of this response quite well: "There is a separate black subculture and social system, it is alleged, which is dependent upon encapsulated social relationships and which provides little basis for movement beyond the ghetto." However, up to this point in this paper we have attempted to demonstrate that the relationships of husband to wife, kin, and friends are typically solid, highly organized, and integrated and quite surprisingly not as different as many might well have expected. If the black ghettoite is not as alienated and atomized as the stereotypes of pathology have so long argued, is it not also possible that he is not socially encapsulated either? Wellman (1971: 622) discovered in Pittsburg, when comparing black and white adolescents, that "the black adolescent . . . cannot be viewed as ghetto-bound in the classical sense. With respect to behavioral measures such as trips outside of their neighborhoods and trips downtown, *more blacks than whites* report that they do such things frequently." He also discovered that blacks were more anxious than whites to meet people from other areas of the city (different from those in their own neighborhoods).

Beyond the Wellman data are other studies which demonstrate means which the black ghettoite has used to break down the contrary stereotypes of atomization and encapsulation. One set deals with the surprising amount of contact black ghetto residents have had with the mass media and another deals with the extent and intensity of participation in formal groups.

THE EXTENT OF CONTACT WITH THE NEWS MEDIA

There is a considerable amount of evidence in the studies we have been reviewing of the extension of ghetto-oriented social integration of black residents, kin, neighbor, and friendship networks to the outside urban environment through the media. For example, Caplovitz and Bradburn (1964: 24) show that black ghettoites watch more television and follow international and local news more closely than do their white counterparts at all class levels. And Podell (1968: 42) points out that 51% of New York City black women on welfare (the poor) read the newspaper daily and another 25% read it a few times a week. Forty-six percent read magazines regularly and the rest irregularly. In our Boston study, we found that the percentage of a relatively low-income group who read the Boston papers daily was similar to that found for a general sample of high school graduates outside ghettos (65%). These poor black women did not appear to differ greatly from other urbanites of roughly the same educational level. Indeed, the sample was substantially more cosmopolitan-oriented than one might expect. Nearly two-thirds were regular readers of the metropolitan dailies, while only a quarter were regular readers of the two local ghetto weeklies (Feagin, 1966).

Moreover, included among the diverse questions asked of the black housewives in this same Boston sample was a time budget schedule. This required the respondent to detail her dominant activities for each of 72 fifteen-minute time segments from sunrise to midnight. Over two-thirds of these housewives had spent one-and-one-quarter hours or more before the TV set on the last weekday preceding the interview; the mean amount of time watching TV for these black housewives was about two hours-and-fifty-four minutes. The lesser significance of radio listening for contemporary urbanites, noted in some recent studies, also seems to be borne out by the Boston data. Overall, then, two-thirds of these black respondents spent at least three-and-one-quarter hours a day in contact with the four types of mass media–radio, TV, magazines, newspapers.

The mass media also provided tenuous links between persons in urban areas, at least in the sense of parasocial ties; that the mass media have become a habit for most urbanites has been borne out in numerous research studies. In comparison with whites studied, however, black men and women do not appear to be very different. They tend to read newspapers, watch TV, and see movies about as often as whites do–particularly those whites of comparable occupational status. The previous data indicating relatively strong primary ties, together with these findings on media contact, also argue against a contention that some neo-Wirthian students of ghettos and slums might well make: that being isolated from primary contacts would force black and other Americans into spending an unusual amount of time absorbed in the mass media. We would speculate further, following Gans (1962), that most blacks watch TV with their friends and relatives.

An argument could be made from these studies that many black ghetto residents know more about the white communities than do their white counterparts. It can at least be said that there is definite evidence that black urbanites expend as much–if not more–energy trying to understand the environment around them, as such information is offered by the various news media. Such behavior does not appear too deviant, anomic, or pathological–in fact, it points to quite the contrary conclusion.

THE EXTENT AND INTENSITY OF INTERACTIONS IN ORGANIZATIONS

In addition, there is substantial evidence extant in studies of urban ghettos which demonstrates that not only are urban blacks just as likely to join at least one voluntary association as urban and suburban whites, but they are also as likely to be as active–if not more active–than their associational counterparts (Stone and Schlamp, 1966: 69-71; Caplovitz and Bradburn, 1964: 19-21). Only a few studies, to our knowledge, have attempted to systematically investigate black voluntary association memberships and provide some statistical detail. A recent article which reviewed this literature discovered that most of these studies found associational participation to be higher among black urbanites than among whites (Curtis and Zurcher, 1971). While it is common knowledge that Myrdal (1964) found the same pattern and then argued that such a pattern was evidence of social pathology among black Americans, we are not at all sure that such inferences of disorganization can be raised from evidence of organization. Our reservations are strengthened by more recent studies which have been more systematic in their approach and more statistical in their detail. One such study by Babchuk and Thompson (1962: 647-655) is devoted to the precise questions of (1) the extent to which blacks affiliate with formal voluntary organizations, and (2) the variation in patterns of membership by social categories. Using a random sample, Babchuk and Thompson found that three-quarters were affiliated with at least one voluntary association, even excluding church and union membership. They conclude that their findings strongly support Myrdal's point that blacks belong to more voluntary associations than whites. A second study of black associational memberships was done in the Highland Park area of Detroit, Michigan. Again emphasizing black-white comparisons, Meadow (1962) found that her small sample of predominantly female heads of household distributed their memberships (including church and union memberships) as follows: (1) no memberships, 32%; (2) one membership, 34%; (3) two memberships, 24%; and (4) three or more, 11%. She found that church or church-related organization memberships accounted for 60% of the memberships of these black urbanites. In our Boston study of predominantly low-income ghetto residents, we found that—with the exception of church participation—their associational ties were not extensive, but the overall picture was broadly similar to white associational participation (Feagin, 1966).

Of the community organizations which blacks are most likely to join, religious affiliation appears to be the highest. To illustrate, 52% of the black husbands in welfare families in California belonged to at least one organization; fully seven in ten of these belonged to religious organizations (Stone and Schlamp, 1966: 69). Attendance in religious and other organizations was also quite high, the median being weekly attendance. The prevalence of religious affiliation may portend the trends of political group formation in terms of black nationalism, civil rights, and community and class struggle, which we have witnessed in the outgrowth of race politics in the past two decades. However, the evidence on this point is equivocal. Marx (1967: 64) points out that "with their stake in the status quo, established religious institutions have generally fostered conservatism, although as the source of humanistic values they have occasionally inspired movements of protest." While Marx's study suggests that religious attachment is not found among civil rights militants, he does find that 66% of black militants consider themselves either "very religious" or "somewhat religious" or possessing a high degree of humanistic values. Thus we believe it valid to assert, at least tentatively, that the historical importance of the church for black Americans, while not always evidenced in organizational strength, has been demonstrated by its influence of humanistic commitment. Again, the evidence is not present to validate this as a solid conclusion, but we offer it as a tenable hypothesis worthy of future study. Finally, in the same vein, we suggest that the evidence of the strength of formal organizations in the black ghetto coupled with the influence of church association and the history of the church as a catalyst of political associations and protest, may form some of the important roots of racial and community political structures. On this hypothetical note, we would like to turn to the last section of this essay which deals directly with political evidence of the new ghetto perspective we are arguing for here.

THE POLITICAL IMPORTANCE OF A NEW GHETTO PERSPECTIVE

In keeping with the preceding section of this essay, we would like to present the argument that the various manifestations of black politics of the sixties have been by and large evidence of highly integrated and socially organized behavior. They do not represent, just as the black family complex, kinship networks, friendship circles, and formal organization attendance have not represented, alignments of atomized and pathological people but rather socially organized and politically aware people who have introduced a variety of new and original political tactics. Indeed, social structure provides the foundation for various political actions. A great deal of the literature on the process of black urbanization and ghetto formation distorts the picture by giving far too much emphasis to the impersonal and disorganizing aspects of the movement of rural blacks to an industrialized urban milieu. There are other features of this process that are equally deserving of attention and have contributed to social and political organization and even to the eventual outbreak of collective protest. As increasing numbers of black migrants from the South entered urban areas in the North, they congregated in relatively confined sectors of the cities (Feagin and Hahn, 1972).[5] Their choices were severely limited by the prevailing discrimination, both individual and institutional, that divided the city into distinctive black and white areas. Thus, geographic residential segregation became one of the most crucial and continuing social facts both for black residents and for cities in which they reside.

According to the available evidence we have surveyed, the social networks and organizations that emerged in urban ghettos have promoted a substantial degree of social cohesion in most ghetto neighborhoods, conventional social pathology theories notwithstanding. Moreover, as a result of the nature of their associations, and as a consequence of the omnipresent discrimination they faced, black urbanites have tended to develop common perspectives. This dual base of social cohesion and a common sense of experience of discrimination has provided a potential nexus around which to advance the collective interests of the black ghetto, or significant segments thereof.

More particularly, we believe it possible to argue that the communal framework we have previously suggested relates to new political patterns. The perceived unresponsiveness of the educational, law enforcement, city government, federal government and other public service structures attendant on urban society has resulted in a variety of forms of black political organization. At the same time, not all ghetto politics is based on reaction to white governmental institutions: nationalistic cries ranging from black power and black nationalism on to variants of black separatism have formed important organizational bases for the black ghetto residents' response.

One example of what makes such organizational patterns both necessary and important forces in community development, not simply passing fads, is the evidence now compiled concerning the level of politically motivated anger directed at agencies of the outside urban establishment which rules over the ghetto. While riot participants still comprise minorities in black communities of this country, it is now estimated that a sizable minority, perhaps as much as a third in many areas have participated in urban riots (Fogelson, 1971). One-fifth of Los Angeles blacks probably participated in the major Watts riot in 1965, and many more had generally favorable feelings about the riot (Sears and McConahay, 1970: 260). While the long-term effectiveness of the riots as a method of promoting black demands is still in question, it does appear as if the rioters achieved at least momentary control over the cities that had for so long ignored them. Until the official law enforcement agencies were able to reassert control, the life of America's major urban centers was ruled from the streets and the political emergence of urban black America was indelibly imprinted on the contemporary consciousness of the vast majority of the white society. As it became apparent that while there was not full-scale revolution this collective violence did provide at least home rule for the instant, the white authorities retaliated with quick and increasingly sophisticated measures which ensured that any attempt to secure permanent political power by rioting would be extraordinarily difficult.

This outbreak of ghetto violence in the late sixties was not a chance accident of history. The riots were not simply reactions to the unresponsiveness and reticence of certain governmental officials at a certain moment in time. They were also responses to a political structure and concomitant administrative network which had historically prevented an emergent minority in its midst from securing the goals it was allegedly guaranteed. In line with this reasoning, Feagin and Hahn (1972) conclude that "Both during the ghetto riots themselves and in the movement for community control that accompanied later riots, black Americans seemed to be pressing a demand for significant powers of self-governance and for the right to control their own lives and destinies."

Into this environment of rioting and propensity to cohere for group goals have come a variety of national political groups and new ghetto organizations. No other minority group with similar discriminatory impediments to full entry into the American mainstream has evidenced as diverse and sophisticated a repertoire of political organizations as have contemporary black Americans. Just as it is basically inaccurate to talk about a single ghetto response or black response to an issue because there is not constant agreement, so, too, it would be the height of folly to equate black political organization with constant agreement among all black political groups, or, more important, political disorganization with group disagreements. The diversity and sophistication of political approaches of various black political interest groups are based upon different strategies and political philosophies; and they are often aimed at different sectors of the American system. From our own normative perspective we see groups such as the Black Panthers and the Black Muslims as not only representatives of more militant organizations but also examples of high levels of social cohesion and racial consciousness. These factors stand as highly politicized symbols of the struggle against the myths and realities attached with construing the black ghettos as encapsulated jungles of alienated and atomized persons.

The Black Muslims have served for some time a dual purpose: first, they meet the needs of working-class blacks and their sense of futility with the recurring failures of integrationist strategies; second, they provide an integrative value through their meetings and procedures. This, coupled with the nationalistic theme of a separate state for black persons (since blacks already know they have the sense of community and capacity to compete and survive), is attractive if one accepts the internal organization of the black ghetto as something other than a pathological state.

At the same time the Black Panthers, while short on actual members, have the same sort of sympathetic support as the riot participants had in Watts. They represent a break with the legend of black docility which adhered to a tradition of accommodation and servility on the part of demanding blacks. Their demands for freedom to determine their own destiny; their demands for "land, bread, housing, education, clothing, justice, and peace" transcend the subordinate position with which other groups were treated as representatives of "socially disintegrated" groups who were to be treated with the paternalistic response with which tragically deprived persons "should" be treated.

If the Panthers and the Muslims stand as examples of groups engaged in highly concerted efforts to raise black consciousness and then to undertake certain political or economic tasks, there are many other groups which undertake at a less practiced and militant level some of these same and even other tasks. There are two basic groups of organizations: first, those which have been, by and large, national in membership, with national goals and identity; second, those organizations comprised of local offices of the national organizations or simply neighborhood or ghetto-wide organizations dedicated primarily to local issues and local community control and development. At the national level the Congress on Racial Equality (CORE), more than any other moderate black political organization, has moved to a separatist posture with much of its leadership arguing for black statehood. Other national groups have taken a variety of strategies aimed at different goals. The Urban League leadership, at the national and local levels, has been most consciously involved in economic activities. The NAACP leadership has tended to be comprised of lawyers at the national level. As a result, the goals and strategies of the NAACP have been most visible and successful in legal battles. Another group with national visibility has been the Southern Christian Leadership Conference (SCLC) with a substantial base in religious organization and its leadership at both the national and local levels has been by and large composed of ministers. Also at the national level, new political visibility has been achieved through the vocal and articulate Black Caucus in the House of Representatives and through black candidates for the nation's highest elective office in the persons of Dick Gregory and, now, Shirley Chisholm. While these latter candidates do not pose serious threats to white control of the White House at this time, they do stand as symbols of the visibility and sophistication of a burgeoning black politics.

In comparison, black organizations at the local level are not concerned with national image or even with national politics to any great degree. Rather they are more concerned with the immediate demands of their constituents in the neighborhoods or the ghetto as a whole. Examples of such organizations are FIGHT in Rochester, New York, which switched from highly overt political tactics to highly sophisticated economic development tactics as the needs of the Rochester black community changed. Other examples of organizations geared for local needs and using different strategies are the Woodlawn organization of Chicago and the Black Stone Rangers. There are also political organizations which are geared to putting black leaders in traditional political office—organizations which have put the Hatchers, and Stokes, and Dellums into office, and, at the same time, have put new ward leaders, the leaders of Model Cities boards, and the like into office. Although there is really no way of telling exactly which organization most closely provided the most substantial impetus to black electoral success overall, it is patently obvious that black political machines are having an important impact. Perhaps the most important evidence of this impact lies in the fact that by 1970 black politicians in the South had been elected to over 700 important political positions at the state and local level (Feagin and Campbell, 1971).

Thus the great variety and strength of these various organizations with different strategies and goals can no longer be denied. They are so strong, visible, and organized that it would seem rather anachronistic to term them as organizational manifestations of allegedly atomized and alienated people who manifest social disorganization and pathology. Furthermore, the sophistication of political alternatives that the black ghetto now represents cannot be interpreted as simple reactions to Whitey. The common themes of black consciousness, cultural pride, and independent competitiveness are now mainstays in all black political organizations. It does not take more than a peripheral look at the position of the black now versus ten years ago—much less thirty or forty years ago—to realize that these themes have been conceived in a new generation of brotherhood. With these themes, generated through the religious base of the integration movement, the conscious-raising of the black power movement, the struggles and failures of the civil rights movements, and the rebirth of pride generated by the separatist groups, has come a more visible and more strongly articulate sense of ethnic identity and political organization.

CONCLUSIONS

In this article we have attempted to place in question the rather long tradition of the social sciences to view the black ghetto of the American city as a pathological place filled with atomized and alienated people estranged from each other as well as the rest of society. In doing so, we have argued:

(1) That for all the protestations to the contrary, much of the literature of the social sciences (and those influenced by the social sciences) continues to adhere to a Wirthian position for ghettos, namely that the emerging urban way of life is characterized by a weakening of friendship bonds, family organization, and breakdown of community organization.

(2) While the literature and empirical hypotheses have revolved around such a background, the evidence has been less than satisfactory as substantiation of such a view of the ghetto. In fact, very few studies have approached research of ghetto organization from any other than a more or less stereotypical view of the ghetto as disorganized and disintegrated due to its pathological nature.

(3) We have reviewed a small but growing body of literature which tends to confirm an alternative view of the black ghetto: that there is a socially significant, healthy, and diverse primary family structure, there is a high level of kinship interaction which is also quite intense, and there is a high level of friendship and neighboring. Hence, there is evidence which suggests that the internal structure and organization of the ghetto is not necessarily an environment which atomizes the person in an alienated state with no friends or organizational support from family and relatives.

(4) At the same time we found growing evidence that the black ghettoite's world is not simply one in which, he, like the turtle, is encapsulated and, when threatened, will not leave his shell. We discovered that he has as much, if not more contact with the news media than white urbanites; he participates as much if not more than white urbanites in voluntary organizations; and he can be as cosmopolitan if not more cosmopolitan than white urbanites when it comes to traveling around the city and interacting with persons of other social identities.

(5) Finally, we have argued that the diversity, sophistication, visibility, and strength of black interest groups and political organizations cannot be dismissed as scattered evidence of disorganized people who are simply reacting to the conditions that suppress them. To categorize black politics as simply a phenomenon based upon black demands made of whites would deny the whole growth of black consciousness and dismiss black politics as having no identity of its own beyond alleged white suppression. There is enough observable evidence to make such a denial quite untenable.

The extent to which we continue to view either the social organization of the black family or the political organization of a black ghetto as deviant or disorganized due to certain evidence of social pathology has very real implications not only for society but also for the social sciences. If we continue to ignore the evidence we have reported and reviewed in this essay, what we really will be seeing is a measure of the threat such evidence is to the selective world of the dominant society. The growth of black political organizations, for example, and the concomitant feelings of internal organization and brotherhood intensifies the centuries-old problem of race-group conflict at the heart of ethnic politics. Edgar Litt (1970: 4) has stated this problem in the following terms:

> The shared symbols, interests, affections, and real or imagined traits which draw some men together into the group or community are the walls which separate those men from others. Hence, the communion that nurtures intragroup cohesion is often the first condition for intergroup conflict. To state it more tersely, for there to be "brothers" there must also be "others."

In view of Litt's comment, it can be asked whether the more evidence both academia and government gather concerning the organizational sense of family, kinship, friendship, and politics, the more academics and government researchers begin to feel like the "others?" It could well be that as the social sciences continue to stereotype such evidence as disorganized and pathological we are not learning so much about the "others" of the black ghetto as we are about the "others" of the dominant white society. To put it another way, we are learning more about the alienation of the observers from the observed than of the alienation of the black ghettoite from his neighbors or even the rest of his urban environment.

NOTES

1. In particular, see a comprehensive list of scholars who have called the Wirthian tenet into question in footnote four of Bell and Boat (1957: 391). See also Janowitz (1952).

2. In his book, Dollard (1957: 433) argued that "the dominant aim of our society seems to be to middle-classify all of its members."

3. Watts et al. (1964: 8; see also 56-57, 90). This study was conducted with a sample of 250 middle-income black families in the Roxbury district of Boston.

4. Caplovitz and Bradburn (1964: 34). These conclusions are an alternative analysis to the one presented by Caplovitz and Bradburn. The two authors couch their analysis in the traditional neo-Wirthian and socially pathological frame of reference and hence explain this reading of adjustment to parenthood on the part of black ghettoites as based on the fact that ghettoite parents are more likely out of work and thus must spend time with their children, or perhaps it is because suburbanites are commuters and cannot see their children. This sort of analysis just does not follow from the data. Ours could be just as accurate or relevant–especially considering the Stack finding.

5. This section of the essay draws heavily from an unpaginated draft of this forthcoming article.

REFERENCES

ALLPORT, G. W. (1958) The Nature of Prejudice. Garden City, N.Y.: Doubleday Anchor.

BABCHUK, N. and R. V. THOMPSON (1962) "The voluntary association of Negroes." Amer. Soc. Rev. 27 (October): 647-655.

BELL, W. and M. BOAT (1957) "Urban neighborhoods and informal social relations." Amer. J. of Sociology 62 (January): 391-398.

BILLINGSLEY, A. (1968) Black Families in White America. Englewood Cliffs, N.J.: Prentice-Hall.

CAPLOVITZ, D. and N. M. BRADBURN (1964) Social Class and Psychological Adjustment: A Portrait of the Communities in the "Happiness" Study. Chicago: National Opinion Research Center.

CLARK, K. (1965) Dark Ghetto. New York: Harper & Row.

CURTIS, R. L. and L. A. ZURCHER, Jr. (1971) "Voluntary associations and the social integration of the poor." Social Problems (Winter): 339-357.

DOLLARD, J. (1957) Caste and Class in a Southern Town. Garden City, N.Y.: Anchor.

FEAGIN, J. R. (1966) "The social ties in Negroes in an urban environment: structure and variation." Ph.D. dissertation. Harvard University.

--- and H. HAHN (1972) "Theories of urban violence: an appraisal," in H. Hirsch and D. C. Perry (eds.) Violence and Politics: A Collection of Original Essays. New York: Harper & Row.

FEAGIN, J. R. and D. CAMPBELL (1971) "Black political power in the South." University of Texas. (unpublished)

FOGELSON, R. M. (1971) Violence is Protest. Garden City, N.Y.: Doubleday.

FRAZIER, E. F. (1939) The Negro Family in the United States. Chicago: Univ. of Chicago Press.

GANS, H. J. (1962) The Urban Villagers. New York: Free Press.
GLAZER, N. and D. MOYNIHAN (1963) Beyond the Melting Pot. Cambridge, Mass.: MIT and Harvard Univ. Presses.
JANOWITZ, M. (1952) The Community Press in an Urban Setting. New York: Free Press.
KATZ, E. and P. F. LAZARSFELD (1964) Personal Influence. New York: Free Press.
LAZARSFELD, P. F. and R. K. MERTON (1963) "Friendship as a social process: a substantial and methodological analysis," in M. W. Riley (ed.) Sociological Research. Vol. I: A Case Approach. New York: Harcourt, Brace.
LITT, E. (1970) Ethnic Politics in America. Glenville, Ill.: Scott, Foresman.
MACK, D. E. (1971) "Where the black-matriarchy theorists went wrong." Psychology Today 4 (January): 24, 86-87.
MARCUSE, H. (1964) One-Dimensional Man. Boston: Beacon.
MARX, G. T. (1967) "Religion: opiate or inspiration of civil rights militancy among Negroes." Amer. Soc. Rev. 32 (February): 64-72.
MEADOW, K. P. (1962) "Negro-white differences among newcomers to a transitional urban area." J. of Intergroup Research 3 (1962): 320-330.
MILLER, S. M. and F. RIESSMAN (1968) Social Class and Social Policy. New York: Basic Books.
MOGEY, J. M. (1956) Family and Neighborhood. London: Oxford Univ. Press.
MYRDAL, G. (1964) An American Dilemma: The Negro Problem and Modern Democracy. New York: McGraw-Hill.
National Advisory Commission on Civil Disorders (1968) Report. New York: Bantam.
OESER, O. A. and S. B. HAMMOND [eds.] (1954) Social Structure and Personality in a City. London: Routledge and Kegan Paul.
PODELL, L. (1968) Families on Welfare in New York City: Kinship, Friendship, and Citizenship. New York: Center for Social Research.
RAINWATER, L. (1966) "Crucible of identity: the Negro lower-class family." Daedalus 95 (Winter): 172-216.
SEARS, D. O. and J. B. McCONAHAY (1970) "Riot participation," in N. Cohen (ed.) The Los Angeles Riots: A Socio-Psychological Study. New York: Praeger.
SINGER, B. D. et al. (1970) Black Rioters: A Study of Social Factors and Communication in the Detroit Riot. Lexington, Mass.: Heath Lexington.
STACK, C. R. (1970) "The kindred of Viola Jackson: residence and family organization of an urban black family," in N. E. Whitten, Jr. and J. F. Szwed (eds.) Afro-American Anthropology: Contemporary Perspectives. New York: Free Press.
STEIN, M. R. (1964) The Eclipse of Community. New York: Harper Torchbook.
STONE, R. C. and F. T. SCHLAMP (1966) Family Life Styles Below the Poverty Level. San Francisco: Institute for Social Science Research.
U.S. Department of Labor (1965) The Negro Family: The Case for National Action. Office of Policy Planning and Research. Washington, D.C. Government Printing Office.
VALENTINE, C. A. (1968) Culture and Poverty: Critiques and Counter Proposals. Chicago: Univ. of Chicago Press.
WATTS, L. G. et al. (1964) The Middle-Income Negro Faces Urban Renewal. Washington, D.C. U.S. Department of Commerce.
Webster's Seventh New Collegiate Dictionary (1965) Springfield, Mass.: G. C. Merriam.
WELLMAN, B. (1971) "Crossing social boundaries: cosmopolitanism among black and white adolescents." Social Sci. Q. 72 (December): 602-624.
WIRTH, L. (1938) "Urbanism as a way of life." Amer. J. of Sociology 44 (July): 1-24.
YOUNG, M. and P. WILLMOTT (1957) Family and Kinship in East London. Baltimore: Penguin.

17

The Barrio as an Internal Colony

MARIO BARRERA
CARLOS MUÑOZ
CHARLES ORNELAS

I. THE BARRIO AS INTERNAL COLONY

□ MEXICAN AMERICANS have often referred to themselves as a forgotten people, and to a large extent this has been true. Within the last few years, various dramatic activities connected with what is loosely referred to among Chicanos[1] as "The Movement" have led to a partial discovery of America's second largest minority by the media, politicians, academics, and government administrators. As in the case of Latin America after the Cuban Revolution, there is now being evidenced a sudden concern for the welfare of Chicanos and a desire to help them overcome their "underdeveloped" status. However, this latest Alliance for Progress venture on the domestic front is likely to produce the same striking non-results as its international counterpart, and if it does it will largely be due to the same sorts of causes. Chief among these will be a mistaken analysis, insufficient funding, a certain amount of self-serving by the earnest benefactors, and a determined unwillingness to see that the problems being treated have, in fact, been largely produced by those same benefactors.

What is needed at the present time is an analysis of the Chicano situation that avoids some of the more blatant distortions that have

---

AUTHORS' NOTE: *We would like to thank Carlos Cortés, Robert Blauner, Itsugi Igawa, and Marcelo Cavarozzi for their comments on an earlier draft of this paper.*

been perpetrated by writers viewing the Chicano reality through lenses colored by American society's dominant myths about ethnic relations. Our concern in this paper is to further such an analysis, focusing on the political dimension.

One beginning point is the recognition that Chicanos are now predominantly an urban people. The popular stereotype of Mexican Americans as overwhelmingly rural continues to be perpetuated by television documentaries, and by the broad coverage given such rurally based movements as César Chávez's United Farm Workers in California and Reies Tijerina's Alianza in New Mexico. Yet the 1960 census shows almost eighty percent of the "Spanish-surname" population in the Southwest living in urban areas, a figure roughly comparable to that of the Anglo population (Grebler et al., 1970: 113). In the cities, most Mexican Americans continue to live in relatively well-defined areas, which are referred to as "barrios."

Although Chicanos are found in many areas of the Midwest and Pacific Northwest, they are concentrated in the five southwestern states, and this concentration increases their political significance. Mexican Americans outnumber all nonwhite groups combined in each of the southwestern states, and in 1960 constituted approximately twelve percent of the population in this area (Grebler et al., 1970: 105). The Chicano presence in such large, politically pivotal states as California and Texas should also be noted.

At the present time there do not exist a large number of studies that deal with Chicano politics in urban areas. The few studies that touch on this problem in recent times are reviewed in part II of this paper. In part III we summarize some of the dominant themes in this literature, and argue that these writings have been influenced by a latent model which we call the "assimilation/accommodation model." In analyzing Chicano politics, these writers dwell on what they perceive as weak leadership and lack of political organization, which they attribute to characteristics of Mexican American culture and social organization. They project solutions based on an analogy with European immigrants, calling for cultural assimilation and the politics of accommodation.

In parts IV and V we develop an alternative perspective, in many cases drawing upon evidence presented in the works reviewed in part II. Our view is that the barrio is best perceived as an internal colony, and that the problem of Chicano politics is essentially one of powerlessness. Powerlessness, in turn, is a condition produced and maintained by the dominant Anglo society through a number of mechanisms, some of which we have begun to identify. We consider the contemporary situation to be a form of internal neocolonialism, characterized by the predominance of relatively subtle and indirect mechanisms.

The concept of internal colonialism has been advanced by others to describe the position of Afro-Americans in the United States, and has also been applied by some Latin American scholars to situations in their countries. The general concept appears to apply to a number of cases, and is valuable in emphasizing the structural similarities and common historical origins of the positions of Third World peoples inside and outside the United States. However, it should not be used to obscure variations in individual cases. The situation of Chicanos is in many ways unique, as we shall describe in part V.

To be colonized means to be affected in every aspect of one's life: political, economic, social, cultural, and psychological. In the present essay we have limited ourselves to discussing primarily the political aspect, not because we consider the other elements to be less important, but because of limitations of time and space. The economic aspect in particular needs to be much more thoroughly researched than it has been to this point. Hopefully it will be possible to put together a more complete picture of internal colonialism in the near future.

The development of an adequate model of the Chicano position is essential if realistic solutions to Chicano problems are to be forthcoming. In the concluding part of this essay we survey some aspects of the contemporary political scene and some of the strategies for change that have been adopted or tried out by various groups. While Chicano urban political groups vary in their analyses and their approaches, their actions can generally be seen as a response to some aspect of the Chicano's colonized status.

II. SURVEY OF STUDIES ON THE POLITICS OF THE BARRIO

RUTH TUCK: *NOT WITH THE FIST*

The first postwar study to deal with an area of Chicano politics was that of Ruth Tuck (1946). The city she studied was San

Bernardino, California, which she labeled "Descanso," and her research appears to have been done during the war. The title of her book is taken from an observation of Charles Horton Cooley:

> wrongs that afflict society are seldom willed by anyone or any group, but are by-products of acts of will having other objects; they are done, as someone has said, with the elbows rather than the fists. There is surprisingly little ill-intent [Tuck, 1946: xix).

In spite of this seeming absolution of "Descanso's" white majority, Tuck documents in considerable detail the pervasive racism of Anglos with regard to Mexican Americans, and the negative social consequences it produced. Among her observations (Tuck, 1946: 10): "Descanso believes with simple, naive fervor that North Europeans and their descendants represent a superior breed of people." "To tell Descanso that it has created caste, or semi-caste, situations for some of its citizens—permanent lower groupings in which they are supposed to stay and against which barriers have been erected to prevent their rise—is to insult its picture of itself as a generous, friendly town" (Tuck, 1946: 44). "Until very recently, the community agreement to keep the Mexican or person of Mexican extraction in his place, economically, was effective for all but occasional members of the group" (Tuck, 1946: 174). She also refers to the deliberate patterns of residential and educational segregation in San Bernardino.

Mexican Americans comprised twelve percent of the population of San Bernardino during the period of which Ruth Tuck wrote. Many had come in the late nineteenth and early twentieth century, in response to the need for labor, and had been encouraged to do so by farmers and industrialists. Of course, Mexicans had lived in the area before Anglos had come to settle. In spite of this relatively long history in the city, Tuck was unable to find very much in the way of stable political organizations among Chicanos. According to her, there have existed Chicano organizations oriented toward particular political issues or that acted in a political manner, but they have been short-lived. One came into existence in 1943 to presecute the city for refusing Mexican Americans the use of the swimming pool. But after winning that battle it disappeared. There was also a chapter of the LULACs (League of United Latin-American Citizens), but it too lasted only a short time.

In describing and explaining this observation, Tuck makes reference to Descanso's "repressive civic policy," but she also devotes considerable attention to discussing Mexican American leadership patterns. According to her, five or six persons are acknowledged as "leaders" by Chicanos because they have some knowledge of Anglo ways and are able to perform as intermediaries with the Anglo society (Tuck, 1946: 137). This leadership is essentially conservative, working within the system to accommodate the desires of the dominant society, and they tend to be drawn from the more "respectable" elements of the barrio.

The barrio leaders in San Bernardino encounter considerable suspicion from barrio people. Some consider that the leaders are overly Anglicized and have turned their backs on their own culture. Accusations of opportunism and feathering one's nest are also frequently heard, and Tuck (1946: 149) notes some tendency for leaders to overconcentrate power in their own hands. However, she puts the major responsibility for the state of barrio politics on Anglo society (1946: 141-142): "The great responsibility of the dominant community lies in the fact that it is like a moon, which pulls the tides of the [barrio's] society this way and that. The most enlightened leadership and the most arduous individual efforts cannot prevail, except in individual and isolated instances, against the forces that arise from without." And she points clearly to the attempts by Anglo society to isolate and co-opt able Chicano leadership (1946: 142): "This effort to cut the Mexican-American of superior background or achievement off from the rest of his group is constant in most Anglo-American circles."

While Tuck's position is one sympathetic to the Mexican American, her perspective is basically assimilationist. As she puts it (Tuck, 1946: xx): "It is the writer's profound hope that this book will play a small part in producing a community where it is no longer necessary to invent terms for splitting up citizens into racial and national islands, where there will be no linguistic devices for emphasizing social isolation and difference, and where the term 'American' will serve for everyone." She thinks of the Chicano as analogous to the European immigrant (1946: 42), and she downplays cultural differences (1946: 117, 119). She is concerned to demonstrate that Mexican Americans *can* assimilate, as other groups have done.

FRANCES WOODS: *MEXICAN ETHNIC LEADERSHIP IN SAN ANTONIO*

Shortly after Tuck's work appeared, a second study of Mexican Americans in a Southwest city was published. This was by Sister Frances Jerome Woods (1949), which had been her Ph.D. dissertation. San Antonio, a good-sized city, was one-third Mexican American at the time. While this work has less scope and depth than Tuck's book, it is more focused on political matters. She employs a rather simplistic twofold typology of leaders based on their methods. One type she calls "Radicals," distinguished by their use of protest and other "militant" tactics, and their emphasis on combatting discrimination against Chicanos. The other type is labeled "Diplomats," and is characterized by more conservative tactics, such as the use of legal channels, and by their stress on Mexican American self-help efforts (Woods, 1949: 74-75, 118). "Radicals" tend to have more popular support; "Diplomats" enjoy more Anglo support (1949: 78).

The Mexican American leadership in San Antonio performs several functions. One is the representative function of expressing the group's values and interests. This is related to their function as intermediaries or mediators between the Anglo and Chicano groups, a role requiring a certain amount of acceptance of their legitimacy by Anglos. A third function has to do with coordination of group efforts, which, according to Woods (1949: 87) is particularly important, given that "the Mexicans tend to be strongly individualistic." In another section of her work, Woods refers to Mexican Americans who act as precinct or campaign managers for Anglos in Chicano areas, but she does not make clear whether she considers this to be a leadership "function."

Woods is also concerned with describing the backgrounds and social characteristics of San Antonio's Mexican American leaders. According to her, they tend to be between 45 and 55 years old, and are generally of "respectable" backgrounds. Traditionally, many have been lawyers, although businessmen have become more important in recent times. There are only a few younger leaders, and fewer still from lower-class backgrounds.

Woods alludes several times to the problem of suspicion of their leadership by Chicanos, and at one point attributes this to the leaders' tendency to make promises which they do not intend to fulfill. This trait, in turn, is traced to the value which Mexican culture puts on courtesy.

In contrast to Tuck, Woods mentions several continuing Mexican American political organizations in San Antonio, and she seems to be aware that many ostensibly nonpolitical organizations play a political role in the community.

Whereas Ruth Tuck was a straightforward assimilationist, Woods speaks favorably of the attempt to merge desirable characteristics of Anglo and Chicano societies. She sees this type of "fusion" as the aim of most Mexican American leaders.

D'ANTONIO AND FORM: *INFLUENTIALS IN TWO BORDER CITIES*

After these two studies in the 1940s, there were no extensive published works on urban Chicano politics until the one of D'Antonio and Form (1965). This is a comparative study of community decision-making and power structure in El Paso, Texas, and Ciudad Juárez, Mexico. Although not focusing on Chicanos as such, there are numerous references to this group, which constituted forty-five percent of El Paso's population in 1960. Their main conclusion with respect to Chicanos is that they were largely excluded from participation in decision-making (1965: 245). They attribute this undemocratic state of affairs to various factors: "citizen apathy" (1965: 245), "the neglect to socialize that part of the Spanish-name population of the nation and community" (1965: 245), and the failure to develop viable political clubs on the part of Chicanos (1965: 246). They also refer to political factionalism and to the Mexican Americans' "relatively low level of internal social integration" (1965: 30).

They do discuss the 1957 city elections, in which a Mexican American, Rivera, won a contested Democratic primary and went on to become mayor, with the solid electoral support of the Chicano community. Yet there were circumstances which rendered this a hollow victory. In order to win, Rivera apparently felt it necessary to run with an all-Anglo slate of city councilmen. Once in office, he proceeded with a great deal of caution in his policies and appointments, to the extent that he had soon won the hearts of the city's business-oriented influentials (D'Antonio and Form, 1965: 142). However, the influentials soon moved to eliminate primaries from the nominating process (1965: 142).

While D'Antonio and Form (1965: 228) attribute the lack of aggressive, issue-oriented Chicano politics in El Paso to factors internal to the Chicano population and to Chicano leaders' "basic identification . . . to American values," they provide sufficient data to support an alternative explanation. Such an explanation would stress intentional efforts by Anglo power holders in El Paso and in the Southwest to prevent a viable Chicano politics from appearing. D'Antonio and Form (1965: 26) refer to the barrier represented by poll taxes, to the exclusion of Mexican Americans from positions in the dominant Democratic Party (1965: 246), and from local government positions (1965: 26). But there are also many more subtle mechanisms that can be noted in their account. One of these is the practice, beginning in 1951 in El Paso, of including one token Chicano on the city council slate of an Anglo candidate for mayor (D'Antonio and Form, 1965: 81, 134). The mechanism of cooptation also appears to have been firmly entrenched. In one instance, they feel that "the decision . . . to nominate José Jiménez for one of the city council posts was only a token step toward recognizing the potential power of the Spanish-name masses, but it also represented an attempt to co-opt ethnic leadership" (1965: 231).

Perhaps the most insidious mechanism, however, might be labeled the "racist mobilization of bias," adapting a term from Bachrach and Baratz (1970).[2] This consists of putting Chicanos on the defensive by constantly suggesting that they are inferior to Anglos. In politics, a typical euphemism that is employed for this purpose is that there are no "qualified" Chicano candidates available (1965: 134). The prevailing attitudes of El Paso's elite can be summed up in the following statement: "How can we hold our heads up in the State of Texas when we have a Mexican mayor (1965: 142)"? The defensive attitude that this kind of racism produces may be noted in a quote attributed to a Mexican American: "If we didn't do a good job, it would not be just another John Smith who had not done a good job as mayor. Ah, no, it would be that José Ramírez, and people with names like Ramírez and Fuentes, are just no good in politics. So we have a special duty to ourselves and to our nationality to perform well" (1965: 138n). The path that this "special duty" took in El Paso was illustrated by the actions of the mayor, "who did not want to deviate from traditional political paths but wanted to demonstrate that 'Mexican-Americans' could do 'as good a job as anyone else in public office.' Operationally, this meant that he proceeded cautiously, made use of impartial fact-finding committees, and avoided purely partisan issues" (1965: 146).

That D'Antonio and Form could include so much evidence for the proposition that Chicano powerlessness was due to Anglo manipulation, and yet put the weight of explanation on Chicano cultural and social characteristics, does not speak particularly well for the ability of "impartial" social scientists to escape society's dominant myths. Perhaps this is related to one of the book's outstanding weaknesses: the almost complete absence of any reference to the history of Chicano-Anglo relations in the Southwest. In our estimation, it is impossible to attribute causes to these complex political patterns, as D'Antonio and Form do, without taking into account that those patterns originated in conquest and have been maintained through force (see, e.g., Carey McWilliams, 1968). More recently, of course, those patterns have been reinforced through the more subtle kinds of political mechanisms that D'Antonio and Form document in their work.

More generally, the prevailing frame of reference which D'Antonio and Form apply to Mexican Americans is similar to that of Ruth Tuck; that is, the analogy with European immigrants. At times they make the comparison explicit (D'Antonio and Form, 1965: 141, 246).

ARTHUR RUBEL: *ACROSS THE TRACKS*

Another study of urban Chicanos appeared (Arthur Rubel, 1966). This was a study of Weslaco ("New Lots" in the book), in the Lower Rio Grande Valley in Texas. During the late 1950s, when Rubel conducted his research, Weslaco had a population of about 9,000 Mexican Americans and 6,000 Anglos, perhaps a less typically urban city due to its relatively small size and agricultural setting.

Echoing themes in other studies, Rubel comments on what he perceives to be weak Mexican American leadership and an absence of instrumental, issue-oriented groups engaged in pressure politics. In seeking explanations for this pattern, Rubel's remarks again have a familiar ring. According to him, "some societies display a certain 'flair' for organizing and proliferating instrumental groups. Clearly, the Chicanos do not." And again, "Organization of groups for the attainment of goals, whether diffuse or particular, is not one of the instrumental techniques made available to them by their culture" (Rubel, 1966: 135). He also refers to personalism, which he sees as the converse of forming instrumental groups (1966: 139). Quite apart from Rubel's tendency to generalize about Chicanos from one study of one town, his interpretations can be challenged on internal grounds. The existence of Chicano defensiveness and fear of consequences can be seen from Rubel's own account (1966: 128) of one political "instrumental group" in Weslaco:

> At a meeting of the League of United Latin-American Citizens (LULAC) held during the campaign, several members urged the local council to support a drive to induce the Chicano voters to purchase poll taxes in the following year. The move was voted down on the strength of Ray's argument that such a drive would be construed by Anglos as having political overtones. The LULAC council, he argued, would no longer be accorded the esteem of Anglos who conceived of the group as an educational and citizenship-oriented organization.

Rubel's emphasis is even harder to explain than that of D'Antonio and Form, since he does have an awareness of the historical setting. Rubel demonstrates in the historical section of his book that he is fully aware of the tactics used by Anglos to deprive Chicanos of the land which they owned, and of the violence used to maintain Anglo control. Even more interesting, however, is his account of the existence of a Democratic "machine" in that area in the early part of the century, based largely on Chicano support (Rubel, 1966: 48-49). For those scholars who make use of analogies with European immigrants in the United States, Rubel's description of the fate of this machine should be instructive. According to him (1966: 46-47, 50), it was destroyed in the 1930s, in the name of the "Good Government League," during a period of heightened anti-Mexican feeling. "By the end of 1930 the Anglo society was assured unassailable superiority in all spheres of activity over the subordinated Mexican-Americans" (Rubel, 1966: 50).

SAMORA AND LAMANNA: *MEXICAN-AMERICANS IN A MIDWEST METROPOLIS*

In 1967 appeared the work of Julian Samora and Richard Lamanna based on research done around 1965. East Chicago, Indiana, is an industrial city in the Chicago urban area, and has a population of about 58,000, of which eleven percent are Mexican Americans.

In contrast to some other writers, Samora and Lamanna noted several political organizations among Chicanos, e.g., the LULACs, the Mexican-American Democratic Organization, and the Latin Civic Political Club. They also comment on the strong sense of Mexican American nationalism, which they see as providing an important source of cohesion in the community. However, they also note that the various organizations do not as a rule work well together, nor do they have really extensive memberships (1967: 98). They do detect a trend towards broader involvement in politics, a more active leadership, and a greater political role for women.

At the time the study was done Chicanos had not achieved a great deal of political representation in East Chicago, although there had been one Mexican American councilman elected. In the authors' estimation, the Kennedy campaigns had been important in politicizing the Chicano community. While they do not make a detailed study of public policy, they do comment on various policy areas as they affect Chicanos. In this connection, they cite the urban renewal program in the area as disruptive and severely criticized among Mexican Americans (Samora and Lamanna, 1967: 104).

Samora and Lamanna attempt to give their study a more general perspective by relating it to ethnic assimilation theory. Although they do not commit themselves normatively, it appears safe to say that they consider assimilation (in the various senses in which they use the term) to be necessary for Mexican American improvement (1967: 125ff).

EDWARD BANFIELD: *BIG CITY POLITICS*

Edward Banfield (1965) briefly covers two cities with large Chicano populations. About Mexican Americans in Los Angeles he has almost nothing to say. His comments about El Paso's Chicanos are either superficial or incredible. He states "In El Paso (but nowhere else in Texas) there has never been 'racial' discrimination against them. The city has never had a 'Mexican' school; 'high-class Mexicans' belong to the best clubs and intermarry with the best Anglo families" (1965: 67). Or, again, "One reason for the Latin's political incapacity is poverty. Another is lack of education . . . but perhaps the Latins' most serious handicap is their persistent attachment to Mexican, rather than North American, cultural standards. Among other things, this leads them to be satisfied with things as they are" (1965: 76).

In a more recent book Banfield (1970: 68) states the latest urban-migrant thesis in its baldest form: "Today the Negro's *main* disadvantage is the same as the Puerto Rican's and Mexican's: namely, that he is the most recent unskilled, and hence relatively low-income, migrant to reach the city from a backward rural area." Apparently Banfield (1965: 67) has forgotten his own words: "El Pasoans think that *most* of the South Side [Mexican American] residents are recent arrivals, but census figures show that this is far from being the case." Indeed, in such Southwest cities as El Paso, San Antonio, Albuquerque, Los Angeles, and San Francisco the Chicano presence is of very long standing. In many such cities Chicanos can trace the history of their barrio to the founding of the city, and it is the Anglos who are the newcomers.

A METHODOLOGICAL NOTE

In making their causal attributions, the authors reviewed here are on uniformly weak grounds methodologically. Most of the works are case studies of one city at one point in time. Banfield (1965) covers more than one city, but he does not proceed in a truly comparative manner, and he discusses Mexican Americans in only one case. D'Antonio and Form (1965) based their book on a comparative study, but their comparative focus is on matters other than Chicano political organization. To our knowledge, then, there are no comparative works on urban Chicano politics. Yet most of the authors cited above do not hesitate to make sweeping generalizations about Chicano politics or to confidently assign causes to the patterns they feel they have detected.

There is also a tendency in these works to make an implicit comparison with Anglo political patterns, but the comparison is with an idealized "textbook" image of Anglo politics as rational, pragmatic, instrumental, and universalistic. A systematic empirical study of Anglo politics in the cities studied may have revealed just as much "factionalism" and "personalism," to cite only two possibilities.

Another possible approach, given the emphasis on culture, would have been to compare Chicano politics with politics in Mexico, where similar cultural attributes presumably exist. Even a superficial comparative study would probably have been enough to eliminate the more simplistic cultural explanations.

In general, we believe it can be said that the observations cited by the authors do not constitute adequate evidence for their conclusions. It is relatively easy to cite observations from these same works that can be used to support very different hypotheses, as we in fact do below (part V). This being the case, it seems likely that their conclusions are in fact drawn from some other source. In the next part we argue that this source is a type of latent model drawn from the prevailing American myths and stereotypes about Mexican Americans.

III. THE ASSIMILATION/ACCOMMODATION MODEL

There are several themes and perspectives which recur in the studies which have just been summarized. In this part we shall indicate some of the most important ones and establish links among them in terms of what we will call the assimilation/accommodation model.

LACK OF REPRESENTATION

One observation on which all the authors seem agreed is that urban Chicanos are unrepresented or underrepresented politically. This nonrepresentation is expressed throughout the range of local offices, from mayor to councilman to grand jury to relatively minor administrative positions, and extends to party offices. This lack of representation is related to the Chicano's disadvantaged socioeconomic position, and often condemns him to the role of victim of such public policies as urban renewal.

WEAK POLITICAL LEADERSHIP AND ORGANIZATIONS: LOW PARTICIPATION

Another common theme in these studies is a perceived lack of strong political leadership, weak or nonexistent political organizations, and generally low rates of participation. Several of the authors refer to factionalism among Chicanos and to a generalized feeling of suspicion toward the leadership. In most cases, the authors argue that it is these various elements which explain the lack of representation. In turn, these elements are explained in terms of two sets of factors, cultural and social-economic.

Cultural factors are given considerable weight by Woods, who refers to the Chicano's "individualism" as a barrier to collective action. This theme is frequently echoed in other studies of Mexican Americans. Sheldon (1966: 127) states that "Individualism is a major characteristic of Mexican culture . . . [whereas] Anglo-Americans have accepted the British tradition of working together to achieve a common goal and rallying round a common cause." As a consequence, "It is not surprising that Mexican-Americans have been unable to put to effective use the tool of the mass voice to promote the common good of their group."[3] Rubel, as we have seen, puts a great deal of weight on the Chicano's "personalism," which he sees as the antithesis of organized, instrumental group action. Banfield believes that Mexican Americans are held back by their fatalism.

Social-economic factors such as low levels of education and income are referred to by such authors as Banfield and D'Antonio and Form to explain the Chicano's political situation. Sheldon (1966: 125) feels that the Chicano's heterogeneity is an important causal element. D'Antonio and Form also perceive a "low level of social integration."

While the existence of past and present discrimination is acknowledged by the authors (e.g., Rubel), and some even feel that this may have something to do with the Chicano's present position, only Ruth Tuck makes a point of stressing Anglo machinations as a major cause of the Mexican American's political nonrepresentation.

THE CHICANO AS THE LATEST URBAN MIGRANT

In striving to conceptualize the situation and prospects of the Mexican American community, social scientists have usually resorted to an analogy with European migrants to eastern U.S. cities. The idea here seems to be that just as various European immigrant groups have come from foreign countries to American cities, Chicanos have recently come to the cities either directly from Mexico, or by a two-step process involving initial settlement in a U.S. rural area. Thus

While the colonial status of a people is a generalized status, affecting all aspects of their existence, we will confine our discussion largely to the political dimension. In political terms, the situation of internal colonialism is manifested as a lack of control over the institutions of the barrio, and as a lack of influence over those broader political institutions that affect the barrio. In essence, then, being an internal colony means existing in a condition of *powerlessness.*[7] One result is that public and private institutions, in their dealings with Chicanos, are able to function in exploitive and oppressive ways, whether by ignorance or intent.

Before going on to describe the model in a more formal way, it is possible to indicate some of the implications of adopting this point of view. One is the rejection of the latest-urban-migrant theme. Rather than looking at Chicanos as a recently arrived group in the process of rapid assimilation into Anglo society, it must be recognized that in many cases Chicanos have existed in urban areas for a long period of time and that their disadvantaged position is actively maintained today through the workings of a set of mechanisms of domination. It is also true that many Mexican Americans have actively resisted assimilation in the interest of maintaining their historic culture and social organization.

Second, disadvantage is explainable not in terms of factors inherent in Chicano culture and social organization, but as a function of external causes.

Third, the analysis of Chicano politics in terms of weak leadership and organization is misleading. This line of reasoning is deficient in several respects: (1) It should be clear from looking at Mexico that there is nothing in the Mexican heritage that precludes effective leadership and strong political organizations from developing; (2) as Tirado (1970) has documented, there have in fact been many Mexican American community organizations that have played a political role, although often not considered political groups by outside observers because they also perform nonpolitical functions. In addition, many such organizations have been forced to adopt a low profile and conceal their political functions because of the threat of reprisal from the dominant society; and (3) the existence of overtly political organizations with a well-defined leadership varies considerably from locale to locale, as the studies reviewed earlier should make clear. One significant factor appears to be the degree of urbanization. An ongoing historical study (Cortés et al., forthcoming) shows considerable organizational variation among four

Southern California cities. Much of the variation is not easily explainable and appears to be the product of complex factors and the specific historical development of the city. Unfortunately, such comparative studies are virtually nonexistent.

Finally, this perspective points to community control rather than individual mobility or the politics of accommodation as a solution to the colonized status of Chicanos.

V. THE INTERNAL COLONIAL MODEL

The concept of internal colonialism has been used by several American writers to refer to the ghetto or the barrio. Generally the term has been employed as an analogy with classic colonialism (Carmichael and Hamilton, 1967; Moore, 1970; Allen, 1970). The approach taken in this paper will be to define internal colonialism as a variety of the more general concept, colonialism, of which classic or "external" colonialism is another variety. We will also differentiate subcategories of internal colonialism.

DEFINING CHARACTERISTICS

Colonialism is a complex phenomenon with many variations. No one disputes that colonialism in its modern usage refers to a relationship in which one group of people dominates and exploits another. Without going into great detail, we can adopt González-Casanova's statement (1969: 128) of colonialism as "a structure in which the relations of domination and exploitation are relations between heterogeneous and culturally different groups." The factor of race and culture is important, as González-Casanova points out, because it differentiates this exploitive relationship from one based on class, or on some other basis.

> The colonial structure and internal colonialism are distinguished from the class structure since colonialism is not only a relation of exploitation of the workers by the owners of raw materials or of production and their collaborators, but also a relation of domination and exploitation of a total population (with its distinct classes,

> proprietors, workers) by another population which also has distinct classes (proprietors and workers) [González-Casanova, 1969: 131-132].

The crucial distinguishing characteristic between internal and external colonialism does not appear to be so much the existence of separate territories corresponding to metropolis and colony, but the legal status of the colonized. According to our usage, a colony can be considered "internal" if the colonized population has the same formal legal status as any other group of citizens, and "external" if it is placed in a separate legal category. A group is thus internal if it is fully included in the legal-political system, and external if it is even partly excluded from equal participation in a formal sense. It may be that the term de facto colony better expresses this distinction, but the term internal is already in common use. At any rate, this definition would classify such groups as the native people of the Union of South Africa as an external colony, even though the dominant population does not have its center in an overseas metropolis. On the other hand, the Black and Chicano communities in the United States are internal colonies, since they occupy a status of formal equality, whatever the informal reality may be. The degree of formal inequality of the external colony may vary considerably, of course, so that some would resemble internal colonies more than would others.

The other important distinction to be made is that between interethnic relations that represent internal colonialism and those that are internal but noncolonial. It is precisely here that we differ from earlier writers, in that we consider Chicanos an internal colony, but not European immigrants. Robert Blauner (1969: 396) has specified four basic components that serve to distinguish the internal colonization complex:

> (1) *Forced entry.* The colonized group enters the dominant society through a forced, involuntary, process.
>
> (2) *Cultural impact.* The colonizing power carries out a policy which constrains, transforms, or destroys indigenous values, orientations, and ways of life.
>
> (3) *External administration.* Colonization involves a relationship by which members of the colonized group tend to be administered by representatives of the dominant power. There is an experience of being managed and manipulated by outsiders in terms of ethnic status.

(4) *Racism.* Racism is a principle of social domination by which a group seen as inferior or different in terms of alleged biological characteristics is exploited, controlled, and oppressed socially and psychically by a superordinate group.

While Blauner's categories are useful in making the distinction we wish to make, it is important to realize that the experience of each ethnic group in the United States has been unique in some respects. Some of his categories need to be qualified in order to fit the specific case of Chicanos.

In the first case—that of forced entry—the basic Chicano-Anglo relationship was formed in a context of conquest and subsequent take-over of the land. While it is true that a large number of Chicanos have come to the United States since that time, and may be said to have come "voluntarily," once in the country they found themselves in a situation that had been structured through violence. So while each individual may not have found himself involuntarily included in the system, the group as a whole did, and the structures and attitudes formed in the earlier period have continued in one form or another into the present.

In a more recent essay, Blauner (1972) has stressed the role of unfree labor as a factor differentiating the experience of Chicanos and other Third World peoples within the United States from the experience of European immigrants. This factor is closely related to the nature of entry into the society, and deserves more extensive treatment than we are able to give here.

Blauner's second point is also applicable, with qualification, to the Chicano experience. While it is not true that Mexican-derived culture and social organization were destroyed to the same extent as that of Blacks brought to this country in slavery, the dominant society has largely destroyed Chicano economic organization, severely limited political organization, and waged a constant attack on Chicano values and other cultural traits through the schools, the media, and other institutions.

Blauner's remarks about external administration are perfectly applicable to Chicanos, and constitute a central aspect of the internal colonial situation.

Racism, as Blauner uses it, is clearly intended to include what has come to be called institutional as well as individual racism, and holds for Chicanos to a certain extent. However, it appears that in the case of Mexican Americans, cultural factors are at least as important as biological ones as a basis of discrimination, and probably more so, particularly in the urban areas. Racism should thus be considered a mixed biological/cultural category.

Combining and extending these various comments, then, we can specify what the status of internal colony means for Chicanos at two different levels. At the institutional or interpersonal level, internal colonialism means that Chicanos as a cultural/racial group exist in an exploited condition which is maintained by a number of mechanisms (to be spelled out below). This relationship is most clearly experienced as a lack of control over those institutions which affect their lives. These institutions are as a rule administered by outsiders, or at best, those who serve the outsiders' interests. This condition of powerlessness is manifested specifically in outside ownership of barrio business (Sturdivant, 1969), in Anglo domination of barrio schools, and in Chicano underrepresentation in every type of public institution. One result of this situation is that the Chicano community finds itself in a general condition of disadvantage: low incomes, poor housing, inadequate health care, low educational level, and so on. It also results in the community finding its culture and social organization under constant attack from a racist society.

At the individual level, the colonized individual finds that because he identifies himself with a particular culture, he is confronted with barriers that prevent him from achieving the economic, social, and political positions which would otherwise be accessible to him (for the economic aspect, see Schmidt, 1970). At the same time, he finds himself under psychological assault from those who are convinced of his inferiority and unworthiness.

For a Chicano, or at least for one sufficiently Caucasian in appearance, it would appear possible to escape his colonial status by completely taking on the culture of the Anglo majority, and renouncing his language, values, behavioral patterns, and self-identification. This would especially seem to be the case in the larger urban areas. While he would not achieve an equal status overnight, he would no longer be confronted with the same barriers. However, this would not produce a noncolonized Chicano, but a noncolonized non-Chicano. Thus the apparent semi-permeability of the colonial barrier for Chicanos is illusory, since there is no escape from the colonial status for an individual *as a Chicano.* If the Chicano community were to take this approach, the result would be cultural genocide. The choice presented to the Chicano community by Anglo society, then, is very clear-cut: colonialism or genocide.

THE POLITICAL DIMENSIONS OF INTERNAL COLONIALISM

Internal colonialism is manifested along many different dimensions: social, economic, political, psychological, cultural, and so on. Within each broad category we can identify further subdivisions. Among the political dimensions are the following.

Political Representation: Parties

While there appear to be no studies of Chicano representation within the major political parties, it is clear to those familiar with the political systems of the southwestern states that Chicanos have not been accorded many significant positions. Since Chicanos have been faithful to the Democratic party in their voting, one would expect to see many Chicanos in high Democratic party posts. Such is not the case. In California, for example, one rarely sees a Chicano name above the urban ward level.[8]

Political Representation: Governmental Bodies

Chicanos find themselves underrepresented at every level of government, from the national and state legislative bodies to county boards of supervisors, to city councils, boards of education, and grand juries (Grebler et al., 1970: 560ff, 222ff; U.S. Commission on Civil Rights, 1970, 1971). A typical situation was pointed up in a special report of the 1970 San Bernardino Grand Jury, based on a state Fair Employment Practices investigation. They refer to "gross inequities" in the hiring of Chicanos and other minorities in the various county administrative offices, and document it office by office. The investigators found that "The vast majority of those interviewed voiced no concern or knowledge of affirmative action, no need to make a concerted effort to increase the utilization of minorities, and no effort to disseminate this policy to employees" (California FEPC, 1970: 59). They also found on the part of numerous department heads attitudes that were either manifestly racist or very good substitutes (1970: 11).

The problem of nonrepresentation is even more acute than statistics show, since in many cases the representatives of the barrio are hand-picked by Anglos on the basis of acceptability to Anglos. It is also true that some Chicanos have collaborated with Anglos in every historical period. These Chicanos were often given some sort of official position, and thus are often cited to prove that Chicanos have "participated" in the system all along. But a close look will usually reveal that these representatives of the Chicano were in fact powerless subordinates of the dominant power structure. As Swadesh (1968: 165) has put it for New Mexico: "Hand-picked Hispanos served in the Legislature as junior partners to those who really held the reins. Their official task was to represent the overwhelming majority of the population, but in practice they helped keep this majority under control." Thus Chicano *interests* are even less represented than Chicano bodies. Again, this points to the centrality of the cultural factor. For a Chicano to self-consciously identify himself with the Chicano community, and with its culture and interests, is to virtually ensure that he will not be among even the token representatives chosen by outsiders.

Contact with Public Agencies

Chicano contact with public agencies is universally described in terms that can only be termed colonial. Easily the best-documented situation is that of the schools' treatment of Mexican Americans (for two recent discussions, see Carter, 1970; Castañeda et al., 1971). Among the many charges that have been listed and documented against the schools are suppression of the Spanish language, channeling Chicano children into vocational training and discouraging them from college-oriented courses, abuse of Anglo-biased testing instruments, underfunding of barrio schools, neglect of Chicano history and other academic areas, the instilling of a sense of cultural inferiority, and, in general, creating a climate that produces progressively worse performance with each passing year and results in a phenomenally high dropout rate.

While more scholarly attention has been given to education than any other area, the public agency does not exist that has not been subjected to harsh criticism by barrio representatives. The police have frequently been accused of brutality, harassment of political organization, and overzealous policing; at times they have been compared to an occupying army (U.S. Commission on Civil Rights, 1970; Morales, 1970). The courts have long been charged with discriminatory treatment (U.S. Commission on Civil Rights, 1970; Lemert and Rosberg, 1948).

The various welfare, housing, health, and antipoverty agencies operating in the cities are characterized as unable to communicate with their clients, as insensitive to the concerns of Chicanos, and as patronizing and paternalistic in their attitudes and practices (Grebler et al., 1970; Samora and Lamanna, 1967; Ornelas and Gonzálex, 1971).

While the list could go on, the view from the barrio, which is to say the view of those with direct experience with these agencies, is generally clear. The kind of treatment meted out to Chicanos, of course, is a function of the overall lack of representation of Chicanos in the political system. The situation is compounded in the American political system by the way in which that system parcels out public responsibilities to private interest groups, and allows such groups to strongly influence or control public policy-making (for a detailed description and critique of this system, see Lowi, 1969). This is so because Chicanos have also been largely excluded from such groups as private real estate boards, which are incorporated officially or nonofficially into the policy-making activities of urban planning commissions, urban renewal agencies, and so on. Unfortunately, Chicanos have only begun to document the often subtle workings of the policy process as it affects Chicanos. For the time being, however, our concern is with the crude fact of external administration and its detrimental consequences for the barrio.

THE MECHANISMS OF POLITICAL DOMINATION

Central to an understanding of internal colonialism is a grasp of the mechanisms by which the colonial situation is maintained over time. While no definitive account is possible at this time, we can begin to list them and to advance some ideas as to their relative importance under various circumstances.

The most direct and obvious mechanisms are those involving force and outright repression. Instances range from the use of the Texas Rangers to repress Mexican American organizational efforts to the widespread Ku Klux Klan anti-Chicano terrorism of the 1920s, often with the complicity of the forces of law and order. Also included in this category would be the various forms of nonviolent reprisals that are taken against Chicanos who dare to break with established political patterns (for a recent case, see Hill, 1963; Grebler et al., 1970: 563).

A second set of mechanisms has the effect of disenfranchising Chicanos. Devices include poll taxes and literacy tests.

Outright exclusion of Chicanos from political parties and governmental bodies has already been discussed. The practice is often justified on the grounds that there are no "qualified" Chicanos available to fill the positions.

Gerrymandering is another mechanism that has found favor with those seeking to minimize Chicano political influence. Generally this consists of splitting Chicano voters into many districts, so that they do not form a sufficiently large group in any one district to elect their own representatives. The case of East Los Angeles is a prominent example (Grebler et al., 1970: 562).

A fifth mechanism involves changing the rules of the game when it becomes apparent that existing rules might lead to greater Chicano political strength. In El Paso, for instance, the political elite successfully carried out a campaign to make city elections nonpartisan after a Mexican American had won the Democratic primary election for mayor.

The mechanism of "divide and conquer" has been attested to by Pauline Kibbe (1964: 227): "When an ambitious and capable Latin American announces for office in opposition to an Anglo incumbent or candidate . . . Anglo politicians follow the tried-and-true formula of 'divide and conquer.' They immediately sponsor the candidacy of another Latin American, preferably a personal enemy of the man who has previously announced, and thereby split the Latin American vote and assure the election of the Anglo candidate." Ruth Tuck has already been cited in connection with Anglo attempts to create divisions among Chicanos.

When outright exclusion of Chicanos is not possible, the resort is likely to be tokenism. This involves minimal representation of Chicanos, with the representatives likely to be carefully selected so as to play the game. Tokenism includes not only "representative tokenism" but also "policy tokenism," in which minor but highly publicized policy concessions are made (Muñoz, 1971: 98-99).

The mechanism of cooptation has already been mentioned several times. By offering limited material and status benefits to those Mexican American individuals who are deemed acceptable on cultural grounds and who are willing to act as information sources and token showpieces, the Anglo elite in effect co-opts them into the structure of domination. Watson and Samora (1954) have carefully described the workings of this mechanism in one Colorado community, and have pointed out its detrimental consequences for the development of more effective Chicano political action.

Of all the mechanisms of domination, however, the racist mobilization of bias may be the most pervasive and the most subtle in its effects. Carlos Cortés (1971) has documented the way in which symbols are manipulated in the media and in the schools to perpetuate the myths of biological and cultural inferiority. In an earlier section of this paper we illustrated the defensiveness this produces in Chicanos, and the way in which this defensiveness prevents Chicanos from acting in the interests of their own community. Any Chicano in public office who self-consciously serves the needs of the Chicano community leaves himself open to charges of "reverse racism" and divisive parochialism. That such charges come from Anglos who have systematically excluded Chicanos for decades and who continue to serve only non-Chicano interests does not render their charges ineffective. To the extent that Chicanos have been manipulated into internalizing the prevailing biases, they are driven into a defensive posture and into seeking Anglo approval for their actions (see, e.g., the statements by Chicano politicians in a California city in Faragher, forthcoming; see also the comments cited by D'Antonio and Form in part II above). This pattern of behavior is undoubtedly reinforced by the Chicano experience of having to function in institutions, such as the schools, dominated by non-Chicano authority figures from whom it is necessary to gain approval (Cortés, personal communication). Thus in this instance, as in many others, the various mechanisms of domination reinforce each other and multiply their effects.

NEOCOLONIALISM

The various mechanisms of domination that have been discussed vary considerably in their degree of subtlety or overtness. The mechanisms of force and repression are unmistakable and relatively uncomplicated; many of the others are more subtle, and thus more difficult to combat. There is a danger in focusing on the cruder measures and underrating the less direct, but often more effective, mechanisms. We are inclined to disagree with Blauner (1969: 404) when he states that "the police are the most crucial institution maintaining the colonized status of Black Americans." In most cases the police do not have to be used, because more sophisticated mechanisms have removed the threat, or, preferably, prevented a threat from arising.

What needs to be recognized is that the American political system has become increasingly sophisticated in its methods of exercising control, and that power holders prefer to use indirect methods wherever possible. In the case of the Chicano, the last one hundred twenty years have seen a change from a relatively direct system of exploitation aimed at depriving him of his land and establishing him in a subordinate status, to a more subtle system functioning to maintain him in that position. At this point it may be well to characterize this change by referring to the earlier situation as "classic" internal colonialism, and applying the term internal neocolonialism to the present-day situation. The basic distinction here is in the nature of the mechanisms of domination that are typical in each case. We do not mean to imply an either/or choice, of course. Thus force and the threat of force continue to play an important role in internal neocolonialism; but for the most part these direct measures tend to be held in reserve today, with the more indirect mechanisms being relied on for the day-to-day maintenance of the system.

Robert Allen (1970) has begun to sketch the outlines of such a model, which he calls domestic neocolonialism, for the case of Blacks in America. While he describes this as a shift toward "an indirect and subtle form of domination," he puts emphasis on the development of a Black bourgeoisie which would act on behalf of the Anglo elite and the status quo, in an American form of indirect rule. While this type of cooptation is undoubtedly important in the workings of a neocolonial form of domination, it should be kept in mind that other mechanisms also play an important role. One such mechanism which we emphasize is the mobilization of bias. It should also be noted that indirect rule is by no means restricted to neocolonial systems, but can be a feature of both classic "external" colonialism and classic internal colonialism, although playing a relatively less important role here.

If it is true that there has been a shift toward neocolonialism during the last century, it also appears to be the case that the process has been carried further in the more urbanized areas. It may be that the greater concentration of population and the greater social

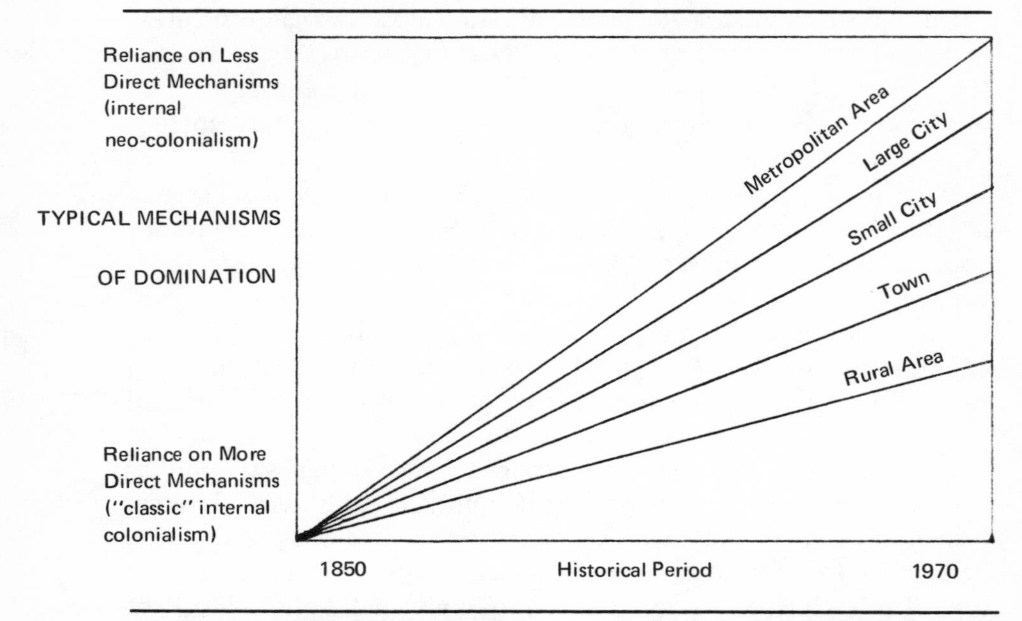

Figure 1: Relationship Between Time, the Urban-Rural Dimension, and the Mechanisms of Domination. The Tendency has been to Move in the Direction of Neo-colonialism, With the More Urban Areas Showing the Greater Shift.

complexity of the city make indirect methods of domination more efficient. We could also hypothesize that it is in the more highly industrialized sectors of society that more sophisticated mechanisms of social control are developed, as in work relations, and that these carry over into the internal colonial relationship.

In addition, the shift over time can be explained in terms of the degree of stability of the relations of domination. In the early stages, a greater amount of force is needed to establish the initial relationship; once the situation is relatively stabilized, more indirect mechanisms are usually sufficient to preserve it, with force playing a secondary role. Of course, if the colonized population again begins to mount a strong challenge, we can expect the system to increasingly revert to the more openly repressive tactics typical of classic internal colonialism.

ASSIMILATION AND CULTURAL DEFENSE

One theme that has been implicit up to this point is the authors' rejection of assimilation as a desirable goal for the Chicano community. The great bulk of the Chicano people have demonstrated through their tenacious cultural defense that they have little enthusiasm for assimilation, however much they may want to participate in the American prosperity that was built on their land and labor, and that of other Third World peoples. Our rejection of cultural assimilation is based on several grounds, among which are a belief in the value of the Chicano culture and a desire to see it develop according to its own internal logic; a distaste for cultural homogenization and a regard for human diversity; and a feeling that assimilation for most Chicanos would involve the trading of a genuine human culture for a bland, dehumanized, consumer-oriented, made-in-America mass culture. From this perspective, the assimilationist approach in America, intellectual or otherwise, is not only an expression of cultural imperialism but in effect an instrument of dehumanization.

VI. THE CONTEMPORARY SITUATION AND PROSPECTS FOR DECOLONIZATION

An assessment of the contemporary scene must begin with an awareness that Chicano activists today are conscious of participating in what is generally referred to as "The Movement." While difficult to characterize in an unambiguous manner, the movement is marked by a self-conscious ethnic pride, a high rate of participation by youth, a developing Chicano nationalist ideology, and an attempt to achieve at least some degree of coordination among the various groups and regions involved. The idea of a Chicano movement is a product of the 1960s, and has continued to gain strength in the seventies. Preexisting political groups have become increasingly conscious of the movement, and some have identified themselves with it in various degrees.

Nevertheless, it is true that coordination is still highly imperfect, and the various political groups differ considerably in the problems which they identify, the manner in which they analyze them, and the tactics which they employ.

One broad category of approaches is fairly traditional, and is primarily aimed at public institutions. Falling within this category are pressure group tactics directed at administrative agencies and seeking to achieve greater attention to Chicano needs. Civil rights efforts fall under this broad heading. Generalized political groups such as MAPA (Mexican American Political Association) and PASSO (Political Association of Spanish Speaking Organizations) as well as more specialized groups such as AMAE (Association of Mexican American Educators) frequently employ such tactics. Actions aimed at the courts are initiated by such groups as the Mexican American Legal Defense and Educational Fund. Also within this general category should be included electoral strategies directed at increasing Chicano influence over elective offices. Traditionally Chicanos have been affiliated with the Democratic party. More recently there has been a tendency toward increased independence from the two major parties. Still, to the extent that this independence is aimed only at producing a kind of "swing vote" leverage for Chicanos, it should be included among this category of approaches.

There is another set of tactics that is directed more at the Chicano community itself, at least in the initial stages. This is generally more "activist" in nature. One such tactic might be termed "organizational," since it is aimed at building multipurpose organizations with a broad base. Such is the strategy currently being pursued by La Raza Unida in the California Inland Empire area (east of Los Angeles). There is also an attempt being launched throughout the Southwest to build an independent Chicano third party, an aspect of the movement which is rapidly gaining importance. This approach received its greatest impetus from the electoral victories of La Raza Unida Party in South Texas, and has since spread to Colorado and California. "Chicano Power" is its rallying cry. A third type of strategy in this general area has to do with the building of alternative institutions, such as schools and business enterprises. The Denver-based Crusade for Justice has made some interesting starts in this direction, as has the Los Angeles community group known as TELACU. An effort to develop a joint Chicano and Indian university has begun in Davis, California, and a Chicano college was recently started in South Texas. Finally, there are a number of groups directing their efforts at communication and increasing Chicano cultural and political consciousness. Among these are the various movement-oriented newspapers and numerous Teatro groups which have followed the original Teatro Campesino formerly associated with the farmworkers' union.

A final set of approaches involve some form of direct action, ranging from peaceful boycotts and picketing to more disruptive sit-ins and demonstrations to force and violence. Such Chicano organizations as the Brown Berets and the Chicano Moratorium Committee have engaged generally in militant but nonviolent forms of direct action. Riots that have resulted from some demonstrations are blamed on the police by Chicano participants. More recently, however, a group identifying itself as the Chicano Liberation Front in Los Angeles has claimed credit for a number of bombings aimed at Anglo institutions.

It should be kept in mind, of course, that organizations often employ more than one kind of tactic, so that it is difficult to make one to one correlations. The most widespread Chicano student organization, MECHA, has used a number of different tactics in its efforts, as have other groups. It is also possible to think of approaches that have not been used extensively by urban Chicanos. There is, for example, no urban counterpart to César Chávez's Farmworkers Union, and there are few Chicano urban organizations that have actively sought long-term coalitions with non-Chicano groups.

At this point it is too early to tell which specific approaches are most likely to produce change in the situation of the Chicano. A variety of types of action is desirable, in that different appeals must be made to different sectors of the Chicano population in order to engage them in political action. It is also true that many of these activities are in essence "probes," designed to test reality in order to find out what works and what does not. Eventually these various activities may contribute to the development of a cohesive ideology which will be useful in clarifying the Chicano situation, specifying goals, and indicating future directions for the movement.

Nevertheless, it is possible now to begin to address the question of what constitutes a solution for Chicanos. To be considered an effective solution, a proposed change must contribute to decolonization—that is, it must enable Chicanos to gain greater control over their environment while maintaining their collective identity. This means, among other things, increasing the range of alternatives open to Chicanos and developing Chicano control over those institutions which most directly affect their lives. Those approaches which seek to incorporate Chicanos into Anglo institutions without making fundamental changes in those institutions will not contribute to decolonization, although they may allow individual Chicanos to increase their social mobility. Thus increasing the number of Chicanos in a school system that functions in such a way as to undermine Chicano culture does not contribute to decolonization. The same thing can be said of incorporating Chicanos into economic institutions that increase Chicanos' social and geographical mobility at the expense of their ties to the barrio.

In part this problem can be attacked through the creation of alternative institutions designed for and controlled by Chicanos. However, there is no way of constructing self-contained Chicano educational, economic, or political systems, so that successful decolonization will depend on producing far-reaching changes in the institutions of the larger society. Since it is doubtful that Chicanos can mobilize the necessary political strength to produce such adjustments on their own, there will eventually have to be coalitions and alliances formed with other groups interested in change. It may well be that a true decolonization for Third World peoples within the United States will require a radical transformation of the structures of this society.

NOTES

1. In the article we have employed the terms "Chicano" and "Mexican American" interchangeably, to refer to the population of Mexican extraction residing in the United States, whether citizens or not, and regardless of place of birth. "Anglo" is used to refer to the entire non-Chicano population, excluding Blacks and other racial minorities. The term "barrio" simply refers to the urban neighborhoods in which Chicanos are concentrated.

2. Bachrach and Baratz (1970: 43) define the mobilization of bias as "a set of predominant values, beliefs, rituals and institutional procedures . . . that operate systematically and consistently to the benefit of certain persons and groups at the expense of others."

3. For general critiques of the tradition of attributing the Chicano's position to his culture, see Romano (1968), Vaca (1970), and Rocco (1970).

4. For a discussion of this concept and its limitations, see Litt (1970).

5. In this connection, see Easton (1965), and Schaefer and Rakoff (1970).

6. Ruth Tuck (1946) relates the man-in-the-street's version of benign neglect in the epilogue to her book, entitled "It Will All Work Out in Time."

7. Kenneth Clark has expressed this succinctly: "Ghettos are the consequence of the imposition of external power and the institutionalization of powerlessness. In this respect, they are in fact social, political, educational, and above all–economic colonies." (Clark, 1964).

8. Eduardo Pérez, Secretary of the California Democratic Party during the administration of Governor Edmund Brown, claims that he was appointed to his position only so that his name would appear on the party literature (Steiner, 1969: 191). Alex García, a California State Assemblyman, has testified that he was ignored by Democratic Party officials when he ran for office. After being elected, he was expected to endorse party positions and accept appointments without being consulted (written testimony to the Democratic Commission on Party Structure and Delegate Selection, June 21, 1969).

REFERENCES

ALLEN, R. (1970) Black Awakening in Capitalist America. Garden City, N.Y.: Doubleday Anchor.

BACKRACH, P. and M. BARATZ (1970) Power and Poverty. New York: Oxford Univ. Press.

BANFIELD, E. (1970) The Unheavenly City. Boston: Little, Brown.

--- (1965) Big City Politics. New York: Random House.

BLAUNER, R. (1972) "Colonized and immigrant minorities," in R. Blauner (ed.) On Racial Oppression in America. New York: Harper & Row.

--- (1969) "Internal colonialism and ghetto revolt." Social Problems (Spring): 393-408.

California Fair Employment Practices Commission, FEPC (1970) Special Report of the San Bernardino County Grand Jury, based on an investigation of the county of San Bernardino.

California Roster of Federal, State, County and City Officials (1970).

CARMICHAEL, S. and C. HAMILTON (1967) Black Power. New York: Vintage.

CARTER, T. (1970) Mexican Americans in School: A History of Educational Neglect. New York: College Entrance Examination Board.

CASTANEDA, A., M. RAMIREZ, C. CORTES and M. BARRERA [eds.] (1971) Mexican Americans and Educational Change. Riverside: University of California.

CLARK, K. (1964) Youth in the Ghetto. New York: Haryou Associates.

CORTES, C. et al. (forthcoming) The Bent Cross: A History of the Mexican American in the San Bernardino Valley.

--- (1971) "Revising the 'all-American soul course': a bicultural avenue to educational reform," pp. 314-340 in A. Castañeda et al. (eds.) Mexican Americans and Educational Change. Riverside: University of California.

D'ANTONIO, W. and W. FORM (1965) Influentials in Two Border Cities. Milwaukee: Univ. of Notre Dame Press.

EASTON, D. (1965) A Systems Analysis of Political Life. New York: John Wiley.

FARAGHER, J. (forthcoming) "Redlands," in C. Cortés et al. The Bent Cross: A History of the Mexican American in the San Bernardino Valley.

GONZALEZ-CASANOVA, P. (1969) "Internal colonialism and national development," pp. 118-139 in I. L. Horowitz et al. (eds.) Latin American Radicalism. New York: Vintage.

GREBLER, L., J. MOORE, and R. GUZMAN (1970) The Mexican-American People. New York: Free Press.

HILL, G. (1963) " 'Latinos' govern tense community." New York Times (September 21).

KIBBE, P. (1964) Guide to Mexican History. New York: International Publications Service.

LEMERT, E., and J. ROSBERG (1948) The Administration of Justice to Minority Groups in Los Angeles County. Berkeley: University of California Publications in Culture and Society, II, no. 1.

LITT, E. (1970) Ethnic Politics in America. Glenview, Ill.: Scott, Foresman.

LOWI, T. (1969) The End of Liberalism. New York: W. W. Norton.

McWILLIAMS, C. (1968) North From Mexico. New York: Greenwood Press.

MOORE, J. (1970) "Colonialism: the case of the Mexican Americans." Social Problems (Spring): 463-472.

MORALES, A. (1970) "Police deployment theories and the Mexican American community." El Grito (Fall): 52-64.

MUÑOZ, C. (1971) "The politics of educational change in East Los Angeles," pp. 83-104 in A. Castañeda et al. (eds.) Mexican Americans and Educational Change. Riverside: University of California.

ORNELAS, C. and M. GONZALEZ (1971) The Chicano and Health Services in Santa Barbara. University of California, Santa Barbara: Chicano Political Communications Project, Report no. 3.

ROCCO, R. (1970) "The Chicano in the social sciences: traditional concepts, myths, and images," Aztlán (Fall): 75-98.

ROMANO, O. (1968) "The anthropology and sociology of the Mexican American: the distortion of Mexican American history." El Grito (Fall): 13-26.

RUBEL, A. (1966) Across the Tracks: Mexican-Americans in a Texas City. Austin: Univ. of Texas Press.

SAMORA, J. and R. LAMANNA (1967) Mexican-Americans in a Midwest Metropolis: A Study of East Chicago. UCLA Mexican-American Project, Advance Report no. 8.

SCHAEFER, G. and S. RAKOFF (1970) "Politics, policy and political science: theoretical alternatives." Politics and Society (November): 51-78.

SCHMIDT, F. (1970) Spanish Surnamed American Employment in the Southwest. Study prepared for the Colorado Civil Rights Commission.

SHELDON, P. (1966) "Community participation and the emerging middle class," pp. 125-158 in J. Samora (ed.) La Raza: Forgotten Americans. Notre Dame: Univ. of Notre Dame Press.

STEINER, S. (1969) La Raza. New York: Harper & Row.

STURDIVANT, F. (1969) "Business and the Mexican-American community." California Management Rev. (Spring): 73-80.

SWADESH, F. (1968) "The Alianza movement: catalyst for social change in New Mexico," pp. 162-177 in Amer. Ethnological Society proceedings of the 1968 Annual Spring Meeting.

TIRADO, M. (1970) "Mexican American community political organization." Aztlán (Spring): 53-78.

TUCK, R. (1946) Not With The Fist: Mexican-Americans in a Southwest City. New York: Harcourt, Brace.

U.S. Commission on Civil Rights (1971) Participation of Mexican Americans in California Government. California Advisory Committee. (mimeo)

--- (1970) Mexican Americans and the Administration of Justice in the Southwest. Washington, D.C.: Government Printing Office.

VACA, N. (1970) "The Mexican-American in the social sciences, 1912-1970." El Grito (part I, Spring: 3-24; part II, Fall: 17-51).

WATSON, J. and J. SAMORA (1954) "Subordinate leadership in a bicultural community: an analysis." Amer. Soc. Rev. (August): 413-421.

WOODS, Sister F. J. (1949) Mexican Ethnic Leadership in San Antonio, Texas. Washington, D.C.: Catholic University of America.

18

The Environment: New Priorities and Old Politics

RONALD O. LOVERIDGE

□ AMID MANY "CRISES," the environment offers exceptional credentials to compete for the nation's attention and resources. At stake is the character, perhaps survival, of urban America; "But if we should falter now," writes Richard Wagner (1971; see also Commoner, 1971), "man's very humanity will become extinguished in a mere struggle for survival." Yet environmental problems pose difficult and troublesome political dilemmas, for the structure, rules, and faith of American politics frequently clash with policy requirements necessary for a tolerable environment.

The problems of the environment raise the question of political responsiveness. No current political theme is more commonplace than the call to examine the nation's priorities and to reallocate its resources to focus on human and environmental problems. Yet extraordinary doubt and cynicism exist when we turn to building livable cities and otherwise straightening out the disordered public scene. Many critics talk at length (Drucker, 1969; Goodwin, 1967; Harrington, 1965, Lowi, 1969) about the inadequate or uneven performance of the American polity. And on environmental matters, the political verdict certainly reads that the governmental response, while noteworthy, has been largely ineffectual.

To evaluate the American political response to the environmental crisis, this chapter will examine the character of environmental problems, their translation into political demands, and the governmental response to these demands. Then, the contradictions between present political practices and requisite solutions to environmental problems will be identified and discussed. Environmental problems represent what economists call externalities—that is, the private market inadequately or erroneously allocates the costs and benefits of environmental goods. As a result, proper allocation of environmental costs and benefits depends on governmental action. Unfortunately, the American political system appears miscast to resolve complex and interrelated environmental problems and to coordinate environmental objectives with the nation's many economic, social, and political interests.

WHAT IS AT STAKE?

For the foreseeable future, the environment will be a preeminent issue, firmly set on the political agenda. In its meteoric climb, the issue of the environment embraced a bewildering mosaic of problems, causes, and solutions. Nevertheless, two widely accepted conclusions (see Dansereau, 1970; Darling and Milton, 1966; Garnsey and Hibbs, 1967) have established its political credentials. Environmental problems are *more complex* and *more serious* than any other domestic policy problems—*complex* because of the fragile and intricately related natural environment and *serious* because they threaten the survival of man.

In addressing the environmental "crisis" in the United States, three definitional emphases can be identified. The first centers on the environment man inhabits and tends to catalogue all aesthetic, social, and physical ills of our society as environmental problems (Gordon, 1963). A second focuses on man's survival and questions his very future unless the problems of pollution and overpopulation are resolved (Ehrlich and Ehrlich, 1970). Usually the earth's environmental support systems are the objects of concern. And a third emphasis stresses the physical environment and deals mainly with pollution, land use, and natural resources. This chapter primarily adopts the third emphasis.

Some critics charge that the environmental crisis is a passing fad, a subjective creation of the media's incessant search for crisis and conflict. Quite the contrary, the problems that comprise the environmental crisis can be identified and measured and their consequences for man are all too real. "It is a state of affairs," writes Harold Sprout (1971: 41-42), "that obtrudes pervasively, inescapably, and increasingly unpleasantly and injuriously into the lives of everyone, old and young, rich and poor—none can escape." In an excellent review, the Council on Environmental Quality (1970: 8-11) offers these introductory descriptions of the major physical problems of the environment:

> Pollution: It is a highly visible, sometimes dangerous sign of environmental deterioration. Pollution occurs when materials accumulate where they are not wanted.... Pollution threatens natural systems, human health, and esthetic sensibilities. Historically, man has assumed that the land, water, and air around him would absorb his waste products. It is clear now that man may be exceeding nature's capacity to assimilate his wastes.
>
> Land Use: Urban land misuse is one of today's most severe environmental problems. Old buildings and neighborhoods are razed and replaced by structures designed with little or no eye for their fitness to the community's needs. A jumble of suburban developments sprawl over the landscape. Unlimited access to wilderness areas may transform such areas into simply another extension of our urban, industrialized civilization.
>
> Natural Resources: Natural resource depletion is a particular environmental concern to a highly technological society which depends upon resources for energy, building materials, and recreation. And the methods of exploiting resources often create problems that are greater than the value of the resources themselves.

Pollution, land use, and natural resources represent serious, perhaps critical, problems. Barry Commoner, Paul Ehrlich, and Kenneth Watt are not alone in crying wolf, for most environmental scientists and many urban writers conclude that unless we begin to control such problems effectively, environmental degradation will result and extinction may follow. For example, Lewis Mumford (1970: 225) observes, "As cultivated areas shrink, as millions of square miles of good soils are turned into sterile expressways, concrete clover leaves, parking lots, and airports, as the air, the water, and the soil become polluted with chemical poisons, nuclear wastes, and inorganic debris, as essential bird and insect species die off, the prospects for human survival on any terms diminish." No matter if we reject apocalyptic views, environmental stakes represent the primary crucible for the future character of urban life (Caldwell, 1967; Ewald, 1968; Herber, 1965).

Environmental problems result from the interplay of basic yet varied causes, many of which can be found in the mainsprings of American society. For example, one cause is the pricing system which fails to take into account the environmental costs and damages of pollution. A second is the belief and commitment to the values of consumer choice, economic development, and personal freedom. Third, population growth places increasing stress on major metropolitan areas, contributing to the pollution of air and water and the misuse of land. Fourth, technology is continually reshaping man's control and impact on his environment. (One product, the automobile, contributes 60% of the air pollution, accounts for over 50,000 annual deaths, consumes yearly one-fifth of our economic resources, requires the work of approximately 15 million Americans for its support, and determines the shape of the American settlement patterns and the character of its recreational activities [Schneider, 1971].) And fifth, environmental policies are largely decided by the ad hoc pressures of organized interests and competing preferences rather than any guiding values of the relationship between man and nature.

The causes of environmental problems thus require solutions that are fundamentally different from other policy issues. The Council on Environmental Quality (1970: 231-241) offers seven present and future environmental needs: a conceptual framework, stronger institutions, financial reforms, pollution control curbs, monitoring and research, a system of priorities, and comprehensive policies. Representative of the solutions most often advocated, the list calls for awesome changes—a new set of values, new public agencies, massive infusion of funds, extraordinary controls on business and consumer activities, extensive applied research, a new yardstick for policy priorities, and comprehensive environmental policies. In effect, the call is for a new conceptual and to some extent, new political framework for the policy process; as Lynton Caldwell (1970: 229) explains, "Environmental administration has become necessary because, without it, contemporary man cannot prevent the destruction of his life-support base. But to protect and maintain this base (the planetary biosphere), public policies and powers will be required that would have been unthinkable in the simpler past."

POLICY DEMANDS–ISSUE OF THE ENVIRONMENT

The issue of the environment now occupies a redoubtable place on the political agenda of the United States, with widespread demands increasing in clarity, intensity, and strength for governmental action on environmental problems. The problems of the environment are publicly perceived and defined as an issue area for which government bears a major responsibility. A variety of environmental groups–ad hoc, single issue, and conservation–are growing in membership and beginning to exert sophisticated and sustained political pressure; many other groups–good government, church, professional, and civic–are with increasing frequency championing environmental causes. Even assorted administrators and politicians cleave to and represent the environment with a fervor formerly reserved for the more traditional shibboleths of politics.

As a major political issue, the environment rose to the forefront in the late 1960s. Before 1965, little public attention was devoted to general environmental questions. Of course, problems of air pollution, water pollution, solid waste, urban sprawl were topics of scientific and, to some extent, administrative concern. Yet environmental problems troubled few people beyond small numbers of conservationists, health officers, and academic investigators; and almost all environmental reformers (Lewis, 1965) saw their strategic mission as public evangelism. Cecile Trop and Leslie Roos (1971: 54) aptly note the formerly low political profile of the environment, "An examination of the 1964 political party platforms reveals that only the Democratic Party gave the environment brief attention, emphasizing conservation of the nation's natural resources and the expansion of recreational facilities. The Republican platform did not mention the environment." Demands for attention to environmental problems began in the middle 1960s and gained striking support from the public, various groups, media, scientists, administrative officials, and politicians. By the end of the decade, the environment entered the center ring in the political arena (Caldwell, 1970: 2):

> Public responsibility for the state of the human environment is becoming implicit in popular movements for environmental quality and in the actions of governments and of some industries. But until the late nineteen sixties, there had been no clear or explicit

> formulation of a public responsibility for the state of the environment. Popular perceptions of the environment have been fragmented and particularized. Fragmented public policies have followed, and deterioration of the environment has been a consequence. If the quality of human life is to be maintained at present levels or is to be improved, public responsibility for the quality of the total environment must become explicit and be made effective. The major task of environmental policy making in the nineteen sixties was to make explicit the general nature and extent of public responsibility. By late 1969, this task was on the way to accomplishment in the United States.

We freely talk about the demands for government action on environmental problems without, however, much explanation or documentation. Charles Jones (1970: 35-42) suggests five stages in a problem becoming a demand: perception, definition, aggregation, organization, and representation. These conditions have largely been met by the issue-area of environment, for environmental problems have been recognized and stated as political issues.

THE PUBLIC

The environment has become a major public issue: abundant evidence points to strong public concern for varied environmental problems and to a demand for their resolution by government. After a review of available public opinion surveys and studies, James McEvoy (1970: 26) concludes, "First of all, Americans are concerned about the quality of their environment. This fact is underlined both by the rapid growth of environment and conservation organizations and by the high level of concern, accompanied by a willingness to increase tax loads, expressed in the survey data. . . . Americans want their environmental problems solved, and they support, in general, technological, legal, and financial solutions to these problems." More than specific surveys and studies, we can all distinguish the signs of a national interest in environmental matters. A range of "unobtrusive measures" (see Webb et al., 1966) can be selected at will: for example, environmental ads by business and industry, ecology curriculum in elementary schools through colleges and universities, purchase of environmental products such as bottled water and special detergents, number of environmental columns by syndicated pundits, jokes by comedians, and action topics of local groups. Interest in the environment has assumed the proportions of a national cause—as expressed by President Nixon in his 1970 State of the Union Message:

> The great question of the seventies is, shall we surrender to our surroundings, or shall we make our peace with nature and begin to make reparations for the damage we have done to our air, to our land, and to our water? Restoring nature to its natural state is a cause beyond party and beyond factions. It has become a common cause of all the people of this country.

GENERAL PUBLIC

The public at large recognizes and defines the environment as a problem area. In the fall of 1969, George Gallup polled a cross-section of the American public and asked, among others, the following question:

> You may have heard or read claims that our natural surroundings are being spoiled by air pollution, water pollution, soil erosion, destruction of wildlife and so forth. How concerned are you about this: deeply concerned, somewhat concerned, or not very concerned?

A majority of Americans (51%) said they were deeply concerned, and another third (35%) replied they were somewhat concerned. In a *Newsweek* (October 6, 1969) cover story on the "Troubled American," Lou Harris conducted a survey of the white population with special attention to the middle-income group—the blue- and white-collar families who make up three-fifths of U.S. whites. When given a lengthy list of problems and asked, "On which problems do you think the government should be spending more money—and on which should it be spending less money?" Harris' Middle America gave the highest priority for spending more money to air and water pollution.

The state of California has been among the most celebrated and popular of the fifty states. Recently Mervin Feld (1971) completed a poll of California residents and found a growing disenchantment with life in California:

> At the root of this growing lack of appeal of California as a place to live are the negative results of a 30 year population explosion, overcrowding, and smog and air pollution in most parts of the state.... The problem of smog and air pollution throughout the state has increased in seriousness today to where it is the chief source of complaint among those who have something less than an enchanted view of California. Overcrowdedness is another drawback.

In a statewide poll for the University of California's Project Clean Air (Gold, 1970), we provided a list of ten problems and asked our respondents to select the three problems they thought were most serious. At least 50% of every age, income, education, and racial category saw air pollution as one of the three most serious problems.

ATTENTIVE PUBLIC

Though the environment has become a major public issue, it is noteworthy that the attentive public—those who pay attention to policy issues—are especially aware of and concerned about environmental matters. First, while estimated at no more than 10% of the population, the attentive public serves, according to Rosenau (1961: 41), "as the critical audience for opinion makers" and sets "the limits within which the opinion-policy relationship operates." Second, the attentive public and the membership rolls of the many interest groups in the United States are largely one and the same (Wright and Hyman, 1958). And if group interests are animating forces in the political process, then the importance of the environmental views of the attentive public are obvious. And third, the attentive public is disproportionately made up of members of the upper-middle class. Because of their time, education, and money, the upper-middle class can amass considerable political resources to achieve their own policy preferences. Thus, the kinds of people most interested now in environmental problems have traditionally been the most effective and influential citizen participants in American politics (Wildavsky, 1967).

GROUPS

Environmental interests are now aggregated and organized. The Council on Environmental Quality estimates that in 1970 there were over 3,100 environmental organizations. Admittedly, many of these are local groups organized to battle a specific environmental threat. However, beyond local groups, state and national groups are rapidly increasing in size. By June 1971, the combined membership in the five largest national environmental organizations–the National Wildlife Federation, the National Audubon Society, the Sierra Club, the Izaak Walton League of America, and the Wilderness Society–totaled over 1.6 million, a one-third increase over the previous year. Most environmental groups are now less concerned with educating the public and more interested in governmental action; these groups have organized letter-writing campaigns, funded full-time lobbyists, instituted court action, and opposed some political candidates.

Besides specific environmental groups, other civic, professional, and scientific groups have taken up environmental causes. For example, in the summer of 1969, the city of Riverside exceeded the state's acceptable air quality level for ozone for sixty consecutive days. Many city groups reacted by using their local and statewide organizations to work for air pollution policy changes–the groups included the Tuberculosis Association, Medical Association, League of Women Voters, Junior Chamber of Commerce, Junior League, City Council (California League of Cities), and Board of Supervisors (California Supervisor's Association). When such groups converge with environmental groups, the resources available to influence environmental policy decisions markedly expand (Loveridge, 1970).

MEDIA

The media are now devoting unprecedented time and space to environmental matters. Many magazines, newspapers, and television stations have assigned sections, writers, or reporters to the environmental beat. If the media tell us what to think about and perhaps what to think period (Cohen, 1963), any number of measures suggest that the media encourage, sustain, and extend the demands for environmental action by government. To illustrate, one index is the number of articles on the environment listed in the *Reader's Guide to Periodical Literature:*

Year	*Number of Articles on Environment*
1950	3
1955	39
1960	23
1965	67
1970	424

And public opinion studies indicate that the people who read magazines are more likely to engage in policy influence attempts. The importance of media attention to the environment should not be underestimated; Lang and Lang (1968: 305) explain:

> The media . . . structure a very real political environment which people can know about only through the media. Information about this environment is hard to escape. It filters through and affects even persons who are not directly exposed to the news or who deny that they are paying a great deal of attention to what the media say. There is something pervasive about the content of the mass media, something that can make its influence cumulative.

By headlining the environment, the media nurture public awareness and concern and foster the political commitments of groups and organizations to environmental issues.

SCIENTISTS

Scientists now quicken and fuel the demand for environmental action by government. The scientific community has devoted increasing attention to studying and reporting environmental problems (Brown, 1971), thereby publicizing and certifying the issue of the environment. Numerous conferences–local, national, even international–have and will be held and most scientific and professional societies have assigned panels and task forces to environmental problems. Prompted and encouraged by changes in the priorities of federal research and development monies, a marked shift in scientific interests has taken place. This new research interest is reflected in many recent environment-related articles, journals, monographs, and books (see Environmental Reporter, n.d.). Colleges and universities have developed new courses, majors, departments, even colleges around the concepts of ecology and environment. In addition to publicizing environmental problems, the research work and other activities of scientists supply data, explanations, and scenarios that certify the credentials and thus legitimate the demands for governmental action.

GOVERNMENT AGENCIES

Environmental agencies exercise considerable political clout in encouraging, aggregating, and representing demands for government action on environmental problems. Many environmental agencies have the visibility, public support, and sense of mission to act as advocates for environmental interests. At the national level, the Council on Environmental Quality and the Environmental Protection Agency represent a conscious federal commitment to environmental matters. Russell Train and especially William Ruckelshaus have been catapulted forward as national advocates of environmental quality and protection. At the state and local levels, the numbers and expenditures of environmental agencies have likewise increased. For example, a recent study (Haskell, 1971) of administrative changes in nine states cited four types of new government institutions in environmental management: consolidated environmental departments (Illinois, Minnesota, Washington, Wisconsin, New York), land use agencies (Vermont, Maine), a waste management agency (Maryland), and state courts have been given a new role in environmental protection through public interest lawsuits by private citizens (Michigan).

ELECTED OFFICIALS

Most elected officials cannot avoid at least the appearance of representing environmental interests in the political arena. Executives and legislators find it necessary to take positions on and often to promote environmental policies and programs. More important, while everyone declares against pollution and for improving the quality of the environment, selected politicians are developing their expertise and building their reputations on environment matters. Elected officials find among their ranks an increasing number who champion environmental reform. President Nixon devoted much of his 1970 State of the Union Message to the environment and his administration has sought to reorganize the federal effort to manage the environment. And Senator Muskie has achieved his greatest legislative successes in the areas of air and water pollution control and has established a record as someone who represents environmental interests. No matter what their personal views, most elected officials must in some form represent the demand for governmental action on environmental problems; politicians have taken up the cause of the environment, at least to the point where environmental problems are recognized and defined as legitimate political issues (Congressional Quarterly, 1970).

POLICY RESPONSE: ENVIRONMENTAL PROBLEMS

The good texts in political science say a legitimate demand will be met in kind by a response. For example, according to Herbert Spiro (1959: 23-26), the five steps of the policy process are the recognition of problem, statement of issue, deliberation, resolution of issue, and solution of problem. Along with many other political scientists, Spiro suggests some balance between demands and the policy response of government. Unfortunately, many environmental problems require policy solutions that conflict directly with revered American political traditions.

The policy response to the environment should not be minimized, especially as a beginning. New regulatory agencies have been organized and established agencies have quickened their activities; increased sums of money have been allocated to resolving environmental problems; many laws have been passed at all levels of government; and politicians have liberally distributed symbolic benefits in terms of promises and expressions of concern. The Council on Environmental Quality (1971: 3-4) states, for example, "If 1970 was 'the year of the environment,' then 1971 may be known as the environmental year of action. . . . Federal, state, and local governments, international organizations, industry, and citizens all have moved vigorously to restore and protect the environment. . . . The Nation has clearly demonstrated it intends to move forward now, with the best knowledge at its command." If we count monies, agencies, laws, or speeches, the response of government to the demand for environmental action has been indeed noteworthy.

Yet, while the response has been promising, the results have been more often than not disappointing. What exists is the familiar gap between stated objectives and actual performance; as diagnosed by Joseph Sax (1971):

> Today the management of environmental controversies is in disarray. Administrative agencies have been gravely deficient, and public confidence in them is eroded. . . . Citizens reach out to the legislatures for help. New statutes are abundant, but their rhetoric far exceeds their effort. . . . Legislators continue to pile more and more burdensome procedures upon agencies whose problems are far deeper than procedural failings. New councils, task forces, and commissions proliferate, but they seem little more than a revival of old institutional mistakes with new names.

The environmental crisis by most accounts continues nearly unabated, despite what the CEQ calls the year of action. The United States has taken some steps forward, some backward, all the time running to stay in the same place. Environmental problems require more than ad hoc, piecemeal, approximate measures—whether they be stronger institutions, more pollution curbs, higher appropriations, or increased research and monitoring. They require a concerted and coordinated effort, a holistic, integrated approach (see Arvill, 1967; Caldwell, 1970; Sprout and Sprout, 1971).

The gap between good intentions and governmental performance can primarily be explained in terms of the contradictions between requisite environmental solutions and the institutions, rules, and values of American politics. Scientists and writers propose in chorus that the American polity must adopt and enforce comprehensive policies that consider environmental problems in a holistic and integrated context. They state that most major environmental problems cannot be resolved independently and apart from the consideration of others, for the environment is made up of complex, highly organized interdependent systems (Wagner, 1971). The governmental response has, however, followed the hallowed tradition of segmental policies and they result in what John Straayer and Roy Meek (1971) call the "iron law of environmental disorder." While rational as incremental decisions, segmental policies often effect or permit wholly undesired environmental conditions—their consequences therefore are an urban environment no one chooses or controls.

THREE OBJECTIVES

In asking why the United States has failed so dismally in handling the unwanted effects of technology, Max Ways (1970) finds the answer in the principle of separation, differentiation, and specialization or in short, fragmentation: "In modern society the principle of fragmentation, outrunning the principle of unity, is producing a higher and higher degree of disorder and disunity." He further concludes that passing new laws setting forth what cannot be done will be largely ineffective unless we innovate new ways of deciding what should be done. To deal with political fragmentation, solutions to environmental problems generally center on three main governmental objectives: *centralization, coordination or planning,* and a *new ethos.* Together, these three objectives represent and propose an integrated and comprehensive governmental response to environmental problems.

The rationale for centralizing environmental policy is based upon several fundamental premises. A tolerable environment can only be achieved by direct, concerted, and intelligent intervention by government. The response of the national government holds the key to most of the issues shaping the physical environment (Grant, 1971), for no other governmental level can deal with society's basic social, economic, and technological forces. To prevent a future shaped by random policies and undirected forces of change, the national government must become the champion of environmental protection and quality. The increasing scale of our society and the power of large organizations, corporate and otherwise, underline the weaknesses and spotlight the limited jurisdictions of state and local governments (Greer, 1968). Moreover, state and local governments are more likely to be open to pressures of economic interests (Crenson, 1971), many of which cancel each other out at the national level. Effective and comprehensive environmental policies require centralized decision-making, beginning at the national level and continuing as guidelines, plans, and programs are implemented at state, regional, and local levels.

Planning and coordination are the watchwords of environmental management. The problems of fragmentation call for new devices, institutional and procedural, to plan and coordinate matters of the environment. Centralized decision-making is of little consequence unless national environmental policies are developed and—most importantly—implemented. A comprehensive approach to the environment requires that one consider the environment as an ecological system and as such, planning and coordination should parallel the integrative network of nature. The failures to plan and coordinate result in fragmented management with such attendant ills as gaps in control and enforcement, duplication of facilities and services, inadequate financial resources, and weak political support for resolution of conflicts.

The third environmental objective is a new ethic—a set of values that establish the priority of the environment as a major factor in the public policy process. Centralized decision-making and planning and coordination will not guarantee an America-the-beautiful. Marshall Goldman (1970), in a superb essay, documents the environmental chaos and disruption that exist in the Soviet Union. Amid the production mission of Soviet bureaucracies and industries, the environment seldom occupies a central role in planning and coordination. To a large extent, the problems of environmental control and enhancement are in fact conceptual; as Rene Dubos (1970) explains, "We can change our ways only if we adopt a new social ethic—almost a new social religion. Whatever form this religion takes, it will have to be based on harmony with nature as well as man, instead of the drive for mastery." A new political ideology is required that stresses ecology rather than competition and thus gives a high priority to the quality and protection of the environment (McHale, 1970). The preamble to the 1969 National Environment Policy Act suggests the values that must be taken seriously for the environment to be cast as a central factor in the policy process:

> To declare a national policy which will encourage productive and enjoyable harmony between man and his environment; to promote efforts which will prevent or eliminate damage to the environment and biosphere and stimulate the health and welfare of man; to enrich the understanding of the ecological systems and natural resources important to the Nation.

POLITICAL FACTS OF LIFE

The three environmental objectives clash, in many ways, with basic American political patterns. In a detailed account of the regulatory process at the national level, Lewis Kohlmeier (1969) examines—among other areas—the regulation of transportation. The federal government closely involves itself in the regulation and planning of transportation, namely through pricing, technological innovation, and allocation of resources. In a policy area ostensibly centralized, planned, and coordinated in the public interest, chaos best describes the result of federal tutelage. The reasons for a chaotic transportation system can be found in the predominant patterns of American politics (Kohlmeier, 1969: 134, 185):

> America's system of highways, waterways, railways and airways today is the product of almost 150 years of patchwork political decision-making. The fragmentation of transport policy is indeed a particularly sharp reflection of the pressures and compromises that always have been the hallmarks of the legislative process in Washington. Fragmentation feeds itself. Each new bureaucracy becomes its own best lobbyist. Each, with the backing of an omnipresent corps of industry lobbyists lending moral and other appropriate support, promotes its own brand of transportation, while the public cost, both economic and social, goes up and up and up . . . [T]he questions of willingness to pay for more and better transportation and of the effectiveness of future federal planning remain unresolved issues that pit the public interest in coordination against private interests in a continuation of uncoordination. The fight is on and the public is not winning.

More generally, government attempts at coordinated planning have a disappointing course record. To some extent, comprehensive planning has failed because it has not been accepted as appropriate government activity. However, even when plans are adopted, they have been notoriously ineffectual; for example, John Reps contends (Feiss, 1968: 218):

> Nowhere in this country can one find a major city or a major sector of an important city, which in the present era has been developed as planned. It is not a case of an occasional departure from an officially adopted plan. It is not even a situation where a majority of cities do not grow as planned. It is, rather, a record of complete and consistent failure.

Planning attempts have resulted in little more than good sources of public information. Largely limited to definitions of hopes, coordinated planning efforts have been axed by the special interests of economic groups and public bureaucracies. The greater the scope, the more likely a plan will fail. The environment mandates massive doses of planning in the uses of air, water, land, and natural resources; and these plans will be difficult to adopt and even more difficult to implement. Calling for a systems approach, whether for environmental or other problems, does not change the existing political format—in the past, the "great game of politics" has aborted coordinated planning attempts.

Jules Feiffer (1970) portrays a dialogue between two men, beginning with "Vietnam is dead as an issue. The real issue is ecology." And the reply, "But what can we do about it?" After several exchanges, the conversation concludes, "We have to control industry." "Oh, you mean socialism." Then the final verdict, "Ecology is dead as an issue." While overstated, Feiffer underscores the differences between environmental objectives and the current political order. A comprehensive policy approach to the environment is designed to respond to a number of factors: the interdependence of air, water, land and resource management; the inability to control the influences of economic consumption and growth or the impact of science and technology upon the environment; the difficulty of setting priorities without knowledge of and controls upon the total system; and the multitude of agencies and levels of government, each with a small piece of the action. However, a centralized, planned, and coordinated approach is confounded and rebuffed by three central patterns of the American political system.

DISPERSED DECISION-MAKING

First, decision-making authority is dispersed. The United States governmental structure is characterized by separation of powers—executive, legislative, and judicial branches, bicameralism, federalism, special districts, and local autonomy. Moreover, most economic decisions are largely left to the private sector. The resulting legal and political system means that enormous stakes exist in the dispersed decision-making arrangements (Lieber, 1970: 282-283):

> Confronted with a wide range of emerging environmental problems, administrative theory's answer to the litany of doom has been an antiquated catechism whose key phrases—regional and coordination—fail to reflect political reality. Consider the following syllogism: 1. Air pollution does not respect political boundary lines. 2. Existing local and state agencies deal with air pollution on a limited, piecemeal, and uncoordinated basis. Therefore—3. An areawide agency covering the entire problem-shed would be the most effective

> unit to deal with the problem. But whether through interstate compact, council of government, or special district, there is hardly a single effective regional air pollution agency functioning today. With the possible exception of the Delaward River Basin Commission, the same could be said for water pollution control. . . . [T]he immense political difficulties in organizing areawide control operations vitiate the apparent administrative advantages attributed to such regional agencies.

The political stakes in existing institutional patterns are illustrated, for example, by the resounding defeats of attempts to form metropolitan governments. Efforts to centralize environmental decision-making directly collide with the ongoing structural arrangements of American politics.

Dispersed decision-making also means that many points exist where critical environment decisions are made—or not made. The opposition to particular environmental decisions have many points of access to ask for modification or reconsideration. Especially for implementing policies, dispersed decision-making presents staggering problems. Often the major political obstacles are not those of adopting big plans but rather the tough and difficult work of implementation; Davies (1970: 201-202) emphasized this issue:

> Given the large number of governments and firms which in effect have a veto power over control efforts, compliance is the most vulnerable part of the pollution control process, the stage at which failure is most likely. Of course, compliance is also the stage which has the most impact on the interests of concerned parties. It does not cost anybody anything to write good laws or set stringent standards if the laws and standards are not to be enforced.

PLURALISM

Second, the dispersion of decision-making authority primes and supports what political scientists call pluralism, a style of politics distinguished by bargaining, coalition-building, and presence of veto groups (Dahl, 1956: 150):

> The making of governmental decisions is not a majestic march of great majorities united upon certain matters of basic policy. It is the

> steady appeasement of relatively small groups. . . . For it [American polity] is a markedly decentralized system. Decisions are made by endless bargaining; perhaps in no other national political system in the world is bargaining so basic a component of the political process. . . . [American polity] does provide a high probability that any active and legitimate group will make itself heard effectively at some stage in the process of decision.

For obvious reasons, the politics of pluralism vitiate the adoption of comprehensive environmental policies. As important, the variety of interests intersecting the dispersed centers of power act to reinforce the status quo, for bargaining emphasized the adjustment of demands within existing group arrangements. Once having achieved special access or modus vivendi in the political arena, interests seek to protect their own prerogatives and privileges and thus support, directly or indirectly, the group arrangements of the status quo (Nimmo and Ungs, 1967).

Quite apart from the adoption of environmental policies, the pluralistic style of American politics tends to undermine the implementation of most environmental programs. The administrative process shares the same characteristics as the legislative process—as Francis Rourke (1969: 103) points out, "Agencies respond to group pressures by modifying existing policies or developing new ones. Bargaining or the adjustment of conflicting interests is as constant a feature of administrative politics as it is of the relations among legislators and legislative committees." Laws do not provide detailed rules but instead offer considerable room for discretion and interpretation by administering agencies. The influence of clientele groups on environmental agencies cannot be overstated. If agencies antagonize their clientele groups, they risk attack and the failure of their mission. Thus, as the reciprocal relations between agencies and clientele groups are fostered, agencies frequently mirror their clientele's interests. The agencies then become protectors of the status quo and use their power to maintain the stakes of the groups they regulate. The politics of regulation too often provide reassuring symbols for the public and concrete benefits for organized interests (Edelman, 1964).

GRASS-ROOTS DEMOCRACY

And third, the objectives of environmental reform contradict fundamental American political values. The concepts of centralization, coordination, and planning are alien notions when compared to orthodox American political virtues such as checks and balances, decentralization, local autonomy, and citizen control. Planning especially is viewed with suspicion, with plaudits instead given to individual and corporate choice. A political ideology which stresses the importance of ecology faces two other problems. As Faltermayer (1968) indicates, our political culture has historically demonstrated little respect for nature and has maintained a belief in the inexhaustability of the nation's resources. While such articles of faith are under some attack, these notions are firmly implanted in the institutions, rules, and laws of the country. And second, there is only partial agreement on the specific values of a new ethic. The relations between man and his environment are immensely complicated and do not lend themselves to the formulation of an operational and systematic ethic. Rather, the ecological ethic will essentially pivot on political choices among values; how, for example, will environmental policy integrate matters of conservation, agriculture, aesthetics, recreation, economic development, human health, and survival?

TOO RESPONSIVE?

Paradoxically, the American political system is perhaps too responsive. The attempts to meet the rising expectations and increasing demands of many individuals and groups for government intervention and resources translate into piecemeal responses to whatever interests place demands on the political agenda. The political system is structured to make short-range responses rather than meet the long and comprehensive objectives of environmental reform. After making a similar argument, Straayer and Meek (1971: 243) conclude that the iron law of environmental chaos may be preferable to the imposition of comprehensive environmental solutions, "there may be something to say for incremental and often muddled planning; after all, it does suggest a society with a rather exciting variety of value systems and considerable freedom to pursue them." To illustrate why at least modified comprehensive solutions are preferable to environmental chaos, let us examine one of the most famous postwar environmental problems–Los Angeles smog.

LOS ANGELES SMOG

Air pollution, the most critical environmental problem for the residents of the Los Angeles basin, pointedly demonstrates the failure of old politics. While responding with alacrity to a powerful demand, old politics has failed because political fragmentation and its constraints work against adequate solutions. Although great public interest exists, active citizen groups plentiful, extensive media coverage unrelenting, excellent research conducted, many laws passed, superb standards adopted, an effective air pollution agency satisfactorily funded and supported, the pollution of Los Angeles air has increased.

Identified and recognized by the general public in the early 1940s and in spite of marked and successful government intervention in controlling emissions, air pollution remains a serious problem. As stated by the *Los Angeles Times,* "Air pollution has long since ceased to be merely an inconvenience in Metropolitan Los Angeles . . . it is an urgent and serious threat to public health that demands bold and concerted action." The incidence of air pollution can be shown by the number of days in 1970 that each of four pollutants exceeded the state of California's health smog standards somewhere in the Los Angeles basin (Mayor's Council on Environmental Management, 1971: 11):

Pollutant	*Number of Days in Excess of State Standards in 1970*
Carbon Monoxide	203
Ozone	241
Nitrogen Dioxide	115
Sulfur Dioxide	95

By any measure, the demand for government action is striking. Los Angeles poll data indicate that for nearly twenty years, the general, attentive, and elite publics have viewed air pollution as one if not the most important problem government should do something about. Sheldon Samuels (1970), former chief of field services for NAPCA's (National Air Pollution Control Administration) Office of Education and Information, says that the Los Angeles basin is probably the best national example of effective community involvement in air pollution control matters: its citizens are—according to Samuels' criteria—aware, concerned, organized, working in a positive climate, confronting the authorities, and focusing their efforts on policy decisions. Protest groups have risen to smite the dragon of air pollution and joined by many other conservation, civic, and professional organizations, they have entered the political arena on behalf of improving air quality. The media have literally campaigned against the problem of air pollution. For example, the *Los Angeles Times* has since 1945 published over 10,000 articles or editorials dealing with air pollution. (On radio and television weather reports, smog levels are noted along with temperature, rain, and wind.) Scientists from the University of California at Los Angeles, University of Southern California, California Institute of Technology, and University of California at Riverside have been the national leaders in air pollution research. Local air pollution control officials have sought to publicize and win support for their activities. And elected officials have long accepted air pollution as a given issue for representatives of Los Angeles. It is difficult to conceive a stronger and more sustained demand for governmental action on a major environmental problem, barring of course a life-and-death crisis of immediate and massive proportions.

The policy response in terms of governmental intervention has also been striking. The Los Angeles County Air Pollution Control District was activated in 1949. It quickly developed into the model for local control of stationary sources. For over fifteen years, the District has had the highest per capita expenditures and the strictest emission standards of any local control program in the country. The County Board of Supervisors has provided the District with the necessary funding and political support. The District's success is nearly uniformly lauded—in the words of then Vice President Hubert Humphrey (Fuller, 1967: 217), "The war Los Angeles is waging against air pollution is already a modern legend. . . . [T]he experience of Los Angeles has shown that local governments can control most sources of pollution if they will." Responding to pressure from Los Angeles, the state of California in 1960 adopted legislation proposed by the only State Senator from Los Angeles County, Richard Richards, and set up the California Motor Vehicle Pollution Control Board. And the MVPC Board and its successor, the Air Resources Board, have in effect all but led the nation in setting standards for auto emissions. Many of Los Angeles leading control officials and research scientists have been at the forefront of state and federal air pollution control efforts. And legislators from the Los Angeles basin have been the primary proponents of tougher standards and enforcement programs at the state level and among the most active at the national level. By conventional measures, the government in the Los Angeles basin has certainly been responsive to the demand for tolerable air.

Despite an extraordinary response by government, however, Los Angeles now produces more smog daily than it did fifteen years ago. While perhaps saving the basin from becoming an inversion cemetery, the state and District controls have not kept pace with the increase in people (1.9 million more) and in motor vehicles (1.7 million more). On an average day in 1970, the county's residents contributed 26.2 million pounds of air pollutants or 1.1 million more pounds than on an average day in 1955. Rather than solving the problem, government intervention has seemingly only restrained and reduced the rate of the deterioration of the air in the Los Angeles basin (Boffey, 1968: 990).

> The Los Angeles Basin which is said to have the most vigorous air pollution control program in the nation, has made little progress in recent years in its struggle to cleanse the air of smog. Though the LA APCD has "pioneered" in imposing stiff curbs on industrial polluters, and the State of California has pioneered in cracking down on motor vehicle emissions, the gains made thus far have been offset by new pollution stemming from new growth of population, automobile traffic, and industry. As a result, air quality in Los Angeles has shown no appreciable improvement in this decade.

Happy days are not here again, for as the Los Angeles Medical Association pointed out (Boffey, 1968: 992), "air pollution is becoming increasingly worse and may lead to great lethality in this community."

The increasing levels of pollution can be explained only in part by the failure to enforce control standards and programs. Many observers now modify their tributes to the Los Angeles APCD. However, their principal criticisms center on its loss of mission or what Anthony Downs (1967) describes as "the law of increasing conservatism": "All organizations tend to become more conservative as they get older . . . " The central villain is the automobile, estimated to account for more than 80% of the air pollution problem in the Los Angeles basin. Here legitimate and serious questions can be raised about implementation problems; however, the level of government now responsible is the state, for it has preempted control of vehicular traffic. The state has relatively stringent automobile emission standards, yet substantial evidence exists that most automobile emission controls deteriorate quite rapidly once upon the California freeways. Professor James Pitts (1971: 2), Director of the University of California Statewide Air Pollution Research Center, stressed this point in testimony before a California Assembly Subcommittee, "But the overriding cause of our atmospheric deterioration seems to be the failure of exhaust control devices to live up to official assurances and public expectations." Yet, even if automobile emissions controls were resolutely enforced, the Technical Advisory Committee to the California Air Resources Board concluded (Pitts, 1971: 11):

> the best estimates indicate that even with the proposed 1975 auto exhaust emission standards, the maximum hourly average oxidant values in Los Angeles will probably not be below 0.25 to 0.35 ppm by 1980. Even by 1985, ten years after these stringent auto emission requirements may be implemented, the oxidant level in Los Angeles will probably be 0.20 to 0.25 ppm.

(The state of California's health-related air quality standard for oxidant is 0.1 ppm per one hour.) In other words, even if all control standards were wholly effective, the problem of air pollution in the Los Angeles basin would remain serious and urgent.

Notable lessons can be found in the failure of the governmental response to control smog in the Los Angeles basin. Local or decentralized efforts are inadequate; while necessary, they are simply not sufficient. No matter what the demand, the local Los Angeles polity cannot resolve the air pollution problem. Its response is inevitably inadequate because control officials lack the resources and jurisdiction to deal with pollution sources, and more important, the requisite solutions. For obvious economic and legal reasons, local officials have almost no control over emissions from vehicular traffic. However, the major reasons for failure rest with their inability to act upon requisite solutions. State experts agree that if the ambient air quality in the Los Angeles basin is to meet state air quality standards, major socioeconomic actions must be taken. For example, Louis Saylor, Director of the State's Department of Public Health, recommended these changes in an interview with John Dreyfuss (1971) of the *Los Angeles Times:*

> In areas such as the Los Angeles Basin, Saylor said, population growth should be limited by sharp restrictions in living-unit construction. He said motor vehicle and aircraft use should be limited, all polluting industries and power plants should be removed from the basin or made emission-free, and a massive nonpolluting public transportation system should be developed. "Smog is slow poison," Saylor said. "To avoid it, people must learn to live differently, They must use their cars far less, and must stop moving to big cities. Factories have to clean up or move out of cities. . . . The State Air Resources Board's technical advisory committee made recommendations similar to ours nine months ago. We think medical evidence supports these recommendations."

These solutions are beyond the will and powers of a local control district, for the local "fix" is limited to relatively technical objectives and does not include even encouraging mixed transportations systems or planning for new industrial sites. The steps suggested by Saylor and others will require, in brief, centralized decision-making, planning and coordination, and a new ethic in order to respond satisfactorily to the physical, economic, and health interactions of pollutants and the socioeconomic and technological measures necessary for their control.

Los Angeles smog illustrates why old politics cannot resolve environmental problems. At its best, fragmented management of the environment simply does not work. When in April 1971, William Ruckelshaus announced new air quality standards for the country, many California control officials said impossible. For example, Haagen-Smit, Chairman of California's Air Resources Board, said he had no idea how Southern California can meet the EPA air quality standards, "How the hell are we going to do it? I don't know." These officials are in fact correct. Without comprehensive policies and new legal, economic, and social devices that can control and when necessary reshape the life style of its residents, the Los Angeles basin will be unable to meet federal air quality standards. At present, air pollution control officials do not exert such controls or influence over the fundamental social and economic patterns of the basin—nor can they expect to in the future. When serious and complex air pollution problems are added to the list of other serious and complex environmental problems, the need for a systems approach becomes clear (Romer et al., 1971: 24-25):

> In common sense terms, the systems approach requires that one consider all relevant aspects of a problem. Planning adds the future dimension—anticipating new problems. The systems approach is particularly suitable to tackling environmental problems. The environment is one ecological system, existing as an equilibrium of differing elements. The techniques of the total systems approach can alone cope with and encompass all these ecological demands.

PROSPECTS FOR THE FUTURE

The environmental prospect requires that the management of man's relationship to his environment become a major task of government. The problems of the environment are real and urgent with possible solutions that are largely matters of public policy rather than questions of science and technology. The task is imposing and fraught with difficulties, for it requires at least some modification of the nation's major political institutions, rules, and values. Yet the failure to act effectively will mean the certain deterioration of urban life and perhaps a fatal weakening of the ecosystems necessary to sustain man himself. The gauntlet has been dramatically thrown, the American polity must successfully respond to the demand to protect and enhance the environment (Falk, 1971).

America has responded to widespread demands for governmental action. On the national level, a National Environment Policy Act has been passed, two important agencies set up—Council on Environmental Quality and Environmental Protection Agency, a new department proposed—Department of Natural Resources, environmental appropriations steadily increased, and numerous environmental laws enacted. At the state and local level, extensive reorganization of environmental agencies has taken place along with increased attention to as well as funding of environmental programs. Nonetheless, the response has, in total, been unsatisfactory. The American polity handicaps and frustrates environmental policies; its dispersed centers of power, pressure group politics, freedom of individual and corporate choice, and values of economic competition and growth intersect and scatter comprehensive environmental solutions. While agreement can be reached on the general mission and purpose of environmental action, attempts to translate such objectives into legislative and especially administrative action dissolve into countless conflicts of interests and values. With its resources staked in the status quo, old politics triumphs over new environmental priorities.

Yet a tolerable environment, urban or otherwise, mandates comprehensive policies; without integrated policy choices and decisions keyed on ecological values, the outlook—no matter good intentions—is one of gloom if not doom. After reviewing the political obstacles, some observers back off and agree with Joseph Gusfield (1971: 61), "We are left with the necessity of solutions to environmental problems in technological terms." Unfortunately, if we make such a choice, we are waiting for Godot. Environmental problems require the intervention of government and the crucial questions become what kind of governmental policies and programs are necessary.

Most comprehensive policies involve, in the last analysis, three objectives: centralized decision-making, planning and coordination, and a new ethic. While the rhetoric of these objectives may seem revolutionary, they essentially call for a more sanely managed central state, a benign form of capitalism, and some controls over technology and science (Ridgeway, 1970). Nevertheless, such reforms collide directly with major American political patterns and as a result, the opposition line-up assumes awesome proportions. Most political pundits say that realistic hopes hinge on gradual but consistent and steady reforms that over time begin to achieve the substance of the requisite environmental solutions. At present, the incremental response appears neither swift enough nor effective enough to provide the necessary controls or directions. Some have argued that before comprehensive environmental policies are possible, the pluralistic and decentralized character of American politics must somehow be reordered, thereby reducing the importance of pressure group politics and state, local, and corporate autonomy. If these changes are in fact requirements, the dice are loaded against the American polity taking a determinate and successful action on major environmental problems.

Whether or not a comprehensive approach can be approximated will depend, of course, on many factors. One necessary condition in a democratic polity will be widespread understanding of the character of environmental and ecological problems. Even though specific problems may often be beyond their comprehension, the public can express a general awareness and concern. However, public interest cannot indefinitely be sustained at a high level of intensity.

Environmental reform will notably be tied to scientific research and to the assumption of a scientific rationale; environmental policies will focus on and be fashioned by scientific knowledge (Lane, 1966; Schooler, 1971). The documentation of problems of pollution, misuse of land, and natural resources may act to counter political objections and obstacles. Yet, public concern and scientific findings–to make a difference–must be translated into the mission and performance of our societal institutions. Here social scientists have a major role to investigate and propose new institutional devices and incentives to control environmental problems. In contrast to most political issues, scientists share a major responsibility for the fate of environmental reform. Without their articulate support, the call for comprehensive environmental policies will only join many other demands that government take massive and conclusive action.

The prospects of the future are uncertain. We still have time and can as many have urged "take sensible advantage of this period of grace." Yet collective self-control is perhaps the most difficult task a democratic society can undertake. Kahn and Weiner (1967) wisely remind us that even the best attempts to plan the future will only meet with partial success. Most environmental scientists argue that we must try, for without attempting to adopt and implement comprehensive policies, we will rush headlong into a shabby hell of our own making. Yet the new priorities of environmental policies clash with the traditions of the "great game of politics" and it is the political game that holds the key to the good environment or man's last moments in the sun (Rienow and Rienow, 1967).

REFERENCES

ARVILL, R. (1967) Man and Environment: Crisis and the Strategy of Choice. Baltimore: Penguin.

BOFFEY, P. M. (1968) "Smog: Los Angeles running hard, standing still." Science 161 (September): 990-992.

BROWN, M. (1971) The Social Responsibility of the Scientist. New York: Free Press.

CALDWELL, L. K. (1970) Environment: A Challenge for Modern Society. Garden City, N.Y.: Natural History Press.

--- [ed.] (1967) Environmental Studies–Papers on the Politics of Public Administration–Environmental Relationships. Vol. I-IV. Bloomington: Indiana Univ. Press.

COHEN, B. (1963) Press and Foreign Policy. Princeton, N.J.: Princeton Univ. Press.

Congressional Quarterly (1970) Man's Control of the Environment: To Determine his Survival . . . or to Lay Waste his Planet. Washington, D.C.: Congressional Quarterly.

Council on Environmental Quality (1971) Environmental Quality: Second Annual Report. Washington, D.C.: Government Printing Office.

--- (1970) Environmental Quality: First Annual Report. Washington, D.C.: Government Printing Office.

CRENSON, M. A. (1971) The Unpolitics of Air Pollution: A Study of Non-Decision Making in the Cities. Baltimore: John Hopkins Univ. Press.

COMMONER, B. (1971) The Closing Circle. New York: Alfred A. Knopf.

DAHL, R. (1956) Preface to Democratic Theory. Chicago: Univ. of Chicago Press.

DANSEREAU, P. [ed.] (1970) Challenge for Survival: Land, Air, and Water in Megalopolis. New York: Columbia Univ. Press.

DARLING, F. F. and J. P. MILTON [eds.] (1966) Future Environments of North America: Transformation of a Continent. Garden City, N.Y.: Natural History Press.

DAVIES, J. C. (1970) The Politics of Pollution. New York: Pegasus.

DOWNS, A. (1967) Inside Bureaucracy. Boston: Little, Brown.

DREYFUSS, J. (1971) "Change in way of life urged in smog battle." Los Angeles Times (July 2): pt. II, 1, col. 5.

DRUCKER, P. (1969) Age of Discontinuity. New York: Harper & Row.

DUBOS, R. (1970) quoted in Newsweek 75 (January 26): 47.

EDELMAN, M. (1964) The Symbolic Uses of Politics. Urbana: Univ. of Illinois Press.

EHRLICH, P. R. and A. H. EHRLICH (1970) Population Resources Environment: Issues in Human Ecology. San Francisco: Freeman.

EWALD, W. (1968) Environment and Policy: The Next Fifty Years. Bloomington: Indiana Univ. Press.

Environmental Reporter (n.d.) Washington, D.C. Bureau of National Affairs.

FALK, R. (1971) This Endangered Planet: Prospects and Proposals for Human Survival. New York: Random House.

FALTERMAYER, E. K. (1968) Redoing America: A Nationwide Report on How to Make Our Cities and Suburbs Livable. New York: Harper & Row.

FEIFFER, J. (1970) Cartoon in Calendar, Los Angeles Times (April 26): 48.

FEISS, C. (1968) "Taking stock: a resume of planning accomplishments in the United States," pp. 214-236 in W. Ewald (ed.) Environment and Change: The Next Fifty Years. Bloomington: Indiana Univ. Press.

FELD, M. (1971) "For Californians, California is losing its popularity." Riverside Press (July 15): pt. C, 5, cols. 1-2.

FULLER, L. J. (1967) Los Angeles County Air Pollution Control Officer, Testimony Before the Subcommittee on Air and Water Pollution U.S. Senate Public Works Committee, Senator Edmund Muskie, Chairman. (February 18).

GARNSEY, M. E. and J. R. HIBBS [eds.] (1967) Social Sciences and the Environment. Boulder: Univ. of Colorado Press.

GOLD, D. (1970) "Public concern and beliefs about air pollution in California: a statewide survey." Project Clean Air: University of California.

GOLDMAN, M. (1970) "The convergence of environmental disruption." Science 170 (October 2): 37-42.

GOODWIN, R. (1967) "The shape of American politics." Commentary 44 (June): 25-40.

GORDON, M. (1963) Sick Cities. New York: Macmillan.

GRANT, D. (1971) "Carrots, sticks, and consensus," pp. 99-115 in L. L. Roos (ed.) The Politics of Ecosuicide. New York: Holt, Rinehart & Winston.

GUSFIELD, J. (1971) "Defusing the population bomb." Transaction 8 (October): 56-61.

GREER, S. (1968) "The shaky future of local government." Psychology Today 2 (August): 64-69.

HARRINGTON, M. (1965) The Accidental Century. New York: Macmillan.

HASKELL, E. (1971) Managing the Environment: Nine States Look for New Answers. Washington, D.C.: Smithsonian Institute. (unpublished)

HERBER, L. (1965) Crisis in Our Cities. Englewood Cliffs, N.J.: Prentice-Hall.

JONES, C. (1970) An Introduction to the Study of Public Policy. Belmont. Calif.: Wadsworth.

KAHN, H. and A. J. WIENER (1967) The Year 2000: A Framework for Speculation on the Next Thirty-Three Years. New York: Macmillan.

KOHLMEIER, L. M. (1969) The Regulators: Watchdog Agencies and the Public Interest. New York: Harper & Row.

LANE, R. (1966) "Decline of politics and ideology in a knowledgeable society." Amer. Soc. Rev. 31 (October): 649-662.

LANG, K. and G. E. LANG (1968) Politics and Television. Chicago: Quadrangle.

LEWIS, H. R. (1965) With Every Breath You Take. New York: Crown.

LIEBER, H. (1970) "Public administration and environmental quality." Public Administration Rev. 30 (May/June): 277-286.

LOVERIDGE, R. (1970) "Types, ranges, and methods for classifying human behavioral responses to air pollution," pp. 53-70 in A. Atkisson and R. Gaines (eds.) Development of Air Quality Standards. Columbus, Ohio: Charles Merrill.

LOWI, T. (1969) The End of Liberalism. New York: Norton.

McEVOY, J. (1970) The American Public's Concern with the Environment: A Study of Public Opinion. Davis, Calif.: Institute of Governmental Affairs.

McHALE, J. (1970) The Ecological Context. New York: Braziller.

Mayor's Council on Environmental Management (1971) "Action on the Los Angeles environment." Los Angeles: City of Los Angeles.

MUMFORD, L. (1970) "Survival of plants and man," pp. 221-235 in P. Danserau, Challenge for Survival: Land, Air, and Water for Man in Megalopolis. New York: Columbia Univ. Press.

NIMMO, D. and T. D. UNGS (1967) American Political Patterns: Conflict and Consensus. Boston: Little, Brown.

PITTS, J. N. (1971) Remarks Before the Subcommittee on Transportation and Natural Resources of the Ways and Means Committee of the Assembly, State of California (March 12): 1-20.

RIDGEWAY, J. (1970) The Politics of Ecology. New York: Dutton.

RIENOW, R. and L. T. RIENOW (1967) Moment in the Sun: A Report on the Deteriorating Quality of the American Environment. New York: Ballantine.

ROMER, R., J. E. FRINK, and C. KRAMER (1971) "Environmental health services: multiplicity of jurisdictions and comprehensive environmental management." Los Angeles: University of California. (unpublished)

ROSENAU, J. N. (1961) Public Opinion and Foreign Policy: An Operational Formulation. New York: Random House.

ROURKE, F. (1969) Bureaucracy, Politics, and Public Policy. Boston: Little, Brown.

SAMUELS, S. W. (1970) "The role of behavioral research in air pollution control." Background Paper of the Symposium on the Role of Perceptions and Attitudes in Decision-Making in Resources Management at the University of Victoria, British Columbia (April 13): 1-26.

SAX, J. L. (1971) Defending the Environment: A Strategy for Citizen Action. New York: Alfred A. Knopf.

SCHNEIDER, K. (1971) Autokind vs. Mankind. New York: Norton.

SCHOOLER, D. (1971) Science, Scientists, and Public Policy. New York: Free Press.

SPIRO, H. (1959) Government by Constitution. New York: Random House.

SPROUT, H. (1971) "The environmental crises in the context of American politics," pp. 41-50 in L. L. Roos (ed.) The Politics of Ecosuicide. New York: Holt, Rinehart & Winston.

--- and M. SPROUT (1971) Toward a Politics of the Planet Earth. Cincinnati: Van Nostrand.

STRAAYER, J. and R. MEEK (1971) "Iron law of disorder," pp. 237-242 in R. Meek and J. Straayer (eds.) The Politics of Neglect: The Environmental Crisis. Boston: Houghton Mifflin.

TROP, C. and L. L. ROOS (1971) "Public opinion and the environment," pp. 52-63 in L. L. Roos (ed.) The Politics of Ecosuicide. New York: Holt, Rinehart & Winston.

WAGNER, R. (1971) Environment and Man. New York: Norton.

WAYS, M. (1970) "How to think about the environment." Fortune 81 (February): 98-101 ff.

WEBB, E., D. CAMPBELL, R. SCHWARTZ, and L. SECHREST (1966) Unobtrusive Measures: Nonreactive Research in the Social Sciences. Chicago: Rand McNally.

WILDAVSKY, A. (1967) "Aesthetic power or the triumph of the sensitive minority over the vulgar mass: a political analysis of the new economics." Daedalus 96 (Fall): 1115-1127.

WRIGHT, C. R. and H. H. HYMAN (1958) "Voluntary association membership in American adults: evidence from national sample surveys." Amer. Soc. Rev. 23 (June): 284-294.

19

Political Patterns in Federal Development Programs

RANDALL B. RIPLEY

□ WHAT IS AND IS NOT "urban" has never been very clear (see Cleaveland, 1969). It could be argued that in a country, the population of which is overwhelmingly urban, every domestic policy has urban implications. Even if that position is not taken, it can at least be argued, and is here, that most public policies in the area of economic and human resource development have substantial, even if not exclusive, impact on urban areas of varying sizes.

This article seeks to explore the involvement of the national government of the United States in promoting economic and human resource development in the last few years, primarily through analyzing four specific programs: Model Cities, the Job Corps, the Appalachia program, and the programs of the Economic Development Administration. The first two are unquestionably urban in focus; the latter two have tangential relations to metropolitan areas and are directly aimed at smaller cities and towns as well as some

---

AUTHOR'S NOTE: *This report stems from an ongoing inquiry into policy-making in the executive branch of the national government of the United States supported by the Mershon Center for Education in National Security at Ohio State University. Both the framework for inquiry suggested in this article and the data on which various summary statements are based were developed through interaction with graduate students, undergraduates, and my colleagues*

more rural areas. No claim is made that these four programs somehow are representative of federal development programs in the last six or eight years; rather it is claimed that these programs have both similarities and differences (and intrinsic interest) that lend themselves to useful analysis.

Specifically, this article will first describe and analyze government at work in the four programs; second, it will propose in brief form an analytical scheme and research strategy for making sense not only of the differences and similarities encountered in the four programs but for pursuing policy research in general.

A COMPARATIVE ANALYSIS OF THE FOUR PROGRAMS

All levels of American government have been involved in promoting economic development since the earliest years of the Republic. More recently, various levels of government have added the development of human resources to their agendas. The range of federal government activity in the economic and human resource areas is great. For example, in April 1970, the Office of Economic Opportunity produced a *Catalog of Federal Domestic Assistance* that took 1,034 pages to describe very briefly 1,019 programs administered by 57 different federal departments and agencies. Most of these programs are designed to promote economic and human resource development in one way or another.

Between 1964 and 1966 the federal government initiated the Job Corps, the Appalachia program, the Economic Development Administration, and the Model Cities program to deal with a number of economic and human resource problems that had come to its attention.[1] In the same years, and during some of the immediately preceding ones, the federal government had also initiated a number

---

in the Department of Political Science at Ohio State. I especially wish to thank three graduate students who have worked closely in this endeavor–Carl Frantz, William Moreland, and Richard Sinnreich–and my administrative assistant, Grace Mullen, for their contributions. For a full report of the data on which the generalizations in this article are based, see Ripley (1972). For a much fuller and somewhat different version of the framework (complete with variables and hypotheses), see Ripley, Moreland, and Sinnreich (1972).

of other related programs: the rest of the poverty war programs, a host of education programs, a number of housing programs, and Medicare, to name only the most visible.

The four specific programs studied here all attacked various aspects of the same general problem. All four programs, however, were structured differently in part. Table 1 summarizes some of these differences.

PROGRAM ORIGINATION

All four of these programs originated primarily within the federal government. In the case of Appalachia, the program originated partially within state governments, and, in the case of the Job Corps, partially within the private sector.

TABLE 1
SELECTED CHARACTERISTICS OF FOUR FEDERAL PROGRAMS AIMED AT ECONOMIC AND HUMAN RESOURCE DEVELOPMENT

Program	Source of Funds	Recipient of Funds	Locus of Planning Initiative	Object of Program
Economic Development Administration	Federal government, some matching funds from local units of government	Economic Development Districts	EDDs, localities, federal government	Construction—tangible development
Job Corps	Federal government	Federal agencies corporations, state agencies, nonprofit organizations	Center contractors, federal government	Education, training, conservation projects
Appalachia	Federal government, some matching funds from local units of government	States, localities, traditional governmental units	State governments, federal government, local governmental units	Construction, education, conservation, health
Model Cities	Federal government	Cities	City governments, Model Cities congressds and various local groups, federal government	Training, education, housing, employment, delivery of services, coordination of governmental efforts

Model Cities developed almost totally as a result of the perceptions—by federal officials and members of the House and Senate—of the weaknesses of existing housing programs. Formulation of the Model Cities program took place almost wholly within the confines of the federal government.

The Economic Development Act was, in many important ways, an extension of the Area Redevelopment Act. The latter act was, of course, the result of Senator Paul Douglas' (D-Ill.) labors at coalition-building basically within Congress and then the assumption of office by friendly Kennedy appointees.

Initial stimulation for the Appalachia program came from the governors of the states in the region and from some of their staff members. Once they had called the attention of the federal government to their problems (as a result of the disastrous floods in the region in 1963) the Appalachian Regional Commission was appointed and it began to generate interest in what became the 1965 Act among officials throughout the federal government also. Again, initiation came from within government—this time at several levels.

Nongovernment academics and publicists were involved in calling the attention of the federal government officials to the problems of unemployed and underemployed youth. Then key governmental officials worked with some individuals without official government positions to produce, first, the juvenile delinquency program, and, second, the Jobs Corps idea itself. Conservation lobbyists also helped shape part of the program in a direction favorable to themselves and their input had been considerable through the years in generating the idea of a Youth Conservation Corps, which was in effect fused into the Job Corps enterprise.

THE PERVASIVENESS OF SUBSIDY

All four programs were sold by their initial proponents as programs of subsidy. This is an accurate portrayal of Appalachia and EDA, but the proponents of Model Cities and the entire poverty program (including Job Corps) also stressed subsidy elements, although there was some recognition that both programs also involved redistribution (on subsidy and redistribution, see Lowi, 1964).

In important ways all four programs were also administered as programs of subsidy. One result of administering a program as subsidy is that eligibility for the benefits of the program tends to be expanded to more areas, units, or people, even if resources for the program are shrinking or at least growing more slowly than the proposed expansion of eligibility would seem to demand. Appalachia grew in its early years into more states (New York and Mississippi) and into additional counties in some of the states where it was initially established (e.g., Alabama). The Job Corps was conceived initially to exclude high school graduates. As administered, however, a sizable proportion of Job Corps enrollees were, in fact, high school graduates. The Economic Development Administration generated activity on the local level that helped qualify increasing numbers of areas for formation as Economic Development Districts and the receipt of aid. Early in the Nixon administration the Model Cities program dropped its limits on the proportion of a city's population that could be included in the "model neighborhood."

All of these eligibility expansions took place when resources were scarce and even when, in the case of Model Cities, Job Corps, and EDA, they began to shrink in absolute terms.

There were also some conflicting elements in the administration of Model Cities and Job Corps. To the extent that administrators made it evident that they felt they were dealing with redistributive programs, forces were generated that led to shrinking resources and even, to some extent, shrinking eligibility. For example, Congress put a ceiling of 45,000 on Job Corps enrollees, although Shriver had initially planned to expand to at least 100,000. Tight screening procedures were also instituted for the Job Corps that, in impact, cut the number really eligible for acceptance.

THE GAP BETWEEN POLICY STATEMENTS AND POLICY ACTIONS

An examination of these four programs underlines the importance of the gap that can develop between policy statements of intent and policy actions to implement those statements. This gap occurs in large part because the legitimation process by which the policy statements were affirmed did not, in these four cases, result in serious or major changes in the proposals that were offered. The rhetoric of these proposals suggested that the programs should be relatively large-scale innovative endeavors. In the case of the Job Corps, only two specific concessions were needed in 1964 to keep the original proposal largely intact in the poverty program: an understanding about the place of women in the program and an understanding about the importance of conservation projects and centers in the program. (The agreement at the beginning of the process that state governors should have a veto on Job Corps centers being located in their respective states can be considered as a third concession.)

The Economic Development Act of 1965 coincided almost exactly with the original administration proposal. The only casualty during the legitimation process was an interest-reduction clause in Title II, but this was more than offset by a doubling of Title I authorization and a 20% increase for technical assistance. The other two changes in what was proposed were more symbolic than real: the amendment of eligibility requirements to assure that every state would have at least one area eligible for aid and the prohibition of loans to industries in areas where existing industries of the same type were already in an unfavorable economic situation.

Appalachia was changed in no significant way from what the administration requested. Model Cities was changed significantly in the sense that long-range authorization was not granted on the scale that the administration desired. But the level of initial authorization was about what was asked for and the administrative arrangements were not altered.

In general, then, the most important changes in the four programs took place after initial legitimation of most of what was asked for by the administration. Either bureaucratic reevaluation or legislative amendment has been the primary source of changes. And since the policy legitimation represents the initial policy statement, and since that policy statement tends to be repeated rather than altered, this means that subsequent actions which deviate from the initial statement will continue to widen the gap between statements and actions. In short, the initial statement, which in these four programs was not altered from a fairly ambitious vision of what the programs could and would do, tends to persist regardless of the shape of subsequent actions, and so a large-scale disjunction between statements and actions becomes normal rather than unexpected. Only in the case of the Job Corps has the Nixon administration substantially and clearly altered the original policy statement, although, as indicated below, its policy actions in all four programs have changed over time.

In none of the four programs have the policy actions been as ambitious as the policy statements. Three major factors help to explain this disjunction between statements and actions. First, if a program is formulated as redistributive but legitimated and administered as subsidy, the gap between statement and action almost immediately begins to appear, either because redistributive goals are not being met (because the legitimation, even though perceived as applying to a program of subsidy, included the rhetoric of redistributive ends) or because there is substantial and successful opposition to proposals by administrators to take truly redistributive actions. Both the Job Corps and Model Cities were caught in this bind. EDA and Appalachia were more consistent programs—that is, they were formulated and legitimated as subsidy—thus the gap between statements and actions was smaller.

Second, the seemingly natural tension between the demands of congressional oversight and the demands of administrative flexibility helps produce a gap between statements and actions. New programs are always likely to run into unforeseen problems that demand changes in pace or direction. Administering agencies thus feel that they need a great deal of procedural and substantive latitude to cope successfully with changing and often unexpected circumstances. Yet the typical congressional committee is concerned with predictability and consistency, because the presence of these conditions enables them to conduct their oversight task with less strain and with more self-assurance that they are protecting their own vision of the public interest. Thus, for example, Congress kept toying with the idea of earmarking poverty funds for specific ends rather than allowing OEO flexibility and for a two-year period instituted such earmarking. Congress became intimately involved in the details of Job Corps recruitment procedures and standards. Congress refused to endorse the long-range nature of the original Model Cities proposal but instead wished to preserve more frequent review.

Third, the pressure for short-term provable results from social programs helps exacerbate the gulf between policy statements and policy actions. In these four programs the policy statements tended to be ambiguous—implying that results would be swift and also implying, when read carefully, that the problems dealt with were large and that a reasonably long period would be needed both for experimentation and for implementation before significant results could be observed. In Congress particularly there is constant interest in measurable and immediate results. Thus Congress (or at least selected members thereof) often began to demand evidence of impact before the program had really begun. Overzealous administrators such as Sargent Shriver often laid this trap for themselves by promising considerably more than they could deliver even if the program worked perfectly and on a reasonable timetable. Thus the rhetoric of part of the policy statements is inflated and when early and inflated demands for results are added to this rhetoric, the policy actions often appear to the participants to be particularly inappropriate, slow, and fruitless. Typically, this helps to produce paring and restricting of the program.

PATTERNS OF CONGRESSIONAL SUPPORT

The party support in 1964-1968 for these programs in Congress as measured by roll call votes, both on final passage of authorization measures and on motions that would either add or subtract specific amounts of money, is summarized in Table 2. All these votes were taken during a period with a Democratic President and a democratically controlled Congress.

TABLE 2
PARTY SUPPORT FOR CREATION AND EXPANSION OF FOUR FEDERAL PROGRAMS, 1964-1968
(percentage pro-creation or pro-expansion)

	Job Corps (1965-1968)			Appalachia (1964-1967)		
	n of Votes	% D Supp.	% R Supp.	n of Votes	% D Supp.	% R Supp.
House votes on final passage	4	81	25	2	79	21
House votes on change in specific monetary items	3	72	21	3	84	17
Senate votes on final passage	4	82	44	3	90	54
Senate votes on change in specific monetary items	6	70	23	0	–	–
	EDA (1965)			**Model Cities (1966-1967)**		
	n of Votes	% D Supp.	% R Supp.	n of Votes	% D Supp.	% R Supp.
House votes on final passage	1	83	25	1	73	16
House votes on change in specific monetary items	1	82	8	3	71	13
Senate votes on final passage	1	95	65	1	81	52
Senate votes on change in specific monetary items	2	88	6	2	80	45

This table will support four generalizations. First, Senate Democrats tended to be more supportive of the programs on both kinds of votes than House Democrats. Second, Senate Republicans tended to be more supportive of the programs on both kinds of votes than House Republicans. Third, Democrats in both houses were more supportive of the programs on both kinds of votes than Republicans in both houses. Fourth, members of all four parties (two in each house) tended to be more supportive of the programs on votes involving final passage than on votes involving changes in specific monetary items.

The rank ordering of preferences among the programs within the four parties is shown in Table 3. This table supports the generalization that all parties tended to be more supportive of programs most clearly engaged in subsidy (specifically, EDA) than of programs most clearly engaged in a manipulative enterprise (specifically, Model Cities). Appalachia attracted less support in some of the parties than Job Corps, and, in one instance, than Model Cities. This was not because it is subsidy but because it is subsidy specifically for one region only and this helped produce some opposition from other regions that are not in a position to share in the benefits. Thus these data support the proposition that a program of widespread subsidy is likely to get stronger support in all four congressional parties than either a program of geographically concentrated subsidy or programs involving redistribution.

Few votes were taken on the four programs in 1969 and 1970. The few that were lend support to the generalization that because the Republican President decided not to kill any of the programs in those years but instead requested continued authorization and funding for all of them (even though on a reduced basis in some cases), the level of Republican support in Congress went up dramatically. For example, only 16% of the House Republicans supported final passage of the original Model Cities bill in 1966; in 1969, 97% of them supported a new authorization bill and in 1970, 85% of them supported another authorization bill. During the Johnson administration only about one-fifth of the House Republicans supported authorization bills for Appalachia; in 1969 over half supported an authorization bill. Democratic support continued to remain higher than Republican support on these votes. An EDA authorization bill passed both houses in 1969 by voice vote. The partisan nature of even the Job Corps was partially muted. The passage of the OEO authorization bill in 1969, a bill supported by

TABLE 3
CONGRESSIONAL VOTING ON FINAL PASSAGE OF FOUR PROGRAM AND INITIAL AUTHORIZATION BILLS, 1965-1968[a] (percentage for final passage)

	House Democrats		House Republicans		Senate Democrats		Senate Republicans	
A. Program Authorization Bills								
Highest support	EDA	83	EDA	25	EDA	95	EDA	65
	Job Corps	81	Job Corps	25	Appalachia	90	Appalachia	54
	Appalachia	79	Appalachia	21	Job Corps	82	Model Cities	52
Lowest support	Model Cities	73	Model Cities	16	Model Cities	81	Job Corps	44
B. Initial Authorization Bills								
Highest support	Job Corps	84	EDA	25	EDA	95	EDA	65
	EDA	83	Appalachia	19	Appalachia	88	Model Cities	52
	Appalachia	81	Model Cities	16	Job Corps	81	Appalachia	42
Lowest support	Model Cities	73	Job Corps	13	Model Cities	81	Job Corps	31

a. The number of votes on which these mean figures are based are reported in Table 2.

Nixon, attracted the support of 94% of the Senate Republicans and 58% of the House Republicans. These figures compare to one-fourth of the House Republicans supporting similar authorization bills in the Johnson administration and less than half the Senate Republicans supporting such bills. Again, the level of Democratic support remained even higher than the increased Republican support. The Job Corps also retained some of its partisanly divisive potential. In two votes on specifics of the program in the Senate in 1969 about three-fourths of the Democrats opposed virtually all of the Republicans. But, in general, these four programs had lost much of their controversial nature by early 1971. They had, in a sense, been cleansed of partisan taint by the support of a moderately conservative Republican administration, although the President himself did not use much of his own personal credit in support of any of them. And, in 1971, he proposed removing any remaining Democratic hue by subsuming all four programs in various revenue-sharing efforts.

Figure 1 compares the level of congressional responses to administration requests for appropriations for the four programs. All the programs have had periods of relative decline in congressional support measured in this way. This decline centered in fiscal years 1967 and 1968. All of the programs have had increasing support during the Nixon administration. This phenomenon, coupled with some of the data on roll call voting, suggests that the support of a moderately conservative administration helps produce a large congressional coalition of mildly conservative Republicans, moderate Republicans, liberal Republicans, and most Democrats willing to give the administration most of its modest monetary requests for these programs.

THE BUDGETARY COMMITMENT

Table 4 summarizes appropriations for the four programs through fiscal 1971. In the budget for fiscal 1972 and in the revenue-sharing messages sent to Congress in March 1971, the Nixon administration has provided some signs about the future of the programs (even if the revenue-sharing proposals fail).

Model Cities is in the most ambiguous situation of all: no new funds have been requested for 1972, although previous appropriations will permit estimated spending (net outlays) of $450 million (although in domestic programs actual outlays are almost never as high as estimated outlays). This request for no new funds came

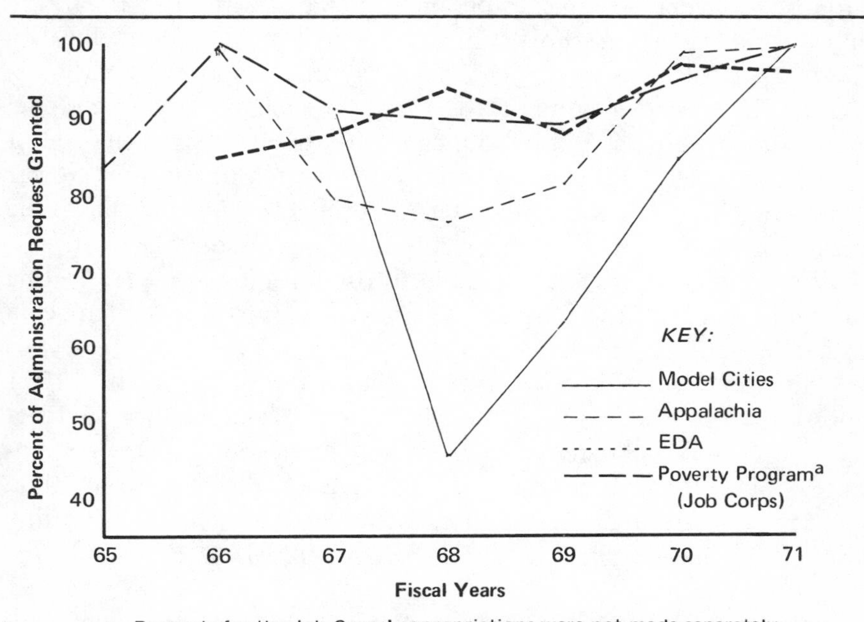

a. Requests for the Job Corps' appropriations were not made separately.

Figure 1: CONGRESSIONAL RESPONSE TO APPROPRIATIONS REQUESTS FOR FOUR PROGRAMS, FISCAL YEARS 1965-1971

TABLE 4

APPROPRIATIONS FOR EDA, JOB CORPS, APPALACHIA, AND MODEL CITIES, FISCAL YEARS 1965-1971

(in millions of dollars)

Fiscal Year	EDA	Job Corps[a]	Appalachia	Model Cities[b]
1965	–	183	–	–
1966	332	310	246	–
1967	296	211	160	–
1968	275	285	127	200
1969	275	280	174	313
1970	271	170(est.)	283	575
1971	252	180(est.)	294	575

a. Appropriations were never made explicitly for the Job Corps. This figure represents the Job Corps allocation from a larger appropriation account (for OEO until fiscal 1970 and for the Labor Department after that).

b. These are appropriations for supplementary grants only.

despite 1971 legislation authorizing $200 million in new appropriations for 1972. In theory, the administration supports the continuance of the programs—at least in its goals—through community development programs developed by states and localities under the proposed revenue-sharing program. In fact, early in 1971 the administration for a short time considered holding Model Cities funds "hostage" in order to spur mayoral support for revenue-sharing. Letters to mayors of 147 cities designated to receive Model Cities funds were drafted announcing a freeze on funds at the end of calendar year 1971. These letters were never sent, however, after representatives of cities protested vigorously. Thus, at best, the future of Model Cities is uncertain. It seems reasonable to suggest that although outlays will continue to rise for a few years (reflecting unspent balances), the program will level off shortly and then begin to decline, at least in relative terms.

The 1972 request for EDA is about the same as the 1971 appropriation: $253 million. Because Congress almost always cuts at least some modest percentage from any domestic request it seems likely that EDA will continue to decline gradually in both absolute and relative terms. If the proposed revenue-sharing for rural community development is passed, EDA will vanish as an independent program.

The Appalachia request for 1972 ($277 million) is smaller than the 1971 appropriation ($294 million) and suggests that although outlays will continue to increase for a few years, the Appalachia program is probably slated to follow the EDA path: gradual decline in both relative and absolute terms. The administration requested abolition of a separate Appalachian program by fiscal 1973 (to be replaced by community development programs based on revenue-sharing), but Congress seems almost certain to keep the program intact.

The administration seems to be planning some modest expansion of the Job Corps from its low points in 1970 and 1971 but it certainly does not seem likely that it will return to the "boom" days of 1966 and 1967, or even the relatively modestly funded days of 1969. In 1969 the Job Corps made obligations of $278.4 million. The obligations in 1970 were $158.2 million and in 1971 are estimated to be $156.2 million. In the fiscal 1972 budget the administration estimates that 1972 obligations for the Job Corps will be $196.1 million. It seems reasonable to expect that it will level off at something like this figure. The Job Corps will also vanish as an independent entity if the President's proposed revenue-sharing in the manpower field were enacted.

When the four programs are viewed together, four general periods of development emerge. Table 5 summarizes these periods.

The Johnson administration was enthusiastic, at least at the rhetorical level, about all four programs. The future of each program was, in turn, portrayed as being one of expansion and increasing effectiveness. This was particularly true in 1964 and 1965 when the Job Corps, EDA, and Appalachia programs were created.

The initial rhetoric was the same for Model Cities in 1966, but with the increase in Vietnam spending beginning in 1965 and 1966, the enthusiasm for all four programs waned in terms of what resources were asked for and allocated to them. Thus levels of attention and interest declined throughout 1966, 1967, and 1968. Only in 1968 did some interest in growth reappear in the Johnson commitment—at least for Appalachia and Model Cities. And the decline of Job Corps and EDA was checked.

The Nixon administration very early charted its course clearly. It continued deemphasizing EDA, immediately made a large cut in Job Corps, supported Model Cities at a modest level mainly as a regulatory program for intergovernmental relations and interagency coordination, and showed some interest in Appalachia. The Nixon administration could naturally be skeptical of all four programs, since they were all part of the "Great Society" and had largely Democratic origins. The one exception was Appalachia, which was

TABLE 5

PERIODS OF DEVELOPMENT FOR EDA, JOB CORPS, APPALACHIA, AND MODEL CITIES PROGRAMS IN JOHNSON AND NIXON ADMINISTRATIONS

	Period 1 Early Johnson Administration (1964-1966)	Period 2 Late Johnson Administration (1966-1968)
EDA	Creation in 1965; enthusiasm for potential impact of program; relatively high level of funding in 1965 and 1966	Gradually declining level of funding in 1967 and 1968
Job Corps	Creation in 1964; great enthusiasm for potential impact of program; relatively high level of funding in 1964 and 1965	Large funding cut in 1966; moderate increase in funding for 1967; small decrease in 1968; original ambitious plans for whole poverty program cut back severely
Appalachia	Creation in 1965; great enthusiasm for potential impact of program; relatively high level of funding in 1965	Substantial cuts in funding in 1966 and 1967; some growth in 1968
Model Cities	Creation in 1966; short-lived enthusiastic and ambitious rhetoric about goals and potential of program	Relatively low level of funding in 1966, 1967, 1968; goals shift from redistribution in society to administrative matters (coordination and strengthening local units of government)

	Period 3 Early Nixon Administration (1969-1970)	Period 4 Projection of Short-Term Future Under Nixon (1971-1972)
EDA	Continued gradually declining level of funding in 1969 and 1970	Continued gradually declining level of funding; no threat to existence of program
Job Corps	Funding severely cut in 1969; original concept of program abandoned; funded at new low level in 1970	Stability at moderately low level of funding; probably no threat to existence of program until after 1972 presidential election
Appalachia	Revival of interest in program; increased funding in 1969 and 1970	Gradually declining funding
Model Cities	Modest cut in proposed funding in 1969; stability of funding at relatively low level in 1970; continued stress on administrative goals	Continuation of program at relatively low and declining level; continued focus on administrative goals

a. Assumes that President Nixon's revenue-sharing proposals encompassing these programs will not be enacted.

certainly important as part of the early Great Society proposals but had a history of Republican support in the affected region. The program was also attractive to the new Republican administration because it stressed local and state initiative, planning, and control and this stress was consonant with Republican political philosophy. EDA was not anathema to Republican thinking, although it seemed to be a program on which some savings could be effected. Model Cities, had it retained its originally redistributive implications, would have been distasteful to the Nixon administration. But since it had become administratively oriented late in the Johnson administration it continued to be attractive to at least some Nixon officials as a tool for accomplishing some of the administrative revamping that had been promised the country in the 1968 campaign. The Job Corps was clearly redistributive and the one program of the four most clearly bearing a "made by Democrats" label. Therefore, it is little wonder that it received attention only in the sense of getting an early and vigorous pruning.

A FRAMEWORK FOR POLICY ANALYSIS

Summary data have been presented on four programs and a series of comparative statements have been made about the programs. This way of commenting on public policy has considerable validity. Yet, to further the comparative enterprise, a more complete analytical framework is needed. The remainder of this article suggests such a framework in general terms and then, in very brief form, indicates the utility of the framework in analyzing the four programs just discussed.

The *governmental policy arena* is the locus of those governmental responses to the environment that include statements of intent against which subsequent actions by the government and results in both government and society can be measured (compare Eyestone and Eulau, 1968: 39; Ranney, 1968: 7; Rosenau, 1968: 222). In the language of psychology, policy-making is the response of the organism we call government to stimuli coming from the environment. Policy is not the *only* response government can make to an environmental stimulus; it is a special kind of response—the central ordering element of which is an explicit statement of intent on the part of government, the direct result of a deliberative process.

The policy arena has three constituent elements: the *policy-relevant environment,* the *policy process,* and *policy responses* (also trichotomized–into *policy statements, policy actions,* and *policy results*).

POLICY-RELEVANT ENVIRONMENT

The concept of environment has always been a troublesome one for those wishing to distinguish between a system and an environment. If a system is viewed as "open," however, the necessity for viewing environment as something outside of the system vanishes (see McClelland, 1961: 414-417; Easton, 1965: 17-21; Katz and Kahn, 1966: ch. 2).

The notion of *policy-relevant environment* is here proposed as useful. This is essentially what decision makers, broadly defined, perceive to be relevant to their endeavors. It must be noted, of course, that parts of the environment not perceived by decision makers may also have a profound impact on policy responses.

The policy-relevant environment includes four main elements. First, there is the perception by policy makers of basic conditions and trends and specific events in the world or in whatever part of the world is relevant to any given policy maker or potential policy maker. Obviously, these conditions, trends, and events do not have a uniform impact on all participants. Some may react strongly to some features while others are oblivious to those features.

Second, the perceptions of need for a given kind of action or inaction on the part of the various participants (both policy makers and potential policy makers) constitutes part of the policy-relevant environment.

A third element in the policy-relevant environment is the perception on the part of the various participants of the potential social impact of various specific courses of action and of the specific techniques appropriate to achieving a given impact. In the domestic sphere some programs are perceived as being largely distributive in impact. Lowi (1964) has summarized the nature of distributive policies: they are

> characterized by the ease with which they can be disaggregated and dispensed unit by small unit, each unit more or less in isolation from other units and from any general rule. "Patronage" in the fullest meaning of the word can be taken as a synonym for "distributive." These are policies that are virtually not policies at all but are highly individualized decisions that only by accumulation can be called a policy. They are policies in which the indulged and the deprived, the loser and the recipient, need never come into direct confrontation.

Governmental techniques appropriate to the achievement of distribution are those of subsidy. Some perceptions are focused on the fact of distribution; some on the specific mechanics of subsidy; and some on the details of how the techniques of subsidy achieve the goal of distribution.

Other programs in the domestic sphere are perceived as being largely regulatory in nature. Lowi (1964) says that

> Regulatory policies are also specific and individual in their impact, but they are not capable of the almost infinite amount of disaggregation typical of distributive policies. . . . Regulatory policies are distinguishable from distributive in that in the short run the regulatory decision involves a direct choice as to who will be indulged and who deprived.

There are specific techniques of regulation that seem most appropriate to achieving regulatory ends. Again, the perceptions of participants can be on the ends, or on the techniques, or on both simultaneously.

Finally, still other domestic programs are perceived as being largely redistributive in nature. Lowi (1964) speaks of redistributive policies as being

> like regulatory policies in the sense that relations among broad categories of private individuals are involved and hence, individual decisions must be interrelated. But on all other counts there are great differences in the nature of impact. The categories of impact are much broader, approaching social classes.

Here the question of who wins and who loses in broader than individual terms becomes paramount. Techniques aimed at manipulation of the environment to produce "winners" and "losers" on a mass basis are appropriate to achieving the end of redistribution. The perceptions of the participants are again divided between the goal of redistribution, the techniques of manipulation, and the link between techniques and goal achievement.

Fourth, the perceived history of governmental attention to a problem or function is part of the policy-relevant environment.

POLICY PROCESS

The *process* through which policies emerge depends in large part on the interrelationships between a variety of participants. The character and behavior of coalitions favoring specific governmental activity is an important part of the process.

Process also includes a number of elements essentially within specific units of the government: specifically, characteristics of and norms about functions and goals in institutionalized policy-making units; and characteristics of and norms about internal decision-making processes (e.g., communications, authority, control, conflict management, change, and innovation) in institutionalized policy-making units.

POLICY RESPONSES

A *policy statement* is a declaration of intent on the part of the government to do something. This declaration is sometimes highly visible—as in a statement by the President or in the passage by Congress of an important new statute. This declaration can also be invisible to most of the public—as in a memo by a bureau chief to a subordinate. Different levels of the government and different parts of the government may make conflicting policy statements simultaneously. Even the same participant in the policy process may make conflicting statements within a relatively short period of time.

Not all statements by governmental actors are policy statements, however. The function of the latter is explicit in its attempt to order the responses of government. Accordingly, the policy statement always implies some choice among alternatives.

In addition to its deliberative source, moreover, a policy statement is identified by its attempt to

(a) guide, induce, or constrain subsequent decisions, statements or actions, with a view to preserving or attaining a desired state of affairs;

(b) delimit the decision latitude of actors subordinate to, or subject to sanction by, the promulgating agency;

(c) provide to interested parties a forecast or expectation of the contingent responses to future stimuli; and

(d) lend an underlying consistency or coherence to a continuity of actions expected to follow the establishment of a rule. This frequently may (but need not) take the form of a justification for the rule and the contingent behavior.

The policy statement is not the totality of policy and, by itself, it is inadequate either to explain government's interaction with environment, or to evaluate its result. But it is nevertheless the central ordering element of the policy arena, for it is the policy statement that links governmental intent to governmental behavior, and, therefore, provides the yardstick against which to judge the former and assess the latter.

Policy action is what the government actually does, as distinguished from what it has said it is going to do (sometimes with many and conflicting voices) in its policy statement.

Just as not all statements can be considered policy statements, so many actions cannot be considered policy actions. Often governmental actors behave in ways that have no bearing—positive or negative—on any previously expressed governmental intentions. Such behavior is essentially irrelevant to the policy arena. However, the results of such actions may subsequently become important as a stimulus to subsequent activity.

A *policy result* is what happens in society or in the government itself as a result of the government's policy statements and actions. A policy result is predicated upon the prior existence of both a policy statement *and* a policy action.

Figure 2 summarizes the analytical framework that is here proposed as useful. The arrows simply indicate that there are two-way interrelationships between all of these elements. The triangular configuration indicates that this framework is not chronologically bound.

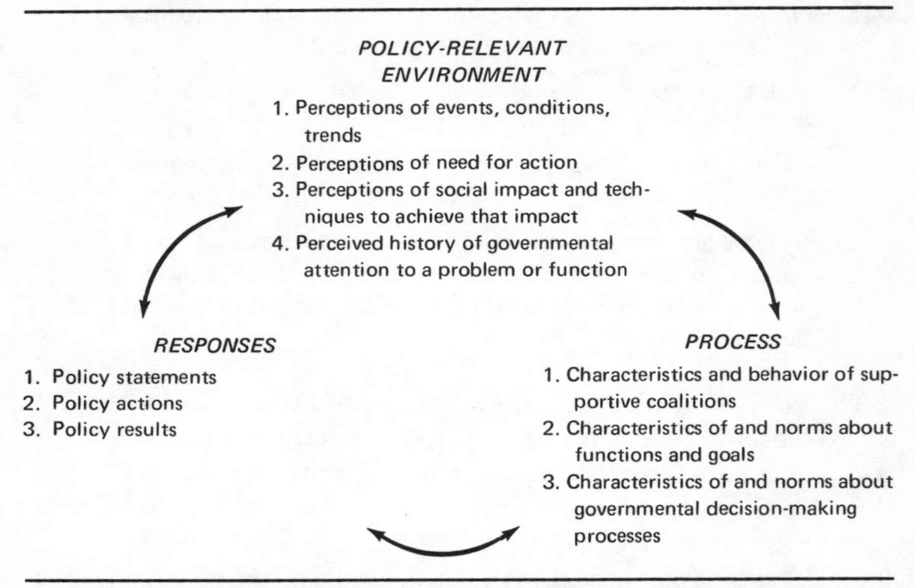

Figure 2: A POLICY ANALYSIS FRAMEWORK FOR FUTURE RESEARCH ON GOVERNMENT PROGRAMS

THE UTILITY OF THE FRAMEWORK

What results stem from the application of this framework in analyzing EDA, Job Corps, Appalachia, and Model Cities? First, a study of these four programs supports the assertions that the policy-relevant environment does help shape both the policy process and policy responses and that policy process and policy responses both have an impact on each other.[2] Second, the specific elements identified in the policy-relevant environment and in the process clusters are helpful in explaining what happened in the specific programs.

ELEMENTS IN THE POLICY-RELEVANT ENVIRONMENT

Perceptions of events, conditions, and trends

In the case of EDA there were widely shared perceptions of administrative and political failings on the part of the Area

Redevelopment Administration, the direct forerunner of EDA. These perceptions resulted in a variety of specific changes in the character and mandate of EDA compared to ARA.

In the case of the Job Corps there were perceptions of the magnitude and seriousness of poverty, particularly among urban black young people, that helped lead to the formulation of proposals for what became the Job Corps. Once the program began, perceptions of trouble (racial incidents and crime in towns near Job Corps centers) at Job Corps centers helped lead to a variety of administrative restrictions and changes designed to suppress further trouble.

In the case of Appalachia disastrous floods in 1957 and particularly in 1963 were perceived in such a way as to point up the general needs of the area. Likewise, perceptions of a special commitment made by John Kennedy to the state of West Virginia during and after his primary campaign there in 1960 helped lead to programmatic initiative designed to meet the problems of the region.

In the case of Model Cities, liberals and conservatives alike shared unfavorable perceptions of federal housing programs before 1966 and this led them both to seek new alternatives. Perceptions of the importance of the urban riots that began in the summer of 1964 also helped spur policy statements and actions in this area.

Perceptions of need for action

Despite the shortcomings of ARA, a large number of policy participants believed in the need for the continuation of a "depressed areas" program of some kind. This helped lead to the specific design of EDA. In the case of the Job Corps a large number of the policy makers perceived a specific need for a program to train young dropouts. In the case of Appalachia there was widespread agreement on the economic and human resource needs of the region. In the case of Model Cities there was considerable concern for the needs of the cities, although shortly after the initiation of the program it became clear that the most widespread agreement was on the need for strengthening local governments and the need for interagency coordination of programs aimed at the cities, not on specific programs with the residents as direct beneficiaries.

Perceptions of social impact and techniques to achieve that impact

There was general agreement that the programs of EDA and Appalachia would be distributive in impact and would involve techniques of subsidy. There was agreement that the programs of the Job Corps would be redistributive in impact and would involve techniques of manipulation (and this agreement helped lead to the decline in the popularity of the Job Corps). There was an early lack of agreement in the case of Model Cities. Some of the liberal proponents urged a view of the program that put it clearly in the redistributive-manipulative arena, but these views were never dominant and thus agreement was reached on the basis of a program of distribution and techniques of subsidy and, increasingly, on a program aimed at regulation of administrative arrangements and relations. All of these perceptions helped shape the specific content of laws and administrative actions in the four programs.

Perceptions of the history of governmental attention

The perceptions of the various facets of the ARA experience help explain what happened in the case of EDA. The perceptions of past federal concern with the job market for relatively uneducated and hard-to-employ youth (the Civilian Conservation Corps and the proposed Youth Conservation Corps) help explain what happened in the case of the Job Corps. The perception that the federal government had ignored the unique problems of Appalachia helped shape what happened in that program. The perception of the existence of an overly large number of categorical grant programs helped condition developments in the Model Cities program.

ELEMENTS IN THE PROCESS

Characteristics and behavior of supportive coalitions

The pro-EDA coalition has been relatively stable and relatively bipartisan in nature. The coalition that first sustained an expansionist Job Corps disintegrated and was replaced by a more conservative, bipartisan coalition that would support only a minimal Job Corps enterprise. The coalition behind Appalachia at first was visible as a Democratic coalition and helped in the creation of a moderately ambitious program; bipartisan elements in the coalition that were originally less visible have now emerged clearly and are helping sustain a growing program even in a Republican administration. The coalition behind Model Cities almost immediately became bipartisan and dependent on moderates and moderate conservatives. This helped shape the nature of the evolution of the Model Cities program.

Characteristics of and norms about functions and goals

The policy makers have been virtually unanimous in agreeing that EDA should pursue a program of subsidy in many different geographical areas. Policy makers agreed that the Job Corps was to seek to transmit skills to young dropouts that would make th employable in decent jobs and there was also agreement that this goal involved the Job Corps in a redistributive enterprise. Policy makers working with the Appalachia program agreed that the goal of the program was the economic rehabilitation of the region through programs of subsidy. Policy makers dealing with Model Cities reached agreement on a series of regulatory goals for their programs. Each of the programs pursued these goals consistently within the constraints of available resources.

Characteristics of and norms about governmental decision-making processes

In the case of the Job Corps, the lack of hierarchical control—except through the personal interest of Shriver—and controversies over screening procedures for applicants, and the distance of enrollees from their homes, help explain some of the problems that have led to a shrinkage of the program. The expectations about the leadership role of the Appalachian Regional Commission, coupled with its initially weak position that was gradually strengthened, help explain the relatively slow start of the programs followed by somewhat more aggressive performance. The great personal interest of the Secretary of the Department of Housing and Urban Development in the Model Cities administration helps explain the continued high visibility of the program despite shrinking funding and low interest in the White House.

CONCLUSION

In this paper four federal programs have been analyzed comparatively, a framework has been presented that is asserted to have some promise for future policy research, and its utility has been hinted at by applying it in brief summary fashion to the four programs. The framework is only a rough beginning and is offered in an exploratory and tentative mood. The conviction that lies behind it is that social scientists have many as yet unrealized contributions to make to the study of public policy. Comparative policy studies demand much theoretical refinement, hypothesis generation, and data base construction. As these tasks proceed, social science will become "relevant" to society in the most meaningful way.

NOTES

1. Because of the summary nature of this article, space limitations, and the desire to promote readability, references to primary sources (mainly government documents) are omitted here. Readers interested in those citations should see Ripley (1972). The most useful secondary references for the four programs, in addition to newspapers, Congressional Quarterly, and National Journal, follow: Davidson (1966); Donovan (1967); Kershaw (1970); Levitan (1969, 1964); Moynihan (1969); Sundquist (1969, 1968); Weeks (1967).

2. The study of these four programs did not lend support to the assertion that process and responses have a direct impact on policy-relevant environment but neither was there any evidence supporting the assertion that there is no impact running from process and responses to policy-relevant environment. Rather, lack of sufficient data simply made a comment on this relationship impossible.

REFERENCES

CLEAVELAND, F. N. [ed.] (1969) Congress and Urban Problems. Washington: Brookings.

DAVIDSON, R. (1966) Coalition-Building for Depressed Areas Bills: 1955-1965. Indianapolis: Bobbs-Merrill.

DONOVAN, J. C. (1967) The Politics of Poverty. New York: Pegasus.
EASTON, D. (1965) A Systems Analysis of Political Life. New York: John A. Wiley.
EYESTONE, R. and H. EULAU (1968) "City councils and policy outcomes: developmental profiles," in J. Q. Wilson (ed.) City Politics and Public Policy. New York: John Wiley.
KATZ, D. and R. L. KAHN (1966) The Social Psychology of Organizations. New York: John A. Wiley.
KERSHAW, J. A. (1970) Government Against Poverty. Chicago: Markham.
LEVITAN, S. A. (1964) Federal Aid to Depressed Areas. Baltimore: Johns Hopkins Press.
--- (1969) The Great Society's Poor Law. Baltimore: Johns Hopkins Press.
LOWI, T. J. (1964) "American business, public policy, case-studies and political theory." World Politics 16 (July): 677-715.
McCLELLAND, C. A. (1961) "Applications of general systems theory in international relations," in J. N. Rosenau (ed.) International Politics and Foreign Policy. New York: Free Press.
MOYNIHAN, D. P. (1969) Maximum Feasible Misunderstanding. New York: Free Press.
RANNEY, A. [ed.] (1968) Political Science and Public Policy. Chicago: Markham.
RIPLEY, R. B. (1972) The Politics of Economic and Human Resource Development. Indianapolis: Bobbs-Merrill.
--- W. B. MORELAND, and R. H. SINNREICH (1972) "Policy-making: a conceptual scheme." Amer. Politics Q. 1 (forthcoming).
ROSENAU, J. N. (1968) "Moral fervor, systematic analysis, and scientific consciousness in foreign policy research," in A. Ranney (ed.) Political Science and Public Policy. Chicago: Markham.
SUNDQUIST, J. L. (1969) Making Federalism Work. Washington: Brookings.
--- (1968) Politics and Policy. Washington: Brookings.
WEEKS, C. (1967) Job Corps: Dollars and Dropouts. Boston: Little, Brown.

20

The Politics of Urbanization

MATTHEW HOLDEN, Jr.

□ IT IS NOW TIME FOR POLITICAL SCIENTISTS to consider urban reform—not as a matter of mere moral preachment, nor in the style of academic Elmer Gantrys, nor even as our equivalent to sociology's "social problems" course, but as a serious matter. Intellectual coherence and practical relevance depend on whether we agree with the dour judgments of Banfield's *Unheavenly City* or with Norton Long, Peter Rossi, Robert Agger or others in the legion of Banfield critics. The question posed by the Banfield debate is whether Manchester liberalism is really appropriate to twentieth-century urbanism: whether it is true that government has so little capacity to do useful things without worse spillovers that it should be restricted to the most limited functions. (Later, we even see that the question of *the capacity of government* arises with respect to the basic function of achieving and maintaining public order.)

The relevance of the reform tradition in political science, as a source of articulating the capacity of government question, is the

---

AUTHOR'S NOTE: *The author gratefully acknowledges the generous support of the Center for the Study of Public Policy and Administration and its director, Professor Clara Penniman, in the preparation of this paper.*

first point considered in this essay. The second question considered is the specific relation of urbanism to political problems, to those political problems about which we ask whether government has a capacity to manage. The third question is a specific discussion of the capacity of government in the urban setting under present conditions.

These three questions, together, go some way to define how we may look at the relation between the public and political situations in cities and how political science may improve its comprehension of the politics of urbanization.

OLD REFORM AND NEW REFORM

Political science has been closely associated with problems of urban reform—perhaps symbolized by the number of named or endowed chairs in political science (such as the Dorman B. Eaton Professorship at Columbia) in honor of some figure in urban reform. There are two reform traditions which I may call old reform and new reform. Old reform came out of the nineteenth century and was aimed chiefly at "good city government." Good city government was never defined very precisely, but it surely meant two things. City governments should deliver services efficiently, in the manner that scholars—who had far more comparative experience than contemporary scholars have—judged British and European cities to do. (How the comparative tradition was lost would be an interesting story in intellectual history itself, but the writers before 1920—such as Goodnow, Rowe, A. Lawrence Lowell, John A. Fairlie, William Bennett Munro, Albert Shaw, et al.—quite commonly built their analyses around comparisons of four countries: the United States, the United Kingdom, France, and Germany.) Central to this, though not sufficient by itself, was an attack on corruption, i.e., on local officials' use of legal authority in order to enrich themselves privately. The objective was to diminish the chances that a Plunkitt might see and take.

Precisely how much corruption there was at the time is uncertain. Most who wrote on the subject were adversaries, with every incentive to dramatize the matter. But there is little doubt that it was very substantial indeed. As a matter of history, the baseline of the concern with reform was the period between the Civil War and 1900. In that period, the scale of government was increasing rapidly, while the procedures for accounting and publishing were still rather crude. With much bigger amounts of money in city treasuries (Holden, 1965) there was more to be taken by outright theft and by disguised theft, such as would occur if a public buying agent permitted a private seller to overcharge the government by, say, 25% and then himself got a rake-off of 10%. Beyond that, with a large number of public decisions to be made on such matters as gas, electric, streetcar and telephone franchises, it was easy for "merit" criteria to be overcome by the criterion of selling to the highest bidder (except that the auctioneer would be the politician in his private interest rather than his public role). Political scientists did not create the old reform, but they were a part of it. At the time, they seemed to come to the problem with an institutionalist bias—although the writings of the same political scientists, such as Goodnow and Munro, indicate a far more sophisticated conception of urban politics than the popularized reform literature would suggest. (This raises the interesting problem about the political scientist as public actor. For one suspects that Goodnow et al. found then what one can find now—that educated and sophisticated nonacademicians are delighted and anxious to have academic authority in support of them, but are reluctant to regard such authority if it does not support their preexistent convictions. This has, at any rate, been my observation in public commissions and organizations, and is compatible with the experience of the economists as reported by Heller.)

In any event, the political science literature associated with the old reform seemed to presuppose formal structure as a major variable in determining the conduct of local government and local officials. But the old reform political science was not simply or naively "formalistic." Nor is it even fair to say that the old reform was simply a commitment to upper-class values and interests. There is merit in that position at least to the degree that the old reform as practical politics more often than not got its support from the parts of the urban population that were better off. It is also true that, in both practical politics and political science, the old reform became more and more focused upon a relatively small set of institutional proposals. These proposals, virtually the program of the National Municipal League, were, if not "conservative," at least not calculated to alter the status quo in favor of the less powerful. But it is not correct to suppose that the political institutions being rearranged were very much more satisfactory to lower-class interests. The "machine" was not an expression of "lower-class interests or values." One might as well call the factory an expression of lower-class interests—if not of lower-class values. After all, it certainly provided the lower-class man his income far more reliably than did any political machine. The machine was an organizational form by which political entrepreneurs—upper class and lower class—used the lower-class population as resources to achieve entrepreneurial dominance and enrichment through the manipulation of government.

Without the businessmen, the machine would have been far less profitable, if not in fact inconsequential. As Lincoln Steffens saw, there was a very close connection between the machine as it existed in American cities, the rapid growth of city government, and the bifurcation of the local elites into "economic" and "political" sectors, and the businessmen's need for some political instrumentality. The businessmen were always accustomed to having governmental decisions made to suit private economic purposes and the former presumption of the nineteenth century was that government ought to aid business (Merrill and Merrill, 1971). The ceaseless beat of the economic upon the political frontier, borrowing as we may from Walton Hamilton, continued to create conditions calling for special favors. The unwelcome novelty was the need to pay for those special favors.

Political corruption, through payments to machine politicians, became the necessary—but despised—technique of big businessmen who needed rapid and favorable decisions. The market atmosphere of competitive, semi-organized local politics was what made this mode of operation necessary. Thus, insofar as reform was supported by upper-class interests—and it was by no means always so—it was a strategy for control without having to pay for agreement on what businessmen conceived they had a right to take for granted.

The old reform had a less well-remembered side of social meliorism. The lay expression of social meliorism, as given by Edwin L. Godkin (1898: 181), probably would have been accepted by social scientists as well:

> It is we who ought to have shown the Old World how to live comfortably in great masses in one place. We have no city walls to pull down, ghettos to clear out, or guilds to buy up, or privileged to extinguish. We have simply to provide health, comfort, and

> education, in our own way, according to the latest experience in science, for large bodies of free men on one spot.

Much of the meliorist vein in the old reform took for granted the possibilities of conservative reform not challenging the existing distribution of property or power. Some aspects of the old reform became more challenging in that sense (Johnson, 1911; Perkins, 1946; Steffens, 1931). But whatever variety one emphasizes, a good deal of reform politics at the time was concerned with sanitation and the housing of the poor; what the critics then deemed useful improvements in the court system; a shift in emphasis from "punishment" to "correction" in the jails; and the development of modes for rapid "Americanization" of the immigrant population.

The normative commitment of political scientists, whether to fiscal probity in government or to social change, was at least as strong as the normative commitment of many current political scientists to such ends as environmental rectification, equality of educational opportunity, or racial equality. Their incentive to hold onto their preferred values was at least as great as is now our own. And their ability to reduce dissonance by ignoring inconvenient evidence was at least as great. Moreover, their scientific criteria were even less well developed, so that they had even more excuse for not knowing when they were wrong.

Consequently, they lost critical detachment about the reform prescriptions they were issuing—as we now show some signs of doing again—with bad results for both their science and their relevance to policy. Quite a long time ago, it became clear that many of the prescriptions were not sufficiently in accord with everyday realities to deal with commonsense experiences.

There is a similar risk in the new reform emergent in political science. New reform chiefly emphasizes the capacity of government to satisfy the claims of certain clearly underprivileged groups: Afro-Americans, Chicanos, Puerto Ricans, Indians. (For some advocates of new reform, the scope also extends to "women," "youth," and possibly "workers," although interestingly it does not extend to Appalachian whites in Northern cities or to "white ethnics"—who are a critical part of the facts covered by the term "workers.") New reform as political science is born more of psychic disillusion than of analysis, although much of it, as far as racial issues were concerned, was catalyzed by the violent experiences of 1967. (One wonders—as suggested before—about political scientists who before had no conception of a political system in which force and violence played key roles.) There is also a certain tendency for new reform to be assimilated retrospectively to the "elitist" side of the "community power" debates of a decade ago. That is, those who are most explicit about their new reform commitments also turn out to interpret the evidence such that: if "pluralism" were to be found, then the interests with which they identify would be served rapidly (by their standard) and satisfactorily (by their standard). Since it is evident that this is not occurring–although criteria for rapidity and satisfaction are not yet available–they conclude that the pluralist hypothesis indicates the realities of community power less adequately than does the elitist hypothesis. This brand of new reform is thus far represented by the large number of academic books that concern themselves with the "unworkability" of American public institutions, and produce analyses suggestive of the "need for radical change" and the fruitfulness of "confrontation," if not indeed of "violence" (e.g., Greenberg et al., 1971; Parenti, 1970). New reform is analytically careless in the implicit–if not explicit–that somehow a pluralistic polity means a Pareto-optimal polity in which everybody wins at once, something actually known only in utopias of welfare economists and others. New reform has, moreover, a certain shallowness in its appreciation of social conflict and the requirements of conflict resolution. New reform may at times overestimate the degree to which "justice" will produce "happiness" and happiness social tranquility; as it may also suffer from the illusion that viable political results can be achieved if the decision makers will attend to those audiences of which new reformers most approve without worrying about all the other pertinent audiences. But it has at least the advantage of forcing us to think more carefully about the capacity of government to deal with "urban problems."

CAPACITY OF GOVERNMENT AND PROBLEMS OF GOVERNMENT

THE MEANING OF URBANISM

To describe, explain, or anticipate the capacity of government to handle urban problems implies for political science more than attention to the wheeling and dealing of politicians who happen to be located in cities. It means some approach to the specifically urban within the generically political, in the manner that urban economics handles its comparable subject matter (Thompson, 1965; Wingo and Perloff, 1968). This emphasis, like the emphasis on reform, is a return to urbanism and political problems.

If political science would describe, explain, or anticipate the capacity of government to handle urban problems, then it would also have within the early tradition (Goodnow, 1909, 1904, 1903, 1898; Wilcox, 1897) which sought to consider the city as "a social fact" substantially independent of any particular system of "national law."

What are the features of the city? It is first of all the human settlement utterly dependent on trade and transport–and subject to extreme privation in a short time if trade and transport break down or are interrupted. Is this obscure? Then ask how long New York City could support itself if all food shipments into the city were interrupted at high noon today–a problem no more fantastic in these days when technological extortion seems possible than it was for Paris at the time of the French Revolution (Cobb, 1970, III: part I) or for Spanish cities at the time of the Civil War (Bell, 1966). The city is also the human settlement with the large population, although how large is always a question. This is a tricky question because it is all confused with the "functions" of the city: communications node or message center (one is genuinely in the country when one reaches the town at which it is not possible to reach more than one television station); labor center to which people gravitate from the countryside; or entrepot through which goods pass for distribution to the hinterland or into which other goods are collected for transmission to other entrepots. But whatever one may say of these functions, the city is surely the human settlement with a large enough population that no one could have an overview of the whole system or personal knowledge of all or even most of the other members.

The city is not merely artificial and large, but it is also dense. Urbanization soon tends to mean that densities not merely increase in a given area, but that the adjacent physical area is also affected and densities increase in the adjacent areas as well. (For these purposes, we may also include the process of "suburbanization" or "metropolitanization.") Densities sometimes increase in the adjacent areas without declining in the original built-up areas, or it could mean that densities increase in the adjacent areas–while declining in the original built-up areas. But in any case, it will mean that the "metropolitan" area as a whole is more dense at any Time 2 . . . Time N than it was at some Time 1. People do not come together merely as discrete human beings, but as members of ongoing sociocultural systems. Insofar as urbanization occurs, therefore, it involves a good deal more than increases in the density of population as measured by the impact of discrete human beings one upon another. It also means that the collectivities have many more interactions with other collectivities, and consequently have many more conflict and boundary adjustment problems.

The city embodies all the problems of politics. This makes it a convenient research situs, which is the point of view from which most urban political research has been conducted. But the city as research situs is not our present interest. *Urbanization* is our present interest. Following the lead of the first generation of political scientists (Goodnow, 1909, 1904, 1903, 1898; Fairlie, 1908; Rowe, 1908) we are interested in what urbanization mandates or implies for politics. (For the moment, we bypass the possibility of considering politics as an independent variable for urbanization, although this was clearly important in the ancient and medieval worlds where city growth often was determined by prior political facts such as the existence of an armed force sufficient to offer protection and thus to act as a magnet for trade and population. The existence of such a city as Brasilia represents a political determination of urban significance in the present world, as does the apparent emergence of the new towns policy in the United States government. But for the moment, we leave aside the political determinants of urbanization to consider only the political implications.)

No one has ever been able to prove in any elegant way that urbanization does have political consequences. But American political history seems to provide clear verification. The politics of the old reform might, on the issues alone, have been appropriate to American small towns. But that politics, weak though it was in the cities, could only function in the cities. And the civil rights movement was visited by the violence of its adversaries everywhere, but it was—and is—in the rural counties where life was roughest.

The small town is actually the whole community functioning as a committee and in committee politics the first rule of thumb is: avoid rocking the boat at all, if you can (Holden, 1965). The committee, or the small town, is the unit in which it is almost assured that the elite will be a very few people, and that the elite will be able to maintain effective surveillance over all potential sources of opposition and criticism and to control most of the economic resources, information sources, symbol sources, and physical force. The small-town person or group that "obstinately" remains dissentient encounters a greater likelihood of penalties that make a difference: the much lessened chance of a home loan or a line of credit, a more severely damaged reputation, severe social isolation, or even physical death. If controversy must arise, it must be resolved and smoothed away as soon as possible; and if controversy persists, then intense factional politics also results (Gunn, 1966; Singham, 1968).

Urbanization—and not merely the wheeling and dealing of some politicians who happened to be in cities—was a very important consideration in the initial days of political science as a discipline. Leo S. Rowe (1908), for example, sought to define "the political influence of cities" from at least two points of view: "the effects of city life upon the political thought and action of the population . . . [and] the influence of the cities upon the political life of the nation." Either emphasis suggests that virtually the same themes are played by Herson (1957) in an iconoclastic article, and by William Anderson (1957) who responded critically to Herson. Herson's readers may have regarded him as an advocate of a "new" posture. But his proposal (Herson, 1957: 344) to focus upon "explanations of urban political behavior . . . [and for] this purpose, to draw upon urban social-psychology" is precisely in tune with Rowe's paper more than a half-century before. William Anderson, the elder statesman to whom Herson's critique was most objectionable, implies an intellectually far-reaching stance in a deceptively simple sentence. "Urban concentration of population is advancing rapidly. It is changing the nature of politics not only in the urban places but throughout the [American] states, the nation and the world." These questions must be understood as variants on the basic practical problems of politics—urban or otherwise—that may be seen as five clusters: personnel, finance, structure, policy, and constitution-making. The problem of structure indicates one facet of urbanization. In a sense, the village—the small unit described above—is also a unit in which there is no need for a "complex" bureaucracy in the sense of an organization with multiple levels or hierarchical authority. Nor, if it be small enough, is there any real need for specialized agencies of representation; e.g., wards.

But the problem of structure also related to the territorial distribution of governments, contingent on the bigger problem of suburbanization. Since urbanization means increased densities, one of its consequences is outward movement from the urban center, as some people and institutions look for new portions of the metropolis in order to pursue activities which they find can no longer be pursued in the central areas equally conveniently. Today, we are in need of much better understanding even of the simple question of why suburbia should have to be regarded as a specially favored place. After all, historically this was not so—and Covington, Kentucky; Newark, Cicero, and the downriver suburbs of Detroit are quite as much suburban as anything populated by the Anglo-Protestant upper classes. Suburban growth as we have seen it in the United States began in the 1890s, but is chiefly a function of the years since 1900. Transportation developments (the suburban railway, e.g., "the Main Line," "the IC," and so on) made it possible for people who wanted to escape the lack of amenities of the existing city to go quite far out in distance and yet retain access to their office, firms, stores, and amusement facilities. This was expedited by the development of the internal combustion engine, the rubber tire, and the hard-surfaced road, for a mass potentiality then existed for suburbanization. The fact that the upper-middle and upper classes were now going into the new kind of suburbs gave the idea of suburbia a certain fashionable snob appeal that could be manipulated. But snob appeal did not make much difference until large numbers of people could make the physical movement. The possibility of the physical movement could not be consolidated until large numbers of people could afford the outlying residence, which is chiefly a function of the growth of home mortgage lending, particularly as inspired and protected by various federal insurance and subsidy programs. Once these three factors—fashion, transport, and finance—were in combination, mass suburbanization was inevitable.

Part of the effort of the mass suburban movement, however, was to develop each little area as a "community" and to screen out residents deemed undesirable from the perspective of that sort of community. This brings us to the practical problem of the very present decade, namely, the conflict between a goal of spreading out various racial and income groups throughout the metropolitan areas, and the metropolitan areas' use of zoning and land use controls as a means of keeping each community isolated and as it was. New suburban governments serve the demands of those people (and institutions) who make the outward movement. Some who cannot (or do not wish to) move outward (or some who find that their interests are more complicated for various reasons), try to secure special governmental structures (e.g., special districts) without moving. In sixteenth-century England, the question was whether the growth of outlying areas could be prohibited—a question raised anew in present land use policy discussions. The failure to prevent such spillover, and the attached failure to prevent the creation of new governmental units "out there," produces "the metropolitan problem" with its various shapes and influences around the world. There are other dimensions of suburbia. Twenty years ago the chief topic that interested political scientists was whether the suburbs' growth meant Republican dominance—an issue answered eventually in the negative, but for which some reexamination may be appropriate—while the chief interests of sociologists seemed to be "conformity," child-rearing, and family life.

Politics is also about "policy" or what "outputs" the governing authorities must produce, and must not produce, and that too has its urban dimensions. The term "urban policy" can be used to mean the policies that urban governments adopt, a usage more or less present in the substantial body of urban policy research that political scientists (and most other social scientists) have recently developed (Eyestone, 1971; Doig, 1966; Lupo et al., 1971). But urban policy may also deal with actual or potential issues that are inherent in urbanization even if, at a given moment, no government is dealing with them. Policy issues can, no doubt, be conceived in many ways, but one convenient way is to focus on the kinds of problems or situations they are expected to affect. This implies some "functional" approach to the study of a specifically urban politics of the sort represented in the unfulfilled work of Delos Franklin Wilcox (1897)—an early political scientist who gave up the discipline to become a public administrator. One may, reasoning "functionally," perceive five major sorts of policies:

(a) those of *order* involving issues and aspirations in which the coercive power of the collectivity is a central resource, and the actual or threatened use of force is an important theme;

(b) those of *space control,* relating literally to the allocation of physical space amongst different claimants;

(c) those of *humane services* ("welfare" in an inclusive sense) designed to convey some minimal benefits to human beings, simply because they are human, not necessarily on raison d'etat or the economic well-being of society;

(d) those of *economic management*, relating to facilitating the production of wealth, governing the distribution of wealth, or simply managing prudently some portion of a wealth already in existence (as is the case in local government fiscal policy); and

(e) those of aesthetics, or the search for amenities beyond the levels of order, or sustenance, or gain.

Here I shall comment only upon three of these: amenities, space control, and order. One of the central facts of social life is the ever-present threat of boredom and irritation—a threat which may well be enhanced by the pace of urban life. Among the functions of politics in society is the discharge of irritation, the relief of boredom, the organization of the responses of individuals into a collective image that makes them all feel better about themselves and their relationship to the world around them. In a sense, one might describe this as politics of collective psychiatry. When, as Fustel de Coulanges says was the case in the ancient city, religion is the centerpiece of urban life, then the equation of politics with collective psychiatry is perfect. But there are many degrees along the gradient: from the bread and circuses of ancient Rome, comprehensible partly in terms of the problems of keeping the attention of the population properly focused in a large city with masses of disaffected and unemployed (Scullard, 1959; McMullen, 1966), to the strong support of the arts in European cities, to the modest support—but not trivial—found in the fact that virtually every large American city invests money in owning art museums. But, on the whole, collective psychiatry in American urban politics is not merely secular, but is "private"—at least as private as the sports stadia and the baseball franchise, disputes over which mobilize far more psychic attention than most disputes over public office. The critical relationship of the collective psychiatric function to the whole pattern of life is perhaps sharply focused by Detroit's wild enthusiasm for the Tigers, after the World Series of 1968, an enthusiasm which—at least temporarily—bridged that city's vast racial chasm as nothing had for a long time.

A second very fundamental urban problem is vaguely cited in our many discussions of planning. As much as latter-day planners wish to talk about social planning, the central core is and must be the allocation and regulation of physical space—a problem dictated by the limited collective space that is inherent in the nature of the city. When Wilcox says that the corporate city must control its streets—"symbols of the city's freedom and channels through which its life currents pass"—he has in a certain sense hit first upon the vitality of traffic control as a form of space allocation. Traffic control is not just some technical problem of engineering logistics, but a vital function (Moynihan, 1966). For if one man might be permitted to hog the road all to himself, he could by that very act, deny freedom of movement to all who should not meet his price. (That is precisely why, before the automobile was common, the urban transit franchise—a license allowing one to hog a certain space, namely, the "right of way"—was so valuable.) Traffic control is but the regulation of space consumed by people in motion. But the equally important spatial allocation—which interacts with traffic—is the regulation of spatial allocation for "permanent" construction, settlement, or use.

I have suggested earlier that the ultimate function of urban government is emergency defense, and this aspect of order goes well beyond "crime control" to absorb collective action to maintain the social unit against natural emergencies. In this sense, the study of American urban government might ultimately be equated with disaster research and particularly might the study of such urban government be equated with the study of the developmental history of city fire departments. Why fire departments? Because, as the histories of bombing raids will show, one of the few things that the discrete human being cannot protect himself against is the burning of his house or his workplace. Wilcox is on target in saying that one of the necessities in a city is the fire force—"a delicate barrier between the city and its destruction." At the time he wrote there was an aesthetic and emotional sense that made this intuitively obvious, for it was only twenty-six years after the Chicago fire (Colbert and Chamberlin, 1971) killed nearly 300 people within the twenty-four hours. But those of us who have seen no such thing, and have merely read of Dresden and Rotterdam, not to mention Hiroshima, cannot fail to appreciate the extraordinary human need for protection against fire.

Wilcox grasps, however, the other side of order—and the side we talk about more often: protective action directed against the exercise of human will, guile, or strength to some socially disapproved purpose of which the commoner will include privations of life, liberty, or property. This raises the matter of policing. Making (or the attempting to make) the thoroughfares safe for those who choose to use them, and moving out from that attempt as well to make private places safe within the limits prescribed by law.

As fire is so critical an urban problem because of the density of building, so policing is an important urban problem because of the density of human beings and of human social systems. Sylvia Thrupp (1962: 99) thus comments on fourteenth- and fifteenth-century London:

> Routine police work taxed all their resources, for a mobile and ingenious underworld made its headquarters in the city, recruited and led by fugitives and outlaws from all parts of the country and from the Continent and by desperadoes who occasionally broke out of Newgate prison.

Policing the village depends on the gossip system, and the all-seeing eyes of the neighbors, and exile today cannot be very different from what it must have been in the sixteenth century. Policing the city has different requirements, for the very fact of urbanization means that individuals lose the capacity to "know their enemies"—at least so far as those enemies who would deprive them of property—and lose some of their prior means for self-protection.

Urban policing depends upon the transmission of messages between parties who may not see each other at all, and who are describing other persons or phenomena whom they have not seen. Imagine the simple problem of the report of a middle-aged housemaid to the police of a young man who has snatched her purse. In an age when nearly everybody was illiterate, photography and rapid transmission of photographs had not developed, people could move no faster on land than they or their horses could run, streets were rough and ill lighted, the problem of policing a city must have been insurmountable. In short, the policy choices available are remarkably contingent on the technological possibilities, and the technological possibilities change the range of policy choices. Contrary to the ideas of technocrats, of course, the application of technology is not an unmitigated good. As a historian of the Boston police force points out, the nineteenth century was not precisely a libertarian period and police brutality not unknown. But it was constrained by the fear of private counterviolence. "After the Civil War, the fact that most policemen carried revolvers tipped the physical balance in their favor, making them more formidable but also increasing the danger of brutality on their part" (Lane, 1967: 187-188).

De Tocqueville (1945: I: 290) being unable to foresee this shift, offered some rather grim estimates that are part of a broader pattern of thought about urbanization.

> In cities men cannot be prevented from concerting together and awakening a mutual excitement that prompts sudden and passionate resolutions. Cities may be looked upon as large assemblies of which all the inhabitants are members; their populace exercise a prodigious influence upon the magistrates, and frequently execute their own wishes without the intervention of public officers.

He thought that the United States could not maintain "republican" government with cities unless it developed an armed force–beyond urban control to keep the city mobs in check. De Tocqueville was absolutely clear in his own mind that the cities represented the future danger. As he saw it, the trouble would come from the free blacks and the European immigrants (1945: I: 289-290):

> As inhabitants of a country where they have no civil rights, they are ready to turn all the passions which agitate the community to their own advantage; thus, with the last few months, serious riots have broken out in Philadelphia and New York. Disturbances of this kind are unknown in the rest of the country, which is not alarmed by them, because the population of the cities has hitherto exercised neither power nor influence over the rural districts.
>
> Nevertheless, I look upon the size of certain American cities, and especially on the nature of their population, as a real danger which threatens the future security of the democratic republics of the New World; and I venture to predict that they will perish from this circumstance, unless the government succeeds in creating an armed force which, while it remains under the control of the majority of the nations, will be independent of the town population and able to express its excesses.

Elwyn Powell (1970: 131-132), for instance, believes that the prediction materialized in the form of the urban police force, armed as an instrument of class warfare. "Urban police forces were developed to cordon off the dangerous classes. Their first mission was riot control; their second the protection of business property; the third and almost incidental, the protection of life and limb of the ordinary citizen." Such a view is not self-evidently wholly wrong. There were riots and the protection of property was the vital factor to begin. But it does not meet de Tocqueville's own tests to describe a local police force as an armed force representative of the nation, but not the populations of the cities.

On the other hand, if one is willing to take a long enough time horizon (always a problem in dealing with social scientists' forecasts), we need not worry about local police at all. The cities did provide tumults sufficient to command national attention and to elicit military intervention at least six times, and in all cases but one were related to racial disturbance: the Draft Riots of 1863 in New York City; the race riots of World War I; the race riots of World War II; the Pullman strike of the 1890s; its related industrial strife; and the disturbances of the 1960s. Incidentally, is it significant or trivial that all these—except the Pullman strike and related violence—occurred at a time of war? Probably not. It is more likely that only in wartime is there a military force ready to be put in. But there is some change in the latter circumstances with the development of a new military capacity for intervention, and the development of more systematic operational plans for domestic violence control. On close reading, de Tocqueville appears to have given us a Delphic forecast—insufficiently operational for us to be sure if it has since been verified or not, important enough that it should not be ignored and can intelligently be debated.

Indeed, the most instructive observation is that de Tocqueville's concern for urban disorder might have been more appropriate both to rural France in the decades preceding his journey and to the United States on several occasions thereafter. As Richard Cobb (1970: 92) points out, urban riots virtually disappear from Paris (though some might be found in other cities in 1810-1812). "Rural disorders, on the other hand, scarcely ever cease; they are endemic in the period 1790-92, during most of the Directory, and they were tensive at least in the northeast in 1801-2, and again in 1812-14." In the United States, the excited assemblies that overawe the magistrates and take the law into their own hands have never been an ordinary and expected part of municipal life. As a more or less routine phenomenon, what de Tocqueville anticipates actually was realized in the mob politics of the rural West and even more the rural South. Moreover, not even de Tocqueville's prescription of military control worked in the rural setting. Indeed, one of the reasons for the failure of Reconstruction was not the deal of 1876-1877, but the fact that for the whole of Grant's administration the government could not command sufficient force in the South to suppress the ex-Rebels in their violent war against the blacks in local politics—almost as if the United States between 1945 and 1949 had lacked the force to suppress Nazis in Germany.

But, in any event, the problem of order is an important test of the capacity of government—a test that governments fail so often (to judge by homicide, armed robbery, and burglary rates) that people apparently expect far less than academic analysis suggests. Politically, law and order may well not be, contrary to much received opinion, a very powerful slogan. Presumably, people might expect mayors to do something about crime as they expect presidents to do something about unemployment and inflation. But no one has ever suggested much of a correlation between crime rates and mayoral defeats or mayoral elections. (Even the election of Frank Rizzo seems contrary to the more extreme statements of law and order politics. Rizzo was elected a Democratic mayor in a city where one expects a Democrat to win, an Italian-surname Democrat in a city where it was about the Italians' turn, and yet elected by a little bit more than 53% of the vote. The victory was respectable, but no more than one would expect—even without law and order.) This does not mean that people do not care. But they also have no great confidence, in reality, that anyone's action is going to make make much difference—and so are prepared for self-protection by simple strategies of avoiding troublesome areas.

The problem of group violence always ran through the American urban experience, and was sharply recalled to American attention by the urban experience of the 1960s. Yet urbanization tends to intensify certain of those relationships, such as ethnic conflict. It does not create such conflict, for that conflict appears to be inherent in cultural diversity. But ethnic factionalism is aggravated by the relationship between urban growth and the mobility of labor from widely diverse places. If an area is growing economically, the demands for labor can seldom be met from a single and homogeneous labor source. Since World War II, the heavy migration of labor from Turkey, North Africa, and Southern Europe into Northern Italy, Germany, the Scandinavian countries, and even Britain provides but the most recent historical examples. The mobility of labor is also due to depressed conditions in the sending areas—as is part of the explanation for the migration of Asian and West Indian peoples to Britain since World War II, of the migration of peasants to the Latin American barrios. The "pull" factors are important but the "push" factors are not trivial. Urbanization, being so dependent on migration from several sources simultaneously, generates all sorts of pressures which translate into a factionalized politics.

These matters of amenities, space, and order refer to political problems that arise in cities. The "politics of urbanization" is a term indicating also how the rise of cities may change the macro-polity. Rowe (1908: 108) expresses an early version of the urban-rural conflict idea in trying to show the implications of his view that city growth is destabilizing for the larger polity.

> With the gradual increase in urban population, there is introduced into the body politic a new influence which acts as a dissolvent of political standards, and which may react unfavorably, at least for a time, upon national life. Openness to new ideas, which characterizes the city population, usually means the absence of profound conviction on any fundamental questions.

When he speaks of "the absence of profound conviction on any fundamental questions," he appears to be recalling the old idea of the urban population as a kind of crowd, changeable and easily moved by fashion. If that were so, then an urban electorate would have been as unsettled, from a political viewpoint, as were the earlier street crowds that affected the stability of governments by their mere physical action (Rude, 1964). "The great political parties in all countries have long recognized this fact which explains their desire to prevent the cities from obtaining a controlling influence in political life" (Rowe, 1964: 108).

This is a quite early version of the urban-rural conflict idea (although an even earlier version can be found in Edwin L. Godkin, 1898). Like so many of his contemporaries, Rowe was an unchristened comparativist, so that his discussion of the idea of deliberate underrepresentation is affected by observations on Britain, France, and Germany (as well as the United States). In the case of the United States, he sought to show arithmetically that the largest cities in six states (California, Illinois, Massachusetts, Missouri, New York, and Pennsylvania) were systematically underrepresented. What he could show (see Table I) is that in four states there was some under-representation in the state senate and this was also trouble in two state houses of representatives. But he could not, on the available

TABLE 1
BIG CITY REPRESENTATION IN FIVE STATE LEGISLATURES: 1900

	Boston	Chicago	Philadelphia	St. Louis	San Francisco
City as % of state population	20	29	20	17	25
Size of senate	38	51	50	34	40
Actual number of city senators	8	15	8	6	9
Theoretical number of city senators	8	15	10	6	10
Size of state house	240	153	204	140	80
Actual number of city members	50	42	36	15	18
Theoretical number of city members	48	44	41	24	26

NOTE: This table is adapted from the same information as provided in Rowe, 1908. The only change is that the descriptive designations have been reworded, the cities listed, in alphabetical order by city, and New York omitted. I have omitted New York because there is a certain confusion in Rowe's original table about the classification of Brooklyn. He appears to count Brooklyn as a separate unit (which it was until 1898), but he also refers to "Greater" New York which is the unit created by the consolidation legislation abolishing Brooklyn. From his information, it is not presently possible to arrive at a resolution of the problem here.

information, show either that the underrepresentation was deliberate or that in most cases it was consequential. In the Missouri case, St. Louis apparently was short nine representatives of a theoretical twenty-four, which in the dynamics of legislative politics might be regarded as a consequential number. But this was the only such case. At any event, Rowe does provide what seems not only an interesting hypothesis, that incidentally may well be the earliest expression in political science of the idea of urban underrepresentation (compare Baker: 1955), but a hypothesis lending itself to examination via time series (although there is no evidence that it has been so examined). The point, incidentally, might be pursued not merely into the legislatures but into the collective decision-making structures of the various private associations that have some real part in the actual governance of American society, or the formation of the ideas by which American society is guided. Thus, it would be very likely that the Episcopal church's general convention or the American Medical and the American Bar associations' houses of delegates would show a profound dissimilarity both in the urban-rural distribution of influence and in the ways in which urban and rural participants would respond to situations. This point is extremely clear in the Democratic party's national convention which (in the years between

1920 and 1928) was a recurrent fight between urban (and ethno-Catholic) and rural (and Anglo-Protestant) interests for dominance.

From the Rowe point of view, the urban population is "unstable," so that increasing its political leverage would add to the political instability of the whole nation. Specifically, he argues that the Parisian dominance in French politics is a source of instability and that French politics would be more unstable if Paris were still more dominant. But, in the hands of Rowe, this is not an anti-urban argument (as it was for some late nineteenth- and early twentieth-century writers). Simply, he maintains, the instability is a natural expression of the fact that new urban values have emerged and are competitive with older rural values, but that no equilibrium has yet been established. Stability will be established, he predicts, when government harmonizes with the new urban values.

Rowe, as most American social scientists, writes from a pro-city view that has prevented their thinking so carefully about the real implications of urban growth for the moral order of the macro-polity. Anxiety on the point is not simply the residue of some nostalgic Jeffersonian ideal. It goes much deeper to the lesson that the Chinese Communists learned. A group may prevent itself being exterminated by retreating into the countryside. But to organize and extend the governance of a nation requires the control of the cities, which became a major item on the Communist agenda (Lewis, 1971).

This is not a new thought. It is precisely because the city is strategic, that when the nation-state (or other macro-polity) precedes urbanization, urban growth is regarded as a threat to the established balance in that macro-polity. (Conversely, some of the most difficult problems in nation-*building* arise when town-based governing elites have to try to establish control over, or contact with, rural populations who are not effectively and normatively members of the system the town-based elites want to see emerge.) The former case is represented in the connections between restraining urbanization and maintaining the existing moral and pragmatic order, as seen by contemporary elites, comes out clearly in the time of Elizabeth I. By that time, villeinage was largely a thing of the past, and most men were legally free. But the moral and pragmatic order of the time clearly depended upon a strong status hierarchy, backed by command over physical force (Stone, 1965). In this time when the aristocrat asserted his claim on the basis of rank, and backed that rank by his own private army of hired thugs, the rural poor can hardly have been even as free as the sharecroppers cited in the novels of Erskine Caldwell or John Steinbeck, let alone those described more coldly in the sociological studies of Charles S. Johnson, Lewis Jones, and others hired by the Committee on Cotton Tenancy (Stokely and Dykeman, 1962).

Those who moved to the city might, on the other hand, not merely make themselves better off economically, but they might lose themselves in the urban mass, in the dense poor regions of London. Thus, the Elizabethan government in 1580 first sought to legislate against new buildings. "It was feared," says George, "that such numbers of people (as new buildings would encompass) would give rise to disorders of all kinds, that the danger of plague would be increased and that provisions would be dearer. The new buildings were resented by the City and feared by the queen who dreaded masterless men and the conspiracies of foreigners" (George, 1951: 68).

The Jeffersonian conception is akin in its difference. For it is the conception not of masterless men in a world that should be ordered, but of men with no incentive to be responsible in a world where freedom depends on independence and responsibility that troubles Jefferson. De Tocqueville (1945), with the history of France before him, had a similar view. He thus thought the absence of a great capital city (New York had 202,000 inhabitants in 1830) one of the "first causes of the maintenance of republic institutions in the United States," and could only see a national police as its defense.

De Tocqueville's particular interpretation may have been in error. But it accentuates the idea of *urban policy* not merely as the policy adopted by some local government in an urban area, but as the calculated response of a higher-level government to the fact of urbanization and to the need to make decisions related to urbanization. In one sense, it could be argued that all national policies of consequence are "urban" in the sense that urban populations are necessarily affected by them. United States government policies with respect to inflation control diminish the home financing market in urban areas, simply because money gets scarce. Some rural policies affect urban areas because they force people off the land (and the question is: are cities adequately prepared to receive them)? From the Roman Republic (Scullard, 1959: 23-43) through the eighteenth-century English enclosures (Trevelyan, 1949: 80-85) through farm mechanization in the South during the 1950s and the 1960s, public and private agricultural policies lead to displacement of poor people into cities. Such policies, one may repeat, may be regarded de facto as urban policies, at least in the sense that they have decided urban effects.

But there are certain other policies that lead, almost inevitably, to decisions meant to affect urban areas as such, and which are necessarily consequent to urbanization. It is likely that if we were able to study the matter, we would find that premodern polities have defined their urban policies chiefly as a matter of control. That is, the policies that they have adopted will have first been policies calculated to guarantee the continued control of the present elites (Thrupp, 1962). Modern urban policies, in democratic states, tend to be defined in economic terms. On the whole, the physical control of cities does not arise as a practical problem, so that national elites have little need to be concerned with that. But they do have need to be concerned with the spatial distribution of economic activities–which is both the essence of modern city planning in local government and of the support of such planning by national government. Decisions about the various clusters of human services (welfare, medicine, and the like) are important and may be debated. But such decisions are hard to make within the framework of general agreement that modern democratic politics requires–and so change will be relatively slow. Urban policies in modernizing states will show a persistent dualism: on the one hand, national elites assuming themselves to be in control will focus on technocratic definitions of goals (e.g., expanding the housing supply rationally). On the other hand, they will recurrently be faced with the problem that they are in effective control, and so must modify their urban policies to the degree required to minimize political resistance.

CURRENT PROBLEMS AND THE CAPACITY OF GOVERNMENT IN THE URBAN UNITED STATES

I have hinted above that–virtues notwithstanding–new reform as political science shares some weaknesses of old reform. New reform provides little or no corrective to the technocratic bias so often found in current policy discussion. That bias assumes that the good results are known or agreed, so that the chief problem is to find "technological short-cuts to social change" (Etzioni and Remp, 1972). This is the "if we can put a man on the moon" idea that makes urban problems managerial more in the sense of Taylor than of Barnard, not to mention Antony Jay. Indeed, new reform is often quite akin to this bias—with a leftward spin.

Old reform has been criticized, fairly, for inadequate exploration of its own suppositions and inadequate understanding of the social world around it. New reform suffers similarly, in ways that impede our understanding of the capacity of government under present urban conditions. There are at least three limitations.

(1) *New reform seems profoundly ill-informed* about the actual history of urban governance and thus about the things that urban government could normally be expected to do. Two illustrations may suffice to make the general point clear. The first is that much current discussion and writing seems to imply that, at some time in the past, the control of urban local government was a major means of social reform benefiting some social interest. Tom L. Johnson apparently sought to use the city government of Cleveland as an agency of social reform (1900-1909), and James Michael Curley apparently tried to mix social reform (or, at least, welfarists populism) with some rather personally self-interested politics in the 1920s. Urban government ordinarily functions in extreme emergency conditions to restore that minimum level of order required for people to do business at all, a level far beneath anything that the ordinary American city experiences, but which can be seen in such drastic circumstances as the San Francisco earthquake, the Chicago fire of 1871 (Colbert and Chamberlin, 1971), or the Draft riots of 1863, or the urban rebellion of 1967 in Detroit. Beyond that, urban government is a local service venture, providing the minimum in tax collection services, water supply operations, and similar functions. Beyond that, urban government is chiefly an enterprise that promises benefits to a very large part of the population, but actually delivers benefits to those whom it employs or those who otherwise make money from it. In that sense, a city government (which may well have 5,000 or 10,000 employees) is like a firm of the same size. Public relations requires it to talk generally about benefits to the community, but the real beneficiaries are the owners, the employee, and those who do business with the owners or the employees. Doubtless, this is an exaggerated representation. But it is meant to convey the idea that except for the direct beneficiaries, those who have had to depend mainly upon local government have gotten relatively little for it. But

the exceptions have been exceptional. Only this kind of historical shallowness could justify the repeated assertions that the machine of the past was so important an instrumentality of upward mobility for a significant part of the population–a proposition used (without the slightest evidence and in defiance of much that might be believed) to justify the argument that something like the machine should be revived in the ghettoized central cities of the present time.

At the same time, much current analysis ignores certain rather obvious factors in the pattern of acquiring and using or not using offices. It is inherent in political logic that many candidates, in closely divided constituencies (e.g., the cities), will be able to come to power only by developing support among those who may not seem their natural supporters. The lesson of Boston political history, indeed, is that the first Irish mayors were not merely dependent on dissident Yankee support, but that the rebellion in Democratic politics which began to bring them to power also was organized by dissident Yankee Democrats out to maintain their own power position (Blodgett, 1966). By this token, it is also comprehensible that the first such mayors should have been extremely careful and cautious, doing nothing to alienate the indispensable Yankee support. Academic purists who criticized Carl Stokes, or criticize Kenneth Gibson for doing the same, evince little except their own historical ignorance and their own political naivete.

In the same way, the historical evidence is compatible with what any reasoned analysis would otherwise indicate that a transfer of office holding is by no means settled by one or two elections (the first Irish mayors had successors in a Protestant throwback, much as Cleveland has had a "white ethnic" throwback after Stokes), and that the rise of electoral politicians from a distressed or depressed population does not make so much immediate and obvious difference in daily life, commonsense outcomes. If one reported a nineteen-year-old, working-class boy, engaged in a drunken fight with policemen, beaten to death by police officers, with an ensuing wave of protests, mass meetings, with a hearing against accused officers by a police board, and the officers in the end were adjudged not culpable, one might be talking about a large American city with a black population today. But the case is, instead, Boston in 1871 and the boy was Irish (Lane, 1967: 188-189). This is neither to condone the persistence of dramatic inequality, nor to make the absurd case that racial dispute is in no important sense different from earlier ethnic disputes. It is merely to argue that social scientists, whatever their moral convictions, need hardly waste time on the supposition that drastic change emanates from change in the officeholding pattern of urban govenments.

(2) *The first point is important and sustains* the second: that new reform is rather too loose in its definition of the problems deemed central, in the evaluation of means of meeting those problems, and in the appraisal of alternative means. If we would anticipate government's capacity to handle problems, we should also have to be clearer about what order of problem we are considering. Thus, the traffic problem—though very important in principle—is not so important as the question: "in a period of mounting polarization, how are opposing and conflicting wishes reconciled or resolved equitably?" Let us leave aside the technical question about whether polarization is mounting, for we may be assured that American cities are sufficiently polarized to be centers of danger now and for some time to come. Thus, whether polarization is mounting may be a secondary issue. Let us also leave aside the question of equitable resolution, simply because the question of equity is one of the more complex philosophical issues.

We could say that the political system would work or be judged to work perfectly if we could muster evidence that the polity was a shared moral order in which all participants preferred the presence and participation of all others, and in which no one preferred the exclusion of any others present. As a practical matter, that is a bit idealized, so we might settle for a community in which all parties would effectively renounce the application of force against any and all other parties. The absolute opposite is the polity in which the war occurs of all against all. The gradations between cannot be worked out in detail here, but the general pattern would be clear.

Presumably in real and current terms, new reform would prefer to see American cities come as close to the norm of community (shared moral order) as possible, so that this urban problem would become an enterprise in social peacemaking. Unfortunately, new reform provides no clue to this. Though new reform means to serve the presently disadvantaged, it does so with naive theories of political change that suggest that politics is *nothing but pure power or a capacity for blackmail.* These theories are grossly incomplete. They obscure the obvious fact that a challenging group—whether a foreign nation or an internal faction—*usually has the power to express its own discomfort or ambitions by creating a sense of disturbance among its adversaries.* This is almost always a prior condition to achieving a new situation which it can find satisfying. But it is not the same thing as having the power to create that new situation. The point is also known in industrial relations, where the union leader's demonstration of his power often requires the calling of a strike. But, in union folklore, "anybody can call a strike; but not anybody can settle a strike."

The new satisfying situation does not automatically or inevitably follow from the fact of disturbance, nor even from specific actions in technical compliance. If it did, party X (defender) would merely adjust to the wishes of party Y (the challenger) and that would be that. Under that theory, Mississippi would have been a tranquil biracial polity before 1900. The Kohler strike would not have occurred. And peace would now reign in Vietnam and the Middle East. Such theories not only ignore the residue of hatred which another party builds up if you compel him to eat dirt. They also ignore the high risk, in any current situation, of mutual alarm, madness, and a politics of self-defeat.

Battle (or its immediate threat) provides its own stress, which exaggerates the chances of a politics of self-defeat. Men often lose sight of what their real objectives are, or decide that they have no choice but to act in ways which defeat the objectives they have in mind, even though they usually will not understand this until after the fact.

The specific reference, then, is that to a considerable degree, the urban problems that most concern us are focused upon the central cities or Standard Metropolitan Statistical Areas, and may be seen largely in ethnic terms. This is not to deny many other factors and inhibitions, but the current disputes most likely to prevent the formation of political markets—let alone communities—and to precipitate empirical action, are expressed in ethnic terms (Holden, 1971). The trouble is that very little of recent research is genuinely pertinent to this problem. The state of black politics in the later sixties gave a certain currency to research on the subject. On the whole, the research on blacks is not—perhaps—trivial, but it is surely quite preliminary, being dedicated to an enormous number of localized studies. They show such things as greater politicization of young blacks compared to older ones, more nationalist politics among the younger, or racial partisanship rather than estrangement from the political system, lower levels of efficacy and higher cynicism among black than white students, but less disillusionment with the political system (due to the very good reason that they had less basis to illusion in the first place). But social scientists have come to this research in the spirit of missionaries looking for primitives to practice upon. So they show not only very little comprehension of the culture, social structure, and institutional matrix pertinent to black politics, but they show even less comprehension of the Euro-American populations whose interdependency with blacks is a vital part of urban conflict and urban conciliation. On the whole, this can be regarded as the naive wanderings of social scientists who are just beginning to have some sense of the shape of the society–and who are still experiencing a culture shock which they write into their own research. But it has not yet enabled social scientists to make very useful or defensible contributions to thought about the problem of community in American cities. At most, it has led them to an infatuation with such an idea as "decentralization"–which is of very little value for this purpose.

The relation of these Euro-American peoples to Afro-Americans is actually rather complex, and it cannot be disposed either by defining the former as selfish racists antagonistic to the latter, or by defining the latter as the most recent urban immigrants who will ultimately be absorbed into the mainstream. That essential fact is the interplay of Afro-Americans and Euro-Americans. The fundamental fact of big city politics now is that nobody can really do much that is constructive and intelligent if he has to face the deepseated and protracted resistance of either Afro-Americans or Euro-Americans.

Broadly speaking, the history of Afro-Americans and Euro-Americans, in relation to each other, has been one of coexistence in the literal sense. That is, they have existed simultaneously in the social and physical space of the city, but have surely had little or no conception of themselves as actual or potential members of a community. They had similar voting behavior, being Republican before Roosevelt and Democrat after. This particular convergence provided part of the foundation for big-city party organization as we know it today. The Afro-Americans and Euro-Americans have also been convergent as the main constituency of the mass industrial unions. For these two reasons they compose the electoral base, but definitely not the leadership, of Democratic liberal politics at the national level.

Their cooperation has always been tacit. Leadership in neither group has ever felt inclined, or free, to make active cooperation with the other a part of its public rationale. Such tacit cooperation, however, has always been the thinnest veneer. Underneath lies a pattern of actual and potential conflict, based upon both the pragmatic and symbolic needs of both sets of people.

(3) *New reform thus fails, as a combination of the* first two difficulties, to have a very good idea of the process of social adjustment and of the institutional relations presently involved in that process in the urban United States. Thus, for the problem of social peacemaking, we have to think anew about what social demands are, how they arise to decision makers, and how decision makers function to translate demands into viable outputs.

(a) The first problem involved is that of getting decision makers' attention. In the urban context, this is particularly difficult for several reasons. In the first place, such evidence as is available indicates that relatively few people make any effort to put their concerns on the agendas of decision makers. But this should, I speculate, be used to argue that people have no concerns and are satisfied. Instead, the backlog of anxiety and discontent, which can at times be translated into political terms, may be substantial. But how are decision makers to know what this anxiety and discontent is? And how are they to know how to respond to it? The answer is that there is no presently available way. Therefore, if no demand is already institutionalized in the habits and commitments of particular organizations, and if key decision makers are not themselves individually committed, the likelihood is that the governmental agencies will never give the signals that they themselves precipitate and create pressure groups and pressure group action.

(b) If decision makers' attention is mobilized, the question is what else can they do? Decision makers' problem include receiving messages and engaging in the continuous process of discovering, manufacturing, and articulating ends-means-ends chains that will work in their circumstances. In this respect, the significance of some recent changes in urban local government has been missed. Black mayors, for instance, are important so much because of basic policies that they will change. They are important because of the symbolic function they serve, and because the serving of that engages the attention of their main audiences—in the way that a completely exclusionary situation would not. But symbolism is not the end. Symbolism is part of the process of activating people for a different level of political participation. But the important urban problem is that at this moment there are few actors with any decisive symbolizing capacity who can link the attention of the Euro-American audiences and who can themselves link, as a matter of pragmatic politics, with Afro-American counterparts.

(c) This last observation indicates something of the sense in which new reform is misleading. The decisive symbolizing capacity in American politics seems to be national. But the distribution of burdens on government makes the handling of some critical problems more local. That is to say, to a considerable degree, money, legal authority, and administrative competence for the governance of urban areas do not inhere in the urban local governments themselves—which is the fallacy of decentralization. Decentralization is tne political strategy of careerists promoting the most easily created institutions in which they can be influential, the political error of desperate claimants who see themselves as having nowhere else to turn, and the intellectual error of a political science suffering from an archaic focus on central city government and the central city government virtually alone—a focus utterly inconsistent with contemporary and emergent institutional realities.

But the capacity to govern a city no longer resides mainly in city hall. The capacity to govern a city's housing supply lies quite as much with suburban zoning officials and with the managers of the private government of real estate who make vital allocations of space by race. The market allocates physical space through an infinite series of individual choices about houses, apartments, offices, stores, and so on. But markets of all kinds depend upon prior political decisions about the rules governing transactions. The general pattern—which may now be altering, but not very rapidly—is that brokers, builders, and bankers constitute an effective private government with fairly well-understood decision rules about the racial allocation of physical space. Racial separation is not simply a function of the prejudices individual buyers, sellers, renters, and landlords. Not only may a white seller be reluctant to sell to a black buyer, but his chances of doing so also are reduced—even when he is unaware of it—if brokers have their own rules about not showing properties to potential buyers if those buyers are black. The market is further controlled by builders' decision about those to whom they will sell their new development properties. It is still further controlled if, in mortgage department of banks and savings and loan associations, favorable or unfavorable decisions on loan applications are contingent on the applicants' race and on the location of the property, as well as on ordinary criteria for credit decisions. At any moment, there will, at least in a large metropolitan area, be some brokers, builders, or bankers who would just as soon not adhere to these particular decision rules. But they have to concern themselves with their own reputations for "sound business practices," and may be penalized by their colleagues' noncooperation or even by inquiries into their "business ethics."

The capacity to govern a city lies not only with the mayor who ostensibly controls a police department, but with a county prosecutor—operating under state law and chosen by a wider constituency than the mayor. The police may investigate or arrest whom they choose, but the prosecutor alone has the right to decide who will and will not in fact be prosecuted in court, as, indeed, only 10% of arrestees on the average are so prosecuted, and he also has the authority to investigate virtually any person or group whom he may choose. The prosecutor has the right to decide what evidence will be submitted, how the case will be conducted, and often the authority to advise the judge on an appropriate sentence for guilty parties. New Yorkers have seen many times the expression of this discretion in the work of the much reputed and recently criticized Frank S. Hogan. And those who have watched the complex world of black politics need only be reminded that Fred Hampton and Mark Clark—the two Chicago Panthers killed December 1969—were shot by policemen under the control of the state's attorney's office. By now we may recognize the peculiar importance of the school government as a separate entity. Capacity is also related to the enormous variety of functional governments or special districts (Bollens, 1957), not only school governments but such entities as the Metropolitan Transit Authority in New York and the Chicago Transit Authority that make critical choices about what parts of the public are to be served how, and which play so acute a rule in public capital investment with all its political consequences (Wood, 1961).

A good deal of the capacity for urban governance is actually private, even in what is theoretically the realm of public government. Each attorney general from Robert F. Kennedy through John Mitchell has emphasized the strength and intractability of organized crime. Organized crime is not some spur-of-the-moment affair, but a form of business conducted beyond the approval of the law. Organized crime could be regarded as a private government that provides commercial services prohibited under law and thus negates or undermines that law. But it is important not to be romantic about this, for to call their activities "commercial services" is to say that they make money out of these activities. It is not to imply that there is inevitably a widespread and legitimate public demand that archaic laws are frustrating. It is likely to involve various forms of private coercion, including some killing, that is formally proscribed by the law—and is thus highly political in the sense of negating the public law. The private government of crime intimidates and punishes both its own recalcitrant members who will no longer obey the law of the (underworld) industry—in defiance of that principle that, in a free society men may withdraw from private activity that they find no longer satisfying. It also punishes law-abiding citizens who have never been part of its system, if those private persons become too troublesome, and thus negates the proffered protections of the law.

Suburbs, special districts, and private governments all are so important a part of the major urban mosaic that it is quite unrealistic to think only—or even primarily—of the capacity of central city governments alone. But it is even less realistic to focus our minds chiefly at the local level. We are still dealing in a political science that takes cognizance only of an old-fashioned world unless, at the minimum, we take account of the significance of higher-level governments as decision makers in urban areas. Under any concept of municipal law, local government is in part executing delegated powers from the state.

As Goodnow (1909: 49-50) asserted:

> If . . . the actual execution of the state will is entrusted to the local community free from any effective state control, such local community may, through its powers of execution, or what are really powers of non-execution, or modification, change the will of the state as expressed by the body representing the state as a whole, so as to adapt it to what are believed to be the needs of the local community.

The "state will" language is archaic, and current generations of political scientists prefer such a term as "public policy." But Goodnow is clearly on target. If (a) in a specific local area the proportion of the politically influential population who prefer policy "A" is relatively large, while (b) in the larger area (e.g., the state) the proportion of politically influential people who favor the contrary policy "B" is large, then (c) locally chosen decision makers who have some opportunities to exercise discretion will tend to be relatively sluggish about putting policy "B" into effect. Unless we are going to go all the way and say that electoral behavior has nothing to do with policy, when we have to suppose that state decision makers would be alert to the internal life of the city for their own purposes. Even though the central cities are now diminishing in electoral importance, relative to the suburbs, the fact remains that the central city will still provide a significant share of the total vote for state candidates—even of the Republican party. (Chicago is unquestionably an extreme case, but it provided 31.0% of the gubernatorial vote and 33.5% of the senatorial vote in 1968. This is a notable drop from the more than 40% that it was providing twenty years ago, but is still a critical factor.)

Even though the central cities may be underrepresented in state legislatures, the existence of any reasonably sized bloc provides the usual incentives for coalition politics—and if legislators from the cities are themselves interested in the problems of those cities, then it will follow that a certain amount of attention must be paid by state officials to city politics. Consequently, state-level decision makers may from time to time intervene in local areas from which they ordinarily abstain, for the purpose of assuring their own favored outcomes.

To a large degree, this means state governors and state administrators, although both legislatures and courts also may be pertinent. The Massachusetts pattern, for instance, seems to exalt the state representative as a major figure in Boston city and neighborhood politics, partly because the change in Boston city government organization some twenty-two years ago downgraded the role of the councilman as a local representative (Gans, 1962: 170). Even as late as 1971 a Boston controversy over school desegregation revealed as well the activity of a bloc of five "state reps" accompaning parents to the school in their protest (Boston Globe, 1971).

Where there are very strong party organizations, operating as such in the legislative process, and where the chief officials of the big cities are key figures in that party-legislative process, as Pennsylvania and Illinois, the legislators are more likely to function as emissaries of the local party leader-municipal officeholder than as a direct voice of the constituents themselves. It is possible, and indeed likely, that where there are ample means for local expression the state legislators may not be so compelled to act as agitator-spokesmen themselves. If there is also no strong party leadership, they obviously will not be emissaries in the same sense as in Illinois and Pennsylvania. But what will they be? Two possibilities remain. One is that they will play indeterminate freelance roles, but it is also possible–particularly if the city is very much dependent on favorable state house decision–the role of local political leader exerting greater control than is normally the case.

The judiciary in its own way plays a rather important role. In the criminal law-civil law area, the judiciary is the mechanism through which important distributions of goods and penalties take place. But the judiciary is also an important controller of some actions of local officials, by virtue of its power to interpret the requirements of statutes in those states with little "home rule" and by virtue as well of its power to interpet the meaning of home rule in those states that have constitutional provisions on the subject.

The exercise of legal control over local officials' discretion is one major lever. This is perhaps most widespread in the laws of the various states as those laws regulate local taxation, spending, and debt practices. J. Allen Smith (1907: 277), a precursor to Beard in the economic interpretation of American politics, refers to this in saying that state restrictions upon city government were calculated to serve the urban property-owning class.

> [Since it] is in the cities that the non-possessing classes are numerically strongest and the inequality in the distribution of wealth most pronounced[, a] municipal government responsive to public opinion might be too much inclined to make the public interests a pretext for disregarding property rights.

As illustration of the argument, which of course his available evidence could not allow him to verify, Smith (1907: 279) chose the case of Colorado.

> which has gone as far as any state in the Union in the direction of municipal democracy, [but where] no franchise can be granted to a private corporation or debt incurred by a city for the purpose of municipal ownership without the approval of the taxpaying electors. When we consider that 72 per cent of the families living in Denver in the year 1900 occupied rented houses, and that the value of two hundred dollars are (sic) exempt from taxation the effect of this restriction is obvious.

In some instances, as in the case of New York City, each city budget is contingent upon the favorable or unfavorable decisions that the city administration gets out of negotiations with the Governor and the legislature. The expenditure decisions that mayors feel called upon to make often require revenues far beyond any that they can legally raise, or beyond any that they feel politically comfortable in raising. Accordingly, mayors find it essential to submit legislative programs to the state calling for increased financial aid, for state assumption of expensive city functions, or for increased taxing powers. Governors, on the other hand, usually prefer to resist these mayoral claims, and to urge upon the city greater "financial self-discipline." In the end, what happens depends on the precise deal, which usually entails some city agreement to a particular change in policy or organization that the state decision makers find preferable (Sayre and Kaufman, 1960). But New York's experience is merely an exaggeration of the national pattern. For the same relationships can be found when Gary and Chicago, respectively, seek Indiana and Illinois state assumption of some or all of their school costs, their welfare costs, and similar costs.

State decision makers act not only through their leverage over money, but also through legal control over local program operations. Exactly how this works depends, often, on the particular state–and on its constitutional and political traditions of home rule or lack of same. In some cases, as the notable Ocean Hill-Brownsville school dispute (Berube and Gittell, 1969), the intervention of the state is the key element. State decision makers participate in the governance of urban areas through their own program operations as in highway policy (Lupo et al., 1971). Historically, this state control element is most clearly expressed in the United States in the form of "ripper" legislation: legislation summarily to "rip out" some function or agency or officer in local government. In the case of police, the state governments have from time to time asserted control over the police forces of Boston, Detroit, New York, Kansas City, and St. Louis. In the Boston and Detroit cases, what was clearly involved was the fact that "that influential class of citizens which had felt left out of politics [in the Boston case due to Irish Democratic control of the electorate] for at least a generation, which could hope for satisfaction only in the [legislature]" (Lane, 1967: 213). In Boston, it related to the details of local politics and patronage, while in the Detroit case it related to doubts about the loyalty of a Democratic police force (during the Civil War), doubts surely made more potent by the fact that the police force made no effort to stop a mob from burning out the black section of the city. It will be very surprising if new forms of ripper legislation do not emerge in those states where blacks begin to acquire control of the principal local elective offices. If, for example, Prudential Insurance Company were to conclude that Newark was becoming intolerable, the company might remove its operations from the city. But, at some point, this abandonment of capital invested would no longer be feasible, and pressure may increase for state takeover of the vital police functions.

These are but rudimentary indications of the possible relevance of state decision makers to urban governance. The federal relationship is even more important. Conventional constitutional and legal doctrine that the federal government takes no official cognizance of the cities is obviously no longer descriptive of anything approaching reality, nor a good basis for estimating what will occur. Instead, since the time that the parents of the present generation of college students were born, the federal role has increased and continues to increase.

The federal role first was giving advice. One of the most radical departures came when Herbert Hoover, as Secretary of Commerce, sponsored the Standard City Planning Enabling Act of 1928. The draft legislation had no binding authority. But it was important nonetheless. Local governments had only seen the Supreme Court hold zoning constitutional two years before. Prior to that time, zoning–an inhibition on one's right to use property–was regarded by many as an unconstitutional taking of property. But the court said otherwise. Once the court said otherwise, however, local governments did not know *how* to plan or to zone–and consultants were not nearly so widespread as now. Thus, Secretary Hoover sponsored the model legislation that the locals might copy. This legislation had no binding effect until adopted by the locals. But it was extremely influential as Charles M. Haar's discussion in "The Master Plan as an Impermanent Constitution" makes clear. This model zoning legislation was revolutionary in its departure from the idea of "uncontrolled" private enterprise (an idea Hoover never accepted anyway), although it placed very little reliance on compulsive public authority. Then President Hoover went another step in appointing the National Commission on Law Observance and Enforcement (the Wickersham Commission). The Wickersham Commission was triggered by the widespread evasion of Prohibition, and associated problems. But its inquiries went into almost every corner of the criminal justice system in a manner that was hardly repeated by a federal domestic policy commission until the President's Crime Commission, the Commission on Civil Disorders, the Commission on Violence, and the Commission on Campus Unrest.

The critical departure, as with so much in American politics, came with the Depression and the financial crisis that the Depression generated for local governments (Terkel, 1970). In the atmosphere of the 1970s, we once more face financial crisis, but the Depression-era crisis was of an altogether different quality and magnitude. Now the problem is to find adequate fiscal devices, what with the constraints of current law and politics, to extract sufficient money from the economy to meet urban public purposes. Then it was to find money. Local treasuries were empty, and there was no way to fill them. Public works had to be suspended. Relief claims grew, but relief resources did not. And policemen, firemen, and teachers–on the presumably secure public payroll–had to take unsecured scrip instead of money. In that situation, the present grants-in-aid programs began to develop, and ballooned thereafter (Sherwood, 1948; Charles, 1963; Wright, 1968).

From the grants-in-aid programs, federal administrators developed influence over the content and manner of local administration which now reaches to the point of (a) setting standards for program administration, (b) developing working relationships and generating new ideas for programs, and (c) developing strong pressure for de facto metropolitan government on a function-by-function basis in the name of rational planning and metropolitan area coordination. The consequence of this development is that there are now very few important policy questions, likely to be controverted, on which a final decision can be reached without some local party involving federal representation as well. The development of public housing programs during the Depression (more as a matter of economic pump-priming than as a matter of housing itself) as well as home loan insurance programs such as the Federal Housing Administration, were essential parts of the next step. And from this came the whole paraphernalia associated with the Department of Housing and Urban Development. The fact that state and local governments experienced various fiscal limitations, and that they often lacked the personnel with the appropriate skills, led to an enhanced federal role. The federal role is always ambiguous, looking weak to the federals and strong to local officials, but whatever its precise strength or weakness, it is vital and presumably permanent (Meyerson and Banfield, 1959; Greer, 1965).

There is still another form of federal intervention which is administrative enforcement (or nonenforcement) of the requirements of constitutional and statutory requirements. Perhaps the most far-reaching of these kinds of interventions is the developing policy of the Department of Justice.

No public data permitting a confident evaluation on this practice are available (letter to author from Department of Justice, November 3, 1971). But the Attorney General made the following observation in a speech to the Associated Press Managing Editors' Association in 1971:

> In less than three years the Department of Justice has obtained indictments or convictions of more than 170 state and local officials or former officials on federal charges or on state charges based on federal information. This figure includes only those offenses connected with organized crime, or those charges that appear to link the defendant with organized crime.
>
> Altogether, these 170 or so individuals represents [sic] officials of 21 cities, 12 counties, and five states. They range from positions of judgeships to state elective officers, from mayors to councilmen, from law enforcement officers to purchasing agents.

Whether this is a large number one cannot say on any available information. Nor can one say, in fact, that the Department of Justice is now more active in these matters than before. Considering the very large number of governmental units, and the much larger number of governmental officials and employees, it is quite possible that the number of actual indictments is really quite small—unless one adopts the presumption that the number of unindicted offenders is many times greater. This, however, is a matter meriting independent examination on better evidence than any publicly available. What is pertinent is that the Department of Justice clearly represents itself as being much more active on this matter than its predecessors have been, and that in some sense the Department of Justice might well be thought a new influence in urban affairs. Such influence is one more aspect of increasing executive leverage in urban affairs.

The federal judicial role is also noticeable, although we do not usually consider the federal courts as urban decision makers. This can now be seen in matters as diverse as police brutality (in a recent Pittsburgh case), governmental structure (as the interpretation of one-man, one-vote widens), housing desegregation (as in the Chicago litigation presided over by Judge Richard B. Austin), and school desegregation (as handled by at least two district judges in metropolitan Detroit). As the federal administrative and judicial role in city politics widens, so also would the congressional role. An occasional congressman has been a powerful local political "boss," as Charles A. Buckley in the Bronx and William L. Dawson in Chicago (Wilson, 1960). And some have gone on to prestigious local offices as Fiorella LaGuardia (Mann, 1965), Sam Yorty, Hugh Addonizio, and John V. Lindsay did. But congressmen have generally stayed out of local politics, except to the extent essential to protect their congressional interests. The more this is so, the more also we should expect congressmen to become involved as intermediaries and as overseers of administration. We see, thus, an emergent possibility for a congressional specialization in urban affairs, comparable to congressional specializations in other subjects. The intermediary role is demonstrated by a hospital strike in Charleston, South Carolina, in 1969. The hospital workers, mostly black, were organized into a custodial staff union under the AFL-CIO and with the support of the Southern Christian Leadership Conference. When they struck, they were fired, but the Department of Health, Education and Welfare intervened in their behalf because the hospital's related medical college was a beneficiary of federal grants. As a result of the HEW pressure, the hospital agreed to rehire the fired workers as part of a strike settlement, but this agreement was then undermined by the subsequent intervention of more conservative members of the South Carolina congressional delegation.

Even nonlocal congressmen begin to show interest if local problems and decisions are particularly visible, and are averse to the national policy decisions that those congressmen prefer. Much criticism of the Office of Economic Opportunity was based upon the fact that OEO sustained local projects that often seemed to defy, or at least to irritate, the best established local political and business leadership. Thus, a Senator McClellan might conduct an intense investigation of the granting of federal funds to juvenile street groups in the Chicago area, an investigation abetted by the Chicago police department which represented the street groups as mere thugs. In the same way, the so-called "Green Amendment" was meant to provide local officeholders with greater influence over OEO's Community Action Agencies.

The Tax Reform Act provides another illustration in that congressional antagonism toward the expenditure of tax-free foundation funds for voter registration campaigns in Cleveland was an important part of the background. And the sponsorship of independent investigations of the composition of the Aid to Families with Dependent Children (AFDC) rolls in New York City came not from New York congressmen, but from the House Ways and Means Committee which is responsible for public assistance legislation.

The institutional pattern reveals one of the recurrent problems of social science: to be dealing with data, however imprecise, that reflect current experience and emergent experience. For, try as they may, social scientists seldom truly escape what may be called "contemporary situational description," although they have more or less profound or penetrating theories to apply to those situations. The politics of urbanization, as it reveals itself presently in the United States, and probably elsewhere as well, calls for a more subtle conception of the institutional capacities and arrangements than any of our past analyses have suggested—and surely more subtle than the prescriptions that new reform presently tends to offer.

New reform faces a more serious problem, inherent in its will toward a prescriptive role. It has to be far more encompassing, far more hard-working, far more knowledgeable about the details that must go into policy as policy makers come to see those details. All in all what actually takes place in urban governance shows how much the specific local area is dependent on what happens at some higher point—even at the instance of someone who has no authoritative right to issue orders and who may conceive himself trapped by the loose and uncoordinated nerve ends of local administration. This aspect of urban governance is best understood, though not well understood by anyone, from the viewpoint of those who, like C. Wright Mills or Albert Vidich and Joseph Bensman, seek to capture the role of the local decision makers in the great world of today.

Francine Rabinovitz (1969) has argued that the growth of urban populations and problems has also called forth a new species of urban experts. But Rabinovitz means planners in the conventional sense, who may draw salaries and write reports, but contribute very little actually to policy makers' understanding or options. It is in this respect that the present interest in urban policy approaches the most trivial.

A few years past, a small number of political scientists studied urban renewal and some continue to study related matters (Rabinovitz, 1969; Ranney, 1969). A small number of political scientists also studied education and some still do. In consequence of the interest in violence, apparently, a number of studies of law enforcement policy have become available although some, as in the work of James Q. Wilson, antedate the violence studies by a good deal. There is new material on transportation (Lupo et al., 1971). Indeed, there is probably no subject on which there is not some piece of work in some library (if only a master's thesis). But the study of urban policy is rather underdeveloped (and, when initiated, more likely to be misled into the socioeconomic determinants versus political determinants argument which is secondary).

From the viewpoint of a reform-oriented political science that does not merely dispense its prejudices, the problem is to make a contribution to the development of policy—which requires knowledge of what one is talking about. This is more important because policy makers want to use social science, but do not know how, and we have not learned how, either. Characteristically, one generation of policy makers picks up the ideas of earlier social scientists who have tried to make their ideas acceptable by piggybacking on the strength of more easily accepted ideas in physical planning and hard policy. This is largely futile, for it means that social science is usually discussing past policy. And we ourselves systematically obstruct our ability to develop useful knowledge, or, rather, to put our knowledge in useful forms. For example, National Academy of Engineering recently proposed a series of telecommunications innovations for "improvement in city life and function." The committee passed over the critical privacy issue, with the observation that it would have to be studied, but the probability is that a genuine meeting of social science and engineering conceptions will depend chiefly on the extent to which social scientists are prepared to take the initiative. We know, of course, that such initiative is contrary to the conventional conception of the study of urban politics.

The contention, however, is that political science-urbanists will, in order to improve both the capacity of the discipline and the discipline's contribution to the capacity of government, have to learn enough to be able to enter the discussion process when ideas are still fluid. Moreover, they will have to learn to be operational in a practical sense, without losing a sense of theoretical guidance and direction. Why is this so? Because political scientists and other social scientists explicitly and implicitly make judgments day by day about the necessity or dispensability, the wisdom or the error, of actions taken by decision makers. Yet when real problems are formulated, they are much more likely to retreat into a more academic stance. Thus, even the simplest problems receive no answers and very little attention. That is, the problems of protecting individuals against privations of life, limb, liberty, and property are surely problems in deterrence—quite as much as any problems in international relations—but is there not a political science (there surely is not a sociology-based criminology) that provides people with useful answers to the policy options (Holden, 1971)? What, for instance, is the relationship between the availability of shelter and the presence of order? Do we expect more or less disorder if vast masses or people must live more or less permanently outdoors, or do people effectively define some turf as their own, even when it is on the sidewalks of Calcutta? Nor is there really a social science attentive to the sources of disorder that lie in collective disagreement, in contrast to ordinary criminal action-despite the violence research?

Even in the relatively conventional sense of policy studies, some extremely critical issues remain off our stage. Who, for instance, anticipated—as was possible—the political-economic importance of cable TV and its pertinence to the cities? So far as I can tell, only Bertram Gross, who organized a series of discussions on the subject when he was at the Wayne State Center for Urban Studies. Lupo et al. (1971) have at least opened up the transportation picture a bit more, after the earlier work that came out of Columbia University, notably by Jameson Doig (1966). But other aspects of transport—closely related to urbanism—as per civil aviation still remain largely beyond us. Civil aviation is an interesting case, the more because it supports and sustains some powerful organizations in the urban governmental setting—the airport authorities, organizations that have often the capacity to make such far-reaching decisions in narrowly technical terms that they have major effects on the local economy, land use and environmental conditions, social organizations, and the capacity of other governmental organizations to handle their responsibilities.

Airport planners, like highway planners, have little incentive to consider anything except the requirements of the airplane, the airlines industry, and the bond industry or the federal government through which the money for investment may be raised. Yet it is a distinct agency of social choice. For it means an impairment, which may be necessary but is usually not debated, of a given local area for the convenience of the airline passengers, often a different sort of person altogether. This impairment is not trivial, for the development of an airport facility may mean that within a radius of several miles, public facilities such as hospitals and schools are rendered inoperable or, if operable, at severely disadvantaged conditions. (One estimate is that within a few miles of the perimeter of John F. Kennedy Airport in New York, there are more than 150 public and parochial schools which—were the land now vacant—should not be constructed on those sites due to airport noise.) What is even more noticeable is that there is still virtually no consideration of the important safety hazards that the airport development involves in the major urban concentration (Cotter, 1968).

What would be the effect of airport expansion (or nonexpansion) on the economy of a region, or subparts thereof? The case by the authority representatives, the transportation administrators, and so on, will emphasize economic growth. But what would be the effects on the ability of people who live within defined portions of the metropolitan region (e.g., the inner city) to serve their own interests reasonably well?

It is inquiry of this sort, with a need to go into technicalities usually bypassed, that is essential if political science is to improve its capacity to relate to real-life decision problems in the urban environment.

REFERENCES

ANDERSON, W. L. (1957) "Municipal government: no lost world." Amer. Pol. Sci. Rev., 51, 3 (September): 776-783.

BAKER, G. E. (1955) Rural vs. Urban Political Power: The Nature and Consequences of Unbalanced Representation. Garden City, N.Y.: Doubleday.

BELL, J. B. (1966) Besieged: Seven Cities Under Siege. Philadelphia: Chilton Books.

BERUBE, M. R. and M. GITTELL [eds.] (1969) Confrontation at Ocean Hill-Brownsville: The New York School Strikes of 1968. New York: Praeger.

BLODGETT, G. (1966) The Gentle Reformers: Massachusetts Democrats in the Cleveland Era. Cambridge, Mass.: Harvard Univ. Press.

BOLLENS, J. C. (1957) Special District Governments in the United States. Berkeley: Univ. of California Press.

Boston Globe (1971) September 15.

CHARLES, S. F. (1963) Minister of Relief: Harry Hopkins and the Depression. New York: Syracuse Univ. Press.

COBB, R. C. (1970) The Police and the People: French Popular Protest 1789-1820. Oxford: Clarendon Press.

COLBERT, E. and E. CHAMBERLIN (1971) Chicago and the Great Conflagration. New York: Viking Press.

COTTER, C. P. (1968) Jet Tanker Crash: Urban Response to Military Disaster. Lawrence: Univ. of Kansas Press.

De TOCQUEVILLE, A. (1945) Democracy in America. New York: Alfred A. Knopf.

DOIG, J. W. (1966) Metropolitan Transportation Politics and the New York Region. New York: Columbia Univ. Press.

ETZIONI, A. and R. REMP (1972) "Technological 'shortcuts' to social change." Science, 175, 4017 (January 7): 31-38.

EYESTONE, R. (1971) The Threads of Public Policy: A Study in Policy Leadership. Indianapolis: Bobbs-Merrill.

FAIRLIE, J. A. (1908) Essays in Municipal Administration. New York: Macmillan.

GANS, H. J. (1962) The Urban Villagers: Group and Class in the Life of Italian Americans. New York: Free Press.

GEORGE, M.D. (1951) London Life in the XVIIIth Century. New York: Alfred A. Knopf.

GODKIN, E. L. (1898) Unforeseen Tendencies of Democracy. Boston: Houghton Mifflin.

GOODNOW, F. J. (1909) Municipal Government. New York: Century.

--- (1904) City Government in the United States. New York: Century.

--- (1903) Municipal Home Rule: A Study in Administration. New York: Columbia Univ. Press.

--- (1898) "The Tweed ring in New York City," pp. 335-353 in Bryce, The American Commonwealth, Vol. II. London: Macmillan.

GREENBERG, E. S., N. MILNER, and D. J. OLSON [eds.] (1971) Black Politics, the Inevitability of Conflict. New York: Holt, Rinehardt & Winston.

GREER, S. A. (1965) Urban Renewal and American Cities: The Dilemma of Democratic Intervention. Indianapolis: Bobbs-Merrill.

GUNN, G. E. (1966) The Political History of Newfoundland, 1832-1864. Toronto: Univ. of Toronto Press.

HERSON, L.J.R. (1957) "The lost world of municipal government." Amer. Pol. Sci. Rev., 51, 2 (June): 330-345.

HOLDEN, M., Jr. (1971) "Law and order in the metropolitan area: issues and options." University Forum Background Paper. Pittsburgh: University of Pittsburgh-Urban Interface Program (Contract No. OEG-2-9-480725-1027, Project No. 80725).

--- (1965) "Committee politics under primitive uncertainty." Midwest J. of Pol. Sci., 9 (August): 235-253.

JOHNSON, T. L. (1911) My Story. New York: B. W. Huebsch.

LANE, R. (1967) Policing the City: Boston, 1822-1885. Cambridge, Mass.: Harvard Univ. Press.

LEWIS, J. W. [ed.] (1971) The City in Communist China. Stanford: Stanford Univ. Press.

LUPO, A., F. COLCORD, and E. FOWLER [eds.] (1971) Rites of Way: The Politics of Transportation in Boston and the U.S. City. Boston: Little, Brown.

MacMULLEN, R. (1966) Enemies of the Roman Order: Treason, Unrest, and Alienation in the Empire. Cambridge, Mass.: Univ. of Harvard Press.

MANN, A. (1965) LaGuardia Comes to Power, 1933. Philadelphia: Lippincott.

MARTIN, R. C. (1957) Grass Roots. New York: Harper & Row.

MERRILL, H. and M. G. MERRILL (1971) The Republican Command 1897-1913. Lexington: Univ. Press of Kentucky.

MEYERSON, M. and E. C. BANFIELD (1959) Politics, Planning, and the Public Interest: The Case of Public Housing in Chicago. New York: Free Press.

MOYNIHAN, D. P. (1966) Traffic Safety and the Health of the Body Politic. Middletown, Conn.: Center for Advanced Studies, Weslayan University.

MUNRO, W. B. (1919) The Government of American Cities. New York: Macmillan.

PARENTI, M. (1970) "The possibilities for political change." Politics and Society, 1 (November): 79-90.

PERKINS, F. (1946) The Roosevelt I Knew. New York: Viking Press.

PERLOFF, H. S. [ed.] (1971) Toward the Year 2000: The Future of the U.S. Government. New York: Braziller.

POWELL, E. H. (1970) The Design of Discord: Studies of Anomie. New York: Oxford Univ. Press.

RABINOVITZ, F. F. (1969) City Politics and Planning. New York: Atherton.

RANNEY, D. C. (1969) Planning and Politics in the Metropolis. Columbus, Ohio: C. E. Merrill.

ROWE, L. S. (1908) Problems of City Government. New York: Appleton.

RUDE, G. F. (1964) The Crowd in History: A Study of Popular Disturbances in France and England, 1730-1848. New York: John C. Wiley.

SAYRE, W. and H. KAUFMAN (1960) Governing New York City. New York: Russell Sage Foundation.

SCULLARD, H. H. (1959) From the Gracchi to Nero: A History of Rome from 133 B.C. to A.D. 68. London: Methuen.

SHERWOOD, R. E. (1948) Roosevelt and Hopkins, An Intimate History. New York: Harper.

SINGHAM, A. W. (1968) The Hero and the Crowd in a Colonial Polity. New Haven: Yale Univ. Press.

SMITH, J. A. (1907) The Spirit of American Government: A Study of the Constitution, Its Origin, Influence, and Relation to Democracy. New York: Macmillan.

STEFFENS, J. L. (1931) The Autobiography of Lincoln Steffens. New York: Harcourt, Brace.

STOKELY, J. and W. DYKEMAN (1962) Seeds of Southern Change: The Life of Will Alexander. Chicago: Univ. of Chicago Press.

STONE, L. (1965) The Crisis of the Aristocracy, 1558-1641. Oxford: Clarendon Press.

TERKEL, L. (1970) Hard Times: An Oral History of the Great Depression. New York: Pantheon.

THOMPSON, W. (1965) A Preface to Urban Economics. Baltimore: Johns Hopkins Press.

THRUPP, S. (1962) The Merchant Class of Medieval London, 1300-1500. Ann Arbor: Univ. of Michigan Press.

TREVELYAN, G. M. (1949) Illustrated English Social History. London: Longmans, Green.

WILCOX, D. F. (1897) Study of City Government: An Outline of the Problems of Municipal Functions, Control and Organization. New York.

WILSON, J. Q. (1960) Negro Politics: The Search for Leadership. New York: Free Press.

WINGO, L. and H. S. PERLOFF [eds.] (1968) Issues in Urban Economics. Baltimore: Johns Hopkins Press.

WOOD, R. C. (1961) 1400 Governments: The Political Economy of the New York Metropolitan Region. Cambridge, Mass.: Harvard Univ. Press.

WRIGHT, D. S. (1968) Federal Grants-in-Aid: Perspectives and Alternatives. Washington: American Enterprise Institute for Public Policy Research.

BIBLIOGRAPHY

BIBLIOGRAPHY

A

ABERBACH, J. (1969) "Alienation and political behavior." Amer. Pol. Sci. Rev. 63 (March): 86-99.

ADRIAN, C. R. and O. P. WILLIAMS (1963) Four Cities. Philadelphia: Univ. of Pennsylvania Press.

AGGER, R. E., D. GOLDRICH, and B. E. SWANSON (1964) The Rulers and the Ruled. New York: John Wiley.

AIKEN, M. (1969) "Community power and mobilization." Annals 385 (September): 76-88.

——— and P. E. MOTT [eds.] (1970) The Structure of Community Power. New York: Random House.

ALFORD, R. R. (1969) "Bureaucracy and participation in four Wisconsin cities." Urban Affairs Q. 5 (September): 5-30.

——— (1967) "The comparative study of urban politics," pp. 246-302 in L. F. Schnore and H. Fagin (eds.) Urban Research and Policy Planning. Urban Affairs Annual Reviews, Vol. 1, Beverly Hills: Sage Pubns.

——— and E. C. LEE (1968) "Voting turnout in American cities." Amer. Pol. Sci. Rev. 62 (September): 796-813.

ALINSKY, S. (1970) The Professional Radical. New York: Harper & Row.

——— (1969) Reveille for Radicals. New York: Vintage.

ALLPORT, W. (1958) The Nature of Prejudice. Garden City, N.Y.: Doubleday Anchor.

ALMOND, G. A. and G. B. POWELL (1968) Comparative Politics: A Developmental Approach. Boston: Little, Brown.

ANTON, T. J. (1966) The Politics of State Expenditures in Illinois. Urbana: Univ. of Illinois Press.

——— (1964) Budgeting in Three Illinois Cities. Urbana: Institute of Government and Public Affairs.

APILADO, V. P. (1971) "Corporate-government interplay: the era of industrial aid finance." Urban Affairs Q. 7 (December): 219-241.

APPLEBY, M. (1969) "Revolutionary change and the urban environment," pp. 216-232 in P. Long (ed.) New Left. Boston: Porter Sargent.

ARVILL, R. (1967) Man and Environment: Crisis and the Strategy of Choice. Baltimore: Penguin.

B

BABCHUCK, N. and A. BOOTH (1969) "Voluntary association membership: a longitudinal analysis." Amer. Soc. Rev. (34): 31-45.

BABCHUCK, N. and R. V. THOMPSON (1962) "The voluntary association of Negroes." Amer. Soc. Rev. 27 (October): 647-655.

BACHRACH, P. (1970) "A power analysis: the shaping of anti-poverty policy in Baltimore." Public Policy 18 (Winter): 179-201.

--- (1967) The Theory of Democratic Elitism. Boston: Little, Brown.

--- and M. BARATZ (1970) Power and Poverty. New York: Oxford.

BAILEY, H. A., Jr. (1968) "Negro interest group strategies." Urban Affairs Q. 4 (September): 26-38.

BAKER, G. E. (1955) Rural vs. Urban Political Power: The Nature and Consequences of Unbalanced Representation. Garden City, N.Y.: Doubleday.

BAKER, J. H. (1971) Urban Politics in America. New York: Scribner's.

BANFIELD, E. C. (1970) The Unheavenly City. Boston: Little, Brown.

--- (1961a) Political Influence. New York: Free Press.

--- [ed.] (1961b) Urban Government. New York: Free Press.

BELL, C. (1971) "Controlling residential development on the urban fringe: St. Louis County, Missouri." J. of Urban Law (48): 309-428.

BELL, W. and M. BOAT (1957) "Urban neighborhoods in informal social relations." Amer. J. of Sociology 62 (January): 391-398.

BELLUSH, J. and S. M. DAVID [eds.] (1971) Race and Politics in New York City. New York: Praeger.

BENTLEY, A. A. (1949) The Process of Government. San Antonio, Texas: Principia Press of Trinity University.

BILLINGSLEY, A. (1968) Black Families in White America. Englewood Cliffs, N.J.: Prentice-Hall.

BLANK, B. D., R. J. IMMERMAN, and C. P. RYDELL (1969) "A comparative study of urban bureaucracy." Urban Affairs Q. 4 (March): 343-354.

BLOOMBERG, W. (1966) "Community organization," pp. 317-358 in H. S. Becker (ed.) Social Problems: A Modern Approach. New York: John Wiley.

BOLLENS, J. C. (1957) Special District Governments in the United States. Berkeley: Univ. of California Press.

——— and H. J. SCHMANDT (1970) The Metropolis: Its People, Politics, and Economic Life. New York: Harper & Row.

BONJEAN, C. M., T. N. CLARK, and R. L. LINEBERRY [eds.] (1971) Community Politics. New York: Free Press.

BOSKOFF, A. and H. ZEIGLER (1964) Voting Patterns in a Local Election. Philadelphia: Lippincott.

BRAGER, G. and H. SPECHT (1967) "Social action by the poor: prospects, problems and strategies," pp. 136-141 in G. Brager and F. Purcell (eds.) Community Action Against Poverty. New Haven: College and Univ. Press.

BRAYBROOK, D. and C. E. LINDBLOM (1963) A Strategy of Decision. New York: Free Press.

BRAZER, H. E. (1959) City Expenditures in the United States. New York: National Bureau of Economic Research.

C

CAHN, E. S. and B. A. PASSETT [eds.] (1971) Citizen Participation: Effective Community Change. New York: Praeger.

CALDWELL, L. K. (1970) Environment: A Challenge for Modern Society. Garden City, N.Y.: Natural History Press.

——— [ed.] (1967) Environmental Studies–Papers on the Politics of Public Administration–Environmental Relationships. V. I-IV. Bloomington: Indiana Univ. Press.

CAMPBELL, A. K. (1971) White Attitudes Toward Black People. Ann Arbor: Institute for Social Research.

——— and H. SCHUMAN (1968) Racial Attitudes in Fifteen American Cities: A Report Prepared for the National Advisory Commission on Civil Disorders. Ann Arbor: Institute for Social Research.

CAMPBELL, A. K. and S. SACKS (1967) Metropolitan America. New York: Free Press.

CAMPBELL, A. K., E. CONVERSE, W. E. MILLER, and D. E. STOKES (1960) The American Voter. New York: John Wiley.

CANTRIL, H. (1965) The Patterns of Human Concerns. New Brunswick, N.J.: Rutgers Univ. Press.

CARMICHAEL, S. and C. HAMILTON (1967) Black Power. New York: Vintage.

CARTER, R. F. and W. G. SAVARD (1961) Influence of Voter Turnout on School Bond and Tax Elections. Washington, D.C.: Government Printing Office.

CARTER, T. (1970) Mexican-Americans in School: A History of Educational Neglect. New York: College Entrance Examination Board.

CASTANEDA, A., M. RAMIREZ, C. CORTES, and M. BARRERA [eds.] (1971) Mexican-Americans and Educational Change. Riverside: Univ. of California.

CATTELL, D. T. (1968) Leningrad: A Case Study in Soviet Urban Government. New York: Praeger.

CLARK, K. (1965) Dark Ghetto. New York: Harper Torchbooks.

CLARK, P. and J. Q. WILSON (1961) "Incentive systems: a theory of organization." Administrative Science Q. (6): 129-166.

CLARK, T. N. (1972) "Urban typologies and political outputs," in B.J.L. Berry (ed.) Handbook of City Classification. New York: John Wiley.

--- [ed.] (1968a) Community Structure and Decision-Making: Comparative Analysis. San Francisco: Chandler.

--- (1968b) "Discipline, method, community structure and decision-making: the role and limitations of the sociology of knowledge." Amer. Sociologist 3 (August): 214-217.

--- (1968c) "Community structure, decision-making, budget expenditures, and urban renewal in 51 American communities." Amer. Soc. Rev. 33 (August): 576-593.

COBB, R. C. (1970) The Police and the People: French Popular Protest 1789-1820. Oxford: Clarendon Press.

COLBERT, E. and E. CHAMBERLIN (1971) Chicago and the Great Conflagration. New York: Viking Press.

COLEMAN, J. S. (1957) Community Conflict. New York: Free Press.

COLEMAN, J. (1966) Equality of Educational Opportunity. Washington, D.C.: Government Printing Office.

CONNERY, R. H. and R. H. LEACH (1960) The Federal Government and Metropolitan Areas. Cambridge: Harvard Univ. Press.

COSTELLO, T. W. (1971) "Change in municipal government: the view from inside." J. of Applied Behavioral Science (March-April): 131-143.

COULTER, P. B. (1970) "Comparative community politics and public policy: problems in theory and research." Polity 3 (Fall): 22-43.

COX, J. L. (1967) "Federal urban development policy and the metropolitan Washington council of governments: a reassessment." Urban Affairs Q. 3 (September): 75-94.

CRAIN, R. L. (1968) The Politics of School Desegregation. Chicago: Aldine.

CRECINE, J. P. (1969) Governmental Problem Solving: A Computer Simulation of Municipal Budgeting. Chicago: Rand McNally.

CRENSON, M. A. (1971) The Unpolitics of Air Pollution: A Study of Non-Decision Making in the Cities. Baltimore: Johns Hopkins Press.

CURTIS, R. L. and L. A. ZURKER, Jr. (1971) "Voluntary associations and the social integration of the poor." Social Problems (Winter): 339-357.

D

DAHL, R. (1961) Who Governs? New Haven: Yale Univ. Press.

——— (1960) "The analysis of influence in local communities," pp. 25-42 in C. Adrian (ed.) Social Science and Community Action. East Lansing: Michigan State University.

——— (1956) A Preface to Democratic Theory. Chicago: Univ. of Chicago Press.

DALAND, R. T. [ed.] (1970) Comparative Urban Research. Beverly Hills: Sage Pubns.

DANSEREAU, P. [ed.] (1970) Challenge for Survival: Land, Air, and Water in Megalopis. New York: Columbia Univ. Press.

D'ANTONIO, W. and W. FORM (1965) Influentials in Two Border Cities. Milwaukee: Univ. of Notre Dame Press.

DAVID, O., M.A.H. DEMPSTER, and A. WILDAVSKY (1966) "A theory of the budgetary process." Amer. Pol. Sci. Rev. (September): 529-547.

DAVIDOFF, P. (1969) "The planner as advocate," pp. 544-555 in E. C. Banfield (ed.) Urban Government. New York: Free Press.

DAVIS, J. C. (1970) The Politics of Pollution. New York: Pegasus.

DAVIS, J. T. (1965) "Middle class housing in the central city." Economic Geography (July): 238-251.

DAVIS, O. and G. H. HARRIS, Jr. (1966) "A political approach to a theory of public expenditures." National Tax J. 19 (September): 259-275.

DAVIS, R. (1967) "The war on poverty: notes on an insurgent response," pp. 159-174 in M. Cohen and D. Hale (eds.) The New Student Left. Boston: Beacon.

DAWSON, R. E. and J. A. ROBINSON (1963) "Interparty competition, economic variables and welfare policies in the American states." J. of Politics (25): 265-289.

DE TOCQUEVILLE, A. (1945) Democracy in America. New York: Alfred A. Knopf.

DOIG, J. W. (1966) Metropolitan Transportation Politics and the New York Region. New York: Columbia Univ. Press.

DOLLARD, J. (1957) Caste and Class in a Southern Town. Garden City, N.J.: Anchor.

DONOVAN, J. C. (1967) The Politics of Poverty. New York: Pegasus.

DOWNES, B. T. [ed.] (1971) Cities and Suburbs. Belmont: Wadsworth.

——— (1969) "Issue conflict, factionalism, and consensus in suburban city councils." Urban Affairs Q. 4 (June): 477-497.

DOWNS, A. (1970) Urban Problems and Prospects. Chicago: Markham.

——— (1967) Inside Bureaucracy. Boston: Little, Brown.

——— (1957) An Economic Theory of Democracy. New York: Harper & Row.

DUDAS, J. J. and D. B. LONGBRAKE (1971) "Problems and future directions of residential integration: the local application of federally funded programs in Dade County, Florida." Southeastern Geographer (November): 157-168.

DYE, T. R. (1966) Politics, Economics, and the Public: Policy Outcomes in the American States. Chicago: Rand McNally.

E

EASTON, D. (1965) A Systems Analysis of Political Life. New York: John Wiley.

EDELMAN, M. (1964) The Symbolic Uses of Politics. Urbana: Univ. of Illinois.

EDGAR, R. E. (1970) Urban Power and Social Welfare: Corporate Influence in an American City. Beverly Hills: Sage Pubns.

ELAZAR, D. J. (1967) " 'Fragmentation' and local organizational response to federal-city programs." Urban Affairs Q. 2 (June): 30-46.

ELLIS, W. (1969) White Ethics and Black Power. Chicago: Aldine.

EULAU, H., B. H. ZISK, and K. PREWITT (1965) "Latent partisan elections: effects of political milieu and mobilization," in H. Zeigler and M. K. Jennings (eds.) The Electoral Process. Englewood Cliffs, N.J.: Prentice-Hall.

EWALD, W. (1968) Environment and Policy: The Next Fifty Years. Bloomington: Indiana Univ. Press.

EYESTONE, R. (1971) The Threads of Public Policy: A Study in Policy Leadership. Indianapolis: Bobbs-Merrill.

--- (1966) "The relation between public policy and some structural and environmental variables in the American states." Amer. Pol. Sci. Rev. (60): 73-78.

--- and H. EULAU (1968) "City councils and policy outcomes: developmental profiles," in J. Q. Wilson (ed.) City Politics and Public Policy. New York: John Wiley.

F

FABRICANT, S. (1952) The Trend of Government Activity in the United States Since 1900. New York: National Bureau of Economic Research.

FAINSTEIN, N. I. and S. S. FAINSTEIN (1971) "City planning and political values." Urban Affairs Q. 6 (March): 341-362.

FALTERMAYER, E. K. (1968) Redoing America: A Nationwide Report on How to Make Our Cities and Suburbs Livable. New York: Harper & Row.

FARKAS, S. (1971) Urban Lobbying: Mayors in the Federal Arena. New York: New York Univ. Press.

FARLEY, R. (1970) "The changing distribution of Negroes within metropolitan areas: the emergence of black suburbs." Amer. J. of Sociology (January): 512-523.

FEAGIN, J. R. and H. HAHN (1972) "Theories of urban violence: an appraisal" in H. Hirsch and D. Perry (eds.) Violence and Politics: A Collection of Original Essays. New York: Harper & Row.

FELDMAN, L. D. and M. D. GOLDRICK [eds.] (1969) Politics and Government of Urban Canada. Toronto: Methuen.

FOGELSON, R. M. (1971) Violence as Protest. Garden City, N.J.: Doubleday.

FRANCIS, W. (1967) Legislative Issues in the Fifty States: A Comparative Analysis. Chicago: Rand McNally.

FRAZIER, E. F. (1939) The Negro Family in the United States. Chicago: Univ. of Chicago Press.

FRIED, R. and F. RABINOVITZ (forthcoming) Comparative Urban Performance. Englewood Cliffs, N.J.: Prentice-Hall.

FRIEDMAN, R. S., B. KLEIN, and J. H. ROMANI (1966) "Administrative agencies and the publics they serve." Public Administration Rev. 26 (September): 192-204.

FRIESEMA, H. P. (1971) Metropolitan Political Structure. Iowa City: Univ. of Iowa Press.

--- (1966) "The metropolis and the maze of local government." Urban Affairs Q. 2 (December): 68-91.

FROMAN, L. A., Jr. (1967) "An analysis of public policies in cities." J. of Politics 29 (February): 94-108.

--- (1966) "Some effects of interest group strength in state politics." Amer. Pol. Sci. Rev. 60 (December): 952-962.

G

GAMSON, W. A. (1961) "The fluoridation dialogue: is it an ideological conflict." Public Opinion Q. 25 (Winter): 526-537.

GANS, H. J. (1962) The Urban Villagers: Group and Class in the Life of Italian Americans. New York: Free Press.

GARNSEY, M. E. and J. R. HIBBS [eds.] (1967) Social Sciences and the Environment. Boulder: Univ. of Colorado Press.

GELB, J. (1970) "Black Republicans in New York: a minority group in a minority party." Urban Affairs Q. 5 (June): 454-473.

GELLHORN, W. (1966) When Americans Complain: Governmental Grievance Procedures. Cambridge: Harvard Univ. Press.

GERWIN, D. (1969) Budgeting Public Funds: The Decision Process in an Urban School District. Madison: Univ. of Wisconsin Press.

GITTELL, M. (1969) Confrontation at Ocean Hill-Brownsville. New York: Praeger.

GLAZER, N. and D. MOYNIHAN (1963) Beyond the Melting Pot. Cambridge: MIT Press and Harvard Univ. Press.

GOODNOW, F. J. (1909) Municipal Government. New York: Century.
——— (1904) City Government in the United States. New York: Century.
——— (1903) Municipal Home Rule: A Study in Administration. New York: Columbia Univ. Press.
GORDON, M. (1963) Sick Cities. New York: Macmillan.
GRANT, D. R. (1965) "A comparison of predictions and experience with Nashville 'Metro.' " Urban Affairs Q. 1 (September): 35-53.
GREBLER, L., J. MOORE, and R. GUZMAN (1970) The Mexican-American People. New York: Free Press.
GREENBERG, E. S., N. MILNER, and D. J. OLSON [eds.] (1971) Black Politics: The Inevitability of Conflict. New York: Holt, Rinehart & Winston.
GREER, S. (1968) "The shaky future of local government." Psychology Today 2 (August): 64-69.
——— (1965) Urban Renewal and American Cities: The Dilemma of Democratic Intervention. Indianapolis: Bobbs-Merrill.
GRIER, G. and E. GRIER (1966) Equality and Beyond. Chicago: Quadrangle.
GURR, T. (1970) Why Men Rebel. Princeton: Princeton Univ. Press.

H

HAGGSTROM, W. C. (1968) "The power of the poor," pp. 457-475 in L. Ferman, J. Kornbluh, and A. Haber (eds.) Poverty in America. Ann Arbor: Univ. of Michigan Press.
HAHN, H. (1971) Urban-Rural Conflict: The Politics of Change. Beverly Hills: Sage Pubns.
——— (1970a) "Civic responses to riots: a reappraisal of Kerner Commission data." Public Opinion Q. 34 (Spring): 101-107.
——— (1970b) "Black separatists: attitudes and objectives in a riot-torn ghetto." J. of Black Studies 1 (September): 35-53.
——— (1970c) "The political impact of shifting attitudes." Soc. Sci. Q. 51 (December): 730-742.
——— (1970d) "Ethos and social class: referenda in Canadian cities." Polity 2 (Spring): 295-315.
——— (1970e) "Correlates of public sentiments about war: local referenda on the Vietnam issue." Amer. Pol. Sci. Rev. 64 (December): 1186-1198.
——— (1968a) "Northern referenda on fair housing: the response of white voters." Western Pol. Q. 21 (September): 483-495.
——— (1968b) "Voting in Canadian communities: a taxonomy of referendum issues." Canadian J. of Pol. Sci. 1 (December): 462-469.
——— and T. ALMY (1971) "Ethnic politics and racial issues: voting in Los Angeles." Western Pol. Q. 24 (December): 719-730.

HAHN, H. and R. J. SCHMIDT (1971) "Policy themes and community attitudes: a research note on the anti-poverty program." Soc. Sci. Q. 52 (December): 672-679.

HAMILTON, H. D. (1970) "Direct legislation: some implications of open housing referenda." Amer. Pol. Sci. Rev. 64 (March): 124-137.

HARTMAN, C. (1970) "The advocate planner: from 'hired gun' to political partisan." Social Policy 1 (July/August): 37-38.

Harvard Law Review (1971) "Exclusionary zoning and equal protection." (May): 1645-1669.

HAWKINS, B. W. (1971) Politics and Urban Policies. Indianapolis: Bobbs-Merrill.

--- (1966) "Public opinion and metropolitan reorganization in Nashville." J. of Politics 28 (May): 408-418.

HAWLEY, A. H. (1963) "Community power and urban renewal success." Amer. J. of Sociology (January): 422-431.

--- and B. G. ZIMMER (1970) The Metropolitan Community: Its People and Government. Beverly Hills: Sage Pubns.

HEIKOFF, J. M. (1967) "Justice, politics and urban planning." Urban Affairs Q. 3 (September): 46-61.

HENNESY, T. (1970) "Problems in concept formation: the ethos 'theory' and the comparative study of urban politics." Midwest J. of Pol. Sci. 14 (November): 537-564.

HERBER, L. (1965) Crisis in Our Cities. Englewood Cliffs, N.J.: Prentice-Hall.

HERSON, L.J.R. (1957) "The lost world of municipal government." Amer. Pol. Sci. Rev. 51, 2 (June): 330-345.

HOFFERBERT, R. I. (1970) "Elite influence in state policy formation: a model for comparative inquiry." Polity (2): 316-344.

--- and I. SHARKANSKY [eds.] (1971) State and Urban Politics: Readings in Comparative Public Policy. Boston: Little, Brown.

HOFSTADTER, R. (1955) The Age of Reform: From Bryan to F.D.R. New York: Alfred A. Knopf.

HOLDEN, M., Jr. (1965) "Committee politics under primitive uncertainty." Midwest J. of Pol. Sci. 9 (August): 235-253.

HOROWITZ, G. (1966) "Conservatism, liberalism, and socialism in Canada: an interpretation." Canadian J. of Economics and Pol. Sci. 32 (May): 143-171.

HORTON, J. E. and W. E. THOMPSON (1962) "Powerlessness and political negativism: a study of defeated local referendums." Amer. J. of Sociology 67 (March): 485-493.

HUNTER, F. (1953) Community Power Structure. Chapel Hill: University of North Carolina.

I

IANNACCONE, L. and D. K. WILES (1971) "The changing politics of urban education." Education and Urban Society 3 (May): 225-244.

IANNACONNE, L. and F. W. LUTZ (1970) Politics, Power and Policy: The Governing of Local School Districts. Columbus, Ohio: Charles E. Merrill.

J

JACOB, H. and M. LIPSKY (1968) "Outputs, structure and process: an assessment of changes in the study of state and local politics." J. of Politics (30): 510-533.

JACOBS, J. (1961) The Death and Life of Great American Cities. New York: Random House.

JANOWITZ, M. (1952) The Community Press in an Urban Setting. New York: Free Press.

JENNINGS, M. K. and D. ZEIGLER (1970) "The salience of American state politics." Amer. Pol. Sci. Rev. 64 (June): 523-535.

--- (1966) "Class, party, and race in four types of elections: the case of Atlanta." J. of Politics 28 (May): 391-407.

JONES, C. (1970) An Introduction to the Study of Public Policy. Belmont, Calif.: Wadsworth.

K

KADUSHIN, C. (1968) "Power, influence and social circles: a new methodology for studying opinion makers." Amer. Soc. Rev. 33 (October): 685-699.

--- (1966) "The friends and supporters of psychotherapy: on social circles in urban life." Amer. Soc. Rev. 31 (December): 786-802.

KAGI, H. M. (1969) "The roles of private consultants in urban governing." Urban Affairs Q. 5 (September): 45-58.

KAPLAN, H. (1967) Urban Political Systems. New York: Columbia Univ. Press.

KATZ, E. and P. F. LAZARSFELD (1955) Personal Influence. New York: Free Press.

KATZNELSON, I. (1971) "Power in the reformulation of race research," pp. 51-82 in P. Orleans and W. R. Ellis, Jr. (eds.) Race, Change and Urban Society. Urban Affairs Annual Reviews, Vol. 5, Beverly Hills: Sage Pubns.

KELLEY, S. (1967) "Registration and voting: putting first things first." Amer. Pol. Sci. Rev. 61 (June): 359-379.

KERSHAW, J. A. (1970) Government Against Poverty. Chicago: Markham.

KEY, V. O., Jr. (1963) Public Opinion in American Democracy. New York: Alfred A. Knopf.

——— (1951) Southern Politics. New York: Alfred A. Knopf.

KRAMER, J. and I. WALTER (1968) "Politics in an all-Negro city." Urban Affairs Q. 4 (September): 65-88.

KRAMER, R. and H. SPECHT [eds.] (1969) Readings in Community Organization Practice. Englewood Cliffs, N.J.: Prentice-Hall.

L

LANE, R. (1967) Policing the City: Boston, 1822-1885. Cambridge: Harvard Univ. Press.

——— (1966) "Decline of politics and ideology in a knowledgeable society." Amer. Soc. Rev. 31 (October): 649-662.

——— (1959) Political Life. New York: Free Press.

LANG, K. and G. E. LANG (1968) Politics and Television. Chicago: Quadrangle.

LAZARSFELD, P. F. and R. K. MERTON (1963) "Friendship as a social process: a substantial and methodological analysis" in M. W. Riley (ed.) Sociological Research. V. I: A Case Approach. New York: Harcourt, Brace.

LAZARSFELD, P. F., B. BERELSON, and H. GAUDET (1948) The People's Choice. New York: Columbia Univ. Press.

LE BLANC, H. L. and D. T. ALLENSWORTH (1971) The Politics of States and Urban Communities. New York: Harper & Row.

LEVITAN, S. A. (1969) The Great Society's Poor Law. Baltimore: Johns Hopkins Press.

——— (1964) Federal Aid to Depressed Areas. Baltimore: Johns Hopkins Press.

LEWIS, J. W. [ed.] (1971) The City in Communist China. Stanford: Stanford Univ. Press.

LEWIS, O. (1968) "The culture of poverty," pp. 405-415 in L. Ferman, J. Kornbluh, and A. Haber (eds.) Poverty in America. Ann Arbor: Univ. of Michigan Press.

LIEBER, H. (1970) "Public administration and environmental quality." Public Administration Rev. 30 (May/June): 277-286.

LINDSAY, J. (1969) The City. New York: New American Library.

LINEBERRY, R. L. (1970) "Reforming metropolitan governance: requiem or reality." Georgetown Law J. 58 (March/May): 675-718.

——— and I. SHARKANSKY (1971) Urban Politics and Public Policy. New York: Harper & Row.

LINEBERRY, R. L. and E. P. FOWLER (1967) "Reformism and public policies in American cities." Amer. Pol. Sci. Rev. 61 (September): 701-716.

LIPSKY, M. (1971) "Street-level bureaucracy and the analysis of urban reform." Urban Affairs Q. 6 (June): 391-410.

——— (1970) Protest in City Politics: Rent Strikes, Housing and the Power of the Poor. Chicago: Rand McNally.

——— and C. NEUMANN (1969) "Landlord-tenant law in the United States and West Germany—a comparison of legal approaches." Tulane Law Rev. 44 (December): 36-66.

LITT, E. (1970) Ethnic Politics in America. Glenview: Scott, Foresman.

LOCKARD, D. (1959) New England State Politics. Princeton: Princeton Univ. Press.

LONG, N. E. (1967) "Political science and the city," pp. 243-262 in L. F. Schnore and H. Fagin (eds.) Urban Research and Policy Planning. Urban Affairs Annual Reviews, Vol. 1. Beverly Hills: Sage Pubns.

LORIMER, J. (1970) The Real World of City Politics. Toronto: James Lewis and Samuel.

LOVERIDGE, R. O. (1971) City Managers in Legislative Politics. Indianapolis: Bobbs-Merrill.

——— (1970) "Types, ranges, and methods for classifying human behavioral responses to air pollution," pp. 53-70 in A. Atkisson and R. Gaines (eds.) Development of Air Quality Standards. Columbus, Ohio: Charles E. Merrill.

LOWI, T. J. (1972) "Why mayors go nowhere." Washington Monthly (January): 55-61.

——— (1969) The End of Liberalism. New York: W. W. Norton.

——— (1964) "American business, public policy, case-studies and political theory." World Politics 16 (July): 677-715.

LUPO, A., F. COLCORD, and E. P. FOWLER (1971) Rites of Way: The Politics of Highway Construction in Boston and the American City. Boston: Little, Brown.

M

McCONNELL, G. (1966) Private Power and American Democracy. New York: Alfred A. Knopf.

McDILL, E. L. and J. C. RIDLEY (1962) "Status, anomia, political alienation, and political participation." Amer. J. of Sociology 68 (September): 205-213.

McEVOY, J. (1970) The American Public's Concern with the Environment: A Study of Public Opinion. Davis, Calif.: Institute of Government Affairs.

MANN, A. (1965) LaGuardia Comes to Power, 1933. Philadelphia: Lippincott.

MARRIS, P. and M. REIN (1969) Dilemmas of Social Reform. New York: Atherton.

MARSHALL, D. R. (1971) "Public participation and the politics of poverty," pp. 451-482 in P. Orleans and W. R. Ellis, Jr. (eds.) Race, Change and Urban Society. Urban Affairs Annual Reviews, Vol. 5. Beverly Hills: Sage Pubns.

MARTIN, R. C. (1957) Grass Roots. New York: Harper & Row.

MASOTTI, L. H. and D. R. BOWEN [eds.] (1968a) Riots and Rebellion: Civil Violence in the Urban Community. Beverly Hills: Sage Pubns.

--- (1968b) "Spokesmen for the poor: an analysis of Cleveland's poverty board candidates." Urban Affairs Q. 4 (September): 89-110.

MAYER, A. C. (1966) "The significance of quasi-groups in the study of complex societies," pp. 97-122 in M. Banton (ed.) The Social Anthropology of Complex Societies. New York: Praeger.

MEADOW, K. P. (1962) "Negro-White differences among newcomers to a transitional urban area." J. of Intergroup Res. 3 (1962): 320-330.

MELTSNER, A. J. (1971) The Politics of City Revenue: The Oakland Project. Berkeley: Univ. of California Press.

--- (1970) "Local revenue: a political problem," pp. 103-136 in J. R. Crecine (ed.) Financing the Metropolis: Public Policy in Urban Economies. Urban Affairs Annual Reviews, Vol. 4. Beverly Hills: Sage Pubns.

--- and A. WILDAVSKY (1970) "Leave city budgeting alone!: a survey, case study, and recommendations for reform," in J. P. Crecine (ed.) Financing the Metropolis: Public Policy in Urban Economies. Urban Affairs Annual Reviews, Vol. 4. Beverly Hills: Sage Pubns.

MEYERSON, M. and E. C. BANFIELD (1959) Politics, Planning, and the Public Interest: The Case of Public Housing in Chicago. New York: Free Press.

MILBRATH, L. (1965) Political Participation. Chicago: Rand McNally.

MILLER, S. M. and F. RIESMAN (1968) Social Class and Social Policy. New York: Basic Books.

MILLS, C. W. (1956) The Power Elite. New York: Oxford Univ.

MINAR, D. W. and S. GREEN [eds.] (1969) The Concept of Community. Chicago: Aldine.

MITCHELL, W. E. and I. WALTER [eds.] (1971) State and Local Finance. New York: Ronald Press.

MOGEY, J. M. (1956) Family and Neighborhood. London: Oxford Univ. Press.

MOORE, C. H. and R. E. JOHNSTON (1971) "School decentralization, community control, and the politics of public education." Urban Affairs Q. 6 (June): 421-446.

MOORE, J. (1970) "Colonialism: the case of the Mexican Americans." Social Problems (Spring): 463-472.

MOORE, V. J. (1971) "Politics, planning, and power in New York State: the path from theory to reality." J. of Amer. Institute of Planners (March): 66-71.

MORSS, E. R. (1966) "Some thoughts on the determinants of state and local expenditures." National Tax J. 14 (March): 95-103.

MOYNIHAN, D. P. [ed.] (1970) Toward a National Urban Policy. New York: Basic Books.

——— (1969) Maximum Feasible Misunderstanding. New York: Free Press.

MULLER, E. (1970) "The representation of citizens by political authorities: consequences for regime support." Amer. Pol. Sci. Rev. 64 (December): 1149-1166.

MUNRO, W. B. (1919) The Government of American Cities. New York: Macmillan.

MURPHY, R. D. (1971) Political Entrepreneurs and Urban Poverty. Lexington, Mass.: Heath-Lexington.

MUSHKIN, S. (1970) "PPB for the cities: problems and next steps," in J. P. Crecine (ed.) Financing the Metropolis: Public Policy in Urban Economies. Urban Affairs Annual Reviews, Vol. 4. Beverly Hills: Sage Pubns.

——— (1969) "PPB in cities." Public Administration Rev. 29 (March/April): 167-178.

MYRDAL, G. (1964) An American Dilemma: The Negro Problem and Modern Democracy. New York: McGraw-Hill.

N

National Advisory Commission on Civil Disorders (1968) Report. Washington, D.C.: Government Printing Office.

O

OESER, O. A. and S. B. HAMMOND [eds.] (1954) Social Structure and Personality in a City. London: Routledge & Kegan Paul.

OLSON, D. J. (1969) "Citizen grievance letters as a gubernatorial control device in Wisconsin." J. of Politics 31 (August): 741-755.

OLSON, M. L., Jr. (1965) The Logic of Collective Action. Cambridge: Harvard Univ. Press.

OSTROM, E. (1971) "Institutional arrangements and the measurement of policy consequences." Urban Affairs Q. 6 (June): 447-476.

P

PALLEY, M. L., R. RUSSO, and E. SCOTT (1970) "Subcommunity leadership in a black ghetto: a study of Newark, New Jersey." Urban Affairs Q. 5 (March): 291-312.

PEATTIE, L. (1969) "Reflections of an advocate planner," pp. 556-567 in E. Banfield (ed.) Urban Government. New York: Free Press.

PETTIGREW, T. F. (1972) "Racism and mental health of white Americans: a social psychological view," in C. Willey, B. Kramer, and Brown (eds.) Racism and Mental Health. Pittsburgh: Univ. of Pittsburgh Press.

––– (1971) Racially Separate or Together? New York: McGraw-Hill.

PIDOT, G. (1969) "A principal components analysis of the determinants of local government fiscal patterns." Rev. of Economics and Statistics 51 (May): 176-188.

PITKIN, H. (1967) The Concept of Representation. Berkeley: Univ. of Calif. Press.

PIVEN, F. F. (1970) "Whom does the advocate planner serve?" Social Policy 1 (May/June): 32-37.

PLUNKETT, T. J. (1968) Urban Canada and its Government. Toronto: Macmillan.

PODELL, L. (1968) Families on Welfare in New York City: Kinship, Friendship, and Citizenship. New York: Center for Social Research.

PRESTHUS, R. (1964) Men at the Top. New York: Oxford University.

PREWITT, K. and H. EULAU (1969) "Political matrix and political representation: prolegomenon to a new departure from an old problem." Amer. Pol. Sci. Rev. 63 (June): 427-441.

R

RABINOVITZ, F. F. (1969) City Politics and Planning. New York: Atherton.

RAINWATER, L. (1966) "Crucible of identity: the Negro lower class family." Daedalus 95 (Winter): 172-216.

RANNEY, A. [ed.] (1968) Political Science and Public Policy. Chicago: Markham.

RANNEY, D. C. (1969) Planning and Politics in the Metropolis. Columbus, Ohio: Charles E. Merrill.

RIDGEWAY, J. (1970) The Politics of Ecology. New York: E. P. Dutton.

RIENOW, R. and L. T. RIENOW (1967) Moment in the Sun: A Report on the Deteriorating Quality of the American Environment. New York: Ballantine.

RIEW, J. (1970) "Metropolitan disparities and fiscal federalism," pp. 137-162 in J. P. Crecine (ed.) Financing the Metropolis: Public Policy in Urban Economies. Urban Affairs Annual Reviews, Vol. 4. Beverly Hills: Sage Pubns.

RIKER, W. (1964) Federalism. Boston: Little, Brown.

RIPLEY, R. B. (1972) The Politics of Economic and Human Resource Development. Indianapolis: Bobbs-Merrill.

--- W. B. MORELAND, and R. H. SINNREICH (1972) "Policy-making: a conceptual scheme." Amer. Politics Q. 1.

ROCCO, R. (1970) "The Chicano in the social sciences: traditional concepts, myths, and images," Aztlán (Fall): 75-98.

ROGERS, D. (1971) The Management of Big Cities: Interest Groups and Social Change Strategies. Beverly Hills: Sage Pubns.

--- (1968) 110 Livingston Street. New York: Random House.

ROSE, H. M. (1972) The Black Ghetto: A Spatial Behavioral Perspective. New York: McGraw-Hill.

--- (1965) "The all-Negro town: its evolution and function." Geographical Rev. (July): 362-381.

--- (1964) "Metropolitan Miami's changing Negro population, 1950-1969." Economic Geography (July): 221-238.

ROSENTHAL, A. (1969) Pedagogues and Power. Syracuse: Syracuse Univ. Press.

ROSENTHAL, D. B. and R. L. CRAIN (1966a) "Structure and values in local political systems: the case of fluoridation decisions." J. of Politics 28 (February): 169-196.

--- (1966b) "Executive leadership and community innovation: the fluoridation experience." Urban Affairs Q. (March): 39-57.

ROSSI, P. H. and R. L. CRAIN (1968) "The NORC permanent community sample." Public Opinion Q. 32 (Summer): 261-272.

ROTH, M. and G. R. BOYNTON (1969) "Communal ideology and political support." J. of Politics 31 (February): 167-185.

ROURKE, F. (1969) Bureaucracy, Politics, and Public Policy. Boston: Little, Brown.

RUBEL, A. (1966) Across the Tracks: Mexican-Americans in a Texas City. Austin: Univ. of Texas Press.

RUDE, G. F. (1964) The Crowd in History: A Study of Popular Disturbances in France and England, 1730-1848. New York: John Wiley.

RUNCIMAN, W. G. (1966) Relative Deprivation and Social Justice. London: Routledge & Kegan Paul.

S

SALISBURY, R. H. (1969)"An exchange theory of interest groups." Midwest J. of Pol. Sci. 13 (February): 1-32.

--- and G. BLACK (1963) "Class and party in partisan and nonpartisan elections: the case of Des Moines." Amer. Pol. Sci. Rev. 57 (September): 584-592.

SALTER, P. S. and R. C. MINGS (1969) "A geographic aspect of the 1968 Miami racial disturbance: a preliminary investigation." Professional Geographer (March): 79-86.

SAMUELSON, P. A. (1954) "The pure theory of public expenditures." Rev. of Economics and Statistics (November): 87-89.

SANDALOW, T. (1970) "Federal grants and the reform of state and local government," pp. 175-194 in J. P. Crecine (ed.) Financing the Metropolis: Public Policy in Urban Economies. Urban Affairs Annual Reviews, Vol. 4. Beverly Hills: Sage Pubns.

SAX, J. L. (1971) Defending the Environment: A Strategy for Citizen Action. New York: Alfred A. Knopf.

SAYRE, W. S. (1970) "The mayor," pp. 563-601 in L. C. Fitch and A. H. Walsh (eds.) Agenda for a City: Issues Confronting New York. Beverly Hills: Sage Pubns.

SAYRE, W. and H. KAUFMAN (1960) Governing New York City. New York: Russell Sage Foundation.

SCHAEFER, G. F. and S. H. RAKOFF (1970) "Politics, policy and political science: theoretical alternatives." Politics and Society (1): 51-77.

SCHATTSCHNEIDER, E. E. (1961) The Semi-Semi Sovereign People. New York: Holt, Rinehart and Winston.

SCHMANDT, H. J. and G. R. STEPHENS (1963) "Local government expenditure patterns in the United States." Land Economics 34 (November): 397-406.

SCHUCTER, A. (1969) White Power, Black Freedom. Boston: Beacon.

SCHUMAN, H. and B. GRUENBERG (1970) "The impact of city on racial attitudes." Amer. J. of Sociology 76 (September): 213-261.

SCOTT, S. and H. NATHAN (1970) "Public referenda: a critical reappraisal." Urban Affairs Q. 5 (March): 313-328.

SEEMAN, M., J. M. BISHOP, and J. E. GRIGSBY, III (1971) "Community control in a metropolitan setting," pp. 423-450 in P. Orleans and W. R. Ellis, Jr. (eds.) Race, Change and Urban Society. Urban Affairs Annual Reviews, Vol. 5. Beverly Hills: Sage Pubns.

SHALALA, D. E. (1971) Neighborhood Governance: Issues and Proposals. New York: National Project on Ethnic America.

SHARKANSKY, I. (1968a) Spending in the American States. Chicago: Rand McNally.

--- (1968b) "Agency requests, gubernatorial support, and budget success in state legislatures," Amer. Pol. Sci. Rev. 57 (December): 1220-1231.

--- and A. TURNBULL, III (1969) "Budget making in Georgia and Wisconsin: a test of a model." Midwest J. of Pol. Sci. 13 (November): 631-645.

SHELDON, P. (1966) "Community participation and the emerging middle class," pp. 125-158 in J. Samora (ed.) La Raza: Forgotten Americans. Milwaukee: Univ. of Notre Dame Press.

SINGER, B. D. (1970) Black Rioters: A Study of Social Factors and Communication in the Detroit Riot. Lexington, Mass.: Heath Lexington.

SLATER, D. (1968) "Decentralization of urban people and manufacturing in Canada." Canadian J. of Economics and Pol. Sci. 27 (February): 72-84.

SMALLWOOD, F. (1966) " 'Game politics' vs. 'feedback politics,' " pp. 313-316 in R. Morlan (ed.) Capitol Courthouse and City Hall. Boston: Houghton Mifflin.

SMITH, B.L.R. and G. R. LA NOVE [eds.] (1971) "Urban decentralization and community participation." Amer. Behavioral Scientist 15 (September-October): 3-129.

SPILERMAN, S. (1970) "The causes of racial disturbances: a comparison of alternative explanations." Amer. Soc. Rev. 35 (August): 627-649.

SPROUT, H. (1971) "The environmental crises in the context of American politics," pp. 41-50 in L. L. Ross (ed.) The Politics of Ecosuicide. New York: Holt, Rinehart & Winston.

SROLE, L. (1956) "Social interaction and certain corollaries: an exploratory study." Amer. Soc. Rev. (21): 709-716.

STACK, C. R. (1970) "The kindred of Viola Jackson: residence and family organization of an urban black family," in N. E. Whitten, Jr. and J. F. Szwed (eds.) Afro-American Anthropology: Contemporary Perspectives. New York: Free Press.

STEFFENS, J. L. (1931) The Autobiography of Lincoln Steffens. New York: Harcourt, Brace.

STEIN, M. R. (1964) The Eclipse of Community. New York: Harper Torchbooks.

STEINER, G. (1966) Social Insecurity. Chicago: Rand McNally.

STEINER, S. (1969) La Raza. New York: Harper & Row.

STONE, C. N. (1965) "Local referendums: an alternative to the alienated-voter model." Public Opinion Q. 29 (Summer): 213-222.

STONE, R. C. and F. T. SCHLAMP (1966) Family Life Styles Below the Poverty Level. San Francisco: Institute for Social Science Research.

SUCHMAN, E. A. and H. MENZEL (1955) "The interplay of demographic and psychological variables in the analysis of voting surveys," in P. F. Lazarsfeld and M. Rosenberg (eds.) The Language of Social Research. New York: Free Press.

SUNDQUIST, J. L. (1969) Making Federalism Work. Washington: Brookings Institution.

––– (1968) Politics and Policy. Washington: Brookings Institution.

SURKIN, M. (1971) "The myth of community control: rhetorical and political aspects of the Ocean Hill-Brownsville controversy," pp. 405-422 in P. Orleans and W. R. Ellis, Jr. (eds.) Race, Change and Urban Society. Urban Affairs Annual Reviews, Vol. 5. Beverly Hills: Sage Pubns.

SWANSON, B. E. (1966) "The concern for community in the metropolis." Urban Affairs Q. 1 (June): 33-44.

T

TENHOUTEN, W. D., J. STERN, and D. TENHOUTEN (1971) "Political leadership in poor communities: applications of two sampling methodologies," pp. 215-254 in P. Orleans and W. R. Ellis, Jr. (eds.) Race, Change, and Urban Society. Urban Affairs Annual Reviews, Vol. 5. Beverly Hills: Sage Pubns.

THOMPSON, W. (1965) A Preface to Urban Economics. Baltimore: Johns Hopkins Press.

THRUPP, S. (1962) The Merchant Class of Medieval London, 1300-1500. Ann Arbor: Univ. of Michigan Press.

TIRADO, M. (1970) "Mexican-American community political organization." Aztlán (Spring): 53-78.

TROP, C. and L. L. ROOS (1971) "Public opinion and the environment," pp. 52-63 in L. L. Roos (ed.) The Politics of Ecosuicide. New York: Holt, Rinehart & Winston.

TRUMAN, D. B. (1951) The Governmental Process. New York: Alfred A. Knopf.

TUCK, R. (1946) Not With The Fist: Mexican-Americans in a Southwest City. New York: Harcourt, Brace.

U

UYEKI, E. S. (1966) "Patterns of voting in a metropolitan area, 1938-1962." Urban Affairs Q. 1 (June): 65-77.

V

VALENTINE, C. A. (1968) Culture and Poverty: Critiques and Counter Proposals. Chicago: Univ. of Chicago Press.

VOSE, C. E. (1966) "Interest groups, judicial review, and local government." Western Pol. Q. 19 (March): 85-100.

W

WAGNER, R. (1971) Environment and Man. New York: W. W. Norton.

WALTER, B. and F. M. WIRT (1971) "The political consequences of suburban variety." Social Science Q. (December): 746-761.

WATSON, J. and J. SAMORA (1954) "Subordinate leadership in a bicultural community: an analysis." Amer. Soc. Rev. (August): 413-421.

WATSON, R. A. and J. H. ROMANI (1961) "Metropolitan Cleveland: an analysis of the voting record." Midwest J. of Pol. Sci. 5 (August): 365-390.

WEAVER, L. (1971) Nonpartisan Elections in Local Government. Detroit: Citizens Research Council of Michigan.

WEEKS, C. (1967) Job Corps: Dollars and Dropouts. Boston: Little, Brown.

WEILER, C. J., Jr. (1971) "Metropolitan federation reconsidered." Urban Affairs Q. 6 (June): 411-420.

WHITELAW, W. E. (1970) "The city, city hall, and the municipal budget," pp. 219-243 in J. P. Crecine (ed.) Financing the Metropolis: Public Policy in Urban Economies. Urban Affairs Annual Reviews, Vol. 4. Beverly Hills: Sage Pubns.

WICKWAR, W. H. (1970) The Political Theory of Local Government. Columbia: Univ. of South Carolina Press.

WILDAVSKY, A. (1967) "Aesthetic power or the triumph of the sensitive minority over the vulgar mass: a political analysis of the new economics." Daedalus 96 (Fall): 1115-1127.

––– (1964) The Politics of the Budgetary Process. Boston: Little, Brown.

WILENSKY, G. (1970) "Determinants of local government expenditures," pp. 197-218 in J. P. Crecine (ed.) Financing the Metropolis: Public Policy in Urban Economies. Urban Affairs Annual Reviews, Vol. 4. Beverly Hills: Sage Pubns.

WILLIAMS, O. P. (1971) Metropolitan Political Analysis. New York: Free Press.

WILLOUGHBY, A. (1969) "The involved citizen: a short history of the National Municipal League." National Civic Rev. 63 (December).

WILSON, J. Q. [ed.] (1968) City Politics and Public Policy. New York: John Wiley.

––– (1961) "The strategy of protest: problems of Negro civic action." J. of Conflict Resolution (September): 291-303.

––– (1960) Negro Politics: The Search for Leadership. New York: Free Press.

––– and E. C. BANFIELD (1964) "Public-regardingness as a value premise in voting behavior." Amer. Pol. Sci. Rev. 58 (December): 876-887.

WINGO, L. and H. S. PERLOFF [eds.] (1968) Issues in Urban Economics. Baltimore: Johns Hopkins Press.

WIRT, F. M. [ed.] (1971) Future Directions in Community Power Research: A Colloquium. Berkeley: Institute of Government Studies, University of California.

WIRTH, L. (1938) "Urbanism as a way of life." Amer. J. of Sociology 44 (July): 1-24.

WOLFINGER, R. E. and F. I. GREENSTEIN (1968) "The repeal of fair housing in California: an analysis of referendum voting." Amer. Pol. Sci. Rev. 62 (September): 753-769.

WOLFINGER, R. E. and J. O. FIELD (1966) "Political ethos and the structure of urban government." Amer. Pol. Sci. Rev. 60 (June): 306-326.

WOLMAN, H. (1971) Politics of Federal Housing. New York: Dodd, Mead.

WOOD, R. C. (1961) 1400 Governments: The Political Economy of the New York Metropolitan Region. Cambridge: Harvard Univ. Press.

WOODS, Sister F. J. (1949) Mexican Ethnic Leadership in San Antonio, Texas. Washington, D.C.: Catholic University of America.

WRIGHT, C. R. and H. H. HYMAN (1958) "Voluntary association membership of American adults: evidence from national survey samples." Amer. Soc. Rev. (23): 284-294.

Y

YATES, D. (1971) Neighborhood Government. New York: Rand Institute.

YOUNG, M. and P. WILLMOTT (1957) Family and Kinship in East London. Baltimore: Penguin.

Z

ZALD, M. N. [ed.] (1967) Organizing for Community Welfare. Chicago: Quadrangle.

--- and R. ASH (1970) "Social movement organizations: growth, decay and change," pp. 516-537 in J. Gusfield (ed.) Protest, Reform and Revolt. New York: John Wiley.

ZEIGLER, H. (1965) "Interest groups in the states," pp. 101-147 in H. Jacob and K. L. Vines (eds.) Politics in the American States. Boston: Little, Brown.

--- and M. A. BAER (1969) Lobbying. Belmont, Calif.: Wadsworth.

ZISK, B. H., H. EULAU, and K. PREWITT (1965) "City councilmen and the group struggle: a typology of role orientations." J. of Politics (27): 618-646.

THE AUTHORS

THE AUTHORS

MARIO BARRERA is Assistant Professor of Political Science and Chicano Studies at the University of California (San Diego). He was previously a research and teaching assistant at the University of California (Berkeley), where he received his Ph.D. in 1970. He is the author of two monographs, *Modernization and Coercion* and the forthcoming *Information and Ideological Change;* he is coeditor of two forthcoming volumes, *Embourgeoisement and Radicalization in Latin America* (with David Apter) and *The Mexican American and Educational Change* (with Alfredo Casteñeda, Manuel Ramirez, and Carlos Cortés). His research interests focus on the political structure of the Chicano community, as well as on power, communication, ideology, and leadership.

SHELDON BOCKMAN is an assistant professor in the sociology department at the University of California (Riverside). He received his Ph.D. in 1968 from Indiana University. His particular fields of interest encompass political sociology and industrial sociology. He is currently engaged in research on community decision-making, and is working on a book which will present an interpretation of the social context of political conduct.

LEWIS BOWMAN is Professor of Political Science and Chairman of the Department of Political Science at Emory University. He is the author of numerous journal articles, including contributions to the *Journal of Politics, Social Forces, Midwest Journal of Political Science, Social Science Quarterly,* and the *American Political Science Review.* He is coauthor of two books: the recent *Comparative Public Opinion and Politics: Essays and Readings* and *American Democracy and the Search for Community* (forthcoming in 1972).

WILLIAM P. BROWNE is an assistant professor of political science at Central Michigan University. He did his graduate work at Washington University (St. Louis), with a Ph.D. dissertation focusing on the relationship between urban interest groups and their member municipalities (a study of such organizations as the U.S. Conference of Mayors, the National League of Cities, the Missouri Municipal League, and the St. Louis County Municipal League).

JEROME M. CLUBB is Director of the Historical Archive at the Inter-university Consortium for Political Research and Program Director of the Center for Political Studies (Institute for Social Research) at the University of Michigan—where he is also a lecturer in history. He is coeditor (with Howard W. Allen) of the forthcoming book, *Partisan Change and Stability in American Political History,* as well as a 10-volume work, *American Elections: County Election Returns for the Offices of President, Senator, Representative and Governor, 1824-1952.*

TERRY N. CLARK is Assistant Professor of Sociology at the University of Chicago, where he directs the Comparative Study of Community Decision-Making (with support from both the National Science Foundation and the Ford Foundation). He is the author of numerous articles in American and European sociology journals, as well as a contributor to the 1968 edition of the *International Encyclopedia of the Social Sciences.* He is editor of the Wiley Series in Urban Research. Dr. Clark is the author of the forthcoming volume *Prophets and Patrons: The French University and the Emergence of the Social Sciences,* coeditor of *Community Politics* (with Charles M. Bonjean and Robert L. Lineberry), and editor of both *Community Structure and Decision-Making* (1968) and *Comparative Community Politics* (forthcoming).

BRYAN T. DOWNES is an associate professor at the University of Missouri (St. Louis). He was previously on the faculty of Michigan State University. He received his Ph.D. from Washington University (St. Louis) in 1966. He is the editor of *Cities and Suburbs: Selected Readings in Local Politics and Public Policy.* His fields of interest include urban government, political leadership, decision-making, and comparative politics.

PETER K. EISINGER is Assistant Professor of Political Science at the University of Wisconsin. His field of specialization is urban politics, and he is currently doing research on decentralization and control-sharing in cities, and on urban political protest behavior. He is the author of a chapter entitled "The Impact of Anti-Poverty Expenditures in New York" which appeared in Volume IV (1970) of this series.

JOE R. FEAGIN is currently an associate professor of sociology at the University of Texas (Austin). He has previously been affiliated with the University of California (Riverside), the University of Massachusetts (Boston), and with the MIT-Harvard Joint Center for Urban Studies. His recent work has been mainly in the areas of race relations and urban problems, including urban political conflict.

EDMUND P. FOWLER is Assistant Professor of Political Science at Glendon College of York University in Toronto. He received his Ph.D. from the University of North Carolina in 1969 with a dissertation on "Government Spending as Response and Initiative." With Robert L. Lineberry he was the author of "Reformism and public policies in American cities," which appeared in the *American Political Science Review.* He has been both participant and observer in Canadian city politics in Toronto, and (with Alan Lupo and Frank Colcord) coauthored *Rites of Way: The Politics of Highway Construction in Boston and the American City.* His special areas of interest include political parties and elections, metropolitan and urban governmental systems.

LEWIS A. FRIEDMAN is at the Department of Political Science, New York University (University Heights campus). He recently completed his graduate work at Michigan State University, focusing on urban politics.

BARRY GRUENBERG is a lecturer in sociology at Wesleyan University. He has previously coauthored an article with Howard Schuman entitled "The Impact of City on Racial Attitudes" which in 1970 won the Gordon Allport prize (awarded by the Society for the Psychological Study of Social Issues) for research in intergroup relations. He is currently completing his dissertation on the social and ecological determinants of American leisure styles.

HARLAN HAHN is Professor of Political Science at the University of California (Riverside), where he is also Chairman of the Urban Studies Program. He is the author of numerous articles in social science journals, dealing with urban violence, voting behavior, the police, and public opinion. Dr. Hahn is the author of a recently published book–*Urban-Rural Conflict: The Politics of Change*–and the forthcoming volumes, *The Politics of Urban Law Enforcement* and *Urban Violence* (with Joe R. Feagin). He edited *Police in Urban Society,* and will serve as editor of the journal *American Politics Quarterly* (commencing publication in January, 1973).

MATTHEW HOLDEN, Jr. is Professor of Political Science at the University of Wisconsin (Madison) and an Associate at the Center for the Study of Public Policy and Administration. He received his Ph.D. from Northwestern University in 1961. His research and writing deals primarily with decision-making behavior (including studies in this area involving environmental pollution), with special interest in urbanization and the politics of urbanization, and the processes of political integration and disintegration. He is the author of the forthcoming book, *The Republic in Crisis,* and has contributed to numerous journals including *American Political Science Review, Western Political Science Quarterly,* and *Urban Affairs Quarterly.*

DENNIS S. IPPOLITO is an assistant professor in the department of political science at Emory University. He is the author of several articles in political science journals and in *Social Science Quarterly,* and is coauthor of the forthcoming book, *American Politics: The Democratic Dilemma.* His research interests focus on survey research –particularly in the areas of political parties, public opinion, and urban problems.

M. KENT JENNINGS is Professor of Political Science at the University of Michigan. He received his Ph.D. in 1961 from the University of North Carolina. He is author of *Community Influentials: The Elites of Atlanta,* and coauthor of *The Image of the Federal Service* and *Source Book of a Study of Occupational Values and the Image of the Federal Service.* He coedited *The Electoral Process* (with Harmon Zeigler), and has authored and coauthored numerous articles–including contributions to *Public Opinion Quarterly, Administrative Science Quarterly,* and *American Political Science Review*–and chapters in many books. His major fields of interest are metropolitan and urban government, political parties, political socialization, and elections.

MARGARET LEVI is currently a graduate student in the Department of Government, Harvard University, where she is working on a dissertation involving police unions and their implications for the analysis of the distribution of wealth and power in the city.

MARTIN L. LEVIN is an associate professor in the department of sociology at Emory University, where he directs the Esso Education Project on "Utilization of Educational Resources." He is also engaged in research on the "Politics of a Mass Transit System" (funded by the Institute for Public Administration and the Department of Transportation). Dr. Levin has authored or coauthored several journal articles (many of which have been subsequently reprinted in books), and coauthored a monograph entitled *Black Youth in a Southern Metropolis.*

ROBERT L. LINEBERRY is Associate Professor of Political Science at the University of Texas (Austin). With Edmund P. Fowler he wrote "Reformism and public policies in American cities," which appeared in the *American Political Science Review.* He coedited (with Charles M. Bonjean and Terry N. Clark) a book entitled *Community Politics: A Behavioral Approach.* Dr. Lineberry has written on metropolitan reform in the *Georgetown Law Journal* and on American urban policies in the *American Political Science Review.* He and Ira Sharkansky coauthored a text, *Urban Politics and Public Policy,* which emphasized a policy approach to urban politics.

MICHAEL LIPSKY is an associate professor of political science at the Massachusetts Institute of Technology. Dr. Lipsky received his Ph.D. in politics from Princeton University in 1967. He is the author of *Protest in City Politics,* editor of *Law and Order: Police Encounters,* and author of "Protest as a Political Resource" in *American Political Science Review,* "Street Level Bureaucracy and the Analysis of Urban Reform" in *Urban Affairs Quarterly,* and other articles. He is currently working on a study (with David J. Olson) on the political response to riots in American cities, which will be published in book form during 1973.

RONALD O. LOVERIDGE is Associate Professor of Political Science and Associate Dean of the College of Social and Behavioral Sciences at the University of California (Riverside). He holds a Ph.D. degree from Stanford and is the author of a book entitled *City Managers in Legislative Politics.* He is particularly interested in the politics of environmental pollution.

CARLOS MUÑOZ is presently an assistant professor of political science at the University of California (Irvine), where he is currently working at developing a radical theory of minority politics. He is in the process of completing his Ph.D. in Political Science (from Claremont Graduate School). He is a member of the National Executive Council of the Caucus for a New Political Science. He has written an article in *Aztlan,* contributed a chapter to *Racism in California,* and is the author of "The Politics of Educational Change in East Los Angeles" (a monograph published by the University of California at Riverside). His current interests include urban politics, public policy, revolution, and violence.

CHARLES ORNELAS is a lecturer in political science at the University of California (Santa Barbara). He is also Director of Centro de Investigaciones, Casa de la Raza, in Santa Barbara, and principal investigator of a study of Chicano Political Communication in the Santa Barbara area. He is presently completing his Ph.D. in political science at the University of California (Riverside). He is the recipient of grants and fellowships from the National Science Foundation, Ford Foundation, and the National Institute of Mental Health. He is the author of several reports, articles, and chapters in books—all dealing with aspects of Chicano studies.

DAVID C. PERRY is an assistant professor of government at the University of Texas (Austin). He is the author of articles in the field of urban studies and public administration. He is presently writing a book on police and metropolitanism and is the coeditor of a forthcoming work on political violence.

THOMAS F. PETTIGREW is a professor in the department of social psychology at Harvard University. He is the author of *A Profile of the American Negro,* and *Racially Separate or Together?* He was a consultant to the U.S. Civil Rights Commission and authored numerous articles which have appeared in such journals as *Integrated Education, Journal of Social Issues, Sociology of Education* and the *Harvard Educational Review.*

RANDALL B. RIPLEY is Professor and Chairman, Department of Political Science, The Ohio State University. He is author of *Power in the Senate, Majority Party Leadership in Congress, Party Leaders in the House of Representatives,* and *Public Policies and Their Politics,* as well as articles in such journals as *American Political Science Review, Journal of Politics, Political Science Quarterly,* and *Public Policy.* He is currently engaged in research on the politics of economic and human resource development, and on policy-making in the executive branch of the U.S. government.

HAROLD ROSE is Chairman of the Department of Urban Affairs at the University of Wisconsin (Milwaukee). He is the author of *The Black Ghetto: A Spatial Behavioral Perspective* and articles in *Economic Geography* and *Geographical Review.*

ROBERT H. SALISBURY is Professor of Political Science at Washington University (St. Louis). He is coauthor of *State Politics and the Public Schools, American Government: Problems and Readings for Political Analysis,* and *Politics in the American States.* He edited *Democracy in the Mid-Twentieth Century* and has published several articles in such journals as *Public Opinion Quarterly, Journal of Politics, American Political Science Review,* and *Midwest Journal of Political Science.*

HOWARD SCHUMAN is Professor of Sociology at the University of Michigan, and research associate in the University's Institute for Social Research. The present chapter is based on data originally gathered for the report "Racial Attitudes in Fifteen American Cities," coauthored with Angus Campbell and submitted to the National Advisory Commission on Civil Disorders in 1968. In addition to further work on racial attitudes presently under way, Dr. Schuman has written on industrial workers and peasants in what is now Bangladesh, and on various aspects of survey methodology and public opinion.

MICHAEL W. TRAUGOTT is Assistant Director of the Historical Archive at the Inter-university Consortium for Political Research and Assistant Study Director at the Center for Political Studies, University of Michigan. He is also an election consultant to the American Broadcasting Corporation. He is currently working (with coauthors Jerome M. Clubb and Erik W. Austin) on two books: *Handbook of Quantitative Methods in Historical Research* and *Computer Applications in History and Political Science.*

HARMON ZEIGLER is Professor of Political Science at the University of Oregon and Program Director of the Center for Advanced Study of Educational Administration (CASEA). He is the author of *The Politics of Small Business, Interest Groups in American Society, The Political Life of American Teachers,* and with various coauthors of *Lobbying, The Irony of Democracy,* and the forthcoming *Politics and the Educational Subsystem.* He edited (with M. Kent Jennings) a volume on *The Electoral Process* and is the author or coauthor of over twenty articles in professional journals. He has pursued his research under fellowships from both the Ford Foundation and the John Simon Guggenheim Foundation.

BETTY H. ZISK is an associate professor of political science at Boston University, where she is engaged as senior investigator in a two-year study of urban bargaining behavior funded by the National Science Foundation. She has contributed several chapters to books as well as articles to such journals as *Western Political Quarterly, Journal of Politics,* and *Simulation and Games.* She is the editor of the book, *American Political Interest Groups: Readings in Theory and Research.*

GREAT CITIES OF THE WORLD

Their Government, Politics, and Planning

Edited by WILLIAM A. ROBSON and D. E. REGAN

A new edition of this remarkable book has long been awaited. Now, after nearly five years' intensive preparation, the third edition is being published. The new edition is produced under the joint editorship of Professor William A. Robson and Dr. D. E. Regan.

The purpose of the book is to describe in depth the local government of a selected group of the world's great cities; to consider the problems these vast communities are confronting and what steps they are taken to overcome them.

Twenty-seven cities are described by contributors who have lived and worked in them, often for most of their lives. Nine cities appear for the first time (Belgrade, Birmingham, Cairo, Delhi, Ibadan, Manila, Mexico City, Pretoria and Warsaw). Thirteen of the eighteen cities previously included are dealt with by new authors in entirely new contributions. (They are Amsterdam, Buenos Aires, Calcutta, Chicago, Johannesburg, London, Montreal, New York, Paris, Rome, Stockholm, Sydney, and Toronto.) The chapters on the remaining five cities (Copenhagen, Los Angeles, Osaka, Rio de Janeiro and Tokyo) have been thoroughly revised by their original authors.

In Part One of the book, entitled 'The Great City of Today,' Professor Robson contributes a masterly synthesis of the material contained in the separate studies. This has been rewritten and is substantially enlarged. Among the items he deals with are: the metropolitan community, special constitutional features, the elected council, the alternative varieties of dealing with executive functions (i.e., city councils, council-appointed executives, elected mayors, elected committees and executives appointed by the central government); also considered are municipal services, the need for integration, expansion attempts, the ad hoc authority for special purposes, politics in great cities, divided civic and political interest, relations with higher authorities, municipal finance, planning the great city, and the metropolitan region of tomorrow.

GREAT CITIES OF THE WORLD was a pioneer work which stimulated the production of a considerable amount of research and writing on metropolitan cities. It remains, however, not only a standard reference work, but the only one which contains studies by experts of more than two dozen metropolitan areas scattered throughout the world.

Political scientists, urban geographers, planners, political leaders, governmental officials (at all levels of government), and public administration specialists will find the new edition of this classic reference tool an important addition to their professional libraries.

ISBN 0-8039-0155-0 (Volume I) **Third Edition** **1,170 pages (both volumes)** **Illustrated** **Clothbound**
0-8039-0156-9 (Volume II) **L.C. 75-167875** **$50.00 (complete set--volumes will not be sold separately)**
Market: U.S.A. **published February, 1972**